# AQUACULTURE

# AQUACULTURE

*Editor*

Hildebert Schmid

scitus
academics

**Aquaculture**

Edited by **Hildebert Schmid**

This Book is Printed on Acid-Free Paper. ♾



ISBN: 9781681179469

Library of Congress Control Number: 2017949086


SCITUS Academics LLC,
616, Corporate Way, Suite 2, 4766,
Valley Cottage, NY 10989

www.scitusacademics.com



Printed in United States of America.

# Preface

Aquaculture is the art, science and business of cultivating aquatic animals and plants in fresh or marine waters. It is the extension of fishing, resulted from the fact that harvests of wild sources of fish and other aquatic species cannot keep up with the increased demand of a growing human population. The past 20 years have seen aquaculture evolve from a traditional practice to a science-based activity and grow into a significant food production sector, contributing more to national economies and providing better livelihoods for rural and farming families. It is very diverse and, contrary to many perceptions, dominated by shellfish and herbivorous and omnivorous pond fish either entirely or partly utilizing natural productivity. The rapid growth in the production of carnivorous species such as salmon, shrimp and catfish has been driven by globalizing trade and favorable economics of larger scale intensive farming. Sustainable development of aquaculture must take into account the societal value of ecosystem services for an efficient and environmentally sound production of food. Although some issues regarding the potential benefits and implementation of sustainable aquaculture remain, the consideration of food security and minimizing ecosystem impacts suggest that the time has come to take action.

This book 'Aquaculture' provides an understanding on a large variety of aquaculture related topics. It aims to describe past trends in capture fishery and aquaculture production, particularly concerning the development of marine capture fisheries, and to compare the current production with fisheries potential. The book presents recent research results on the interaction between aquaculture and environment, and includes several case studies all over the world with the aim of improving and performing sustainable aquaculture. An essential requirement for ensuring sustainable fisheries and aquaculture through good policies and management will be the provision of objective information on the state of fisheries and aquaculture. This book will be equally useful for students and professionals in aquaculture and biotechnology.

# Table of Contents


Chapter 1 Sustainable Technique for Selected Live Feed Culture 1

Chapter 2 Microarray-Predicted Marker Genes Molecular Pathways Indicating Crowding Stress in Rainbow Trout (*Oncorhynchus Mykiss*) 33

Chapter 3 Early Development and Allometric Growth Patterns in Burbot *Lota Lota* L. 61

Chapter 4 Biofloc Technology: An Emerging Avenue in Aquatic Animal Healthcare and Nutrition 79

Chapter 5 Seaweed Aquaculture in Norway: Recent Industrial Developments and Future Perspectives 95

Chapter 6 People Matter in Animal Disease Surveillance: Challenges and Opportunities for The Aquaculture Sector 121

Chapter 7 Growth Phase Significantly Decreases The DHA-To-EPA Ratio in Marine Microalgae 155

Chapter 8 Impacts of Aquaculture on Habitats and Best Management Practices (BMPs) 169

Chapter 9 Environmental and Physiological Factors Shape The Gut Microbiota of Atlantic Salmon Parr (*Salmo Salar* L.) 241

Chapter 10 **Impacts of Rearing Techniques on Growth, Survival and Bacterial Microbiota of Carpet Shell Clam (*Ruditapes Decussatus*) Larvae** 265

Chapter 11 **Application of Biological Agents in Abalone Aquaculture: A South African Perspective** 281

Chapter 12 **Fisheries and Aquaculture in Bangladesh: Challenges and Opportunities** 313

Chapter 13 **Use of Probiotics in Aquaculture** 325

Chapter 14 **Time-To-Response Toxicity Analysis as a Method for Drug Susceptibility Assessment in Salmon Lice** 341

Chapter 15 **Nutrient Profiles of Rotifers (*Brachionus* Sp.) and Rotifer Diets from Four Different Marine Fish Hatcheries** 357

Chapter 16 **The Role of Shore Crabs and Mussel Density in Mussel Losses at a Commercial Intertidal Mussel Plot After Seeding** 377

**Index** 397

# CHAPTER 1

# Sustainable Technique for Selected Live Feed Culture

**Zaleha Kassim[1], Akbar John[3], Lim Keng Chin[2], Nur Farahiyah Zakaria[4] and Nur Hidayah Asgnari[1]**

[1]*Department of Fisheries, Faculty of Fisheries and Aqua-Industry, University Malaysia Terengganu, Terengganu, Malaysia*

[2]*Institute of Tropical Aquaculture University Malaysia Terengganu, Terengganu, Malaysia*

[3]*Kuliyyah of Science, International Islamic University Malaysia, Kuantan Campus, Malaysia*

[4]*Unit of Farmer's Organisation of South Johor Bahru, Johor, Malaysia*

## 1. INTRODUCTION

Sustainability in the aquaculture industry depends on several factors including the minimum production cost in comparison to the yield, unexpected environmental conditions which affect the farm and practices in the farm management itself. These factors are inter-connected and always incur a synergistic effect on the issue of sustainability. Live feeds as the fundamental needs for larval rearing and fry production have to be prioritised for sustainable farming activity. Dependency on imported sources of live feeds or inert feed will increase the production cost. Thus, the continued activity of screening, stocking and maintaining some local species as an option for live feed production is economically necessary.

Live feeds are an important basic diet for newly-hatched fish and shrimp larvae as they still have an incomplete digestive system and are lacking in enzymes. They are still at a very young stage to generate their own required nutrients or convert them from any pre-cursor obtained from a diet. They need a ready-made diet with readily available nutrient to be absorbed through their digestive system. There have been many species suggested or tested for their potential as live feed. All test animals were mostly zooplankton in nature and must meet the requirement as live feed. They must be in a compatible size with the mouth size or gape of the larvae predator, or they cannot be swallowed. Since larvae are still weak to track down the food, the wave created by the prey will be a great help, thus 'active' swimming prey is preferred. The most important role of a prey is the ability to supply energy and other nutrients which are essential for the larval survival and growth. Live feeds, as the starter diet in larval rearing and fry production must be continuous in supply. Good, nutritious and compatible-size prey must be able to reproduce fast to meet the requirement and adaptable to a simple mass-production technique.

## 1.1. Copepods As Live Feeds

The conventional live feed, brine shrimp and rotifers, are considered unsuitable as live feed if compared to copepods in term of nutritional value. *Artemia* sp. is deficient in polyunsaturated fatty acids (PUFAs), thus it needs to be enriched before feeding [1, 2]. Similarly, rotifer have poor nutritional value and are small in size [3]. On the other hand, copepod diets were proven to increase the growth of larval marine fish compared to diet of rotifers *Brachionus plicatilis,* [4,5] or *Artemia* [4]. The potential use of copepods as live feed due to their excellent fatty acid content has been highlighted by using an example, a paracalanid [6]. They improved the quality of the cultured organism, particularly the larval stages. The superiority of copepods over other live feed such as brine shrimp and rotifers was further confirmed [7]. They have the appropriate ratio of docosahexaenoic acid (DHA) to eicosapentaenoic acid (EPA) which will improve the growth and survival of fish larvae if compared to the conventional live feeds [8]. Nonetheless, it is reminded that copepods could be better used as supplement rather than a sole diet in larval rearing, unless they are used for some high-valued commercial reef fish due to the high operation cost [9]. The possibility of using copepods, particularly the harpacticoids as alternative live feed in fish larval rearing has been stressed [10] and it is confirmed that the nutritional quality of harpacticoid copepods as live feed is extremely high [11,12]. *Macrosetella gracilis,* a planktonic harpacticoid copepod, is also reported to have better diet quality when compared to *Artemia* [13].

Despite these positive findings, rotifers and *Artemia* continue to be the live feeds of choice in commercial hatcheries, because copepods are not currently cultured at sufficient densities to be economically efficient on a commercial scale [14, 15, 16]. In term of culture condition, it was found that the optimum condition for the high production of a tropical harpacticoid copepod, *Pararobertsonia* sp., was at salinity 35psu and temperature of 25°C [17]. The fluctuation in salinity, pH and temperature in the culture vessel would definitely influence the reproduction and population growth of copepods such as the harpacticoids if kept in small containers [18]. Nonetheless, a strategy to produce harpacticoid in large quantities for hatchery use by using a tray-culture method has been suggested [19].

Another copepod group, a Cyclopoida, *Apocyclops dengizicus* was found to increase the survival and growth of *Panaeus monodon* postlarvae when used as live feed [20]. Cyclopoids are omnivorous, and can be fed a mixture of feeds, mainly phytoplankton or a combination of phytoplankton, yeast or other feeds [7]. As for *Apocyclops panamensis*, there is a report on a successful technique for outdoor ponds [15]. Information on the use of copepods in aquaculture, particularly from the tropical *Apocyclops* sp., is still scarce. The species reported in abundance and potentially exploited as live feeds for shrimp post-larvae in Malaysia for example is *A. dengizicus*. A new species, *Apocyclops ramkhamhaengi*, has been described [21] and added to the present report of 3 species of *Apocyclops* recognised from Asia: *A. dengizicus* (Lepeshkin), *A. royi* (Lindberg), and *A. borneonensis* (Lindberg). This new species is found in abundance in eastern Thailand water and has yet to be determined for its potential in aquaculture. Planktonic copepods such as cyclopoids feed on other plankton including planktonic microalgae. To maintain planktonic copepods in the hatchery or aquaculture ponds, a continuous supply of their diet, particularly the microalgae, will definitely be required.

## 1.2. Microalgae

Microalgae are a diverse group of unicellular autotrophs inhabiting almost all aquatic water bodies. Microalgae are rich in many specific and attractive compounds [22] and their nutritional values for aquaculture had been highlighted [23]. Production of microalgae is mandatory in the hatchery as it is a basic and nutritious diet for live feed, specifically the zooplankton. However, its mass production is generally costly due to huge manpower, space requirements and operation which usually related to the cost of the energy used. A good strategy in manipulating the culture environment, particularly during the production process of microalgae would scale down the operational cost.

Light plays a fundamental role in the development of microalgae through photosynthesis. It is one of the major environmental factors which control the performance of microalgae phototrophic growth and productivity [24, 25, 26]. Light may either be natural or supplied by fluorescent tubes giving the maximum effective radiation which can be absorbed by the pigments of the microalgae. Light intensity plays a vital role, but the requirements vary with the culture depth or volume as well as the density of the algae in the culture. At a higher volume, light intensity must be increased to enable it to penetrate through the culture. However, an extreme light intensity may result in photo-inhibition which reduces the photosynthetic rates and growth [27,28]. Furthermore, overheating due to artificial or natural illumination should be avoided in microalgal culture. The most often employed light intensity is 1000 lux which is suitable for Erlenmeyer flasks but 5000-10000 lux is needed for a greater volume of microalgal culture [29, 30]. The duration of illumination can be varied where photosynthesis of microalgae can be enhanced or increased in the light/dark (LD) cycle (usually 12:12 or 14:10 LD, maximum 16:8 LD). For some microalgae, photosynthesis rate could also be increased exponentially with increasing light/dark frequencies where a longer period of dark in relation to the light period can further increase photosynthetic efficiencies but not vice versa [31]. The illuminations also affect the nutrient utilisation in the culture vessel [32].

A cost-effective and nutritionally-adequate alternative to costly maintenance of live microalgae is the production of moist-microalgae concentrates. It is seen to simplify hatchery procedures and has shown promising potential in the aquaculture industry [33,34].The storing of microalgae concentrates in moist form under low temperature can preserve their high nutrient composition and excellent cell viability [35,33]. Juvenile pacific oyster (*Crassostrea gigas*) fed with different algal pastes had shown significant improvement in growth rate than oyster fed with other diets [36]. Concentrates of *Chaetoceros muelleri*and *Tetraselmis pseudonana* refrigerated for 6 weeks at 4°C were found to promote similar survival rates of the tiger prawn *Penaeus monodon* larvae fed with live microalgae [37]. Supplementation of microalgae concentrates to bivalves, oysters and scallops have also recorded the same extent of growth rates as live microalgae [38,34]. It has been documented that most of the demand for mariculture feed in Japan is supplied with live and fresh microalgae which is thickly concentrated and readily stored at 2-4ºC for 1-8 weeks with good shelf life [39].

Preparation of concentrated condition of microalgae usually involves centrifugation technique. Nonetheless, although this technique has been successfully applied and utilised for preparing microalgae concentrates, it poses some limitations. First, the process involves exposure of cells to high gravitational and shear forces deteriorating the cell structure with the leaking of nutritional contents. Second, centrifuging large volumes of cultures is time-consuming and requires expensive equipments. Several alternative procedures, less damaging to the cells, which can be applied are filtration [40], foam fractionation [41] and flocculation [33, 34, 38]. Previous studies have observed the excellence of ultrafiltration technique in preserving and retaining the cellular structure and properties of fragile algal cells with little loss of material [42, 43].

The level of natural resources exploitation for aquaculture purposes is commonly high. Coastal land and mangroves forests always become the target area for brackish-water aquaculture ponds. The water source of this area, which is always from the nearby river estuary and lagoon, is also used as the live feeds (zooplankton and microalgae) source. Nonetheless, the supply is always seasonal and could become unavailable unexpectedly due to many factors and natural phenomena. This chapter aims to discuss the possible ways to produce local live feeds, a marine microalgae species and a planktonic copepod, sustainably using a simple technique for larval-rearing purposes. Maintaining local species is hypothesised to be more economical and practical. The usage of the microalgae as an enrichment element for live feed copepods will be proved.

## 2. METHODOLOGY

### Experiment 1: Production of *Chlorella vulgaris* Concentrate Isolated from Bidong Island and Assessment as Copepod Diet

Seawater samples were obtained from Bidong Island, Terengganu. The collection was made by lowering a Niskin water sampler to a required depth, following the light-penetration depth. Concentrated water samples were then transferred into chilled, white-plastic containers and brought back to the laboratory for microalgae isolation process. Successive plating out on agar plates was performed in order to select the desired marine *Chlorella* colonies. Monospecific colonies were then transferred into trial culture tubes before scaling up into larger volumes of Erlenmeyer flasks.

The microalgae was then cultured for the preparation of moist concentrates in the temperature controlled room (20±2°C) using the standard batch culture method. Triplicate of actively-growing starter cultures were inoculated into 30 litres acrylic tanks enriched with Conway medium under constant illumination (cool-white type; 110 watts). All cultures were started with an initial inoculum of $2x10^6$ cells $mL^{-1}$. Cultures were aerated continuously using humidified filtered air. Evaporation in the culturing tanks was kept to a minimum by covering the top of the tanks. Cellular density of microalgae cultures was monitored daily using a Neubauer haemocytometer [29]. Scanning electron microscopic observation was also done to determine the ultra structure of the cell. Measurement of radius and height of the target microalgae cells was done under the advanced research microscope (Model Nikon Eclipse 80-i, Japan) and twenty individual cells were measured for the calculation of cell biovolume to avoid biasing results. Cell biovolume was calculated as assumed round-shape volume with the following formula proposed by Sun and Liu [44]:

$$\text{Cell volume} = 4/3\pi R^3 \quad (1)$$

Where, $\pi$= 3.142, R= radius of cell

Specific growth rate was calculated from the expression as proposed [45] which is shown below:

Specific growth rate ($\mu$) = $\ln(F_1/F_0)/t_1\text{-}t_0$ (2)

Where, $\mu$ = specific growth rate, $F_1$= biomass at time harvest, $t_1$ and $F_0$= biomass at time zero, $t_0$.

Doubling time was computed based on the formula as proposed [45] which is shown below:

Doubling time($\tau$)=$\log(2)/\mu$ (3)

Where, $\tau$ = doubling time, $\mu$ = specific growth rate.

All microalgae cultures were grown to late-logarithmic phase for the preparation of concentrates via ultrafiltration technique. The concentrated aqueous suspensions of microalgae were filtered through a membrane filter (0.1µm pore size) to remove access water from the suspension without rupturing the microalgae, thereby obtaining the microalgae concentrate or paste. Cell viabilities of microalgae concentrates were assessed using Eosin dye as a viability assay on the basis of its penetration into non viable-cells based on the expression as proposed [46]:

$$\text{Cell viability}\left(\%\right) = \frac{\text{Viable cells x 100}}{\text{Total cells}} \quad (4)$$

The harvesting efficiency or percentage recovery (%) was evaluated by comparing the remaining total number of cells in the concentrate with the total number of cells before filtration with the following expression:

Harvesting efficiency/Percentage recovery(%) = $C_B/C_A$x 100 (5)

Where, $C_B$ = total number of cells before filtration, $C_A$ = total number of cells after filtration

Microalgal concentrates were compared to live cultures of the same algae as food for marine copepods. Copepods were obtained from existing culture in UMT's laboratory. Two different sets of cultures were done using a Petri dish where each of them was fed with live and microalgae concentrate respectively. Individual copepods were counted daily under the Leica stereo microscope before being fed (1 drop). The maximum specific growth rate (K) was calculated [47] as shown below:

$K=\ln (X_1/X_0)/t_1-t_0$ (6)

Where, K = specific growth rate, $X_1$= the number of copepods at harvest time, $t_1$ and $X_2$= the number of copepods at time zero, $t_0$

The doubling time was computed as:

Doublingtime(т)=log(2)/K

Where, т = doubling time, K = specific growth rate.

## Experiment 2: Effects of Photoperiod and Culture Size on *Chlorella vulgaris* Stock Growth

Pure strains of *Chaetoceros* sp. and *C.vulgaris* were obtained from the microalgae maintenance laboratory at Institute of Tropical Aquaculture, Universiti Malaysia Terengganu, Malaysia, while the pure strain of *Nannochloropsis* sp. was obtained from the Fisheries Research Centre, Pulau Sayak, Kedah, and was maintained in autotrophic conditions in liquid and semisolid agar with Conway media [48]. This axenic culture was stored at 25±2$^0$C for 3 days which served as an inoculum for further experiments.

Microalgae were grown in autotrophic conditions as a monospecific axenic culture in different volumes (250mL, 500mL and 2000mL) containing Conway media. 25mL of pure strain with the cell density of ~2 x $10^6$ cells $mL^{-1}$ were transferred to each Erlenmeyer culture flask and kept at complete illumination provided by luminescent tubes (1000 Lux). Carbon source was provided by bubbling sterile 2% (v/v) $CO_2$ in air through the cultures. Culture flasks were maintained at a constant temperature (22$^{\circ\circ}$C± 1°C) with the pH range of 7-8 and salinity of ~35ppt in an air-conditioned laboratory over 2-3 weeks. Daily cell count was calculated using a haemocytometer. To determine the effect of different photoperiods, microalgae cultured in a 2 litre flask containing Conway media was treated at different photoperiods (light/dark) (24:0, 12:12 and 8:16) in replicates and cell count was achieved as mentioned above. Growth curve for each species of algae was constructed and One-way ANOVA with Dunnett's post-test was performed using Graph-Pad Prism.

Mean cell count and specific growth rate were calculated using the formula $X=\frac{\Sigma\chi_i}{n}$ and $SGR=Ln\,\frac{W_2/W_1}{t_2-t_1}$ respectively (where, X = mean cell count; χi = total number of cells; 'n'= number of cell counts; SGR = specific growth rate; $W_1$= Initial cell density, $cell^{-1}$; $W_2$= Cell density at late exponential phase, cell$^{-1}$; $t_1$ = Time at initial cell density, cell$^{-1}$; $t_2$ = Time at late exponential phase, cell$^{-1}$).

## Experiment 3: Low-Cost Commercial Fertiliser for Mass Culture of Marine *Chlorella vulgaris*: Manipulation of N:P:K Ratio

An investigation was made to see the adaptability of the local marine *C. vulgaris* to the natural conditions in an aquaculture farm. This means that they need to adapt to different fertilisers other than Conway media, different salinity regimes and uncontrolled temperatures. Preparation of NPK-based fertiliser was made by manipulating the ratio of nitrogen, phosphorus and potassium source as summarises in Table 1. Each of the different N:P:K ratio treatments was prepared in triplicate. Source of nitrogen was obtained by using urea fertiliser.

Culture containers were well-cleaned with bleach and rinsed thoroughly before filling up with 1L of the farm water (salinity of between 20-25ppt). The marine *C.vulgaris* concentrate was prepared and 1mL of it was inoculated into the container and 1mL of the fertilizer was added. The containers were vigorously aerated to provide required quantity of oxygen and to keep cells and media in suspension. The containers were kept in the open under 100% outdoor light exposure.

**TABLE 1.** Type and fertiliser used in N:P:K ratio for mass culture of marine *Chlorella vulgaris*

| N:P:K ratios | | | Type and fertiliser used | | |
|---|---|---|---|---|---|
| N | P | K | Urea (g) | P+ (mL) | Potash+(g) |
| **1** | **1** | **1** | **0.98** | **0.98** | **0.98** |
| 15 | 15 | 15 | 14.7 | 14.7 | 14.7 |
| 8 | 8 | 2 | 7.84 | 7.84 | 1.96 |
| 16 | 8 | 6 | 15.68 | 7.84 | 5.88 |
| 12 | 6 | 4 | 11.76 | 5.88 | 3.92 |
| 12 | 8 | 4 | 11.76 | 7.84 | 3.92 |

Sampling of microalgae cells was done daily and counting was carried out using a Neubauer Hemocytometer covered with glass slide under a compound microscope.

The growth rate, divisions per day, and generation time or doubling time was calculated following [49]

Growth rate; K' = Ln $(N_t/N_o)/(t_2 - t_1)$ (8)

Divisions per day; Div.day$^{-1}$ = K'/Ln2 (9)

Generation time(days); Gen't = 1/Div.day$^{-1}$ (10)

Generation time (hours); Gen't = 24(1/Div.day$^{-1}$) (11)

Where, No and Nt = final and initial populations at time t1 and time t2, respectively.

Since sample was collected daily, therefore, t2 – t1 = 1.

## Experiment 4: Egg Production, Growth and Development of *Apocyclops ramkhamhaengi* Fed on Marine *Chlorella vulgaris*

Detailed observation on the reproduction performance of a zooplankton depending solely on a *C.vulgaris* diet was planned to prove the important role played by this local microalgae in live-feed production. Samples of copepods were collected from Sungai Semerak (N 05° 51.737, E 102° 30.809'), Tok Bali, Kelantan using a zooplankton net. This area receives sea water from the South China Sea, which is near to the Thailand coast where the copepod species was first identified and reported. Live copepods were maintained and adapted to the laboratory environment. Sand-filtered sea water from the Marine Hatchery, Universiti Malaysia Terengganu was diluted with deionised water to be at salinity of 25ppt and was further filtered through a GFC membrane filter and then autoclaved at 121°C for 15 minutes [11]. Salinity was measured using a portable hand-refractometer (ATAGO, Japan). Microalgae diet for the copepod was prepared from the marine algae *C.vulgaris* stocked at the Marine Hatchery. The microalgae were cultured in 29-31ppt Conway medium with 24h-light, room temperature of 25-27°C and continuous aeration for 7days. The cell concentration in each 500ml conical flask was determined by using Neuber haemocytometer (0.25mm$^2$ x 0.1 mm) under a compound microscope. The algal production was done weekly and supplied to *A. ramkhamhaengi* culture.

The investigation on the reproduction performance started with fifteen gravid females of *A.ramkhamhaengi* placed into two sets of triplicate of 250mL beakers. The diet constituted,1mL of Baker's yeast (0.02g/L) and 1mL of *C.vulgaris* at density 1x10$^6$cells/mL which were introduced into both sets of the beakers and covered with parafilm layer to avoid contamination. Three subsamples (approximately 1mL) from each beaker of the cultures were observed daily. The number of the copepods at all stages, including nauplii, copepodite, adult and gravid female, were counted under a dissecting microscope (Leica ZOOM 2000) and then returned to the culture. Changing of approximately 80% of the culture medium was done every alternate day by passing the copepods culture through 100 and 40 microns nylon net which would retain all stages of copepods (the smallest size of 60 microns) but remove most of the waste.

The population growth of *A. ramkhamhaengi* was studied for 30 days. The specific growth rates (K) of all stages of the copepods in both diets given were calculated by using the formula [50]:

$$K = \frac{\ln Nt - \ln No}{t} \tag{12}$$

Where, t is the culture days, No and Nt is the number of copepods at the initial and final selected time interval. The doubling time (Dt) was calculated by dividing $\log_e 2$ by the population growth rate (K) of all stages of *A. ramkhamhaengi* in both diets given:

$$Dt = \frac{\log_e 2}{K} \tag{13}$$

Although cyclopoid copepods are known to suspend in water column, *A. ramkhamhaengi* showed its adaptability to swim on the near bottom of its culture vessel. The culture for this experiment was started by introducing a gravid female on the experimental petri dish. The adult was removed after the eggs hatched and the nauplii were monitored until they reached copepodite-v stage and were ready to mate. Adult females and males from the culture were prepared for the experiment. A pair of male and female was put into each set of glass Petri dish filled with 15mL sea water. The use of a Petri dish instead of a beaker eased the daily observation of different stages of the copepod in the population. The cultures were maintained at room temperature of 25-27°C without additional oxygen supply or aeration. Observation was done twice per day under a dissecting microscope (Leica ZOOM 2000) before feeding to avoid the disturbance of the diet materials during individual or population counting. The culture medium was changed approximately 80% daily, and culture containers were subsequently changed every 4days. Daily feeding was done in the morning and evening by dropping 1mL of $1 \times 10^6$ cells/ml *C.vulgaris* into the culture. The time taken for the females to become gravid was based on the observation recorded twice per day (morning/evening). Once the females become gravid, the male broodstocks were removed, and the female were left alone inside the Petri dish in order to determine the number of eggs per female from its first copulation. Observation on the development time from nauplii to adult, maturation time and generation time of *A. ramkhamhaengi* were recorded coupling with the numbers of offspring produced and percentage of hatching.

# 3. RESULT AND DISCUSSIONS

## Experiment 1: Production of *Chlorella vulgaris* Concentrate Isolated from Bidong Island and Assessment as Copepod Diet

The ultra-structure of the *C.vulgaris* isolated and cultivated in this study is shown in Figure 1. The scanning electron micrographs displayed the characteristic features of green single cells with spherical shape and possession of rigid cell wall. There are some differences found in the present specimen if compared to some other established species. The outer shell is rough if compared to the latest SEM of *C.sorokiana* [51].The feature is almost the same as found in SEM of *C.vulgaris* [52]. In terms of size, the specimen was found to be in between the size of marine *C.vulgaris* (2.1μm) and estuarine *C.vulgaris* (2.3μm) from Korean waters [53].

The cell has an average cell biovolume of 5.26±0.87 $\mu m^3$. The cell densities changed following the culture period in both culture of concentrates (paste) and live condition (although they both started at the same density). Nonetheless, they followed more or less the same growth patten. The variation in cell densities during the experimental period is shown in Figure 2. Cell density of *C.vulgaris* increased rapidly to 227.22±0.87 $\times 10^6$ cells $mL^{-1}$ prior to stationary phase. After that, the cell densities maintained at this point for ten days before decreasing significantly thereafter (Figure 2). The average specific growth rate (SGR, μ) achieved during the exponential phase was 0.660±0.001 $day^{-1}$ with the doubling time (T) of 0.580±0.004 hour which then decreased drastically to 0.126±0.001 $day^{-1}$ during the retardation phase with the doubling time of 2.420±0.019 hour before the death phase. Based on cell density and growth rate observed, the following growth phase is described:

i. Exponential (log) phase (days 0-6),

ii. Declining of relative growth rate phase (days 6-12),

iii. Stationary phase (days 12-22),

iv. Death phase (days 22-26).

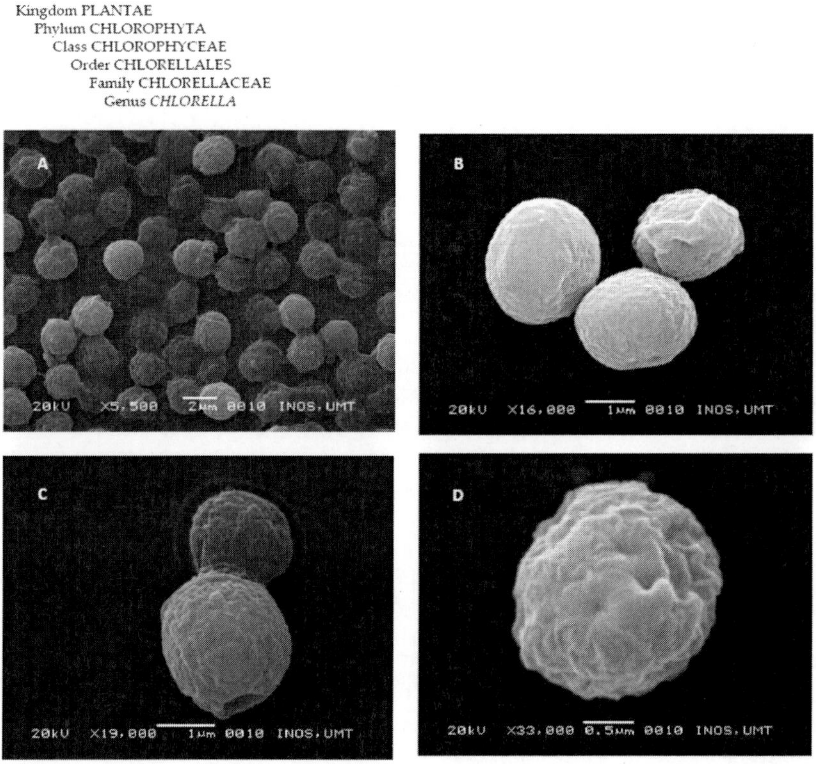


**FIGURE 1.** The scanning electron micrographs of *C.vulgaris* under different views and various magnifications. A, Cell aggregation. B, *C.vulgaris* cells under different magnification C. Cell divison in *C.vulgaris*. D, Single cell and cell wall structures.

*C.vulgaris* paste was successfully concentrated from the pure culture isolated from Bidong Island. This concentrate contains cell density of approximately 58.46±2.44 x$10^9$ cells $mL^{-1}$ - 227.22±0.82 x$10^6$ cells $mL^{-1}$. The present result also showed that this *C.vulgaris* concentrate can still be inoculated after refrigeration for a duration of six weeks and exhibited similar growth characteristics as the live culture (Figure 2). The cells had very high viability even after 6 weeks of storage in chilling conditions (4ºC) as shown in Figure 3. It is interesting to note that the paste had recorded the highest cell viabilities of 99.51±0.57% and continued to display slow and steady decrement of cell viabilities to 83.28±0.58% on the sixth week of storage. Microscopic examination also indicated that the cells were in single forms without any aggregation occurring and can be readily dispersed in seawater medium as single suspension of cells upon inoculation (Figure

1A). The harvesting efficiency of the ultrafiltration technique using membrane filter had recorded a very high percentage recovery of 93.14±1.35% showing the effectiveness of this technique for harvesting and concentrating the microalgae biomass.

Copepod species, *Apocyclops* sp., showed a higher population density when fed with *C.vulgaris* paste (60±4.36 individual $mL^{-1}$) than with the live culture (14.33±0.58 individual $mL^{-1}$) (Figure 4). In addition, the copepod populations fed with this concentrate exhibited a higher instantaneous growth rate, K=0.455±0.008 $day^{-1}$ and faster doubling time (0.662±0.012 hour) than live culture which recorded an instantaneous growth rate of 0.296±0.005 $day^{-1}$ and doubling time of 1.108±0.016 hour. Results from one-way ANOVA test has shown that there is a significant difference ($P<0.005$) between the copepod population densities fed with the *C.vulgaris* concentrate and live culture.

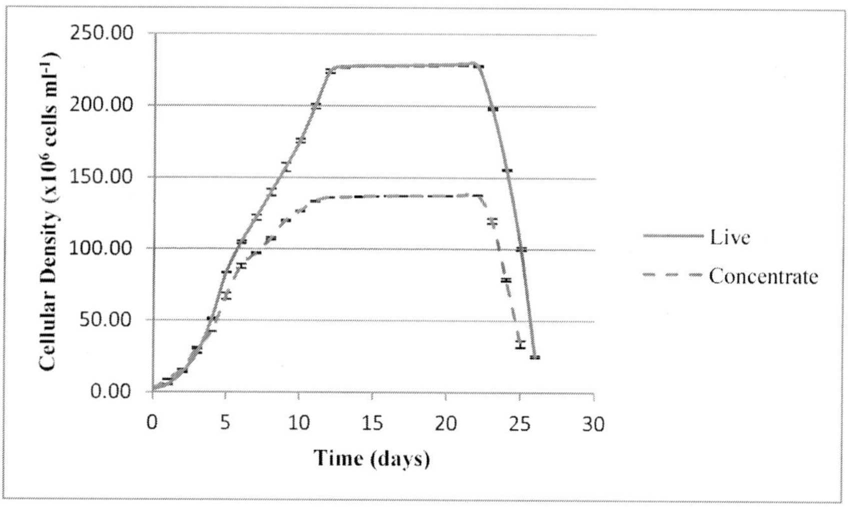


**FIGURE 2.** Cellular densities of live *C.vulgaris* (30L) and concentrate/paste (after reinoculation in 5L) cultured under laboratory conditions with Conway medium. Data are mean value and standard deviation of 3 repetitions.

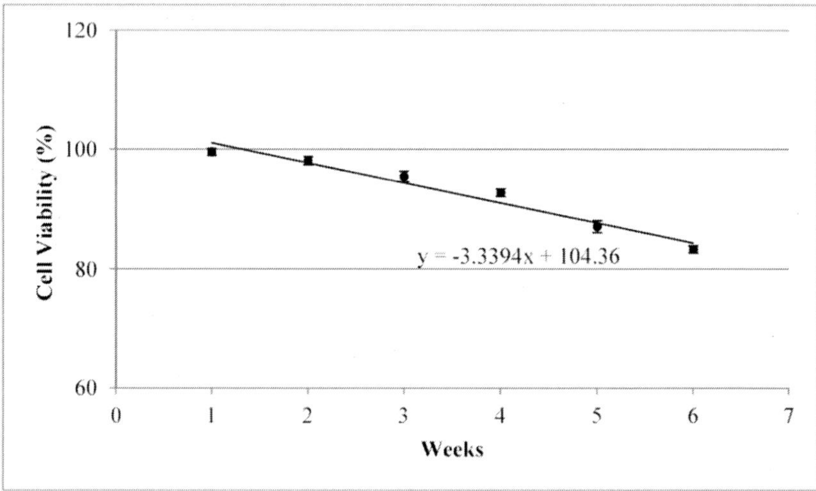


**FIGURE 3.** Variations in the cell viabilities of *C.vulgaris* microalgae concentrate over 6 weeks of storage in chilling condition at 4º C. Data are mean value and standard deviation of 10 repetitions.

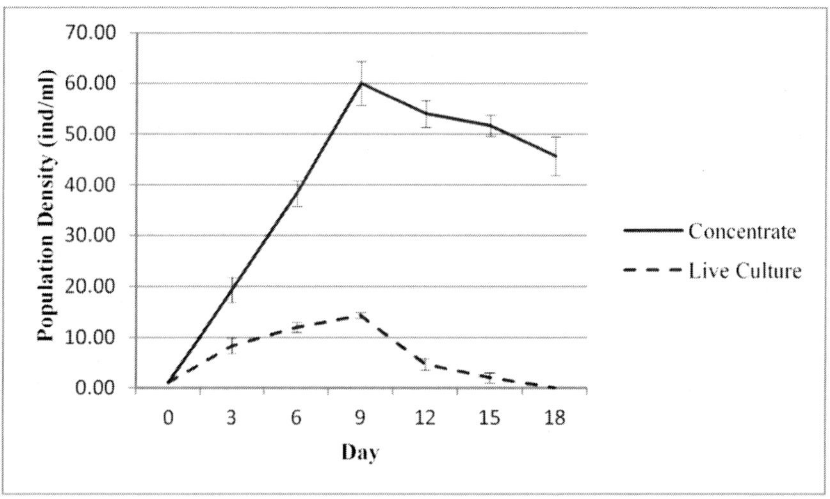


Data are mean value and standard deviation of 3 repetitions.

**FIGURE 4.** Population densities of *Apocyclops* sp. fed with live culture and microalgae concentrate of *C.vulgaris*.

*C.vulgaris* isolated from Bidong Island had exhibited an extremely rapid growth rate. This might be attributed to its relatively small cell volume with an average of 5.26±0.87 $\mu m^3$. This is indeed much smaller than the cell sizes recorded for some other species; *Chaetoceros calcitrans* (8µm diameter, volume 276.95 $\mu m^3$) and *Isochrysis galbana* (4µm diameter, volume 33.49 $\mu m^3$) [54]. Small size species of microalgae grow faster with a rapid growth rate. This is due to the reason that the greater surface-to-volume ratio of smaller size cells facilitates assimilation of nutrients at a relatively faster rate. In addition, the smaller size cells may achieve high density because they occupy less space. Apart from that, it was also cultured with optimal values for all environmental factors in the laboratory, thus promoting favourable environmental conditions for the cells to grow to extremely high density. In this high-density culture, the possibilities of contamination were excluded. The sudden collapse in the growth rate after day six could be mainly the result of the depletion of the nutrient in the culture. Growth rate declines and growth of microalgae ceases when the nutrient in shortest supply relative to the metabolic needs of algal population [27,55]. The populations of *C.vulgaris* cells then entered the stationary phase of the growth cycle and collapsed after day twenty two. The long stationary phase of this culture might indicate that contamination was absent during the culture period. It has been reported that this stationary phase can last for several weeks if there is no contamination in the culture [30].

The ultrafiltration technique which was used to concentrate the *C.vulgaris* cells in this study can be applied to concentrate a range of other microalgae species used as aquaculture feeds. Concentrating and storing the microalgae concentrate in moist form preserves its high nutritional value through maintaining excellent cell viability [33, 35]. The cells were readily re-suspended upon dilution in sea water with high cell viability which was proven by their ability to be inoculated even after storage for a duration of 6 weeks. The efficiency of ultrafiltration through this study was ≥90% which is very comparable to the reported efficiency of ≥80% for flocculation technique by Knuckey *et al.* [34]. There has been no comparative assessment of concentrates prepared by ultrafiltration with those prepared by centrifugation. However, from a practical and theoretical point of view, it is proven that the centrifugation method possesses some disadvantages due to its exertion of shear gravitational forces rupturing the microalgae cell structure during harvesting procedure. This reduces their nutritional values due to leaking of nutritional contents. On the other hand, microalgae concentrates prepared by ultrafiltration are not subjected to the same gravitational forces during harvesting. As reported earlier [38], the major production cost of centrifuged concentrates may exceed US$10,000 (RM32,620) which is unaffordable for small-scale hatcheries and is likely to be limited to larger hatcheries with specialised equipments or facilities specifically set up to produce microalgae concentrates to hatcheries.

Advantages of the ultrafiltration technique used in this study is that it is a relatively simple, inexpensive and volume-independent process which can be readily adopted by small-scale hatcheries to prepare their own microalgae concentrates on site.

The use of *C.vulgaris* concentrate as diet for cyclopoid copepods increases population density, instantaneous growth rate as well as doubling time and it was proven as a better diet than the *C.vulgaris*live cultures. This might be possibly due to the significantly higher cellular density of the microalgae concentrates. The rates of ingestion and egg production in copepods are dependent on the quantity of the provided microalgae [56, 57] implying that quantity of food is the most important factor regulating the productivity of copepod culture. Other studies have also demonstrated that the rate of egg production of calanoid copepod, *Acartia tonsa,* increases with increasing food concentrations [58, 59]. Essential substances such as cholesterol, HUFA and PUFA are present or exist abundantly in microalgae, and, copepod production is positively related to the lipid levels or DHA: EPA ratio in the diet [60]. Thus, microalgae concentrate could be useful as a replacement for live or fresh microalgae. This is extremely important as a stable and continuous supply of live feed for aquaculture hatcheries must always be provided.

## Experiment 2: Effects of Photoperiod and Culture Size on *C.vulgaris* Stock Growth

It is very important for hatcheries to be able to maintain the stock for microalgae for their sustainable live-feeds supply. Batch cultures need to be maintained under optimal environmental conditions and in a suitable culture vessel which will not affect the cell density and quality. Comparison on the effect of photoperiod and culture sizes between *C.vulgaris* and other microalgae was made to investigate the adaptability of the species to simple stock handling in the laboratory or hatchery conditions. No significant difference in the cell density was noted in *Nannochloropsis* sp. (Figure 5A) and *C.vulgaris*(Figure 5C) cultures grown in different volume flasks while *Chaetoceros* sp. Figure 5B showed significant variation in cell-density level at similar culture conditions ($P < 0.001$). However, stationary phases of all cultured species were achieved earlier in 250mL flask compared to the cultures in 500mL and 2L flasks. All cultures showed greater response towards daylight variations whereby higher cell density was noted in culture flasks exposed to continued illumination (24:0 L/D), and it was followed by 12:12 L/D and 6:18 L/D condition. *Nannochloropsis* sp. (Figure 5D) responded less towards the treatment compared to the other 2 species (Figure 5E & F) which clearly showed a specific response towards culture conditions.

It is well-documented that, in natural conditions, microalgae growth is not curtailed by ambient environmental conditions because the growth rate is just enough for species survival. However, their multiplication rate is highly influenced by various environmental parameters. In an *In vitro* setup, the proper maintenance of optimum culture condition triggers the metabolic pathway of target species in a unidirectional fashion to achieve high cell density. In this study, a higher cell density of microalgae in low volume flask culture together with early stationary phase was observed could be used to obtain continuous harvest of selected microalgae.

Highest cell density and specific growth rate were recorded in selected species cultured in 250mL culture flask compared to the cultures in 500mL and 2000mL flasks (Table 2). The highest cell density was achieved during the end of the log phase. Cell density of early stationary phase, which is the end of the log phase for *Nannochloropsis* sp., *Chaetoceros* sp. and *C.vulgaris* was achieved on the $10^{th}$ day of culture in 250mL. After the $10^{th}$ day, density decreased and the lowest level was different for different species and different culture volumes. Similar results were also noted for the specific growth rate values. Significant variation in cell density and specific growth rate was observed between the cultures in different sizes of culture flask ($P < 0.05$ or $P < 0.001$). This observation might probably be due to the light-penetration efficiency in the culture flask. Similar observation was noted for the culture of *Nannochloropsis* sp. in 2000mL flasks which produced greater cell density compared to the culture in 20L carboys [61]. The effect of light saturation could decrease in the denser culture and the average irradiance in the culture reduced due to absorption from other cells [62]. The large volume culture needed higher light intensity to allow light penetration while the smaller volumes were less affected by the light penetration.

**TABLE 2.** Cell density and specific growth rate of selected microalgae cultured at different flask volumes.

| Microalgae | Flask volume (mL) | Early stationary phase | Cell density ($\times10^6$ cells $mL^{-1}$) | Specific growth rate (K') |
|---|---|---|---|---|
| *Nannochloropsis sp* | 250 | Day 10 | $112.5\pm2.36^{a}$ | $0.52\pm0.01^{a}$ |
| | 500 | Day 11 | $110.17\pm1.77^{a}$ | $0.46\pm0.04^{b}$ |
| | 2000 | Day 13 | $92.2\pm0.87^{a}$ | $0.34\pm0.01^{c*(ac)}$ |
| *Chaetoceros sp.* | 250 | Day 10 | $12.460\pm0.018^{a}$ | $0.203\pm0.002^{a}$ |
| | 500 | Day 12 | $10.889\pm0.013^{b**(ab)}$ | $0.145\pm0.001^{b*(ab)}$ |
| | 2000 | Day 14 | $8.225\pm0.001^{c**(ac\ \&\ bc)}$ | $0.037\pm0.003^{c**(ac);\ *(bc)}$ |
| *C vulgaris* | 250 | Day 10 | $117.53\pm0.84^{a}$ | $0.4749\pm0.0007^{a}$ |
| | 500 | Day 11 | $91.0\pm0.55^{a}$ | $0.4081\pm0.0002^{b}$ |
| | 2000 | Day 14 | $86.13\pm0.81^{a}$ | $0.3166\pm0.0007^{c*(ac)}$ |

*Note:* Data represented in Mean ± SD. [*] and [**] indicates significant difference at $P < 0.05$ and $P < 0.001$(respectively) level between different superscripts depicted for each species.

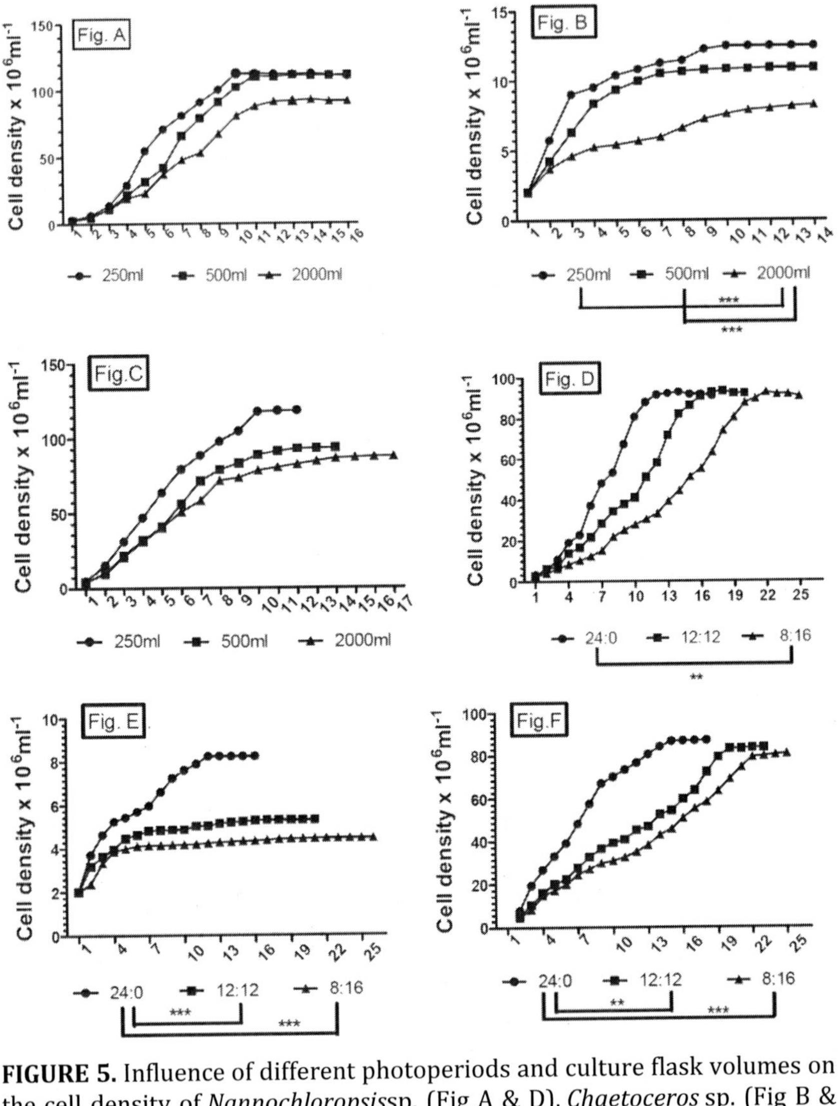


**FIGURE 5.** Influence of different photoperiods and culture flask volumes on the cell density of *Nannochloropsis*sp. (Fig A & D), *Chaetoceros* sp. (Fig B & E) and *C.vulgaris* (Fig C & F) respectively. X-axis shows the days of culture. Data represented as 250ml, 500ml and 2000ml are culture flask volumes while 24:0, 12:12 and 8:16 are photoperiods (Light : Dark phase). [***] and [**] are significantly different at P < 0.001 and P < 0.05 level in the culture conditions respectively.

**TABLE 3.** Cell density and specific growth rate of selected microalgae cultured at different photo periods.

| Microalgae | Photo period (Light : Dark phase) in hours | Early stationary phase | Cell density ($x10^6$cells $ml^{-1}$) | Specific growth rate (K') |
|---|---|---|---|---|
| *Nannochloropsis* sp. | 24:0 | Day 13 | $112.5 \pm 2.36^{a}$ | $0.34 \pm 0.01^{a}$ |
| | 12:12 | Day 17 | $110.17 \pm 1.77^{a}$ | $0.25 \pm 0.01^{b}$ |
| | 8:16 | Day 23 | $92.2 \pm 0.87^{a}$ | $0.19 \pm 0.02^{c* (ac)}$ |
| *Chaetoceros* sp. | 24:0 | Day 12 | $8.225 \pm 0.001^{a}$ | $0.129 \pm 0.003^{a}$ |
| | 12:12 | Day 17 | $5.293 \pm 0.009^{b** (ab)}$ | $0.061 \pm 0.002^{b* (ab)}$ |
| | 8:16 | Day 22 | $4.453 \pm 0.003^{c** (ac)}$ | $0.037 \pm 0.003^{c** (ac)}$ |
| *C.vulgaris* | 24:0 | Day 14 | $86.60 \pm 0.17^{a}$ | $0.3170 \pm 0.0001^{a}$ |
| | 12:12 | Day 19 | $83.04 \pm 0.19^{b** (ab)}$ | $0.2313 \pm 0.0001^{b}$ |
| | 8:16 | Day 21 | $79.23 \pm 0.21^{c** (ac)}$ | $0.2010 \pm 0.0001^{c* (ac)}$ |

*Note:* Data represented in Mean ± SD. [*] and [**] indicates significant difference at $P < 0.05$ and $P < 0.001$ (respectively) level between different superscripts depicted for each species.

Highest cell density and specific growth rate were recorded in all cultured species that were exposed to continued illumination (24:0. L/D) followed by 12:12 and 6:18 L/D respectively. Early stationary phases differed for *Nannochloropsis* sp. (day 13), *Chaetoceros* sp. (day 12) and *C.vulgaris.* (day 14) respectively while the corresponding specific growth rate was also highest under 24hours illumination. Significant variation in both growth parameters was observed between the cultures exposed to different photoperiods ($P < 0.05$ or $P < 0.001$) (Table 3). Photosynthetic efficiency of microalgae can be enhanced by sudden alteration between light and dark phase [63]. During this process, the fast reduction of e-acceptors, Qa and Qb, associated to photosystem II (PSII) followed by their oxidation in the dark period will take place that will ultimately maximise the proton-accepting capacity of PSII during sudden irradiant of light [64].

*C.vulgaris* proved its adaptability to different culture volumes and lighting periods with good growth performance comparable to *Nannochloropsis* sp. and better than *Chaetoceros* sp. The cells responded positively towards continuous illumination of light by producing higher cell density and specific growth rate in the culture media. It was also noted that the culture in the low-volume flask produced an early stationary phase due to high penetration of light and continuous sharing of available nutrients in the media for faster growth and survival. On the other hand, *C.vulgaris* consistently grew at significant cell densities even in larger volume containers and shorter period of illumination than dark condition (comparable to *Nannochloropsis* sp. and better than *Chaetoceros* sp.). In another study to analyse the effect of photoperiod to the cellular essential fatty acid in these species, the photoperiod of 12:12h L/D regime is recommended for the fast and economical technique for batch culture production [65]. A better ratio of essential fatty acid accumulated in *C.vulgaris* exposed in the 12:12h if compared to 24:0 or 8:16 L/D photoperiod.

## Experiment 3: Low-cost Commercial Fertiliser for Mass Culture of Marine *Chlorella vulgaris*: Manipulation of N:P:K Ratio

*C.vulgaris* showed its adaptability to grow well when fertilised with a low-cost commercial N:P:K plant fertiliser (Figure 6). Duration of the log phase for *C.vulgaris* varied among treatments. The 12:6:4 and 12:8:4 ratios had a result of 3 days while the longest period was in the 15:15:15 treatment. Combined applications of urea, P+ and K+ (N:P:K; 12:6:4) produced the highest cell number ($4.0\times10^6$ cells $mL^{-1}$) during log period at 5 days while N:P:K; 15:15:15 (control) produced highest cell number ($4.16\times10^6$cells $mL^{-1}$) at 7 days of log period. Different ratios of N:P:K, 12:8:4, 8:8:2 and 16:4:6 resulted in decrease of cell density, $3.3\times10^6$, $3.0\times10^6$ and $2.7\times10^6$ cells $mL^{-1}$, respectively.

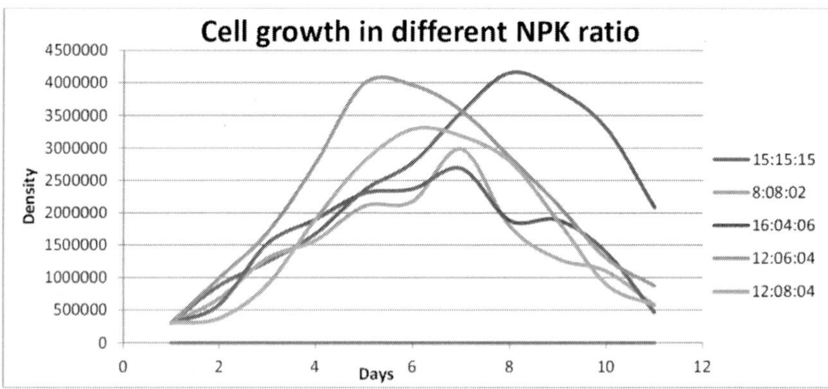


**FIGURE 6.** Density (cells $mL^{-1}$) of *C.vulgaris* cultured with commercial fertiliser of different N:P:K ratios

**TABLE 4.** Growth rate of *C.vulgaris* cultured with commercial fertiliser of different N:P:K ratios

| | | Days in Log Phase | | | | | |
|---|---|---|---|---|---|---|---|
| NPK ratio | 2nd | 3rd | 4th | 5th | 6th | 7th | |
| | | Growth rate; K' | | | | | Average |
| 15:15:15 | 1.07 | 0.35 | 0.30 | 0.34 | 0.17 | 0.25 | 0.41 |
| 8:08:02 | 0.81 | 0.65 | 0.20 | 0.29 | | | 0.49 |
| 16:04:06 | 0.65 | 0.96 | 0.23 | 0.19 | | | 0.51 |
| 12:06:04 | 1.19 | 0.53 | 0.50 | | | | 0.74 |
| 12:08:04 | 0.20 | 0.87 | 0.77 | | | | 0.61 |

**TABLE 5.** Division per day of *C.vulgaris* cultured with commercial fertiliser of different N:P:K ratios

| | | Days in Log Phase | | | | | |
|---|---|---|---|---|---|---|---|
| NPK ratio | 2nd | 3rd | 4th | 5th | 6th | 7th | |
| | | Division/ Day; Div. day-1 | | | | | Average |
| 15:15:15 | 1.54 | 0.50 | 0.44 | 0.48 | 0.25 | 0.36 | 0.59 |
| 8:08:02 | 1.16 | 0.94 | 0.28 | 0.42 | | | 0.70 |
| 16:04:06 | 0.94 | 1.39 | 0.33 | 0.28 | | | 0.73 |
| 12:06:04 | 1.72 | 0.77 | 0.72 | | | | 1.07 |
| 12:08:04 | 0.29 | 1.25 | 1.11 | | | | 0.89 |

The 12:6:4 ratios showed the best average (74%) growth rates of natural increase at log phase. The second was 12:8:4 with 61% average growth rate.15:15:15 NPK ratio showed the lowest average growth rate of 41% (Table 4). The *C.vulgaris* cell in 12:6:4 NPK ratio recorded an average division per day by 107% which was the best compared to others. In 12:8:4 ratios the average cell division was 89% and decreasingly followed by 16:4:6 and 8:8:2 for 73% and 70% respectively. The control ratio which was 15:15:15 showed the lowest average division which was 59% (Table 5).

**TABLE 6.** Generation time (days) for *C.vulgaris* cultured with commercial fertiliser of different N:P:K ratios

| | Days in Log Phase | | | | | | |
|---|---|---|---|---|---|---|---|
| NPK ratio | 2nd | 3rd | 4th | 5th | 6th | 7th | |
| | | Generation time (days); Gen't | | | | | Average |
| 15:15:15 | 0.65 | 2.00 | 2.28 | 2.06 | 4.07 | 2.80 | 2.31 |
| 8:08:02 | 0.86 | 1.06 | 3.55 | 2.38 | | | 1.96 |
| 16:04:06 | 1.06 | 0.72 | 3.04 | 3.62 | | | 2.11 |
| 12:06:04 | 0.58 | 1.31 | 1.39 | | | | 1.09 |
| 12:08:04 | 3.41 | 0.80 | 0.90 | | | | 1.70 |

**TABLE 7.** Generation time (hour) for *C.vulgaris* cultured with commercial fertiliser of different N:P:K ratios

| | Days in Log Phase | | | | | | |
|---|---|---|---|---|---|---|---|
| NPK ratio | 2nd | 3rd | 4th | 5th | 6th | 7th | |
| | | Generation time (Hour); Gen't | | | | | Average |
| 15:15:15 | 15.56 | 47.99 | 54.84 | 49.55 | 97.66 | 67.32 | 55.49 |
| 8:08:02 | 20.60 | 25.55 | 85.22 | 57.05 | | | 47.11 |
| 16:04:06 | 25.44 | 17.28 | 73.01 | 86.79 | | | 50.63 |
| 12:06:04 | 13.97 | 31.34 | 33.37 | | | | 26.22 |
| 12:08:04 | 81.76 | 19.14 | 21.54 | | | | 40.82 |

Measurement of generation time for *C.vulgaris* is summarised in Table 6 and Table 7. *C.vulgaris*cultured with fertiliser of the ratio 12:6:4 only took 1.09 days (26.22 hour) to complete one generation of replication, the shortest time compared to other treatments. The longest generation time was when using the15:15:15 ratio which was completed in 2.31 days (55.49 hour). The other three intermediate treatments recorded 1.70 (40.82 hour), 1.96 (47.11 hour) and 2.1 days (50.63 hour) for 12:8:4, 8:8:2 and 16:4:6 respectively. When comparing the performance by using all of the growth parameters, N:P:K; 12:6:4 ratio gave the best result with average growth rate per day (74%), maximum growth rate $day^{-1}$(107%), maximum cell density ($4.0\times10^{6}$ cell/mL), division's $day^{-1}$ (107%) and generation time (1.09 day; 26.22 hour). *C.vulgaris* in control treatment (15:15:15) exhibited the poorest growth performance. Nonetheless, it is interesting to note that they experienced longer log period which could give more time for reproduction activity, thus the density did not decrease drastically as when cultured using other ratios. The fluctuation of temperature and different salinities could be the reason why cell densities were not as high as the first and second experiment.

Numerous nutrient media have been use for the culture of pure *Chorella* sp. Most of those were for laboratory use and/or for low-grade production of algae. Majority of these media are composed of pure nutrients (N-8). Commercial fertilisers are least considered for *Chlorella* culture because of the conception that they do not provide required nutrients for algal growth and are mostly suitable for crop (land) agriculture. Nevertheless, it has been proved that the commercial plant fertiliser could support a freshwater *Chlorella* [66]. The use of N:P:K fertiliser could be a better choice if compared to the organic fertiliser. Organic matter has its own limitations and depends on the microbial activity to release the inorganic nutrient and it cannot be compared to the performance of pure nutrients. Despite good growth performance, the short period of the log phase when *C.vulgaris* is cultured using N:P:K; 12:6:4 need specific and efficient up-scaling or harvesting method, indicating that other ratios such as 15:15:15 could be a better choice.

## Experiment 4: Egg Production, Growth and Development of *Apocyclops ramkhamhaengi* Fed on *C.vulgaris*

Different diets gives significantly ($P<0.05$) different densities of *A. ramkhamhaengi*. The mean gravid production of *A. ramkhamhaengi* fed on *C.vulgaris* and Baker's yeast was highest on 23$^{rd}$ day with 1.11ind./ml and 0.67ind./ml respectively. The production peaked on 11$^{th}$, 23$^{rd}$, and 26$^{th}$ day and on 17$^{th}$, 20$^{th}$ and 23$^{rd}$ day when fed with *C.vulgaris*. and Baker's yeast respectively (Figure 7). In this 30 days culture, the highest mean population density of *A. ramkhamhaengi* fed with *C.vulgaris* was recorded on the 9$^{th}$ day with 3.31ind./ml and when fed on Baker's yeast was on day 19$^{th}$ with 1.83ind./ml (Figure 8). *A. ramkhamhaengi* fed on *C.vulgaris* showed the higher instantaneous growth rate (K) than when fed on Baker's yeast (Table 8). The period taken to double their population (Dt) was shorter in *A. ramkhamhaengi* fed *C.vulgaris* (8days) than Baker's yeast (11days).

**TABLE 8.** The instantaneous growth rate and doubling time of *A. ramkhamhaengi* fed on different diets.

| Diets | Instantaneous growth rate (K) | Doubling Time(Dt)(day) |
|---|---|---|
| *C.vulgaris* | 0.1150 | 8 |
| Baker's yeast | 0.0756 | 11 |

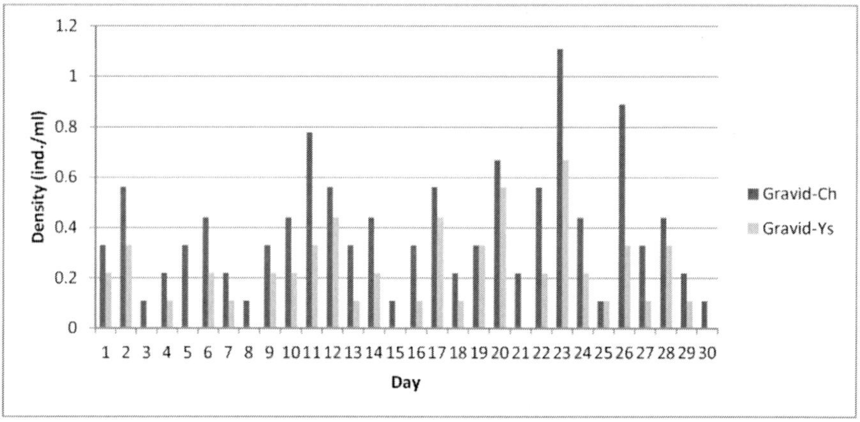


**FIGURE 7.** Mean density of gravid female of *A. ramkhamhaengi* fed with *C.vulgaris* (Gravid-Ch) and Yeast (Gravid-Ys) in 30 days.

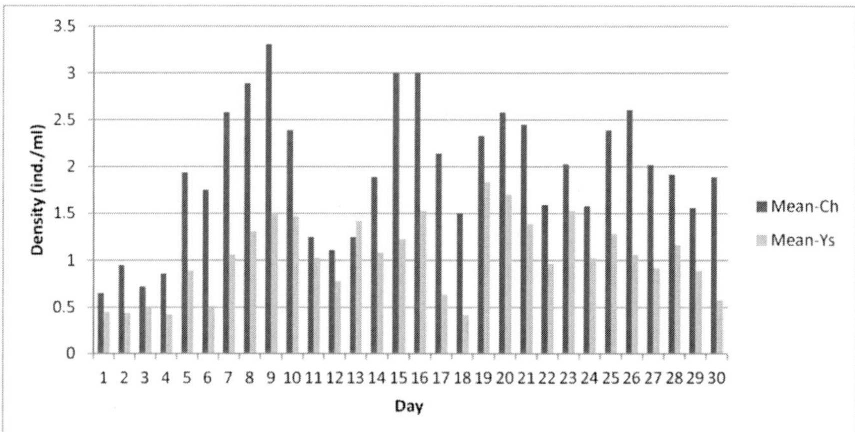


**FIGURE 8.** Mean total density of *A. ramkhamhaengi* fed with *C.vulgaris* (Mean-Ch) and Yeast (Mean-Ys) in 30 days.

The development times for nauplii, copepodite, adult and gravid female were observed separately using the copepod culture fed on *C.vulgaris*. The longest period was at copepodite stage (7.33 ± 2.08days) and the shortest period was the naupliar stage which needed only 1.33 ± 0.58 days (Figure 9). The mean number of eggs produced was 21.33 ± 1.53. Hatching percentage of the three individuals of *A.ramkhamhaengi* was 96.82 ± 2.77 % (Table 9). Maturation time which is the time between the appearance of eggs and their hatching time was 1.33 ± 0.58 days. The time taken to become gravid female from the produced nauplii was about 20.67 ± 3.51days and it is known as generation time.

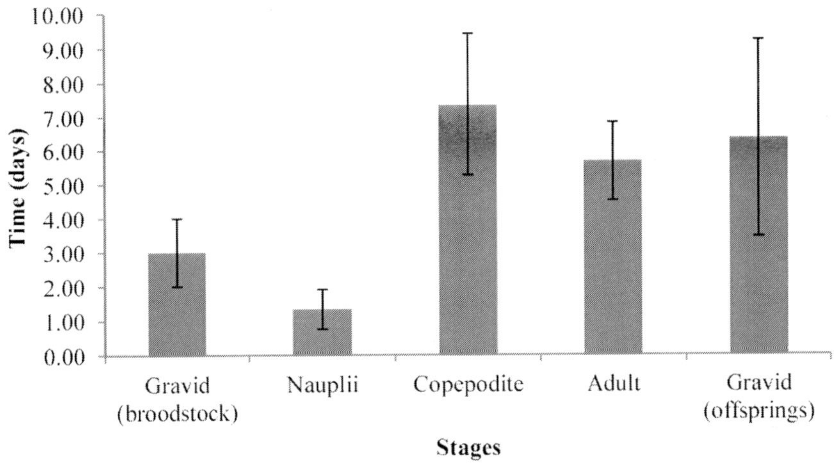


**FIGURE 9.** Mean development time in each stage in *A. ramkhamhaengi* life cycle

**TABLE 9.** The number of observation (N), mean and standard deviation (SD), minimum and maximum of total number of eggs, percentage of hatching of *A. ramkhamhaengi*, maturation time and generation reared in laboratory-controlled condition.

| Parameter | N | Mean ± SD | Minimum | Maximum |
|---|---|---|---|---|
| Number of eggs | 3 | 21.33 ± 1.15 | 20 | 22 |
| % hatching | 3 | 96.82 ± 2.77 | 95 | 100 |
| Maturation time(days) | 3 | 1.33 ± 0.58 | 1 | 2 |
| Generation time(days) | 3 | 20.67 ± 3.51 | 17 | 24 |

Production of the gravid females and the population density obviously increased when *A.ramkhamhaengi* was fed on *C.vulgaris* as compared to Baker's yeast. This finding is in agreement with the previous study [20] on the population growth and production of *A. dengizicus* fed on different diets. Population reached its peak in term of total density for several periods in the 30day culture condition indicates the existence of different populations. These populations reached their peak density in accordance with the diet taken where *Chlorella*-fed population were found to grow faster than those fed on Baker's yeast. It seemed that *A. ramkhamhaengi* has the potential to become more nutritional when enriched, thus growing faster when fed on a good-quality diet such as microalgae if compared to Baker's yeast. The nutritional value has been shown to increase when cyclopoid nauplii stage such as in *A. panamensis* was offered an enriched diet [15]. Although other microalgal diets such as *Tetraselmis*sp. and *Isochrysis* sp. could be a better choice for *Chlorella* sp. [67], at least the present finding is able to prove the potential of the species to reproduce and grow when fed on the marine *C.vulgaris* as used in this study.

A female of *A. ramkhamhaengi* fed on *C.vulgaris* could produce between 20 and 22 eggs with about 97% hatching success. This is more than what has been reported before for *A. panamensis* [68], and it could be related to many environmental factors and culture procedure. Environmental parameters such as temperature, food availability and predation were reported to influence the life-history strategy in copepods [69]. Binary diet of *Nannochloropsis* sp. and T-ISO improved the hatching rate by 88.1 ±2.1 % in a calanoid copepod, *Acartia sinjiensis* [70]. The brackish water cyclopoid, *A. ramkhamhaengi* has shown its potential to be cultured and reproduced under controlled conditions. The population adapted very well to the introduced diet, a marine *C.vulgaris* and a common Baker's yeast. *Chlorella*-fed population of *A. ramkhamhaengi* grow faster and need fewer number of days to double its population than those fed on Baker's yeast.The number of eggs produced was 21.33 ± 1.15 eggs at the maturation time of 1.33 ± 0.58 days and generation time was 20.67 ± 3.51 days. The species show great potential to be cultured together with *C.vulgaris* for hatchery and farm use. A more comprehensive study is essential to investigate the reproductive biology of this species, particularly in a large-scale production system to verify its suitability in aquaculture.

## 4. CONCLUSION

*C.vulgaris* isolated from Bidong Island exhibited a rapid growth rate under optimum environmental conditions in the laboratory culture and was able to achieve an extremely high density when cultured in bigger containers.

The photoperiod of 24:0 proved to be the best condition for cells growth but 12:12 L/D photoperiod could be the more economical. The high-density culture could be harvested using a relatively cheap, inexpensive and simple ultrafiltration technique for other use or reinoculation. This will save the space and long period of maintaining live algae for unexpected use. The cells collected using a ultrafiltration technique showed high viability and long shelf life when kept in 4°C refrigerator. The product is called as *C.vulgaris* paste or concentrate which could be used to enrich or maintain the zooplankton live feeds for aquaculture purposes. The *C.vulgaris* also showed its best growth performance when cultured using a common commercial plant fertiliser with certain ratio of N:P:K. This was shown by their ability to perform cell division and grow and easily adapted to certain ratio such as 15:15:15 and 12:6:4. Nonetheless, the cells density is very much lower than those cultured with the specific chemical fertilizer, Conway media. This problem could be overcome by further investigation on their ion requirement when cultured openly in hatchery or ponds. The suitability of C.vulgaris as enriched diet for a zooplankton potentially used as live feed, A. ramkhamhaengi was proved by the population increase and reached high individual density with good reproduction performance. Maintaining local species of microalgae and zooplanktons in hatchery and ponds will definitely support the continued supply of live feeds for larval rearing and the aquaculture industry.

## 5. ACKNOWLEDGEMENTS

The authors are grateful to the Ministry of Science, Technology and Innovation (MOSTI) for the funding given under the ABI-MOSTI grant (2009-2011), "Mass fry production technology for grouper (Epinephelus sp.)", the Ministry of Education Malaysia for the Knowledge Transfer Programme grant (2012-2013) Using Microalgae and Copepod Live Feeds for Brackish Water Aquaculture Farm, UMT-PPKJBS.

## REFERENCES

1. Sorgeloos, P.The use of brine shrimp Artemia in Aquaculture. The brine shrimp, Artemia. 1980. Proceedings the International Symposium on the Brine shrimp, *Artemia salina*. Universa Press, Wetteren, Belgium. Pp. 25- 46.
2. Leger, P., Bengtson, D. A., Simpson, K. L. & Sorgeloos, P. The use and nutritional value of Artemia as a food source. Marine biology and oceanography: an annual review 1986; 24, 521–623.
3. Chu, K. H. & C. K. Shing. Feeding behaviour of the shrimp, *Metapenaus ensis*, on Artemia nauplii. Aquaculture 1986; 58, 175-184.

4. Kuhlmann, D. G., Quantz and Witt, U. Rearing of turbot larvae (*scopthalmus maximus*) on cultured food organism and postmetamorphosis growth on natural and artificial food. Aquaculture 1981; 23: 183-196
5. Watanabe, T., and Kiron, V. 1994. Prospects in larval fish dietetics. Aquaculture 1994; 124: 223-251.
6. McKinnon, A.D., Duggan,S., Nichols, P.D., Rimmer, M.A., Semmens, G. and Robino, B.The potential of tropical paracalanid copepods as live feeds in aquaculture. Aquaculture 2003; 223(1-4), 89-106.
7. Stottrup, J.G. Review on status and progress in rearing copepods for marine larviculture and advantages and disadvantages among calanoid, harpacticoid and cyclopoids copepods. 2006. VIII Symposium Internacional Acuicola 15-17 Noviembre, Universidad Autonoma de Nuevo Leon, Mexico.
8. Rajkumar & Kumaraguru vasagam. Suitability of the copepod, *Acartia clausi* as a live feed for Seabass larvae (*Lates calcarifer* Bloch): Compared to traditional live-food organisms with special emphasis on the nutritional value. Aquaculture 2006; 261(2), 649-658.
9. Schipp, G. 2006. The use of Calanoid Copepods in semi-intensive, tropical marine fish larviculture. VII Simposium Internacional de Nutricion Acuicola. 15-17 Noviembre. Universidad Autonoma de Nuevo Leon, Monterrey, Nuevo Leon Mexico.
10. Zaleha,K., Ibrahim, B., Akhbar John, B., Kamaruzzaman, B.Y. Generation times of some marine harpacticoid species in laboratory condition. Journal of Biological Sciences 2012; 12(8), 426-432. DOI: 10.3923/jbs.2012.433.437
11. Carli, A., Mariottini, G. L. & Pane, L. Influence of nutrition on fecundity and survival in *Tigriopus fulvus* Fischer (Copepoda, Harpacticoida). Aquaculture 1995; 134,113-119.
12. Cutts, C.J. Culture of harpacticoid copepods: potential as live feed for rearing marine fish. Advances in Marine Biology 2003; 44:295-316.
13. Ananth, S. and Santhanam, P. Laboratory culture and biochemical profile of marine copepod *Macrosetella gracilis* (Dana). Aquaculture 2011; 12(1): 49-55.
14. Rhodes. A. 2003. Methods for high density batch culture of *Nitokra lacustris*, a marine harpacticoid copepod.The Big Fish Bang. In: Howard I. Browman, A., Skiftesvik, B. (eds.) Proceedings of the 26th Annual Larval Fish Conference. Institute of Marine Research, Bergen, Norway.
15. Sumiarsa, G.S. Cyclopoid copepod nauplii Apocyclops panamensis: mass production method in outdoor ponds and their fatty acid profiles.VDM Verlag Dr.muller GmhH & Co. KG. Germany. 2011.
16. Drillet,G., Frouël,S., Sichlau, M.H., Jepsen, P.M., Jonas K., Højgaard, J.K., Joarder,A.K., Hansen, B.W. Status and recommendations on marine copepod cultivation for use as live feed. Aquaculture 2011; 315:155-166
17. Zaleha, K., Farahiyah, I. J. Culture and growth of a marine harpacticoid, *Pararobertsonia* sp in different salinity and temperature. Sains Malaysiana 2010; 39(1): 135–140
18. Kassim, Z.,Ibrahim, B. Culture of Harpacticoid Copepods: Understanding the Reproduction and Effect of Environmental Factors. In: Muchlisin, Z. (ed.) Aquaculture. Rijeka: Intech; 2012.p343-360.
19. Rhodes. A. Methods for high density batch culture of *Nitokra lacustris*, a marine harpacticoid copepod.The Big Fish Bang. In: Browman, H.I., Skiftesvik, A.B. (eds.). Proceedings of the 26th Annual Larval Fish Conference. Institute of Marine Research, Postboks 1870 Nordnes, N-5817, Bergen, Norway. ISBN 82-7461-059-8. 2003
20. Farhadian, O., Yusoff, M.F., Arshad, A. Population growth and production of *Apocyclops dengizicus*( Copepoda: Cyclopoida) Fed on Different Diets. Journal of the World Aquaculture Society, 2008; 39(2), 384-396. DOI: 10.1111/j.1749-7345.2008.00172.x

21. Chullasorn, S., Kangtia, P., Pinkaew, K.,Ferrari, F.D. *Apocyclops ramkhamhaengi* sp. nov. (Copepoda: Cyclopoida) in a Culture Originating from Brackish Waters of Chang Island, Trat Province,Thailand. Zoological Studies 2008; 47(3),326-337.
22. Apt, K. E., Behrens, P. W. 1999. Commercial developments in microalgal biotechnology. *Journal of Phycology, 35*(2), 215-226. doi: 10.1046/j.1529-8817.1999.3520215.x
23. Geudes, A.C., Malcata, F.X. Nutritional values and uses of microalgae in aquaculture. In: Muchlisin, ZA (Ed); InTech Open Acess Publisher
24. Muller-Feuga, A., Moal, J. and Kaas, R. 2003. The microalgae of aquaculture. In: Stottrup, J.G and Mcevoy, L.A (eds.) Live Feeds in Marine Aquaculture. Blackwell Publishing Ltd., USA, p206-252.
25. Mata, T.M., Martins, A.A., Caetano, N.S. 2010. Microalgae for biodiesel production and other applications: a review. Renewal Sustainable Energy Review 2010; 14 : 217-232.
26. Chen, C.Y., Yeh, K.L., Aisyah, R., Lee, D.J. and Chang, J.S. 2011. Cultivation, photobioreactor design and harvesting of microalgae for biodiesel production: a critical review. *Bioresource Technology*, 102, pp. 71-81.
27. Lee, Y.K. and Shen, H. 2004. Basic culturing techniques for microalgae. In: *Handbook of Microalgal Culture, Biotechnology and Applied Phycology* (ed. A. Richmond). Blackwell Publishing Ltd., USA, pp. 40-56.
28. Chen, C.Y., Yeh, K.L., Aisyah, R., Lee, D.J. and Chang, J.S. 2011. Cultivation, photobioreactor design and harvesting of microalgae for biodiesel production: a critical review. *Bioresource Technology*, 102, pp. 71-81.
29. Becker, E.W., 1995. Microalgae biotechnology and microbiology. *Cambridge University Press*, Great Britain, pp.56- 62.
30. Borgne, Y.L. 1996. Culture of microalgae. In: *Aquaculture* (ed. G. Barnabe). *Ellis Horwood Ltd.*, England, pp. 197-206.
31. Grobbelaar, J.U., Nedbal, L. and Tichy, V. 1996. Influence of high frequency light/dark fluctuations on photosynthetic characteristics of microalgae photoacclimated to different light intensities and implications for mass algal cultivation. *Journal of Applied Phycology*, 8, pp. 335-343.
32. Meseck S.,L., Alix J.,H., Gary, H., Wikfors G.,H., 2005. Photoperiod and light intensity effects on growth and utilization of nutrients by the aquaculture feed microalga Tetraselmis chui (PLY429). *Aquaculture* 246: 393–404.
33. Heasman, M., Diemar, J., O'Connor, W., Sushames, T. and Foulkes, L. 2000. Development of extended shelf-life microalgae concentrate diets harvested by centrifugation for bivalve mollusks- a summary. *Aquacult. Res.*, 31, pp. 637-659.
34. Knuckey, R., Brown, M., Robert, R. and Frampton D. 2006. Production of microalgal concentrates by flocculation and their assessment as aquaculture feeds. *Aquacult. Eng.*, 35, pp. 300-313.
35. Montaini, E., Zitelli, G.C., Tredici, M.R., Molina Grima, E.M., Fernandez Sevilla, J.M. and Sanchez Perez, J.A. 1995. Long-term preservation of *Tetraselmis suecica*: influence of storage on viability and fatty acid profile. *Aquaculture*, 134, pp. 81-90.
36. McCausland, M.A., Brown, M.R., Barrett, S.M., Diemar, J.A. and Heasman, M.P. 1999. Evaluation of live and pasted microalgae as supplementary food for juvenile Pacific oyster (*Crassostrea gigas*). *Aquacult. Res.*, 174, pp. 323-342.
37. D'Souza, F.M.L., Knuckey, R.M., Hohmann, S. and Pendrey, R.C. 2002. Flocculated microalgae concentrates as diets for larvae of the tiger prawn *Penaeus monodon* Fabricius. *Aquaculture Nutrition*, 8, pp. 113-120.
38. Brown, M. and Robert, R. 2002. Preparation and assessment of microalgal concentrates as feeds for larval and juvenile Pacific oyster (*Crassostrea gigas*). *Aquaculture*, 207, pp. 289-309
39. Iwamoto, H. 2004. Industrial production of microalgal cell-mass and secondary products-Major industrial species: *Chlorella*. In: *Handbook of Microalgae culture, Biotechnology and Applied Phycology* (ed. Richmond, A.). Blackwell Science Ltd., Oxford, UK, pp. 255-263.

40. Rossingol, N., Vandanjon, L., Jaouen, P. and Quemeneur, F. 1999. Membrane technology for the continuous separation microalgae culture medium: compared performances of cross-flow microfiltration and ultrafiltration. *Aquacult. Eng.*, 20, pp. 191-208.
41. Csordas, A. and Wang, J.K. 2004. An integrated photobioreactor and foam fractionation unit for the growth and harvest of *Chaetoceros* sp. in open systems. *Aquacult. Eng.*, 30, pp. 15-30.
42. Petrusevski, B., Bolier, G., Van Bremen, A.N. and Alaerts, G. J. 1995. Tangential flow filtration: a method to concentrate freshwater algae. *Water Res.*, 29, pp. 1419-1424.
43. Rossingol, N., Vandanjon, L., Jaouen, P. and Quemeneur, F. 1999. Membrane technology for the continuous separation microalgae culture medium: compared performances of cross-flow microfiltration and ultrafiltration. *Aquacult. Eng.*, 20, pp. 191-208.
44. Sun, J. and Liu, D.Y. 2003. Geometric models for calculating cell biovolume and surface area for phytoplankton. *J. Plankton Research.*, 25, pp. 1331-1346.
45. Muller-Feuga, A. 2000. The role of microalgae in aquaculture: situation and trends. *J. Appl. Phycol.*, 12, pp. 527-534.
46. Baraniak, B., Niezabitowskab, M., Pieleckic, J. and Wojcika, W. 2004. Evaluation of usefulness of Magnafloc M-22S flocculant in the process of obtaining protein concentrates from peas. *Food Chem.*, 85, pp. 251-257.
47. Omori, M. and Ikeda, T. 1984. Methods in Marine Zooplankton Ecology. *Wiley*, New York, 322 pp.
48. Usha, T.L., Sarada, R., and Ravishankar, G.A. 2002. Effect of culture conditions on growth of green alga *Haematococcus pluvialis* and astaxanthin production. Acta Physiologiae Plantarum. 24 (3): 323-329
49. Fogg and Thake 1987. Algal Cultures and Phytoplankton Ecology. University of Wisconsin Press.
50. Abu-Rezq, T. S., Yule, A. B. & Teng, S. K. 1997. Ingestion, fecundity, growth rates and culture of the harpacticoid copepod, *Tisbe furcata*, in the laboratory. *Hydrobiologia, 247*, 109-118.
51. Imase, M., Ohko, Y., Takeuchi, M., Hanada, S. 2013. Estimating the viability of Chlorella exposed to oxidative stresses based around photocatalysis. International Biodeterioration and Biodegradation. 78: 1-6
52. Dwaish, A.S., Mohammed, D.Y., Jawad, A.M., Al-kubaicy, A.A. Determine the Uptake of Lead in Chlorella vulgaris Isolated from Tigris River in Baghdad City. International Journal of Scientific and Engineering Research, 2011, 2(9): 1-4
53. Bae, J.H. and Hur, S.B. 2011. Selection of suitable species of Chlorella, Nannochloris and Nannochloropsis in high- and low temperature seasons for mass culture of the rotifer *Brachionus plicatilis*. Fisheries and Aquatic Sciences, 14(4):323-332.
54. Phatarpekar, P.V., Sreepada, R.A., Pednekar, C. and Achuthankutty, C.T. 2000. A comparative study on the growth performance and biochemical composition of mixed culture of *Isochrysis galbana* and *Chaetoceros calcitrans* with monocultures. *Aquaculture*, 181, pp. 141-155.
55. Barsanti, L. and Gualtieri, P. 2006. Algae: anatomy, biochemistry and biotechnology. *CRC Press*, New York, 301 pp.
56. Stottrup, J.G. and Jensen, J. 1990. Influence of algal diet on feeding and egg-production of the calanoid copepod *Acartia tonsa* Dana. *J. Exp. Mar. Biol. Ecol.*, 141, pp. 87-105.
57. Stottrup, J.G. 2003. Production and nutritional value of copepods. In: *Live Feeds in Marine Aquaculture* (eds. Stottrup, J.G and Mcevoy, L.A.). *Blackwell Publishing Ltd.*, USA, pp. 145-205.
58. Kiorboe, T. 1997. Population regulation and role of mesozooplankton in shaping marine pelagic food webs. *Hydrobiologia*, 363, pp. 13-27.
59. Dam, H.G., Peterson, W.T. and Bellantoni, D.C. 1994. Seasonal feeding and fecundity of the calanoid copepod *Acartia tonsa* in Long Island Sound: is omnivory important to egg production? *Hydrobiologia*, 292/293, pp. 191-199.

60. Payne, M.F. and Rippingale, R.J. 2001. Intensive cultivation of the calanoid copepod *Gladioferens imparipes*. *Aquaculture*, 201, pp. 329-342.

61. Hernandez, D.E. and Martinez. S.F. 2001. The effect of the culture conditions on the growth and lipid contents of two strains of Nannochloropsis sp. to be used in aquaculture. Hidrobiologia 11(2):163-168.

62. Sandnes, J.M., K°allqvist, T., Wenner, D. and Gislerød. H.R. 2005. Combined influence of light and temperature on the growth rates of *Nannochloropsis oceanica*: linking cellular responses to large-scale biomass production. Journal of Applied Phycology 17: 515–52.

63. Nedbal, L., Tichyacute, V., Xiong, F. and Grobbelaar. J. U. 1996. Microscopic green algae and cyanobacteria in high frequency intermittent light. Journal of Applied Phycology, 8(4/5), 325-333.

64. Matthijs, H.C.P., Balke, H., van Hes, U.M., Kroon, B.M.A., Mur, L.R. and Binot. R.A. 1996. Application of light emitting diodes in bioreactors: Flashing light effects and energy economy in algal culture (Chlorella pyrenoidosa). Biotechnology and Bioengineering, 50(1), 98-107.

65. Lim, K.C and Zaleha, K. 2013. Effect of Photoperiod on the Fatty Acid Composition of Three Tropical Species of Marine Microalgae. *Malaysian Applied Biology Journal,* 42(1): 41-49

66. Ashraf, M., Javaid, M. Rashid, T., Ayub, M., Zafar, A., Ali, S., Naeem,M. 2011. Replacement of expensive pure nutritive media with low cost commercial fertilizers for mass culture of freshwater algae, *Chlorella vulgaris. International Journal of Agriculture and Biology*, 13: 484–490

67. Lee, K.W., Park, H.G., Lee, S.& Kang, H. 2006. Effects of diets on the growth of the brackish water cyclopoid copepod *Paracyclopina nana*Smirnov. *Aquaculture, 256,* 346-353. http://dx.doi.org/10.1016/j.aquaculture.2006.01.015

68. Phelps, R.P., Sumiarsa, G.S., Lipman, E.E., Lan, H.-P., Moss, K.K., & Davis, A.D. 2005. Intensive and extensive production techniques to provide copepod nauplii for feeding larval red snapper *Lutjanus camperchanus*. In: Copepods in Aquaculture (Ed. By C.-S. Lee, P.J. O'Bryen, N.H. Marcus), pp. 151-168. Blackwell Publishing Ltd. Oxford, UK.

69. Da Grac, M., Melao, A.G., & Rocha, O. 2004. Life history, biomass and production of two planktonic cyclopoid copepods in a shallow subtropical reservoir. *Journal of Plankton Research, 26(8),* 909-923.

70. Milione, M., & Zeng, C. 2008. The effects of temperature and salinity on population growth and egg hatching success of the tropical calanoid copepod, *Acartia sinjiensis. Aquaculture,275,*116-123.

# CHAPTER 2

# Microarray-Predicted Marker Genes Molecular Pathways Indicating Crowding Stress in Rainbow Trout (*Oncorhynchus Mykiss*)

**AlexerRebl[a],ManuelaZebunke[b],reasBorchel[a,c],RalfBochert[d],MariekeVerleih[a],TomGoldammer[a*]**

[a] *Institute for Genome Biology, Leibniz Institute for Farm Animal Biology (FBN), Wilhelm-Stahl-Allee 2, 18196 Dummerstorf, Germany*

[b] *Institute for Behavioural Physiology/Institute for Genetics Biometry, Leibniz Institute for Farm Animal Biology (FBN), Wilhelm-Stahl-Allee 2, 18196 Dummerstorf, Germany*

[c] *Department of Biology, Sea Lice Research Centre (SLRC), University of Bergen, Thormøhlensgate 55, 5008 Bergen, Norway*

[d] *Institute for Fisheries, State Research Institute for Agriculture Fisheries Mecklenburg-Western Pomerania (LFA-MV), Südstraße 8, 18375 Born, Germany*

## ABSTRACT

Stocking density is considered as a key factor determining the productivity of fish aquaculture systems. The transcriptomic response to crowding stress is, however, still poorly investigated. We aimed to identify potential biomarker genes via microarray analyses to gain insight into molecular pathways modulated by density-induced stress in farmed rainbow trout *Oncorhynchus mykiss*. Transcriptome profiling in the liver, kidney, gills was complemented by analysis of classical plasma parameters behavioural observations. Individuals of two trout strains were exposed for nine days to definite stocking densities, ~ 10 kg/m³ (low density); ~ 30 kg/m³ (moderate); ~ 60 kg/m³ (elevated); ~ 120 kg/m³ (high). Only the PYGMA gene, encoding the key enzyme of glycogenolysis, was upregulated in gills kidney, while about 90 genes were differentially regulated in the liver of trout exposed to high density compared with the moderate-density group. Pathway enrichment analyses predicted the upregulation of HIF1A signalling in liver, contributing to glucose homeostasis during stress conditions. This is consistent with the increased plasma glucose levels detected in crowding-stressed trout. The mTOR PI3K/AKT signalling pathways were impaired in liver at higher stocking densities indicating stress at the cellular level. Contrary to our expectations, protein ubiquitination cholesterol biosynthesis pathways were found to be underexpressed. Cholesterol is a precursor of the stress hormone cortisol, whose plasma concentrations were indeed within the normal physiological range in all treatment groups. In addition, we did not observe any aggressive interactions, which are known to be triggered through cortisol release. Taken together, we identified a relatively modest number of rainbow trout-specific crowding markers suggesting that farmed rainbow trout have gained enhanced stress-tolerant phenotypes through decades of selective breeding.

*Statement of relevance:*The comprehensive recording of animal-based biomarkers is becoming increasingly important for the further development of the aquaculture industry with importance for fish well-being consumer opinion. Previous studies on crowding stress in rainbow trout have used single or multi-gene approaches to assess the relevance of putative biomarkers, but transcriptomic approaches to define crowding stress-induced pathways in salmonids are overdue. The present study compares the transcriptomic profiles of rainbow trout exposed to different stocking densities correlates these data sets with classical plasma parameters behavioural observations.

**Keywords:** Glucose metabolism Glycogenolysis Salmonid fish Stocking density Welfare Transcriptome

# 1. INTRODUCTION

Aquaculture might provide a real solution to the ever-growing consumption of seafood provided that fish production is sustainable eco-friendly (Thurstan Roberts, 2014; Williams 2000). Approximately 25% of EU-consumed fish is produced in European aquaculture, mainly rainbow trout Atlantic salmon (European Commission, ). However, market surveys indicate the need to encourage a more positive profile in consumer attitudes towards fish farming, particularly regarding environmental concerns animal welfare (Stubbe Solgaard Yang, ). Intensive fish farming has been of particular concern for decades, stocking density (SD) is considered as a major variable in intensive aquaculture that affects the physiology, behaviour, hence well-being of farmed fish (reviewed by Ellis 2002). Inadequately high or low SDs activate persistent stress responses with possible deleterious consequences for growth, disease resistance, reproduction (Barton, 2002; Huntingford 2006). Stress responses of fish are detectable at three levels (Barton Iwama, 1991; Hemre Krogdahl, 1996). The hypothalamic-pituitary-interrenal (HPI) axis initiates the primary response, releasing corticosteroid hormones into the bloodstream (Hemre Krogdahl, 1996; Mommsen 1999). The secondary response is mainly characterised by the mobilisation of energetic reserves to restore homeostasis (Silbergeld, 1974; Vijayan Moon, 1992). The tertiary response refers to 'whole-animal performance' such as behaviour, disease resistance or growth (reviewed by Barton, 2002; Ellis 2002; Wendelaar Bonga, 1997).

The correlation between well-being stocking densities in aquaculture has been exhaustively analysed reviewed in the past (Adams 2007; Alanara Brannas, 1996; Ellis 2002; Gornati 2004; Laursen 2013; Martins 2012; Melotti 2004; North 2006; Person-Le Ruyet 2008; Turnbull 2005). Nevertheless, concrete minimum maximum density levels await definition (Laursen 2013): 40 kg/m$^3$ has previously been suggested as an upper limit for stocking density in husbry of rainbow trout *Oncorhynchus mykiss*(FAO 2005-2017; FSBI, 2002), whereas several other studies demonstrated that densities exceeding 100 kg/m$^3$ do not obviously impair the well-being of farmed trout (Boujard 2002; Person-Le Ruyet 2008).

New digital technologies based on the internet of things cloud computing of 'big data' will in future allow the continuous high-throughput detection of biomarkers in farmed fish, their waste products their artificial environment to regulate husbry conditions as needed to maintain aquaculture facilities in optimal ranges. However, these technologies require the knowledge of sensitive biomarkers for compromised well-being of farmed fish. Well-established less exploited salmonid-specific biomarkers for crowding are summarised in Table 1. This list reflects, on the one h, the complexity contradictory nature of currently utilised marker sets , on the other h, points moreover to the high variability of individual approaches regarding the definition of high stocking density (ranging from 30 to 120 kg/m$^3$) the duration of the experiment (up to nine months). Moreover, our literature survey revealed that crowding stress-induced pathways have been rarely suggested for salmonids so far, though numerous authors described distinct phenotypic implications such as impaired immunological parameters (Yarahmadi 2016, 2014), altered apparent feed conversion rates (Person-Le Ruyet 2008) reduced body condition factors (North 2006) due to overcrowding. In this context, reference should be made to the studies of Rexroad colleagues identifying transcriptional patterns related pathways in response to acute confinement stress (Liu 2014; Sánchez ).

**Table 1.** Overview about selected physiological parameters altered in response to density stress in salmonid fish species *Oncorhynchus mykiss* (*O. myk*), *Salmo salar* (*S. sal*), *Salvelinus fontinalis* (*S. fon*), *Coregonus maraena* (*C. mar*).

| Species | Stocking density [kg/m³] | Duration of challenge | Putative stress indicator | Direction[a] | Tissue | Reference |
|---|---|---|---|---|---|---|
| *O. myk* | 40, 80 | 9 months | Fin damage | Up | Dorsal, caudal, pectoral fin | North et al., 2006 |
| *O. myk* | 75, 120 | 84 days | Fin damage | Down | Dorsal, pectoral fin | Person-Le Ruyet et al., 2008 |
| *S. sal* | 30 | 10 weeks | ALP, alkaline phosphatase | Up | Blood | Liu et al., 2015a |
| *O. myk* | 45 | 1 month | Cholesterol | Down | Blood plasma | Yarahmadi et al., 2014 |
| *O. myk* | 40, 80 | 9 months | Cortisol | Down | Blood plasma | North et al., 2006 |
| *O. myk* | 120 | 2 weeks | Cortisol | Up | Blood plasma | Person-Le Ruyet et al., 2008 |
| *S. sal* | 30 | 10 weeks | Cortisol | Up | Blood plasma | Liu et al., 2015a |
| *O. myk* | 40, 80 | 1 month | Cytokines (IL1B, CXCL8, TNF, IFNG) | Down | Head kidney | Yarahmadi et al., 2016 |
| *S. fon* | 120 | 1 month | FBP1, fructose-1,6-bisphosphatase | Up | Liver | Vijayan et al., 1990 |
| *O. myk* | 20, 25, 30 | 2 months | G6PD, glucose-6-phosphate dehydrogenase | Up | Kidney | Aksakal et al., 2011 |
| *S. fon* | 120 | 1 month | G6PD | Down | Liver | Vijayan et al., 1990 |
| *O. myk* | 30 | 2 months | G6PD | Down | Liver, muscle, gill | Aksakal et al., 2011 |
| *S. fon* | 120 | 1 month | GK, glycerol kinase | Up | Liver | Vijayan et al., 1990 |
| *S. fon* | 120 | 1 month | Glucose | Down | Blood plasma | Vijayan et al., 1990 |
| *O. myk* | 45 | 1 month | Glucose | Up | Blood plasma | Yarahmadi et al., 2014 |
| *O. myk* | 70 | 2 days | Glycogen | Down | Liver | Conde-Sieira et al., 2010 |
| *S. fon* | 120 | 1 month | Glycogen | Down | Liver | Vijayan et al., 1990 |
| *S. fon* | 120 | 1 month | GPD1, glycerol-3-phosphate dehydrogenase 1 | Up | Liver | Vijayan et al., 1990 |
| *O. myk* | 25, 30 | 2 months | GR, glutathione reductase | Down | Liver, muscle, gill | Aksakal et al., 2011 |
| *S. fon* | 120 | 1 month | HADH, hydroxyacyl-CoA dehydrogenase | Up | Liver | Vijayan et al., 1990 |
| *S. fon* | 120 | 1 month | HK1, hexokinase | Down | Liver | Vijayan et al., 1990 |
| *O. myk* | 25, 30 | 2 months | HSP70, heat shock protein, 70 kDa | Up | Muscle | Aksakal et al., 2011 |
| *S. sal* | 30 | 10 weeks | IgM, immunoglobulin M | Down | Blood | Liu et al., 2015a |
| *O.myk* | 70 | 2 days | Lactate | Up | Blood plasma | Conde-Sieira et al., 2010 |
| *C. mar* | 100 | 9 days | LECT2, leukocyte cell-derived chemotaxin 2 | Up | Liver | Korytář et al., 2016 |
| *S. sal* | 30 | 10 weeks | Maleic dialdehyde | Up | Blood plasma | Liu et al., 2015a |
| *S. fon* | 120 | 1 month | PFK, phosphofructokinase | Up | Liver | Vijayan et al., 1990 |
| *O. myk* | 100 | 1 month | PKF | Up | Liver | Trenzado et al., 2006 |
| *O. myk* | 20, 25, 30 | 2 months | PGD, phosphogluconate dehydrogenase | Up | Kidney | Aksakal et al., 2011 |
| *O. myk* | 25, 30 | 2 months | PGD | Down | Liver, muscle, gill | Aksakal et al., 2011 |
| *C. mar* | 60, 100 | 9 days | SAA, serum amyloid A | Up | Kidney | Korytář et al., 2016 |
| *O. myk* | 45 | 1 month | Triglyceride | Down | Blood plasma | Yarahmadi et al., 2014 |

[a] Relative to control group.

The objective of the present study was to discover novel animal-based biomarkers evaluate well-known existing ones based on transcriptome profiling, to derive inferences about the onset of suggestive pathways. To this end, we profiled the gene expression in three tissue types, liver, kidney, gills. Kidney liver are evolutionarily well-conserved key organs of the stress response: The kidney secretes hormones stimulating stress response pathways, while the liver, as the dominant metabolic organ, provides the required energy (reviewed by Harper Wolf, 2009). The gills are 'a frequent target for stress responses' (Harper Wolf, 2009) have been investigated in numerous studies concerning stress in teleost fish (Amaeze ; Lam 2014; Rebl 2013). The investigation of classical plasma parameters behavioural observations complemented our gene expression analyses.

# 2. MATERIALS METHODS

## 2.1. Fish experimental facilities

Rainbow trout of two different strains, the German Born strain (Rebl 2012; Verleih ) the North American Troutlodge strain (http://www.troutlodge.com/; WA, US; Silverstein 2004) were bred from eggs (Born), 'eyed'- stage eggs (Troutlodge). For SD experiments, trout were placed separately into 400-L glass tanks for 9 d until blood tissue sampling was conducted. Born trout were 10 months in age averaging 26.8 ± 1.4 cm in length (mean ± stard deviation STD), 280.3 ± 48.8 g in weight. Troutlodge trout were 11 months in age averaged 27.5 ± 2.1 cm in length, 335.4 ± 66.7 g in weight.

We investigated four stocking density conditions according to the literature (cf. review from Ellis 2002) practical experience of the Institute of Fisheries, LFA M-V (Suppl. Fig. 1a). An uncrowded 'moderate' density (MD) of 30 kg/m$^3$ was assumed as the optimal stocking density for trout. Ten kilograms per cubic meter was considered as low density (LD), 60 kg/m$^3$ was considered as elevated density (ED), 120 kg/m$^3$ served as high density (HD). Born Troutlodge trout, respectively, were romly assigned to identical glass tanks with MD (34 30 individuals), ED (64 60 individuals), HD (140 120 individuals). One LD tank harboured five individuals each of the Born the Troutlodge strains. Adipose fins were clipped to mark the Born trout.

Each tank (0.74 m length × 0.58 m width × 0.72 m height) held a volume of ~ 310 L brackish water from the Darss-Zingst Bodden chain (2.5–6 practical salinity units) in a recirculating aquaculture system (RAS) with a water exchange rate of approximately 60 turnovers per day. The water inflow was situated on the upper left-h side; the run-off was located ground-proximate on the right-h side of each tank. Drum filter Moving Bed Biofilm reactors in combination with UV radiation precleared the water. Water quality (including $NH_4^+$, $NO_3^-$) was monitored in all tanks throughout the trial the observed values were consistently in optimal ranges, i.e. temperature (16.4–19.8 °C), dissolved oxygen (9.8–10.5 mg/L), pH (7.4–7.8). Two centrally located luminaires above each tank mimicked natural lighting conditions (light:dark period 17.5:6.5) including 0.5 h dimming in the morning evening. Trout were fed commercial dry pellets by automatic feeders located centrally above each tank distributing the food 12 h/d. Throughout the experiment, no mortality or disease occurred among the rainbow trout no technical problems were registered.

## 2.2. Behavioural observations

The behaviour of the Born Troutlodge trout was recorded separately by cameras placed in front of the MD, ED, HD tanks for 4 h on day 8 (until the start of dark period) for another 4.5 h on day 9 (beginning at the start of the light period) after transfer of trout to experimental tanks. The LD tank contained both Born Troutlodge trout was thus analysed without discriminating the strains.

For behavioural analysis, the video recordings were examined for aggression (e.g. charging, chasing, nipping/biting, fighting). Regarding the spatial distribution of the trout within the tank, each tank was subdivided into six equally sized rectangles: One vertical line divided the tank into 'left' 'right', two horizontal lines into 'top', 'centre' 'bottom'. In an interval of ten minutes, stills of the video were examined the position of all visible fish within the respective rectangles was noted. Fifty-three time points were analysed in total per tank. A $Chi^2$-Test (SAS 9.2, proc. Freq) was used to test for equality of spatial distribution of trout across the 53 time points (under the null hypothesis of equal distribution). The test was calculated independent of the strain as well as separately for each strain. Due to the low total number of fish in the LD treatment (n = 10) an accordingly low number of visible fish (n = 2 on average), the counting of fish in the LD treatment was not regarded as meaningful was therefore excluded from further behavioural analysis.

## 2.3. Sampling nucleic acid preparation

Rainbow trout were romly sampled using h nets. Anaesthesia of fish with phenoxyethanol prior to blood tissue sampling was conducted in compliance with the terms of the German Animal Welfare Act (§ 4(3) TierSchG). Nine days after the start of the experiment, blood tissue samples (entire liver; gill filaments; head trunk kidney) were taken from ten Born ten Troutlodge trout kept at MD, ED, HD as well as five Born five Troutlodge trout kept at LD. Each fish was weighed measured.

Blood was taken from the caudal vein of each trout using 5-mL plastic syringes filled with 500 µl 0.5 M EDTA (pH 8.0) solution as anticoagulant. Gill, kidney, liver tissues of the same animals were isolated, snap-frozen in liquid nitrogen stored at – 80 °C until nucleic acid extraction. TRIzol reagent (Invitrogen, Carlsbad, CA, US) was used to extract total RNA from the tissue samples. The RNeasy Mini Kit (Qiagen, Hilden, Germany) in combination with the RNase-free DNase Set (Qiagen) allowed us to obtain high-quality RNA suitable for microarray hybridisations quantitative real-time PCR (qPCR) assays. DNA was isolated from liver samples using the DNeasy

Blood Tissue Kit (Qiagen) to determine the genetic sex of the examined tout exploiting primer sequences adapted from Brunelli et al. (2008).

Nucleic acid concentrations were determined by measuring the 260:280 nm absorbance ratio in a NanoDrop 1000 Spectrophotometer (NanoDrop Technologies/Thermo Fisher Scientific, Darmstadt, Germany). RNA was quality-checked using an Agilent 2100 Bioanalyzer (Agilent Technologies, Waldbronn, Germany) revealing RIN values ranging from 8.5 to 10.

## 2.4. Plasma cortisol glucose analyses

Blood samples were centrifuged (4 °C, 1700 rcf) the supernatant was kept on ice until analysis of blood plasma parameters. The levels of plasma cortisol glucose have traditionally been recorded as indicators of stress in fish (Silbergeld, 1974; Sumpter 1986). Cortisol concentrations were determined using a Cortisol ELISA assay (DRG Instruments GmbH, Marburg, Germany) plasma glucose concentrations were measured using a colorimetric assay (Glucose Assay Kit II; BioVision, Milpitas, California, US). Both analyses were performed according to the manufacturer's instructions using a Beckman Coulter DTX 800/880 Series Multimode Detector (Beckman Coulter, Brea, CA, US) measuring the absorbance at 450 nm.

## 2.5. cRNA synthesis, array hybridisation data analysis

Five individual RNA samples from the same tissue (liver, gills, kidney), strain (Born or Troutlodge) stocking density (MD, ED, HD) were pooled. Sample labelling was performed following the One-Color Microarray-Based Gene Expression Analysis protocol, version 6.6. Briefly, 100 ng of each total RNA sample was amplified labelled with the fluorescent dye cyanine 3 (Cy3) using the Agilent Low Input Quick Amp Labeling Kit (Agilent Technologies). Yields of Cy3-labelled cRNA the dye-incorporation rate were measured. The hybridisation procedure was performed using the Agilent Gene Expression Hybridisation Kit (Agilent Technologies). To this end, an equal amount of fragmented Cy3-labelled cRNA (0.6 µg) in hybridisation buffer was hybridised at 65 °C for 17 h to 8 × 60 K Agilent-049158 Salmon Oligo Microarrays (Agilent Technologies; GEO platform: GPL21057) using Agilent's recommended hybridisation chamber oven. Following hybridisation, the microarrays were washed once with the Agilent Gene Expression Wash Buffer 1 for 1 min at room temperature, followed by a second wash with preheated Agilent Gene Expression Wash Buffer 2 at 37 °C for 1 min.

Fluorescence signals of the hybridised Agilent Microarrays were scanned at 3-µm resolution using a G2505C Microarray Scanner System (Agilent

Technologies). Agilent Feature Extraction Software (FES) 10.7.3.1 was used to read out process the microarray image files using default settings. FES corrected the background based on a two-sided Student's *t*-test. The features having passed this quality control were further analysed using the limma package of R version 3.1.1/Bioconductor suite (Smyth 2005; http://www.bioconductor.org/packages/devel/bioc/vignettes/limma/inst/doc/usersguide.pdf); all customized scripts used for calculation are available from the corresponding author. Following quantile normalization, pairwise comparisons between Born Troutlodge trout were employed to compare transcript abundances under MD HD conditions. The focus of our study was on common physiological features characterising crowding stress impaired well-being in rainbow trout, rather than teasing apart strain-specific responses. To this end, the treatments were adjusted for differences between the strains according to the mathematical formula $Var(X - Y) = Var(X) + Var(Y) - 2\ Cov(X, Y)$, where $X$ denotes the mRNA concentrations at ED or HD $Y$ represents mRNA concentrations at MD (control). 'Strain' was considered as a blocking factor accounting for treatment variability between Born Troutlodge to reduce sources of variability thus allow greater precision (Ritchie ). To control the false discovery rate, p-values were adjusted according to Benjamini Hochberg (1995). Only genes with a corrected p-value (or q-value) of $< 0.05$, an absolute fold change of $> 2.0$ were deemed differentially expressed (DE) genes considered for further evaluation. The full complement of microarray data has been deposited in the NIH/NLM Gene Expression Omnibus (GEO accession: GSE74332). Principal component analysis was performed using Rosetta Resolver Data Reduction (Weng 2006). Comparisons of gene expression across different stocking densities were performed using Venn Diagrams (Hulsen 2008). DE genes were re-annotated using the Basic Local Alignment Search Tool (BLAST). We considered only transcripts with unique BLAST results (coverage sequence identity of $> 80\%$ E-value $< 1 \times 10^{-4}$) joined redundant probes representing identical transcripts to avoid repeated counting of the same gene.

A functional analysis was performed using the Ingenuity program (Ingenuity Pathway Analyses/Qiagen) to evaluate global functional networks canonical pathways of interacting genes other functional groups. Benjamini-Hochberg multiple-testing was utilised as a cut-off criterion. Pathway enrichment analyses were collated with DAVID-generated lists of KEGG pathways (Huang 2008); canonical pathways of mammalian diseases were excluded. Enriched pathways are indicated in the following by italic face.

The determination of the hypothetical chromosomal location of DE genes from rainbow trout was conducted using the comparative analysis of chromosome sequences from salmon trout (Lien 2016; Phillips 2009).

### 2.6. Confirmation of DE genes by qPCR

Quantitative PCR was performed to validate array-predicted mRNA abundances. Gene-specific primers were designed (Pyrosequencing Assay Design software v.1.0.6; Biotage, Uppsala, Sweden) to amplify target fragments between 151 181 bp (Table 2) with efficiencies > 92% ($r^2 > 0.99$). *RPS5* (ribosomal protein S5) *RNA18S* (18S ribosomal RNA) were chosen as suitable reference genes (with averaged GeNorm stability value M = 0.37, coefficient of variation, CV = 0.13). The qPCR analyses were conducted with the LightCycler 96 System (Roche, Mannheim, Germany) using the SensiFAST SYBR No-ROX Kit (Bioline, Luckenwalde, Germany). Melting curve analyses validated the amplification of each particular target fragment. Amplicons were visualised on 3% agarose gels in order to evaluate product size quality. qBase$^{+}$ software (Biogazelle, Ghent University, Belgium) supported the calculation of qPCR data. Pearson's correlation test (http://www.socscistatistics.com/tests/pearson/) was used to assess the concordance between the microarray qPCR results.

**Table 2.** Gene-specific qPCR primers used in this study.

| Gene symbol | GenBank accession # | Sense (5′-3′) | Antisense (5′-3′) | Amplicon length [bp] |
|---|---|---|---|---|
| *COX* | NM_001185033 | TACAGAGTTTGCAGGTAGACCC | ACGTGGTGCTCAAATGCTTTCTT | 176 |
| *EIF4E* | NM_001160620 | GGGGAGGCTTTCGATGACTATA | GGATTCCCAAGCGCTCCTTATA | 154 |
| *FDPS* | XM_014201927 | TAGTTTGGACTGTGAGTGATCCT | GATGACATGATATTCCCAGTTTGA | 156 |
| *G6PC* | AB889398 | AACCTGGTCTTCAAGTGGATTCT | TGTAGTAGACGCCTGCAGCTC | 172 |
| *HSP47* | NM_001124234 | ATGCATCGCACCGGTCTCTATA | ATCTTGCCCATCCAGGTTTCCA | 181 |
| *HSP90AA1* | NM_001124591 | GCTGGGTCTGGGAATCGATGA | ATACAGGTGTGTAGGTTTGAGAC | 151 |
| *PGM3* | NM_001139891 | ATAGACAGCCGGGTGAAGAAGT | CGGTTTAGGCTCCTCTCCTAC | 180 |
| *RPS5* | NM_001160519 | ATGACATCTCACTGCAGGATTAC | ATCAGCTTCTTGCCGTTGTTGC | 171 |
| *RNA18S* | FJ710873 | ACATCTAAGGGCATCACAGACC | TTCTTAGTTGGTGGAGCGATTTG | 169 |

## 3. RESULTS

### 3.1. Born Troutlodge strains regulate common biofunctions to cope with crowding

Microarray technology was used to record the transcriptional changes induced by different SDs in the liver, gills kidney of rainbow trout. The tissue-specific response patterns to different SDs are reflected by the principal component analysis (Suppl. Fig. 1b). Pairwise comparison of microarray datasets from both trout strains evinced a relatively high number of common DE genes, but also exclusive DE genes present in only one of the two datasets (Fig. 1a, b). In liver tissue of both strains kept separately at ED HD, Born trout showed fewer DE features (ED: 208, HD: 281; compared with MD) than Troutlodge trout (ED: 338, HD: 643; compared with MD). Assigning these genes with altered mRNA levels to IPA biofunctions revealed that two common pathways were significantly downregulated in both strains, the *protein ubiquitination pathway* (Born: $p = 1 \times 10^{-6}$, 10 genes; Troutlodge: $p = 1 \times 10^{-3}$, 9 genes; 4 genes shared), the *cholesterol biosynthesis pathway* (Born: $p = 6 \times 10^{-4}$, 3 genes; Troutlodge: $p = 2 \times 10^{-4}$, 4 genes; 2 genes shared).

**a** Born

| | ED *vs.* MD | HD *vs.* MD | | |
|---|---|---|---|---|
| | LIVER | LIVER | GILLS | KIDNEY |
| ↑ | 54 | 97 | 38 | 30 |
| ↓ | 154 | 184 | 42 | 21 |

**b** TLodge

| | ED *vs.* MD | HD *vs.* MD | | |
|---|---|---|---|---|
| | LIVER | LIVER | GILLS | KIDNEY |
| ↑ | 100 | 235 | 4 | 4 |
| ↓ | 238 | 408 | 2 | 46 |

**c** Rainbow trout (Born + TLodge)

| | ED *vs.* MD | HD *vs.* MD | | |
|---|---|---|---|---|
| | LIVER | LIVER | GILLS | KIDNEY |
| ↑ | 13 | 39 | 2 | 1 |
| ↓ | 127 | 210 | 0 | 0 |

**d** Rainbow trout (Born + TLodge)

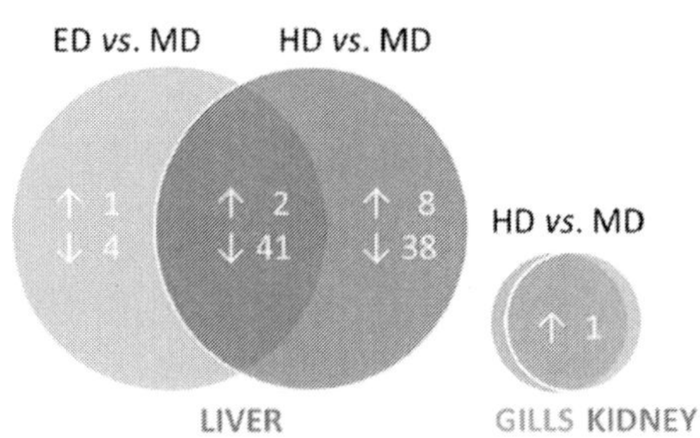


**Figure. 1**. DE genes ($p < 0.05$; $q < 0.05$; $-2.0 > FC < 2.0$) in liver, gills, kidney tissue of rainbow trout from (a) strain Born, (b) strain Troutlodge (TLodge), (c) both strains together kept at ED or HD compared with MD conditions. Up-(↑) downregulated (↓) genes are presented as a function of tissue type SD. (d) Venn diagrams illustrating the common exclusively expressed genes excluding non-annotated /or duplicate features; the gene

numbers are hence different to those listed in (c), although the original data sets were identical. Numbers in the overlap represent DE genes derived from different comparisons. Upregulated genes are listed in Table 2.

Distinct sets of DE genes were also detected in gills kidney of both trout strains, although the total number of DE genes was clearly lower (Fig. 1a, b). Strikingly, 80 DE features were regulated in gills of Born trout, whereas only 6 features were regulated in Troutlodge trout. In contrast, the number of DE features in kidney was similar (about 50). The *protein ubiquitination pathway* (Born: $p = 6 \times 10^{-3}$, 3 genes; Troutlodge: $p = 1 \times 10^{-2}$, 1 gene) was the prominent pathway downregulated in gills of Born Troutlodge strain; the *complement system* (Born: $p = 3 \times 10^{-2}$, 1 gene; Troutlodge: $p = 2 \times 10^{-6}$, 3 genes) was downregulated in the kidney of both strains.

## 3.2. The majority of biological functions were downregulated in rainbow trout at higher densities

Given a clear intersection of DE gene sets in Born Troutlodge, we aimed at the detection of potential strain-independent biomarker genes with relevance for rainbow trout in general. The blocking module included in the Bioconductor package limma extracted those genes that were differentially expressed under ED or HD in both strains (hereinafter collectively referred to as 'rainbow trout') compared with MD conditions, reducing the number of putative indicator genes even further (Fig. 1c, d).

This analysis revealed that most of DE genes in the liver of both strains were lower transcribed under ED (45 annotated genes) HD conditions (79 annotated genes), as elucidated previously for the Born Troutlodge strains in separate calculations. These downregulated genes are mainly involved in five hepatic pathways (Table 3): *protein ubiquitination pathway* ($p < 3 \times 10^{-2}$), *cholesterol biosynthesis* ($p < 3 \times 10^{-4}$), *mTOR signalling* ($p < 1 \times 10^{-2}$), *retinol biosynthesis* ($p < 9 \times 10^{-3}$), *PI3K/AKT signalling* ($p < 2 \times 10^{-2}$). Additionally, the transcript level of *GCLM*, the first rate-limiting enzyme in *glutathione biosynthesis* ($p < 2 \times 10^{-2}$), was reduced. The genes that were more strongly transcribed in liver under ED (3 annotated genes; cf. Table 4) HD conditions (10 annotated genes), were mainly of regulatory nature involved in the control of cell growth (*KLF11, GDF11*) or proliferation (*SOSTDC1*). Notably, *HIF1A signalling* ($p = 0.02$) was the only stress-induced hepatic pathway comprising upregulated positive factors (*ARNT, G6PC*) a downregulated inhibitory factor (*HSP90AA1*) under HD conditions compared with MD (Fig. 2a).

**Table 3.** Relevant ingenuity-predicted canonical pathways downregulated in rainbow trout held at ED or HD compared to MD conditions.

| Canonical pathway (number of included genes) | p-Value | Stocking density | Involved genes |
|---|---|---|---|
| Protein ubiquitination pathway (255) | 1.23E-05 | HD vs. MD | *DNAJC3, -11, HSP90AA1, GRP78, HSPA9, HSPD1, PSMD2, -11* |
| | 2.09E-02 | ED vs. MD | *DNAJC3, HSP90AA1, HSPA9* |
| Cholesterol biosynthesis (28) | 2.20E-04 | HD vs. MD | *FDPS, IDI1, MVD1* |
| | 4.88E-07 | ED vs. MD | *FDFT1, FDPS, IDI1, MVD1* |
| mTOR signalling (188) | 1.21E-03 | HD vs. MD | *EIF3J, EIF4A1, EIF4E, EIF4G1, PPP2R1A* |
| | 9.30E-03 | ED vs. MD | *EIF4A1, EIF4G1, PPP2R1A* |
| Retinol biosynthesis (33) | 8.54E-03 | HD vs. MD | *LIPG, RDH12* |
| | 2.63E-03 | ED vs. MD | *LIPG, RDH12* |
| PI3K/AKT signalling (123) | 1.53E-02 | HD vs. MD | *EIF4E, HSP90AA1, PPP2R1A* |
| | 3.27E-02 | ED vs. MD | *HSP90AA1, PPP2R1A* |
| Glutathione biosynthesis (3) | 1.26E-02 | HD vs. MD | *GCLM* |
| | 4.64E-03 | ED vs. MD | *GCLM* |

**Table 4.** Genes significantly upregulated ($q < 0.05$) in rainbow trout held at ED or HD compared to MD.

| Gene symbol | Gene product | Fold change | Function |
|---|---|---|---|
| Upregulated in liver under ED conditions | | | |
| *SOSTDC1* | Sclerostin domain-containing protein 1 | 3.2 | Regulation of cellular proliferation, differentiation, apoptosis |
| *KLF11* | Krüppel-like factor 11 | 2.0–2.1 | Inhibition of cell growth; induction of apoptosis |
| *ID3* | Inhibitor of DNA-binding protein 3 | 2.0 | Inhibition of DNA-binding interacting Helix-loop-helix proteins |
| Upregulated in liver under HD conditions | | | |
| *ZCCHC3* | CCHC domain-containing zinc finger 3 | 2.8–4.8 | Unknown function |
| *ARNT* | Aryl hydrocarbon receptor nuclear translocator | 2.1–2.9 | Expression of stress response and xenobiotic metabolism genes |
| *KLF11* | Krüppel-like factor 11 | 2.0–2.7 | Inhibition of cell growth; induction of apoptosis |
| *SOSTDC1* | Sclerostin-domain containing protein 1 | 2.7 | Regulation of cellular proliferation, differentiation, apoptosis |
| *CC058* | CC058 protein | 2.3 | Unknown function |
| *HAMP* | Hepcidin antimicrobial peptide | 2.2 | Maintenance of iron homeostasis |
| *CYP24A1* | Cytochrome P450, family 24, subfamily A, polypeptide 1 | 2.1 | Synthesis of cholesterol/xenobiotic metabolism |
| *GDF11* | Growth differentiation factor 11 | 2.1 | Regulation of cell growth and differentiation |
| *G6PC* | Glucose-6-phosphatase, catalytic subunit | 2.1 | Regulation of glucose homeostasis |
| *TRIM39* | Tripartite motif-containing protein 39 | 2.0 | Release of cytochrome *c*, promotion of apoptosis signalling |
| Upregulated in kidney under HD conditions | | | |
| *PYGM* | Glycogen phosphorylase, muscle | 4.2 | Glycogenolysis |
| Upregulated in gills under HD conditions | | | |
| *PYGM* | Glycogen phosphorylase, muscle | 6.7 | Glycogenolysis |

The elevated transcript levels of *PYGM* were the only feature regulated in kidney (4.2-fold) gills (6.7-fold) of rainbow trout kept at HD compared to MD (with $q \leq 0.04$) (Table 4).

## 3.3. Higher densities increase blood glucose levels, but do not affect cortisol levels in trout

In Born Troutlodge trout exposed to HD, the glucose concentration was significantly ($p < 0.05$) increased over trout kept at MD (Fig. 2b). Plasma cortisol levels did not show any significant differences between the SD conditions (data not shown).

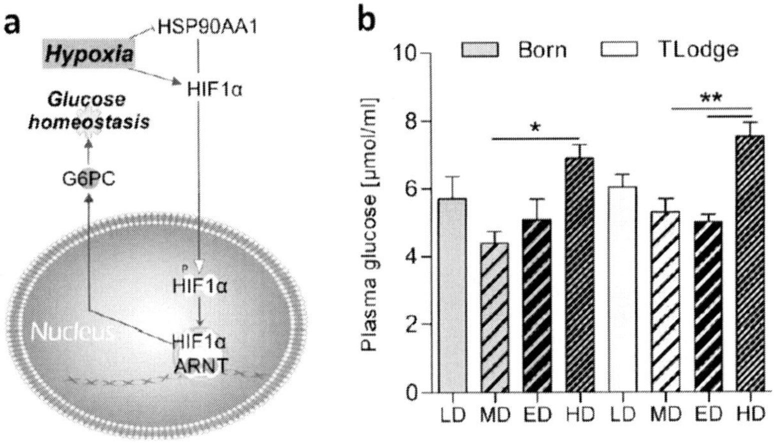


**Figure. 2.** (a) Stress-induced interaction network *HIF1A signalling* involving three genes (*ARNT, G6PC, HSP90AA1*) regulated in trout kept at HD compared with those at MD, as suggested by the Ingenuity Pathway Analyses tool (with $p = 0.02$). This pathway eventually regulates plasma glucose concentrations. (b) Plasma glucose concentrations [µmol/mL] in Born (blue bars) Troutlodge trout (green bars) kept at density treatments of ~ 10 kg/m³ (LD), ~ 30 kg/m³ (MD), ~ 60 kg/m³ (ED) ~ 120 kg/m³ (HD), as indicated on the abscissa. Significant differences are marked with asterisks, *($p < 0.05$); **($p < 0.01$) as evaluated using one-way ANOVA Tukey's tests.

## 3.4. Microarray qPCR confirm the different expression of selected marker gene cidates

Quantitative RT-PCR validated array-predicted expression differences of selected genes involved in the above mentioned pathways biofunctions, i.e. *G6PC* (*HIF1A signalling*), *HSP90AA1* (*protein ubiquitination pathway*), *FDPS* (*cholesterol biosynthesis*), *EIF4E* (*mTOR signalling*), as well as the genes encoding PGM3, HSP47 (SERPINH1), COX (Fig. 3). Student's *t*-test validated a significantly different expression ($p < 0.05$) for all selected genes, except for *COX* ($p = 0.06$). Accordingly, COX profiles showed no correlation (Pearson product-moment correlation coefficient, $r < 0.3$), whereas profiles across the other six target genes revealed a high concordance ($r = 0.86$).

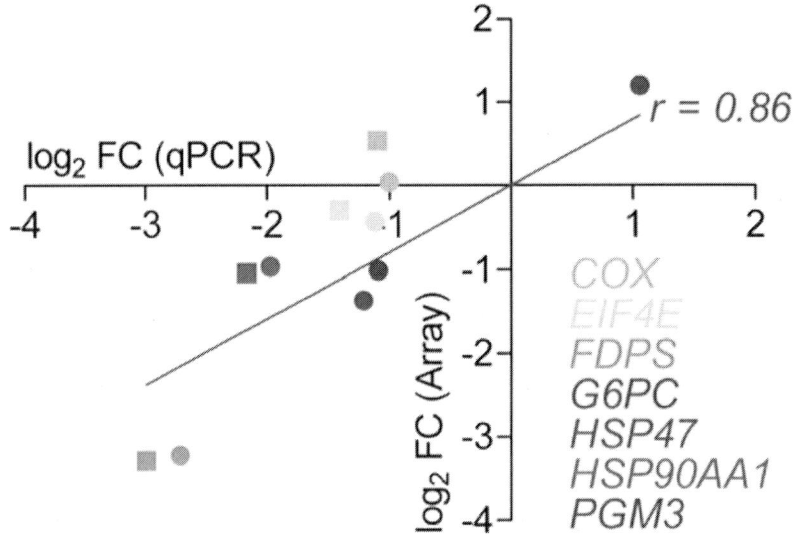


**Figure. 3.** Concordance of array (ordinate) qPCR (abscissa) results for selected genes. The logarithm base 2 of fold-change values ($\log_2$ FC) were plotted for ED vs. MD (squared symbols) HD vs. MD (circular) comparisons. Pearson's correlation is indicated.

## 3.5. Increased stocking is reflected by altered behaviour, although in a strain-dependent manner

In addition to the analysis of molecular characteristics attending crowding stress, we recorded the behaviour of the rainbow trout to determine potential interrelations between transcriptional behavioural features. Video recordings did not reveal aggressive interactions. However, we observed a significantly unequal distribution of trout at LD ($\chi^2 = 32.9$, $p < 0.001$), MD ($\chi^2 = 42.2$, $p < 0.001$) ED ($\chi^2 = 23.3$, $p < 0.001$), but not at HD ($\chi^2 = 4.0$, $p = 0.56$). Notably, this distribution pattern was only evident among BORN trout (Fig. 4, upper panel) indicating that individuals of this strain were mainly residing at the bottom (MD) as well as in the lower central parts of the tanks (ED). In contrast, the Troutlodge strain showed a more or less equal distribution across all SDs (Fig. 4, lower panel).

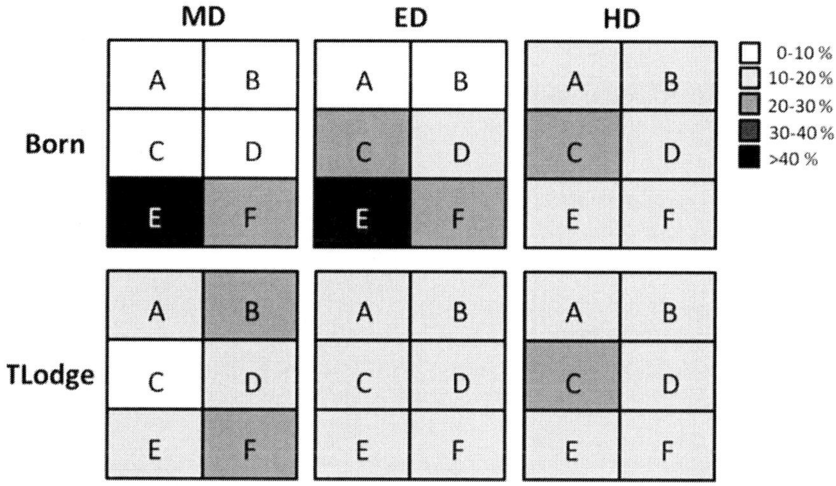


**Figure. 4.** Spatial distribution of rainbow trout from Born (upper panels) Troutlodge strain (TLodge; lower panels) in relation to SD (MD, ED, HD). Shaded fields indicate the percentage of visible fish (average of 53 observation points) across the front of the experimental tanks divided into six rectangles (A–F).

# 4. DISCUSSION

## 4.1. Transcriptome profiling reveals a surprisingly short list of genes indicating crowding stress in trout

Crowding has been judged as one of the most influential stressors affecting fish physiology thus the status of well-being in aquaculture (Barton, 2002). Against this background, it is even more astonishing that microarray technology allowing for the simultaneous recording of > 42,500 sequence features, revealed unexpectedly short lists of differentially regulated genes in rainbow trout from high-versus moderate-density groups (cf. Suppl. Fig. 1b). In kidney gills, crowding induced only the *PYGMA* gene encoding a glycogen phosphorylase. In contrast, almost 100 genes were differentially regulated in the liver of trout kept at high versus moderate density. Strikingly, only little more than one tenth of those features were upregulated under HD conditions. The differentially regulated genes were predicted to contribute to seven hepatic pathways, which are all known to play specific roles during stress responses. These included the overexpressed *HIF1A signalling pathway* as well as the underexpressed *mTOR PI3K/AKT signalling pathways*. This is in line with previous reports demonstrating that ER stress impairs mTOR PI3K/AKT activities (Proud, 2004; Qin 2014), but induces HIF1A (hypoxia-inducible factor 1, subunit alpha) to restore homeostasis (Rahman Thomas, 2012; Werno 2008).

## 4.2. Carbohydrate-metabolising pathways indicate short-term crowding stress in trout

The restoration of homeostasis requires large amounts of glucose as a key fuel, thereby restricting its availability for other vital functions. HIF1A, in concert with ARNT (aryl hydrocarbon receptor nuclear translocator) (Jiang 1996) induces the expression of *G6PC* (Gautier-Stein 2012). This gene encodes the catalytic subunit of glucose-6-phosphatase, a key enzyme regulating blood glucose levels, in rainbow trout as in other organisms (Vernier Sire, 1976) having previously been identified as an indicator for hling crowding stress in rainbow trout (Liu 2014). Not only the blood glucose level, but also the mRNA abundances of *ARNT G6PC* genes were indeed increased in the livers of rainbow trout kept at HD compared with MD in the present study. Interestingly, the genes *ARNT* (its salmon ortholog LOC106568252 maps to chromosome ssa13) *G6PC* (salmon LOC106601296 maps to ssa03) are presumably located on chromosome 12 of rainbow trout, which had been found to harbour the most significant quantitative trait loci for stress response (Rexroad 2013) 16 single-nucleotide polymorphisms significantly associated with cortisol response to crowding (Liu b).

Concomitantly with increased expression of *ARNT G6PC*, *PGM3* transcript levels were found to be decreased. The encoded phosphoacetylglucosamine mutase is indispensable for glycogen biosynthesis breakdown. Glycogenolysis might even be a systemic physiological response, as *PYGM* was the only gene that was upregulated in the kidney gills of rainbow trout kept at HD compared with MD. The encoded glycogen phosphorylase catalyses the conversion of glycogen to glucose-1-phosphate (Mehrani Storey, 1993).

An enhanced glucose level is generally accepted as an indicator of impaired well-being in trout (Trenzado 2006; Yarahmadi ), though few reports document that glucose levels may decrease (Vijayan 1990) or remain unaltered (Laidley Leatherl, 1988; Melotti 2004) in salmonids with increasing SD (cf. Table 1). In addition, reduced glycogen content has been found in rainbow trout (Conde-Sieira 2010; Trenzado 2006) brook trout (Vijayan 1990) held at HD between 70 120 kg/m$^3$ compared with an LD group (10–30 kg/m$^3$) strongly indicating that the breakdown of glycogen indicates (crowding) stress in salmonid fish. Altogether, the abovementioned findings in the context of relevant literature reveal that regulated features of carbohydrate metabolism, in particular those involved in glucose provision the dissociation of glycogen, are strong indicators of crowding stress.

Several enzymes of the cytochrome P450 family (CYP) have also been suggested as indicator genes for stress in diverse fish species (Gornati 2004; Nematollahi 2012; Rahman Thomas, 2012; Scholz 1997). Genes coding for CYP members have been proven to contribute to a broad panel of functions, particularly in regard to the oxidation of many endogenous xenobiotic substrates (reviewed by Frye, 2004). While the level of cytochrome *c* oxidase (COX)-encoding transcripts was predicted to be decreased, *CYP24A1* mRNA abundance was increased in the liver of rainbow trout kept at HD compared with those at MD. The encoded mitochondrial 25-hydroxyvitamin D3-24-hydroxylase plays a role in calcium homeostasis through vitamin D3 metabolism in salmonid fish (Lock 2007). Calcium is an essential second messenger promoting repair or death responses (reviewed by Cerella 2010). In this respect, it has been suggested that stress conditions compromise cell survival trigger apoptotic signalling (Calduch-Giner 2010), which is indeed reflected by the set of upregulated genes comprising a higher proportion of cell death-related genes (*KLF11, SOSTDC1, TRIM39*). As apoptosis-related genes generally influence a plethora of physiological events, the altered transcription level of cell death-related genes seems to be a less robust indicator of impaired well-being. In contrast, glycogen- glucose-metabolising pathways (including the key enzymes PGM3, PYGM, G6PC, respectively) seem to be valuable, informative indicators for short-term (> 1-month period) crowding stress in rainbow trout.

## 4.3. Cortisol HSPs are not suitable biomarkers for non-acute, short-term crowding stress in trout

*Cholesterol biosynthesis* was among the top pathways underexpressed in rainbow trout kept at HD ED compared with those at MD. *FDPS*, one of the most strongly downregulated genes in trout at ED HD, has been identified as the rate-limiting enzyme in the mammalian biosynthesis of cholesterol (Ashby Edwards, 1989), which is the precursor metabolite for the production of cortisol. The main function of this slow-acting steroid hormone is to increase blood glucose levels in response to hypoglycaemia (reviewed by Verberne 2014). As stated above, glucose levels were indeed increased in stressed trout, whereas cortisol levels were similar in all SD groups examined in the present study. This can likely also be attributed to reduced cortisol synthesis in rainbow trout kept under high-density husbry in comparison to moderate husbry conditions, as predicted by IPA.

At first glance, these data seem in conflict with the plethora of studies on stress physiology in fish that refer to plasma cortisol levels as an indicator of the severity of certain stressors (reviewed by Wendelaar Bonga, 1997). A previous long-term study on the impact of SD demonstrated, on the other h, that higher cortisol levels were recorded in trout kept at 10 kg/m$^3$ (comparable to the LD treatment in the present study) than in those kept at 80 kg/m$^3$ (intermediate between the ED HD treatments in the present study) (North 2006). Further studies have documented that cholesterol levels increase (Person-Le Ruyet 2008; Pickering Pottinger, 1987; Pickering 1991), decrease (Leatherl Cho, 1985; North 2006), or remain constant (Kebus 1992; Leatherl, 1993; Procarione 1999) with increasing SD in rainbow trout farming (cf. Table 1). In this context, it should be noted that cortisol levels not only adapt to persistent stressful conditions (Aerts ), but also depend on a broad spectrum of endogenous environmental factors, such as seasonality (North 2006), time of day (López-Patiño 2014), water quality (Person-Le Ruyet 2008), rearing temperature, colour of the tank, food, heritability age of fish (Martinez-Porchas 2009). Relevant literature indicates moreover that higher concentrations of cortisol indicate sudden, short-term stress exposure (reviewed by Barton, 2002; Wendelaar Bonga, 1997), while the level of glucose rises days after stress induction (Hemre Krogdahl, 1996).

It has also been shown that lower densities provoke the establishment of territorial hierarchies (North 2006; Pottinger Carrick, 2001), which in turn influence the level of cortisol (Fox 1997). To document potential crowding-related expressions of hierarchy during the present study, the behaviour of trout was monitored. We did not detect any aggressive behaviour, but we did find an unbalanced spatial distribution of individuals from one of the two rainbow trout strains: While trout were more or less equally distributed in HD tanks, we noted crowding in the lower parts of MD tanks. Laursen et al. (2013) found in a previous study that a few dominant rainbow trout occupy one tank of a two-tank system, while all other trout are crowded in the adjacent tank. Nevertheless, our observation of a crowding-induced hierarchical formation is apparently only a strain- not a species-specific feature. This is in line with other authors, who describe clear differences in behaviour between strains of rainbow trout based on their individual selection history (Campbell ; Millot 2014; Trenzado 2006).

In addition to *cholesterol biosynthesis* the synthesis of the antioxidants retinol glutathione, the downregulation of a further metabolic pathway in liver of trout exposed to higher SD was predicted, namely *protein ubiquitination*. This pathway comprises a considerable number of heat-shock proteins (HSPs) that are downregulated in crowding-stressed trout, such as *HSP47/SERPINH1*, *HSP90AA1*, *HSPA5/GRP78*, *HSPA9*, *HSPD1* related factors. HSPs are generally considered as characteristic marker genes for stress responses, also in rainbow trout, but rather indicative of denatured proteins accumulating upon exposure to thermal (Rebl 2013; Verleih ) or chemical stressors (Kilemade Mothersill, 2001; Koskinen 2004). Although it has been reported that *HSP70* mRNA levels increased in rainbow trout kept at densities of 30 kg/m$^3$ (Aksakal ) as well as 100 kg/m$^3$ (Gornati 2004) compared with the respective control groups, HSPs have not been judged as reliable 'indicators of hatchery stress' in fish (Zarate Bradley, 2003). Their concentration has also been found either to remain at similar levels (Liu 2014; Zarate Bradley, 2003) or to decrease after exposure of fish up to densities of 46 kg/m$^3$ (Alves 2010; Salas-Leiton 2010). A reason for the dynamic, condition-specific variable expression of HSPs may lie in their pleiotropic functional spectra. HSPs contribute to numerous cellular signalling pathways including *HIF1A* signalling (Gradin 1996; Fig. 4a), *toll-like receptor chaperone-mediated autophagy* (Xu ) or *glucocorticoid receptor signalling* (Kirschke 2014; Pratt Dittmar, 1998) dem therefore tight regulation (Wang ).

## 4.4. Evaluation of biomarkers for crowding stress is affected by various animal-based environmental factors

This study presents several genes pathways that are affected by stocking density in rainbow trout farming, thus extends the rich repertoire of different biomarkers for crowding identified in previous challenge experiments (summarised in Table 1). Nevertheless, we others have found that the direction of perturbation of certain biomarkers may be contradictory — with various contributory causes. One certain cause is the duration of stress exposure. The experiments referred to in Table 1 for instance, span a timeline from two days up to nine months. Such variability in temporal scope has been shown to produce variation in biomarker measurements (reviewed by Barton, 2002; Ellis 2002). The present study challenged fish for nine days is therefore presumably situated between acute chronic stress exposure (Pickering 1991). This transitional phase, however, may be characterised by a lower number of detectable biomarkers associated indicative pathways.

Further factors influencing the expression of biomarkers include conditions in the husbry facilities, such as water quality, temperature, nutrition (reviewed by Barton, 2002; Ellis 2002) the size of the experimental tanks. Our experiments were conducted in smaller-scale tanks, oriented to previous studies on crowding stress in fish (Calduch-Giner 2010; Conde-Sieira 2010; Laursen 2013; Procarione 1999; Trenzado 2006; Yarahmadi 2014; see also Table 1), but it should not be ignored that these tanks do not represent volumes used in commercial aquaculture facilities.

Moreover, this study emphasises that 'biomarkers in aquaculture' are generally not only species-specific, but may also vary between strains or breeds. Decades of domestication caused directional genetic changes have inevitably altered the behaviour of rainbow trout (Tymchuk 2009). Contemporary, intensive targeted selection has promoted a rainbow trout phenotype that is clearly distinguished from the wildtype, as it is characterised by enhanced stress tolerance (Barton Schreck, 1987; Jentoft 2005; Trenzado 2006) — at least towards crowding (Rexroad 2012). Our parallel study on another salmonid fish demonstrates that comparatively new aquaculture species have a low level of stress tolerance: We assessed the transcriptional impact of similar SDs on the maraena whitefish *Coregonus maraena* found that almost 200 genes were differentially expressed ($q < 0.05$, $-2.0 > FC > 2.0$) in the kidney of whitefish kept at high versus moderate density (Korytář 2016). By comparison, only one single parameter was found to be differentially expressed in the kidney of trout under comparable husbry conditions (present study). Additionally, a high number of pathways indicating cellular stress immunosuppression in the liver of whitefish are contrastive to the relatively few findings in the liver of rainbow trout in this study. However, it should not be overlooked that the *complement system* (Köbis ) was in part compromised in the kidney of trout (from both strains) kept under stress conditions. Complement function has been found to be suppressed in several fish species upon crowding challenge (Mauri ), also in rainbow trout (Yarahmadi 2016). Future studies involving transcriptomic approaches might investigate whether prolonged crowding challenge might lead to severely impaired immune functions, eventually culminating in a significantly elevated susceptibility towards pathogens.

# 5. CONCLUSIONS

Our data confirm previous investigations revealing that SDs up to 120 kg rainbow trout per cubic meter do not seem to cause severe physiological dysfunctions, at least in the first days of exposure to crowding. However, this assertion is based on an experimental setup that does not meet commercial aquaculture stards depends on various definite parameters, not least on two robust rainbow trout strains, constantly high water quality, sufficient food supply. In addition, the investigated time point of nine days does not provide information about sudden, acute or long-term, chronic stress responses. Subsequent transcriptomic studies may thus address the impact of prolonged exposure to crowding stress, recording the time course of affected networks to indicate, in particular, early signs of stress-induced immunosuppression at the mRNA level.

# AUTHORS' CONTRIBUTIONS

TG conceived of the study coordinated the experiments. RB organized the production of fish provided experimental facilities. MV, MZ, AR, AB, contributed to the design of the study performed the laboratory experiments. AR AB performed gene profiling assays analysed data. All authors were involved in sampling in the interpretation of experimental data. MZ AB determined the distribution of fish in the individual tanks. AR wrote the manuscript. All authors have red approved the final .

The following are the supplementary data related to this .

# ACKNOWLEDGEMENTS

We are indebted to I. Hennings, B. Schöpel, L. Falkenthal for expert technical assistance we acknowledge A. Tuchscherer for providing statistical support (SAS software). Many thanks go to an anonymous referee for constructive comments that greatly contributed to improve the manuscript. Funding for this work came from the European Fisheries Fund (EFF) the Ministry of Agriculture, the Environment Consumer Protection Mecklenburg-Western Pomerania, Germany (EFF pilot project 'BORN-Forelle' grant no. VI-560/7308-4). M. Zebunke was funded by the German agriculture network PHÄNOMICS (funded by the Federal Ministry of Education Research; grant no. 0315536G).

# REFERENCES

1. Adams,C., Adams, J. Turnbull, A. Bell, J. Bron, F.A. Huntingford**Multiple determinants of welfare in farmed fish: stocking density, disturbance, aggression in Atlantic salmon (*Salmo salar*)**Can. J. Fish. Aquat. Sci., 64 (2007), pp. 336-344, 10.1139/f07-018
2. Aerts, J. Aerts, J.R. Metz, B. Ampe, A. Decostere, G. Flik, S. de Saeger**Scales tell a story on the stress history of fish** PLoS One, 4 (), e0123411, 10.1371/ journal. pone.0123411
3. Aksakal, E. Aksakal, D. Ekinci, O. Erdoğan, Ş. Beydemir, Z. Alım, S.B. Ceyhun**Increasing stocking density causes inhibition of metabolic–antioxidant enzymes elevates mRNA levels of heat shock protein 70 in rainbow trout** Livest. Sci., 141 (1) (), pp. 69-75, 10.1016/j.livsci..07.006
4. Alanara A. Alanara, E. Brannas**Dominance in dem-feeding behaviour in Arctic charr rainbow trout: the effect of stocking density** J. Fish Biol., 48 (2) (1996), pp. 242-254, 10.1111/j.1095-8649.1996.tb01116.x
5. Alves R.N. Alves, O. Cordeiro, T.S. Silva, N. Richard, M. de Vareillesa, G. Marinob, P. Di Marcob, P.M. Rodriguesa, L.E.C. Conceição**Metabolic molecular indicators of**

**chronic stress in gilthead seabream (*Sparus aurata*) using comparative proteomics** Aquaculture, 299 (2010), pp. 57-66, 10.1016/ j.aquaculture. 2009.11.014

6. Amaeze,N.H. Amaeze, S. Schnell, O. Sozeri, A.A. Otitoloju, R.I. Egonmwan, V.M. Arlt, N. R.Bury**Cytotoxic genotoxic responses of the RTgill-W1 fish cells in combination with the yeast oestrogen screen to determine the sediment quality of Lagos lagoon, Nigeria** Mutagenesis, 30 (1) (), pp. 117-127, 10.1093/mutage/geu032
7. Ashby, M.N. Ashby, P.A. Edwards**Identification regulation of a rat liver cDNA encoding farnesyl pyrophosphate synthetase** J. Biol. Chem., 264 (1) (1989), pp. 635-640 (PMID: 2909544)
8. Barton, B.A. Barton**Stress in fishes: a diversity of responses with particular reference to changes in circulating corticosteroids** Integr. Comp. Biol., 42 (3) (2002), pp. 517-525, 10.1093/icb/42.3.517
9. Barton, B.A. Barton, G.K. Iwama**Physiological changes in fish from stress in aquaculture with emphasis on the response effects of corticosteroids** Annu. Rev. Fish Dis., 1 (1991), pp. 3-26, 10.1016/0959-8030(91)90019-G (2MB)
10. Barton, B.A. Barton, C.B. Schreck**Influence of acclimation temperature on interrenal carbohydrate stress responses in juvenile chinook salmon (*Oncorhynchus tshawytscha*)** Aquaculture, 62 (3–4) (1987), pp. 299-310 (808KB)
11. Benjamini, Y. Benjamini, Y. Hochberg**Controlling the false discovery rate: a practical powerful approach to multiple testing** J. R. Stat. Soc. Ser. B, 57 (1) (1995), pp. 289-300 www.jstor.org/stable/2346101
12. Boujard, T. Boujard, L. Labbe, B. Auperin**Feeding behaviour, energy expenditure growth of rainbow trout in relation to stocking density food accessibility** Aquac. Res., 33 (15) (2002), pp. 1233-1242, 10.3923/pjbs.2007.964.967
13. Brunelli, J.P. Brunelli, K.J. Wertzler, K. Sundin, G.H. Thorgaard**Y-specific sequences polymorphisms in rainbow trout Chinook salmon** Genome, 51 (9) (2008), pp. 739-748, 10.1139/G08-060
14. Calduch-,J.A. Calduch-Giner, G. Davey, A. Saeraila, B. Houeix, A. Talbot, P. Prunet, M.T. Cairns, J. Pérez-Sánchez**Use of microarray technology to assess the time course of liver stress response after confinement exposure in gilthead sea bream (*Sparus aurata* L.)** BMC Genomics, 11 (2010), p. 193, 10.1186/1471-2164-11-193
15. Campbell, J.M. Campbell, P.A. Carter, P.A. Wheeler, G.H. Thorgaard**Aggressive behavior, brain size domestication in clonal rainbow trout lines**Behav. Genet., 45 (2) (), pp. 245-254, 10.1007/s10519-014-9696-0
16. Cerella, C. Cerella, M. Diederich, L. Ghibelli**The dual role of calcium as messenger stressor in cell damage, death, survival** Int. J. Cell. Biol., 2010 (2010), p. 546163, 10.1155/2010/546163
17. Conde-Sieira, M. Conde-Sieira, A.J. Aguilar, M.A. López-Patiño, J.M. Míguez, J.L. Soengas**Stress alters food intake glucosensing response in hypothalamus, hindbrain, liver, Brockmann bodies of rainbow trout** Physiol. Behav., 101 (4) (2010), pp. 483-493, 10.1016/j.physbeh.2010.07.016
18. FAO, 2005–2017**Cultured aquatic species information programme. *Oncorhynchus mykiss*. Text by Cowx, I.G** FAO Fisheries Aquaculture Department [online] (2005–2017) Rome. Updated 15 June 2005. [Cited 9 February 2017] http://www.fao.org/fishery/culturedspecies/Oncorhynchus_mykiss/en
19. European Commission/Eurostat Eumofa, European Commission/Eurostat Eumofa**Aquaculture – Facts Figures** http://ec.europa.eu/fisheries /cfp/ aquaculture/facts/index_en.htm ()
20. Ellis, T. Ellis, B. North, A.P. Scott, N.R. Bromage, M. Porter, D. Gadd**The relationships between stocking density welfare in farmed rainbow trout** J. Fish Biol., 61 (3) (2002), pp. 493-531, 10.1111/j.1095-8649.2002.tb00893.x
21. Fox, H.E. Fox, S.A. White, M.H. Kao, R.D. Fernald**Stress dominance in a social fish** J. Neurosci., 17 (16) (1997), pp. 6463-6469 (PMID: 9236253)

22. Frye, R.F. Frye**Probing the world of cytochrome P450 enzymes** Mol. Interv., 4 (3) (2004), pp. 157-162, 10.1124/mi.4.3.5
23. FSBI, FSBI**Fish welfare (briefing paper 2)** Fish Soc. Br. Isles (2002), pp. 1-21 www.fsbi.org.uk/publications/briefing-papers/
24. Gautier-Stein, A. Gautier-Stein, M. Soty, J. Chilloux, C. Zitoun, F. Rajas, G. Mithieux**Glucotoxicity induces glucose-6-phosphatase catalytic unit expression by acting on the interaction of HIF-1α with CREB-binding protein** Diabetes, 61 (10) (2012), pp. 2451-2460, 10.2337/db11-0986
25. Gornati, R. Gornati, E. Papis, S. Rimoldi, G. Terova, M. Saroglia, G. Bernardini**Rearing density influences the expression of stress-related genes in sea bass (*Dicentrarchus labrax*, L.)** Gene, 341 (2004), pp. 111-118, 10.1016/j.gene.2004.06.020 (896KB)
26. Gradin K. Gradin, J. McGuire, R.H. Wenger, I. Kvietikova, M.L. Fhitelaw, R. Toftgård, L. Tora, M. Gassmann, L. Poellinger**Functional interference between hypoxia dioxin signal transduction pathways: competition for recruitment of the Arnt transcription factor** Mol. Cell. Biol., 16 (10) (1996), pp. 5221-5231, 10.1128/MCB.16.10.5221
27. Harper Wolf, C. Harper, J.C. Wolf**Morphologic effects of the stress response in fish**
28. ILAR J., 50 (4) (2009), pp. 387-396, 10.1093/ilar.50.4.387
29. Hemre Krogdahl, G. Hemre, A. Krogdahl**Effect of hling fish size on secondary changes in carbohydrate metabolizm in Atlantic salmon, *Salmo salar* L** Aquac. Nutr., 4 (1996), pp. 3-6, 10.1111/j.1365-2095.1996.tb00067.x
30. Huang, D.W. Huang, B.T. Sherman, R.A. Lempicki**Systematic integrative analysis of large gene lists using DAVID bioinformatics resources** Nat. Protoc., 4 (1) (2008), pp. 44-57, 10.1038/nprot.2008.211
31. Hulsen, T. Hulsen, J. de Vlieg, W. Alkema**BioVenn - a web application for the comparison visualization of biological lists using area-proportional Venn diagrams** BMC Genomics, 9 (2008), p. 488, 10.1186/1471-2164-9-488
32. Huntingford, F.A. Huntingford, C. Adams, V.A. Braithwaite, S. Kadri, T.G. Pottinger, P.Soe, J.F. Turnbull**Current issues in fish welfare** J. Fish Biol., 68 (2) (2006), pp. 332-372, 10.1111/j.0022-1112.2006.001046.x
33. Jentoft, S. Jentoft, A.H. Aastveit, P.A. Torjesen, Ø. ersen**Effects of stress on growth, cortisol glucose levels in non-domesticated Eurasian perch (*Perca fluviatilis*) domesticated rainbow trout (*Oncorhynchus mykiss*)** Comp. Biochem. Physiol. A Mol. Integr. Physiol., 3 (2005), pp. 353-358, 10.1016/j.cbpb.2005.06.006 (119KB)
34. Jiang, B.H. Jiang, E. Rue, G.L. Wang, R. Roe, G.L. Semenza**Dimerization, DNA binding, transactivation properties of hypoxia-inducible factor 1** J. Biol. Chem., 30 (1996), pp. 17771-17778, 10.1074/jbc.271.30.17771
35. Kebus, M.J. Kebus, M.T. Collins, M.S. Brownfield, C.H. Amundson, T.B. Kayes, J.A. Malison**Effects of rearing density on the stress response growth of rainbow trout** J. Aquat. Anim. Health, 4 (1) (1992), pp. 1-6, 10.1577/1548-8667(1992)004<0001:EORDOT>2.3.CO;2
36. Kilemade Mothersill, M. Kilemade, C. Mothersill**Heat shock protein 70 levels in rainbow trout primary epidermal cultures in response to 2,4-dichloroaniline exposure: a novel in vitro aquatic toxicity marker** Environ. Toxicol., 16 (3) (2001), pp. 253-259 (PMID: 11409197)
37. Kirschke, E. Kirschke, D. Goswami, D. Southworth, P.R. Griffin, D.A. Agard **Glucocorticoid receptor function regulated by coordinated action of the Hsp90 Hsp70 chaperone cycles** Cell, 7 (2014), pp. 1685-1697, 10.1016/j.cell.2014.04.038 (3MB)

38. Köbis, J.M. Köbis, A. Rebl, C. Kühn, T. Korytář, B. Köllner, T. Goldammer **Comprehensive comparative transcription analyses of the complement pathway in rainbow trout** Fish Shellfish Immunol., 42 (), pp. 98-107, 10.1016/j.fsi.2014.10.032 (3MB)

39. Korytář, T. Korytář, M. Nipkow, S. Altmann, T. Goldammer, B. Köllner, A. Rebl**Adverse husbry of maraena whitefish directs the immune system to increase mobilization of myeloid cells proinflammatory responses** Front. Immunol., 7 (2016), p. 631, 10.3389/fimmu.2016.00631

40. Koskinen H. Koskinen, P. Pehkonen, E. Vehniäinen, A. Krasnov, C. Rexroad, S. Afanasyev, H. Möl sa, A. Oikari**Response of rainbow trout transcriptome to model chemical contaminants** Biochem. Biophys. Res. Commun., 320 (2004), pp. 745-753, 10.1016/j.bbrc.2004.06.024 (543KB)

41. Laidley Leatherl, C.W. Laidley, J.F. Leatherl**Cohort sampling, anaesthesia stocking-density effects on plasma cortisol, thyroid hormone, metabolite ion levels in rainbow trout, *Salmo gairdneri* Richardson** J. Fish Biol., 33 (1) (1988), pp. 73-88, 10.1111/j.1095-8649.1988.tb05449.x

42. Lam, S.H. Lam, E.Y. Lui, Z. Li, S. Cai, W.-K. Sung, S. Mathavan, T.J. Lam, Y.K. Lp**Differential transcriptomic analyses revealed genes signaling pathways involved in iono-osmoregulation cellular remodeling in the gills of euryhaline Mozambique tilapia, *Oreochromis mossambicus*** BMC Genomics, 15 (1) (2014), p. 921, 10.1186/1471-2164-15-921

43. Laursen, D.C. Laursen, P.I. Silva, B.K. Larsen, E. Höglund**High oxygen consumption rates scale loss indicate elevated aggressive behaviour at low rearing density, while elevated brain serotonergic activity suggests chronic stress at high rearing densities in farmed rainbow trout, *Oncorhynchus mykiss*** Physiol. Behav., 122 (2013), pp. 147-154, 10.1016/j.physbeh.2013.08.026 (395KB)

44. Leatherl, J.F. Leatherl**Stocking density cohort sampling effects on endocrine interactions in rainbow trout** Aquac. Int., 1 (2) (1993), pp. 137-156, 10.1007/BF00692617

45. Leatherl Cho, J.F. Leatherl, C.Y. Cho**Effect of rearing density on thyroid interrenal gl activity plasma hepatic metabolite levels in rainbow trout, *Salmo gairdneri* Richardson** J. Fish Biol., 27 (5) (1985), pp. 583-592, 10.1111/j.1095-8649.1985.tb03203.x

46. Lien, S. Lien, B.F. Koop, S.R. Sve, J.R. Miller, M.P. Kent, T. Nome, T.R. Hvidsten, J.S.Leong, D. R. Minkley, A. Zimin, F. Grammes, H. Grove, A. Gjuvsl, B. Walenz, R.A. Hermansen, K. v on Schalburg, E.B. Rondeau, A. Di Genova, J.K. Samy, Vik J. Olav, M.D. Vigel, L. Caler, U. Grimholt, S. Jentoft, D.I. Våge, P. de Jong, T. Moen, M. Baranski, Y. Palti, D .R. Smith, J.A.Yorke, A.J. Nederbragt, A. Tooming-Klunderud, K.S. Jakobsen, X. Jiang, D. Fan, Y. Hu, D.A.Liberles, R. Vidal, P. Iturra, S.J. Jones, I. Jonassen, A. Maass, S. W. Om holt, W.S. Davidson**The Atlantic salmon genome provides insights into rediploidization** Nature, 533 (7602) (2016), pp. 200-205, 10.1038/nature17164

47. Liu ,B. Liu, Y., Wang X, 2015**The effect of stocking density on growth seven physiological parameters with assessment of their potential as stress response indicators for the Atlantic salmon (*Salmo salar*)** Mar. Freshw. Behav. Physiol., 48 (3) (), pp. 177-192, 10.1080/10236244..1034956

48. Liu, S. Liu, G. Gao, Y. Palti, B.M. Clevel, G.M. Weber, C.E. Rexroad III**RNA-seq analysis of early hepatic response to hling confinement stress in rainbow trout** PLoS One, 9 (2014), e88492, 10.1371/journal.pone.0088492

49. Liu, S. Liu, R.L. Vallejo, G. Gao, Y. Palti, G.M. Weber, A. Hernández, C.E. Rexroad 3rd**Identifi cation of single-nucleotide polymorphism markers associated with cortisol response to crowding in rainbow trout** Mar. Biotechnol. (NY), 17 (3) (), pp. 328-337, 10.1007/s10126-015-9621-4

50. Lock, E.J. Lock, R. Ørnsrud, L. Aksnes, F.A.T. Spanings, R. Waagbø, G. Flik**The vitamin D receptor its lig 1alpha,25-dihydroxyvitamin D3 in Atlantic salmon (*Salmo salar*)** J. Endocrinol., 193 (3) (2007), pp. 459-471, 10.1677/JOE-06-0198

51. López-Patiño, M.A. López-Patiño, M. Gesto, M. Conde-Sieira, J.L. Soengas, J.M. Míguez **Stress inhibition of melatonin synthesis in the pineal organ of rainbow trout (*Oncorhynchus mykiss*) is mediated by cortisol** J. Exp. Biol., 217 (Pt 8) (2014), pp. 1407-1416, 10.1242/jeb.087916

52. Martinez-Porchas, M. Martinez-Porchas, L.T. Martinez-Cordova, R. Ramos-Enriquez**Cortisol glucose: reliable indicators of fish stress?** Pan. Am. J. Aquat. Sci., 4 (2) (2009), pp. 158-178

53. Martins, C.I. Martins, L. Galhardo, C. Noble, B. Damsgård, M.T. Spedicato, W. Zupa, M.Beauchaud, E. Kulczykowska, J.C. Massabuau, T. Carter, S.R. Planellas, T. Kristiansen**Behavioural indicators of welfare in farmed fish** Fish Physiol. Biochem., 38 (1) (2012), pp. 17-41, 10.1007/s10695-011-9518-8

54. Mauri,

55. Mauri, A. Romero, L. Acerete, S. Mackenzie, N. Roher, A. Callol, I. Cano, M.C.Alvarez, L. Tort**Changes in complement responses in Gilthead seabream (*Sparus aurata*) European seabass (*Dicentrarchus labrax*) under crowding stress, plus viral bacterial challenges** Fish Shellfish Immunol., 30 (1) (), pp. 182-188, 10.1016 /j.fsi.2010.10.006 (747KB)

56. Mehrani Storey, H. Mehrani, K.B. Storey**Control of glycogenolysis effects of exercise on phosphorylase kinase cAMP-dependent protein kinase in rainbow trout organs** Biochem. Cell Biol., 71 (11 – 12) (1993), pp. 501-506, 10.1139/o93-072

57. Melotti, P. Melotti, A. Roncarati, L. Angellotti, A. Dees, G.E. Magi, C. Mazzini, C. Bianchi, R.Casianco**Effects of rearing density on rainbow trout welfare, determined by plasmatic tissue parameters** Ital. J. Anim. Sci., 3 (4) (2004), pp. 393-400 Millot 2014

58. S. Millot, S. Péan, L. Labbé, T. Kerneis, E. Quillet, M. Dupont-Nivet, M.L. Bégout**Assessment of genetic variability of fish personality traits using rainbow trout isogenic lines** Behav. Genet., 44 (4) (2014), pp. 383-393, 10.1007/s10519-014-9652-z

59. Mommsen, T.P. Mommsen, M.M. Vijayan, T.W. Moon**Cortisol in teleosts: dynamics, mechanisms of action, metabolic regulation** Rev. Fish Biol. Fish., 9 (3) (1999), pp. 211-268, 10.1023/A:1008924418720

60. Nematollahi, M.A. Nematollahi, H. van Pelt-Heerschap, W. Atsma, J. Komen**High levels of corticosterone, gene expression of star, cyp17a2, hsd3b, cyp21, hsd11b2 during acute stress in common carp with interrenal hyperplasia** Gen. Comp. Endocrinol., 2 (2012), pp. 252-258, 10.1016/j.ygcen.2012.01.023 (446KB)

61. North, B. North, J. Turnbull, T. Ellis, M.J. Porter, H. Migaud, J. Bron, N.R. Bromage**The impact of stocking density on the welfare of rainbow trout (*Oncorhynchus mykiss*)** Aquaculture, 255 (1–4) (2006), pp. 466-479, 10.1016/j .aquaculture.2006.01.004 (237KB)

62. Person-Le Ruyet, J. Person-Le Ruyet, L. Labbé, N. Le Bayon, A. Sévère, A. Le Roux, H. Le Delliou, L. Quéméner**Combined effects of water quality stocking density on welfare growth of rainbow trout (*Oncorhynchus mykiss*)** Aquat. Living Resour., 21 (2) (2008), pp. 185-195, 10.1051/alr:2008024

63. Phillips, R.B. Phillips, K.A. Keatley, M.R. Morasch, A.B. Ventura, K.P. Lubieniecki, B.F. Koop, R.G. Danzmann, W.S. Davidson**Assignment of Atlantic salmon (*Salmo salar*) linkage groups to specific chromosomes: conservation of large syntenic blocks corresponding to whole chromosome arms in rainbow trout (*Oncorhynchus mykiss*)** BMC Genet., 10 (2009), p. 46, 10.1186/1471-2156-10-46

64. Pickering Pottinger, A.D. Pickering, T.G. Pottinger**Crowding causes prolonged leucopenia in salmonid fish, despite interrenal acclimation** J. Fish Biol., 30 (6) (1987), pp. 701-712, 10.1111/j.1095-8649.1987.tb05799.x

65. Pickering, A.D. Pickering, T.G. Pottinger, J.P. Sumpter, J.F. Carragher, P.Y. Le BailEffects **of acute chronic stress on the levels of circulating growth hormone in the rainbow trout, *Oncorhynchus mykiss*** Gen. Comp. Endocrinol., 83 (1) (1991), pp. 86-93, 10.1016/0016-6480(91)90108-I (710KB)

66. Pottinger Carrick, T. Pottinger, T. Carrick**Stress responsiveness affects dominant-subordinate relationships in rainbow trout** Horm. Behav., 40 (3) (2001), pp. 419-427, 10.1006/hbeh.2001.1707 (119KB)

67. Pratt Dittmar, W.B. Pratt, K.D. Dittmar**Studies with purified chaperones advance the understing of the mechanism of glucocorticoid receptor–hsp90 heterocomplex assembly** Trends Endocrinol. Metab., 9 (6) (1998), pp. 244-252 (PMID: 18406276) 10.1016/S1043-2760(98)00059-9(619KB)

68. Procarione, L.S. Procarione, T.P. Barry, J.A. Malison**Effects of high rearing densities loading rates on the growth stress responses of juvenile rainbow trout** N. Am. J. Aquac., 61 (2) (1999), pp. 91-96, 10.1577/1548-8454(1999)061< 0091:EOHRDA>2 .0.CO;2

69. Proud, C.G. Proud**The multifaceted role of mTOR in cellular stress responses** DNA Repair (Amst), 3 (8–9) (2004), pp. 927-934, 10.1016/j.dnarep.2004.03.012 (115KB)

70. Qin, L. Qin, Z. Wang, L. Tao, Y. Wang**ER stress negatively regulates AKT/TSC/mTOR pathway to enhance autophagy** Autophagy, 6 (2) (2014), pp. 239-247, 10.4161/auto.6.2.11062

71. Rahman Thomas, M.S. Rahman, P. Thomas**Effects of hypoxia exposure on hepatic cytochrome P450 1A (CYP1A) expression in Atlantic croaker: molecular mechanisms of CYP1A down-regulation** PLoS One, 7 (7) (2012), e40825, 10.1371/journal.pone.0040825

72. Rebl, A. Rebl, M. Verleih, J.M. Köbis, C. Kühn, K. Wimmers, B. Köllner, T. Goldammer **Transcriptome profiling of gill tissue in regionally bred globally farmed rainbow trout strains reveals different strategies for coping with thermal stress** Mar. Biotechnol. (NY), 15 (4) (2013), pp. 445-460, 10.1007/s10126-013-9501-8

73. Rebl, A. Rebl, M. Verleih, T. Korytář, C. Kühn, K. Wimmers, B. Köllner, T. Goldammer**Identification of differentially expressed protective genes in liver of two rainbow trout strains** Vet. Immunol. Immunopathol., 145 (1–2) (2012), pp. 305-315, 10.1016/j.vetimm..11.023 (1MB)

74. Rexroad, C.E. Rexroad 3rd, R.L. Vallejo, S. Liu, Y. Palti, G.M. Weber**QTL affecting stress response to crowding in a rainbow trout broodstock population** BMC Genet., 13 (2012), p. 97, 10.1186/1471-2156-13-97

75. Rexroad, C.E. Rexroad 3rd, R.L. Vallejo, S. Liu, Y. Palti, G.M. Weber**Quantitative trait loci affecting response to crowding stress in an F(2) generation of rainbow trout produced through phenotypic selection** Mar. Biotechnol. (NY), 15 (5) (2013), pp. 613-627, 10.1007/s10126-013-9512-5

76. Ritchie, M.E. Ritchie, B. Phipson, D. Wu, Y. Hu, C.W. Law, W. Shi, G.K. Smyth**Limma powers differential expression analyses for RNA-sequencing microarray studies** Nucleic Acids Res., 43 (7) (), p. e47, 10.1093/nar/gkv007

77. Salas-Leiton, E. Salas-Leiton, V. Anguis, B. Martín-Antonio, D. Crespo, J.V. Planas, C.Infante, J.P. Cañavate, M. Manchado**Effects of stocking density feed ration on growth gene expression in the Senegalese sole (*Solea senegalensis*): potential effects on the immune response** Fish Shellfish Immunol., 28 (2) (2010), pp. 296-302, 10.1016/j.fsi.2009.11.006 (409KB)

78. Sánchez, C.C. Sánchez, G.M. Weber, G. Gao, B.M. Clevel, J. Yao, C.E. Rexroad III**Generation of a reference transcriptome for evaluating rainbow trout responses to various stressors** BMC Genomics, 12 (), p. 626, 10.1186/1471-2164-12-626

79. Scholz, S. Scholz, I. Behn, H. Honeck, C. Hauck, T. Braunbeck, H. Segner**Development of a monoclonal antibody for ELISA of CYP1A in primary cultures of rainbow**

**trout (*Oncorhynchus mykiss*) hepatocytes** Biomarkers, 2 (5) (1997), pp. 287-294, 10.1080/135475097231517

80. Silbergeld, E.K. Silbergeld**Blood glucose: a sensitive indicator of environmental stress in fish** Bull. Environ. Contam. Toxicol., 11 (1) (1974), pp. 20-25 (PMID: 4433778)

81. Silverstein, J.T. Silverstein, C.E. Rexroad, T.L. King**Genetic variation measured by microsatellites among three strains of domesticated rainbow trout (*Oncorhynchus mykiss*, Walbaum)** Aquac. Res., 35 (1) (2004), pp. 40-48, 10.1111/j.1365-2109.2004.00979.x

82. Smyth, G.K. Smyth, R. Gentleman, V. Carey, S. Dudoit, R. Irizarry, W. Huber**Limma: linear models for microarray data** Bioinformatics Computational Biology Solutions Using R Bioconductor, Springer, New York (2005), pp. 397-420, 10.1007/0-387-29362-0_23

83. Stubbe Solgaard Yang, H. Stubbe Solgaard, Y. Yang**Consumers' perception of farmed fish willingness to pay for fish welfare** Br. Food J., 113 (8) (), pp. 997-1010, 10.1108/00070701111153751

84. Sumpter, J.P. Sumpter, H.M. Dye, T.J. Benfey**The effects of stress on plasma ACTH, α-MSH, cortisol levels in salmonid fishes** Gen. Comp. Endocrinol., 62 (3) (1986), pp. 377-385, 10.1016/0016-6480(86)90047-X (952KB)

85. Thurstan Roberts, R.H. Thurstan, C.M. Roberts**The past future of fish consumption: can supplies meet healthy eating recommendations?** Mar. Pollut. Bull., 89 (1–2) (2014), pp. 5-11, 10.1016/j.marpolbul.2014.09.016 (522KB)

86. Trenzado, C.E. Trenzado, A.E. Morales, M. de la Higuera**Physiological effects of crowding in rainbow trout, *Oncorhynchus mykiss*, selected for low high stress responsiveness** Aquaculture, 258 (1–4) (2006), pp. 583-593, 10.1016 /j.aquaculture.2006.03.045 (200KB)

87. Turnbull, J. Turnbull, A. Bell, C. Adams, J. Bron, F. Huntingford**Stocking density welfare of cage farmed Atlantic salmon: application of a multivariate analysis** Aquaculture, 243 (1–4) (2005), pp. 121-132, 10.1016/j.aquaculture.2004.09.022 (363KB)

88. Tymchuk, W.E. Tymchuk, D. Sakhrani, R.H. Devlin**Domestication causes large-scale effects on gene expression in rainbow trout: analysis of muscle, liver brain transcriptomes** Gen. Comp. Endocrinol., 164 (2009), pp. 175-183, 10.1016/j.ygcen.2009.05.015

89. Verberne, A.J.M. Verberne, A. Sabetghadam, W.S. Korim**Neural pathways that control the glucose counterregulatory response** Front. Neurosci., 8 (2014), p. 38, 10.3389/fnins.2014.00038

90. Verleih, M. Verleih, A. Borchel, A. Krasnov, A. Rebl, T. Korytář, C. Kühn, T. Goldammer**Impact of thermal stress on kidney-specific gene expression in farmed regional imported rainbow trout** Mar. Biotechnol. (NY), 17 (5) (), pp. 576-592, 10.1007/s10126-015-9640-1

91. Vernier Sire, J.M. Vernier, M.F. Sire**Evolution of the glycogen content of glucose-6-phosphatase activity in the liver of Salmo gairdneri during development** Tissue Cell, 8 (3) (1976), pp. 531-546, 10.1016/0040-8166(76)90011-2 (5MB)

92. Vijayan, M.M. Vijayan, J.S. Ballantyne, J.F. Leatherl**High stocking density alters the energy metabolism of brook charr, *Salvelinus fontinalis*** Aquaculture, 88 (3–4) (1990), pp. 371-381, 10.1016/0044-8486(90)90162-G (888KB)

93. Vijayan Moon, M.M. Vijayan, T.W. Moon**Acute hling stress alters hepatic glycogen metabolism in food-deprived rainbow trout (*Oncorhynchus mykiss*)** Can. J. Fish. Aquat. Sci., 49 (11) (1992), pp. 2260-2266, 10.1139/f92-247

94. Wang, B. Wang, Z. Chen, F. Yu, Q. Chen, Y. Tian, S. Ma, T. Wang, X. Liu**Hsp90 regulates autophagy plays a role in cancer therapy** Tumor Biol., 37 (1) (), pp. 1-6, 10.1007/s13277-015-4142-3

95. Wendelaar Bonga, S.E. Wendelaar Bonga**The stress response in fish** Physiol. Rev., 77 (3) (1997), pp. 591-625

96. Weng, L. Weng, H. Dai, Y. Zhan, Y. He, S.B. Stepaniants, D.E. Bassett**Rosetta error model for gene expression analysis** Bioinformatics, 22 (9) (2006), pp. 1111-1121, 10.1093/bioinformatics/btl045

97. Werno, C. Werno, J. Zhou, B. Brüne**A23187, ionomycin thapsigargin upregulate mRNA of HIF-1alpha via endoplasmic reticulum stress rather than a rise in intracellular calcium** J. Cell. Physiol., 215 (3) (2008), pp. 708-714, 10.1002/jcp.21351

98. Williams, M.J. Williams, J.D. Bell, M.V. Gupta, M. Dey, M. Ahmed, M. Prein, S. Child, P.R.Gardiner, R. Brummett, D. Jamu**Responsible aquaculture can aid food problems** Nature, 406 (6797) (2000), p. 673, 10.1038/35021300

99. Xu, C. Xu, J. Liu, L.-C. Hsu, Y. Luo, R. Xiang, T.H. Chuang**Functional interaction of heat shock protein 90 Beclin 1 modulates toll-like receptor-mediated autophagy** FASEB J., 25 (8) (), pp. 2700-2710, 10.1096/fj.10-167676

100. Yarahmadi, P Maindare H.K. Miare, S. Fayaz, C.M. Caipang**Increased stocking density c auses changes in expression of selected stress- immune-related genes, humoral innate immune parameters stress responses of rainbow trout (*Oncorhynchus mykiss*)** Fish Shellfish Immunol., 48 (2016), pp. 43-53, 10.1016/j.fsi..11.007

101. Yarahmadi, P. Yarahmadi, H.K. Miare, S.H. Hoseinifar, N. Gheysvi, A. Akbarzadeh**The effects of stocking density on hemato-immunological serum biochemical parameters of rainbow trout (*Oncorhynchus mykiss*)** Aquac. Int., 23 (1) (2014), pp. 55-63, 10.1007/s10499-014-9797-z

102. Zarate Bradley, J. Zarate, T. Bradley**Heat shock proteins are not sensitive indicators of hatchery stress in salmon** Aquaculture, 223 (1–4) (2003), pp. 175-187, 10.1016/S0044-8486(03)00160-1

# CHAPTER 3

# Early Development and Allometric Growth Patterns in Burbot *Lota Lota* L.

**Krzysztof Kupren, Izabela Trąbska, Daniel Żarski, Sławomir Krejszeff, Katarzyna Palińska-Żarska, Dariusz Kucharczyk**

*Department of Lake and River Fisheries, University of Warmia and Mazury in Olsztyn, Oczapowskiego 5, 10-095 Olsztyn, Poland*

## ABSTRACT

This study analyzed the morphological development and allometric growth patterns of *Lota lota* L. (burbot) larvae reared under controlled laboratory conditions. From hatching to day 50, twenty larvae were sampled each [between 1 and 14 days post-hatch (DPH)] or every second day (between 14 and 50 DPH) and measured under a stereoscopic microscope using analytic software. Based on the external morphology, the different stages during early development of burbot were identified: yolk sac larva 0–8 DPH [3.92–4.37 mm total length (TL)]; preflexion larva 9–26 DPH (4.57–12.06 mm TL); flexion larva (between notochord degradation and its replacement with rays) 28–34 DPH (14.00–16.34 mm TL) and postflexion larva/juvenile 36–50 DPH (18.20–29.27 mm TL). Allometric growth patterns of some parameters (e.g., total length, head length, body length, tail length, head depth, body depth, eye diameter) were modeled by a power function and described by the growth coefficient. Organogenesis and changes of body proportions in burbot larvae were more rapid and complex during the yolk sac and preflexion phase of development as larvae developed most of their sensorial, feeding, respiratory and swimming systems and after notochord flexion, when most morphological changes were related to the progressive transformation from pelagic larva to demersal larva/juvenile.

**Keywords: Allometric growth Larval development Morphological characteristics Burbot** *Lota lota*

# INTRODUCTION

Burbot *Lota lota* L. is the only freshwater representative of the cod family (Gadidae). This species is widely distributed throughout the Holarctic zone and typically inhabits cold rivers and lakes (McPhail and Paragamian 2000). Burbot is presently considered to be one of the most endangered freshwater fish species in many parts of Europe (van Houdt et al. 2003, 2005; Paragamian et al. 2008; Hardy et al. 2008). The principle causes of this include the following: disturbances in habitat, inhibition of spawning migrations, overfishing, pollution and climate warming. Burbot is a highly valued fish species and also is a promising candidate for intensive aquaculture (Wocher et al. 2013).

Information on larval development of fish species is essential for a better understanding of their early life stages, priorities during early growth and size-related adaptations, and it also offers insights into fish biology and taxonomy (Gisbert 1999). There is much published data concerning various aspects of early life stages of burbot based on both field and laboratory studies. The published papers investigated the behavior and ecology of larval and juvenile burbot (Ryder and Pesendorfer 1992; Wang and Appenzeller 1998; Fischer 1999; McPhail and Paragamian 2000; Hofmann and Fischer 2001; Donner and Eckmann 2011) and also concerned identification of the basic rearing conditions, such as optimum temperature and photoperiod (Wolnicki et al. 2002; Harzevili et al. 2004, Lahnsteiner et al. 2012) and the type of food supplied (Harzevili et al. 2003; Żarski et al. 2009; Wocher et al. 2013). However, a detailed course of morphological changes during larval stages, in combination with an analysis of inflexion points in growth curves, has not been previously reported. The determination of morphological features and relative growth patterns during early development are also very important: They can contribute to fisheries management and aquaculture by characterizing normal growth patterns under specific conditions and help to optimize rearing protocols (Peña and Dumas 2009).

The aim of the present study was to describe the morphological development and allometric growth patterns for burbot larvae reared under controlled conditions.

# MATERIALS AND METHODS

Larvae were obtained by artificial spawning of fish originating from the Odra River (northeastern Poland). Spawning was stimulated only through manipulation of thermal conditions, according to the methodology described by Żarski et al. (2009). Eggs were obtained from three females of an average weight of 720 ± 108 g (mean ± SD) and were fertilized with semen from three males with an average weight of 332 ± 79 g. For the first month, the eggs were incubated in Weiss jars, at 3–4 °C. After that period, until the moment of hatching, the water temperature was kept as 6 °C. At the moment when the first hatched larvae were observed, eggs were transferred to water at 10 °C in order to synchronize hatching (Żarski et al. 2009). Newly hatched larvae were placed in a semi-closed laboratory recirculating system where they were stocked in two 40-L aquariums. The initial stocking density of larvae was 60 ind. $L^{-1}$. Mortality was recorded daily. Survival was 29.79 ± 1.45 % (mean ± SD) to 50 days after hatching. The system was equipped with biological filtration, aeration and automatic temperature regulation (±0.1 °C). For the first week of the study, the water temperature in the tanks was 10 °C. Following swim bladder inflation, the water temperature was then gradually increased from 10 to 16 °C over 3 days, and these conditions were constant until the end of the experiment. Such a thermal regime has been recommended as optimal for burbot larvae (Wolnicki et al. 2002; Harzevili et al. 2004). Specimens were exposed to a 12 L:12 D photoperiod. The dissolved oxygen and ammonia contents were >80 % saturation and <0.01 mg $L^{-1}$, respectively. Larvae were fed three times a day ad libitum with freshly hatched *Artemia* sp. nauplii (San Francisco origin). They were randomly sampled (total $n = 20$) daily from hatch to 14 DPH (days post-hatch) and every second day from 14 to 50 DPH. Sampled larvae were subjected to overexposure in an anesthetic solution (MS-222 at a dose of 150 mg $L^{-1}$) and digital photographs of each specimen were taken using digital image analysis software ProgRes®CapturePro 2.5 (Jenoptic, Germany). After this, larvae were fixed in a 4 % phosphate-buffered formaldehyde solution for further detailed examinations. Developmental stages were identified according to Kendall et al. (1984) and Dunn and Matarese (1984) and divided into four periods: yolk sac larva, preflexion larva, flexion larva and postflexion larva. The term flexion larva concerned a period of development between notochord degradation and its replacement with rays. Thresholds during early development, such as filling of the swim bladder and exogenous feeding, were considered to be achieved when at least 50 % of the specimens represented this particular stage. On each digital photograph, several body proportions associated with feeding and locomotion were measured to the nearest 0.01 mm. These measurements were snout length (SNL); head length (HL), trunk length (TRL); tail (post-anal) length (TAL);

eye diameter (ED); head depth (HD); body depth (BD); tail depth at anus level (TAD); mouth (maxilla) length (ML) and total length (TL) (Fig. 1). All measurements were taken along lines parallel or perpendicular to the horizontal axis of the body (Gisbert 1999). Dead and abnormal specimens (presence of malformations) were excluded from the analysis. In yolk sac larvae, the diameter (maximum and minimum) of the ellipsoidal yolk sac was also measured and the volume ($mm^3$) calculated using the following formula: $V = 0.1667 \times LH^2$; where $H$ is the minimum diameter and $L$ is the maximum diameter of the yolk sphere (Heming and Buddington 1988).

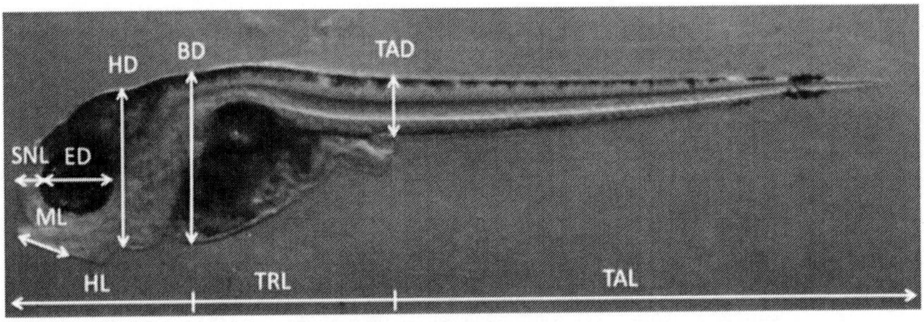


**Figure. 1** Morphometric characters measured in the burbot larvae. *BD* body depth, *ED* eye diameter, *HD* head depth, *ML* mouth length, *SNL* snout length, *TAD* tail depth at anus level, *TAL* tail length, *TL* total length, *TRL* trunk length

## Data analysis

Growth patterns during the developmental stages were modeled by a power function of TL, and the allometric patterns were described by the growth coefficient (i.e., power function exponent) in the equation $Y = aX^b$, where $Y$ is the dependent variable (measured character) and $X$ is the independent variable (TL), $a$ is the intercept and $b$ is the growth coefficient. Isometric growth occurred at $b = 1$ and allometric growth was positive when $b > 1$ and negative when $b < 1$. In addition, linear regressions were performed on log-transformed data (TL as independent variable) and the inflection points were calculated. Inflection points are the $X$ values where the slope of growth changes. The inflection points were calculated according to van Snik et al. (1997). The $x$–$y$ data set was sorted according to increasing $x$, being TL. Regression lines were calculated for $x_{min}$ until $x_{intermediate}$, and for $x_{intermediate}$ until $x_{max}$, $x_{intermediate}$ varied iteratively from $x_{min} + 2$ to $x_{max} - 2$. Also, $t$ tests were performed to check whether the growth coefficients ($x_{min}$ $x_{intermediate}$) and ($x_{intermediate}$ $x_{max}$)

differed significantly. The $x_{intermediate}$ value that iteratively showed the largest *t* was defined as the inflexion point.

No specific experiments were done to the test behavior and swimming ability of larvae. Only general observations were made in the rearing tanks during daylight and night-dark conditions and the distribution of larvae in the water column was noted.

# RESULTS

## Development

Between the day of hatching and 50 DPH, the TL of the burbot larvae body (TL) grew continuously and increased by 25.35 mm (mean length was 3.92 and 29.27 mm in 0 DPH and 50 DPH, respectively) (Fig. 2). Four larval development stages were observed after hatching: yolk sac larvae, preflexion larvae, flexion larvae and postflexion larvae. At hatching (0 DPH), larvae measured 3.92 ± 0.09 mm TL (mean ± SD) and had the head bent downwards without hatching glands (Fig. 3a). Larvae had a large yolk sac (volume 0.20 ± 0.08 $mm^3$, mean ± SD) containing one oil droplet in the anterior lower part of the yolk sac. A finfold covered the body from the dorsal area of the trunk to the ventral area of the yolk sac and was higher in the dorsal part of the trunk and narrowed in the caudal region. Auditory capsules and otoliths were visible in the posterior head region. The gut was an undifferentiated tube in the dorsal posterior region of the yolk sac. Melanophores were observed along the dorsal part of the body and in the posterior part of the yolk sac. The eyes were darkly pigmented. Larvae lay on the bottom and occasionally made rapid movements with the entire body and swam up from the bottom. At 3 DPH, the head was almost straight, the mouth was in more anterior position and open. On day 4 (4.27 ± 0.15 mm TL), the first larvae initiated gas bladder inflation which lasted for the next 3 days (4.34 ± 0.11 mm TL). By 8 DPH, 4.38 ± 0.14 mm (TL), the larvae had almost completely resorbed the yolk sac (0.01 ± 0.04 $mm^3$). The pectoral fins were well developed, and the larvae swam very well in the water column (Fig. 3b).

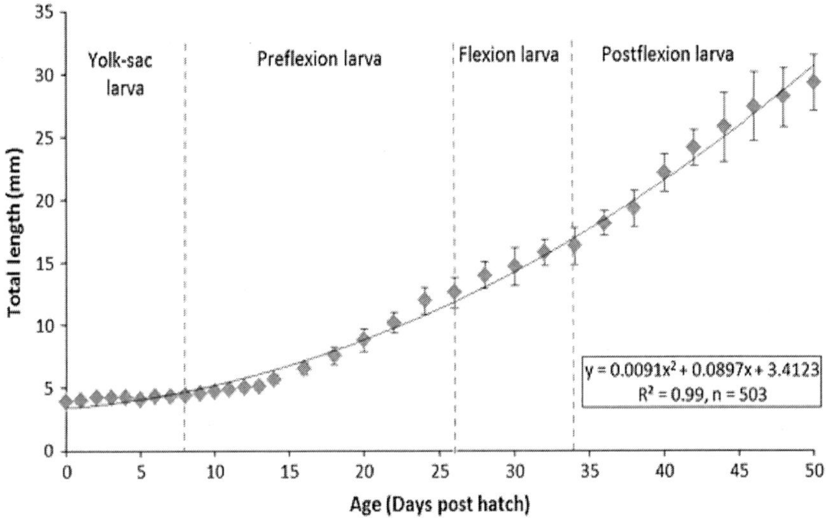


**Figure. 2** Changes in total length (TL) from hatching to 50 DPH. SD (standard deviation) is represented by *whiskers*

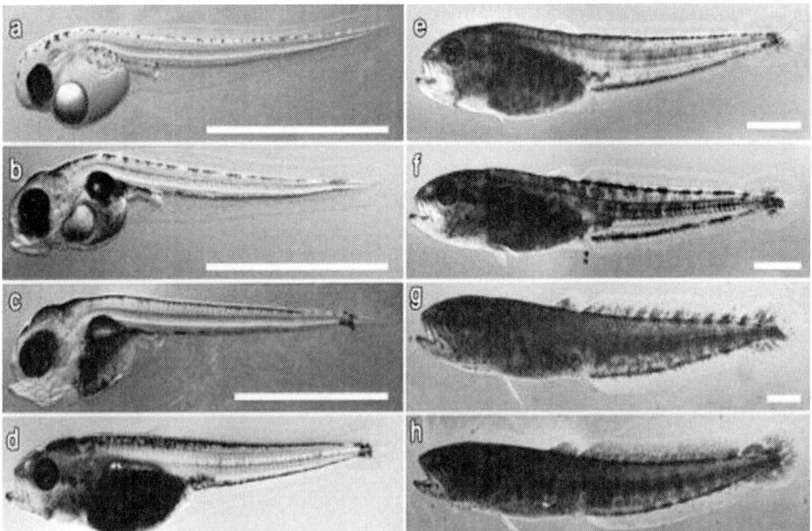


**Figure. 3** Larval development of *Lota lota*. **A** Yolk sac stage (3 h); **B** yolk sac stage (the gas bladder filled) (8 DPH); **C** preflexion (exogenous feeding) larvae (13 DPH); **D** preflexion larva, (22 DPH); **E** flexion larva (26 DPH); **F** flexion larva (34 DPH); **G** postflexion larva (42 DPH); **H** postflexion larva/juvenile (48 DPH). *Scale bars* 2 mm

The preflexion stage began at the first feeding on day 9 (4.57 ± 0.13 mm TL) and ended around day 26 (12.6 ± 1.26 mm TL). At the onset of this stage, the yolk sac was resorbed but a small oil globule was still present in the anterior part of the trunk. A coiled digestive tract was evident and the anus was opened. The number of melanophores increased on the dorsal and lateral parts of the body. On day 13 (5.12 ± 0.36 mm TL), the oil globule was depleted (Fig. 3c). At 18 DPH (7.58 ± 0.70 mm TL), the dorsal and caudal finfold began to differentiate, starting with an indentation behind the future first dorsal fin and a slight narrowing of the finfold near the caudal peduncle. At 20 DPH (8.82 ± 0.92 mm TL), the first rays appeared in the ventralmost part of the caudal fin. Two days later (22 DPH, 10.22 ± 0.80 mm TL), rays also appeared on the dorsal side of the notochord tip and also in the dorsal and anal fins. At the same time, a single barbel bud located on ventral side of the lower jaw appeared (Fig. 3d). On the last day of this phase (26 DPH, 12.60 ± 1.26 mm TL), a few small teeth were visible for the first time in the mouth. Most of the larvae were still homogeneously distributed in a water column, but the largest individuals began to spend more and more time near the bottom of the tanks.

The flexion stage started from day 28 (14.00 ± 1.07 TL) and continued until day 34 (16.34 ± 1.45 mm TL). On day 28, most of the larvae showed a slight upward turn of the notochord tip (Fig. 3e). During the next few days, a shortening of the notochord was observed. On day 34, the notochord was shortened and the empty place was filled by caudal rays. Pigmentation on fins was restricted to the bases (Fig. 3f). Most of the time, the larvae stayed near the bottom of the tanks.

The postflexion stage was observed from day 36 (18.12 ± 0.99 mm TL). The finfold was completely absorbed at 42 DPH (24.10 ± 1.43 mm TL) (Fig. 3g). Elongated pelvic, anal and two dorsal fins with pigmented rays were observed. On 46 DPH (27.39 ± 2.18 mm TL), the second rays of the pelvic fins were visibly elongated and melanophores covered most of the fins (Fig. 3h). The body shape of larvae and the pigmentation pattern on the lateral side of the body were similar to those of the adult. Young burbots showed a completely demersal behavior.

The main events during larval development of the burbot are summarized in Fig. 4.

## Allometric growth

Allometric growth equations between nine measured body segments and TL during the larval development stage (0–50 DPH) are presented in Figs. 5 and 6.

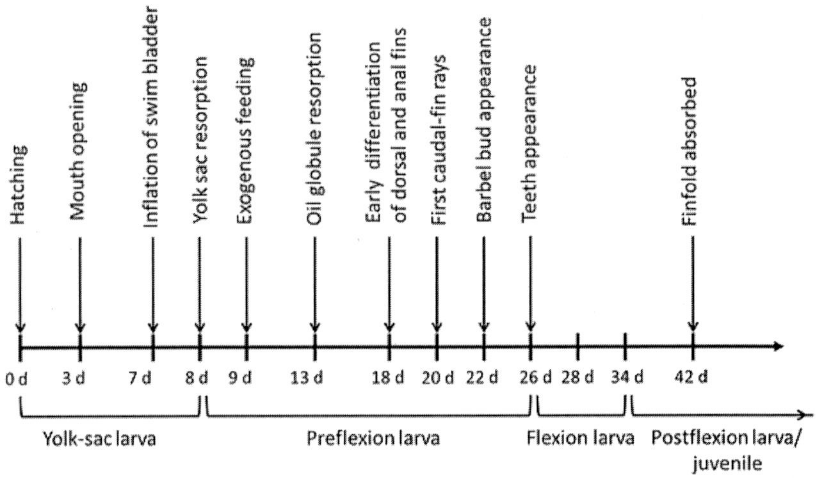


**Figure. 4** The main events of larval development in burbot

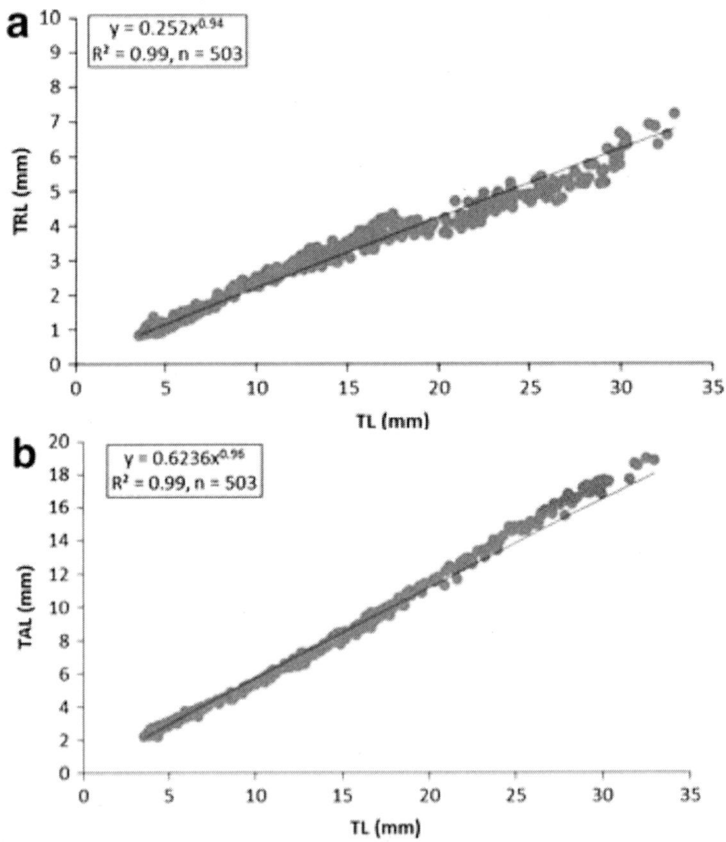

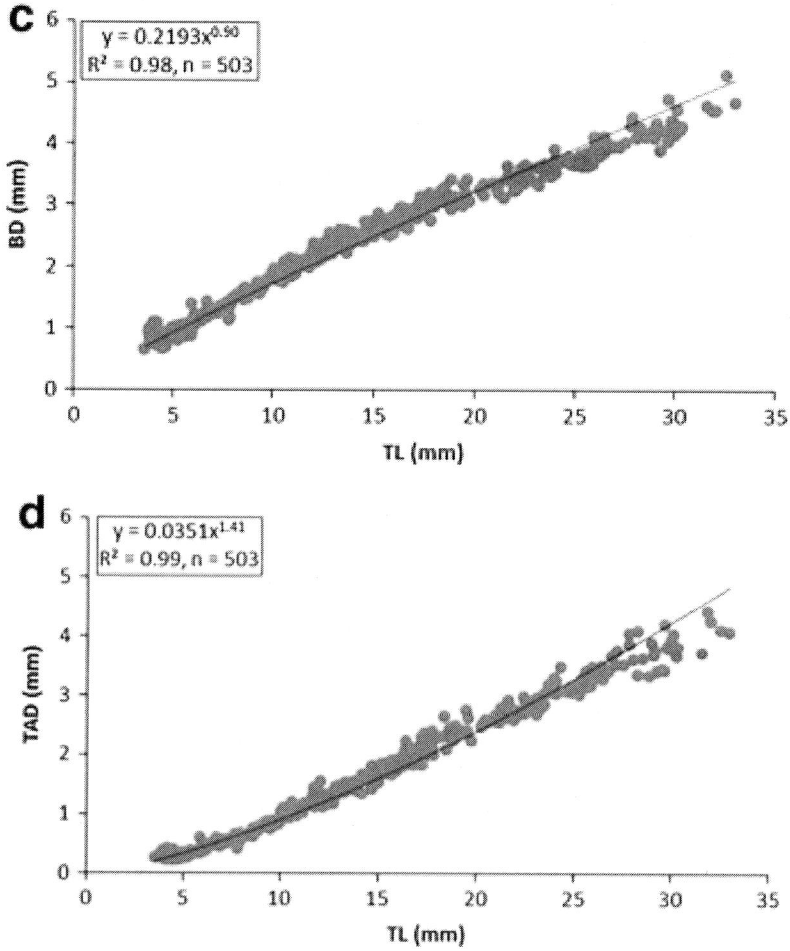


**Figure. 5** Allometric growth equations between measured body proportions and total length during burbot development. The *dashed line* represents the inflection point of growth. *TRL* trunk length, *TAL* tail (post-anal) length, *BD* body depth, *TAD* tail depth at anus level, *TL* total length

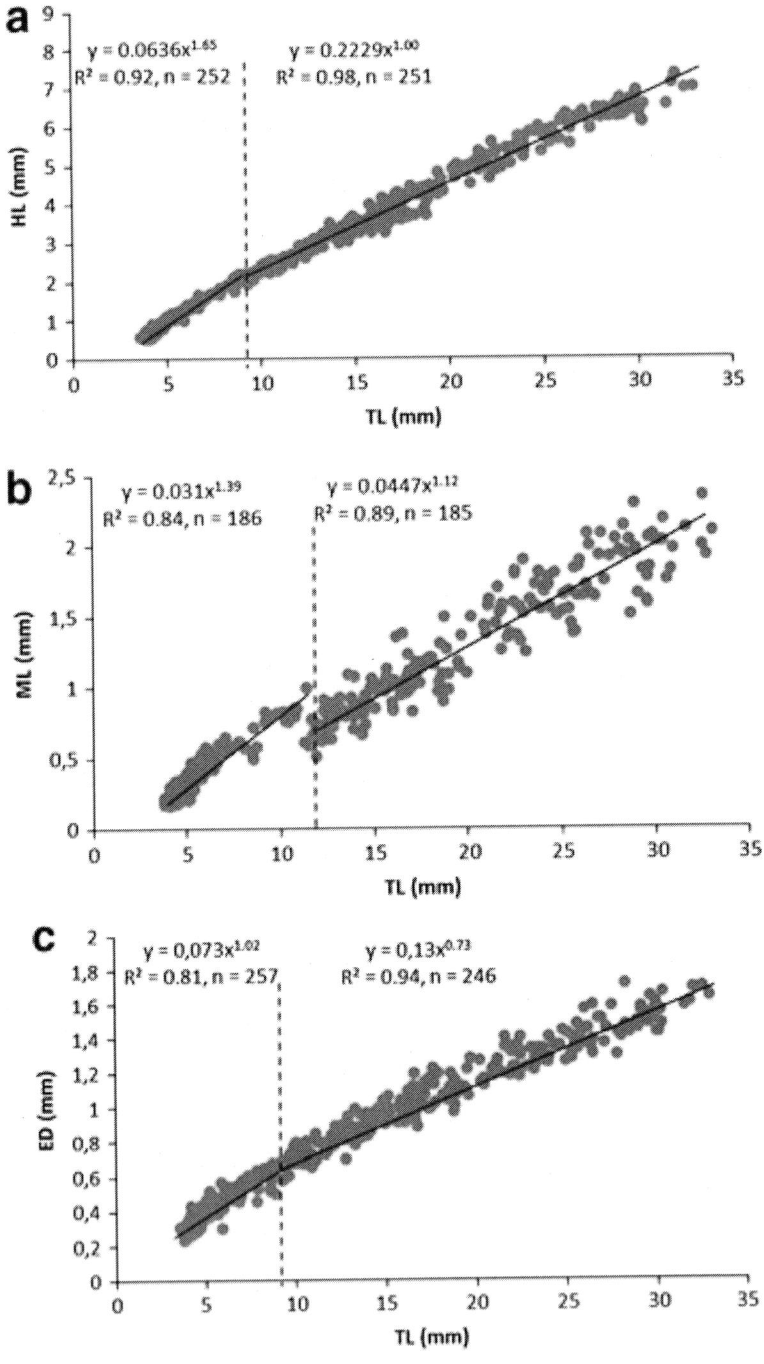
a
y = 0.0636x^1.65
R² = 0.92, n = 252
y = 0.2229x^1.00
R² = 0.98, n = 251
HL (mm)
TL (mm)
b
y = 0.031x^1.39
R² = 0.84, n = 186
y = 0.0447x^1.12
R² = 0.89, n = 185
ML (mm)
TL (mm)
c
y = 0,073x^1.02
R² = 0.81, n = 257
y = 0,13x^0.73
R² = 0.94, n = 246
ED (mm)
TL (mm)

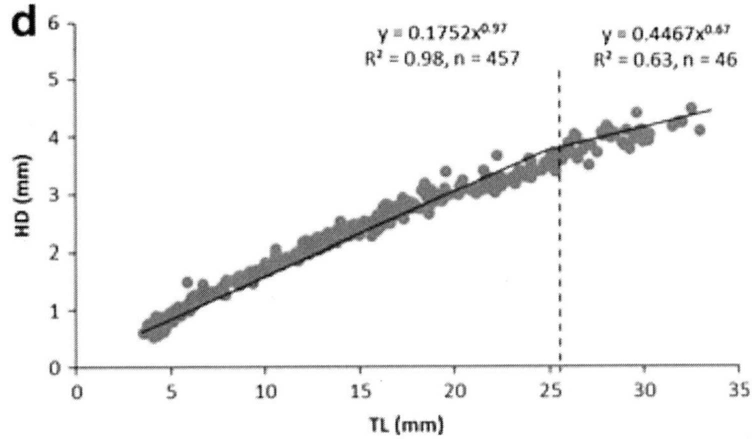


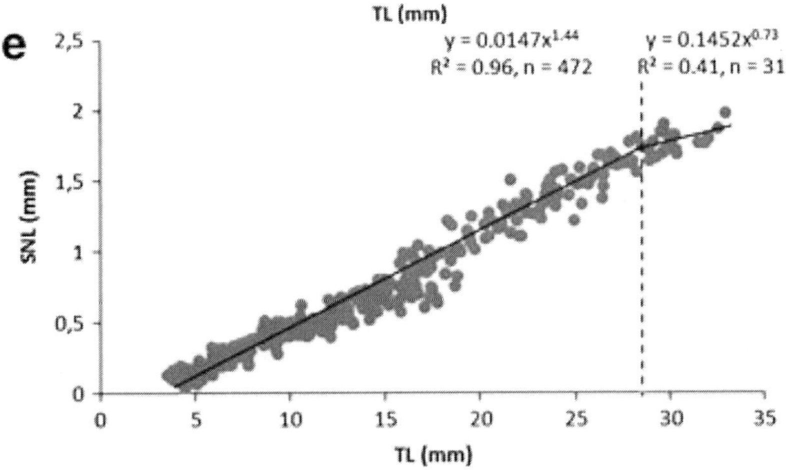


**Figure. 6** Allometric growth equations between measured body proportions and total length during burbot development. The *dashed line* represents the inflection point of growth. *HL* head length, *ML* mouth (maxilla) length, *ED* eye diameter, *HD* head depth, *SNL*snout length, *TL* total length

Growth of the TRL and TAL was near-isometric and monophasic ($b = 0.94$, $R^2 = 0.99$ and $b = 0.96$, $R^2 = 0.99$, respectively) from hatching to 50 DPH (Fig. 5a, b). Growth of BD was monophasic and negatively allometric throughout the study ($b = 0.90$, $R^2 = 0.98$) (Fig. 5c), while the growth coefficient of TAD was positively allometric ($b = 1.41$, $R^2 = 0.99$) (Fig. 5d). Growth of HL was positively allometric from hatching to 20 DPH ($b = 1.65$, $R^2 = 0.92$) and isometric from 22 to 50 DPH ($b = 1.00$, $R^2 = 0.98$) with an inflexion point situated at a TL of 8.64 mm ($P < 0.001$) (Fig. 6a).

Growth of mouth length (ML) (measured from the time of mouth opening at 3 DPH) was positively allometric and biphasic, with faster growth during 4–24 DPH ($b = 1.39$, $R^2 = 0.84$) and slower growth during 26–50 DPH ($b = 1.12$, $R^2 = 0.89$) with an inflexion point situated at a TL of 11.76 mm ($P < 0.001$) (Fig. 6b). Further, a single inflexion point was also found in the growth curve of the eye. This inflexion was situated at a TL of 8.87 mm from where the rate drastically decreased (20 DPH, slope = 1.02, $R^2 = 0.81$–0.73, $R^2 = 0.94$, $P < 0.001$) (Fig. 6c). A similar growth model was characteristic for HD and SNL (slope = 0.98–0.67 and 1.44–0.63, respectively) with inflexion point at a TL of 25.10 mm (44 DPH, $P < 0.001$) and 27.85 mm (46 DPH, $P < 0.001$), respectively (Fig. 6d, e)

# DISCUSSION

Observations on the early development of other cods (Fahay and Markle 1984; Dunn and Matarese 1984; Auditore et al. 1994) agree with the pattern described in the present study, thus confirming the basic uniformity of development in this group of fishes (Dunn and Matarese 1984). However, variations exist among species as to the relative timing of ontogenetic events and structure formation in response to different environmental conditions, e.g., temperature and food availability. Burbot follows a typical course of Gadidae ontogeny through a series of larval and juvenile stages: yolk sac larva, preflexion larva, flexion larva, postflexion larva, pelagic juvenile and demersal juvenile (Dunn and Matarese 1984; Auditore et al. 1994). Instead of true notochord flexion, which is characteristic for different teleost groups (van Snik et al. 1997; Osse and van den Boogaart 2004), there is a shortening of the notochord in burbot and other gadoid larvae (Auditore et al. 1994).

The sequence of important life events described in the present study is similar to the descriptions of this species from field-collected material (Ryder and Pesendorfer 1992; Ghan and Sprules 1993; Fischer 1999; Donner and Eckmann 2011). In lakes, the newly hatched burbot migrate upward from the deep spawning areas to the water surface to fill their swim bladders to gain buoyancy and start exogenous feeding. Next, the burbot larvae go through the pelagic phase, swim in small schools high in the water column, take vertical migrations and feed on copepods and cladocerans during the daytime (Ryder and Pesendorfer 1992; Ghan and Sprules 1993; Donner and Eckmann 2011). Upon reaching about 15 mm, the burbot gradually undergo a habitat shift. Most of the time, they stay in the lower part of the water column near the bottom (pelagic juvenile) and when they reach about 30 mm in TL, they become solitary and benthic (demersal juvenile) and feed primarily at night, almost exclusively on amphipods (Ryder and Pesendorfer 1992; Wang and Appenzeller 1998; Fischer 1999). In our experiment, fish which were initially homogeneously distributed in a water column with a good swimming ability from about 26 DPH (12.6 mm TL, end of preflexion period) gradually started to spend more time near the bottom of tanks. Such behavior was concomitant with the development of specific body structures such as teeth and barbel. This course of larval development has also been partially confirmed in several rearing experiments under laboratory conditions. Wolnicki et al. (2002), Harzevili et al. (2004), Lahnsteiner et al. (2012) and Wocher et al. (2013) found a similar TL at chosen larval stages to be important from a practical (in hatchery) point of view (e.g., yolk sac absorption or the start of exogenous feeding).

The changes in functional development and, consequently, in body form, which are especially dramatic during the few first days after hatching, lead to the formation of characteristic allometric growth patterns. The chronology of allometric changes is related to the chronology of important early life history events (Gisbert 1999). The HL growth of burbot larvae was positively allometric from hatching to day 20 (8.64 mm TL) and, during this time, structures and organs related to feeding and swimming developed sharply and prepared the fish for the pelagic stage. The early differentiation of the nervous (development of the midbrain and hindbrain), vision (positive allometric growth of the eyes) and feeding systems (positive allometric of the mouth) contribute to increasing the probability of zooplankton detection (Fuiman 1983; Koumoundouros et al. 1999). The development of gill arches and filaments would allow switching from temporary (cutaneous) to branchial respiration, resulting in a better oxygen supply and an increase in swimming activity (Gisbert et al. 2002). The swimming ability of burbot larvae to disperse actively in a water column to search for food, avoid potential predators and enhance their survival was progressively improved after swim bladder inflation. Swimming ability was concomitant with rapid development of the tail region (isometric growth of tail length and allometric growth in tail height), rapid development of the pectoral fins and differentiation of the primordial finfold into unpaired fins. The reported positive allometric growth of anterior and posterior parts of the body during the first stages of development in burbot larvae was not surprising, as it is common for many fish species to develop the anterior and posterior body regions prior to the abdominal part of body (Fuiman 1983; Osse et al. 1997; van Snik et al. 1997; Koumoundouros et al., 1999; Gisbert 1999; Gisbert et al. 2002). This growth pattern could also be interpreted as an adaptation to reduce and optimize the energy costs of larval transport (Osse and van den Boogaart 1995; 2004; van Snik et al. 1997)

The decrease in head height growth detected at 25.10 mm TL (44 DPH) and snout length at 27.85 mm TL (46 DPH) appears to be associated with the transition from a pelagic larva/juvenile into a benthic juvenile. This transition resulted in clear changes in proportions of the head, which became more elongated and flattened like in older stages. Other morphological changes related to the adaptation to a new (benthic) way of life are, for instance, fast growth of pelvic fins and an intensive increase in body pigmentation. Similar changes are characteristic for cod *Gadus morhua* L. and haddock *Melanogrammus aeglefinus* L. (Auditore et al. 1994). Another species which also shows transformation following the pelagic phase is the Californian halibut *Paralichthys californicus*, which undergoes progressive transition from a symmetrical (pelagic) larva into a bilaterally asymmetrical (benthic) juvenile. This transition results in profound changes in the organization of the splanchnocranium, especially the jaws and visceral arch skeleton (Gisbert et al. 2002).

In conclusion, organogenesis and differentiation in burbot larvae were more rapid and complex during the yolk sac and preflexion phase of development as larvae developed most of their sensorial, feeding, respiratory and swimming systems and, after notochord flexion, when most morphological changes were related to a progressive transformation from pelagic larva/juvenile to demersal larvae/juvenile. Besides these, all significant and rapid changes in body proportion concerned mainly the head region. Other body parts showed monophasic growth. The lack of inflexion points between 11.76 mm and 25.10 mm (24–44 DPH, flexion and most of the postflexion period) and the near-isometric growth of most body proportions during this period probably reflect long pelagic and prolonged transformation phases (between larval and juvenile stages) related to a gradual habitat shift from a pelagic to a demersal way of life. Such a growth timetable is in contrast with the general patterns in many teleost larvae, in which inflexion points occur within a more or less narrow TL interval. These occur most commonly during the flexion stage and are accompanied by the rapid development of fin rays, changes in body shape and a change in locomotive ability and feeding techniques (Kendall et al. 1984; van Snik et al. 1997; Gisbert 1999; Geerinckx et al. 2008; Peña and Dumas 2009).

The analysis of growth patterns in combination with observed changes in morphological development provides valuable information on the changes in functional demands throughout ontogeny. These models can be useful for both comparative studies of ontogeny in other species as well as for aquacultural applications by characterizing normal growth patterns. Moreover, the results presented in this study may be useful for aquaculturists since an understanding of developmental and growth patterns may lead to designing and providing more suitable rearing conditions for particular life stages.

## ACKNOWLEDGMENTS

This study was cofinanced by the project "Innovations in finfish aquaculture with special reference to reproduction" (acronym: InnovaFish), Operational Programme "Sustainable Development of the Fisheries Sector and Coastal Fishing Areas 2007–2013" (OR14-61724-OR1400003/09/10/11).

## REFERENCES

1. Auditore PJ, Lough RG, Broughton EA (1994) A review of the comparative development of Atlantic cod (*Gadus morhua* L.) and haddock (*Melanogrammus aeglefinus* L.) based on an illustrated series of larvae and juveniles from Georges Bank. NAFO science council studies, Northwest Atlantic Fisheries Organisation, pp 7–18
2. Donner MT, Eckmann R (2011) Diel vertical migration of larval and early-juvenile burbot optimises survival and growth in a deep, pre-alpine lake. Freshw Biol 56:916–925. doi:10.1111/j.1365-2427.2010.02536.x
3. Dunn JR, Matarese AC (1984) Gadidae: development and relationships. In: Moser HG, Richards WJ, Cohen DM, Fahay MP, Kendall AW, Richardson WJ (eds) Ontogeny and systematics of fishes. Amer Soc Icht and Herp, Spec Pub, vol 1, pp 283–299
4. Fahay MP, Markle DF (1984) Gadiformes: Development and relationships. In: Moser HG, Richards WJ, Cohen DM, Fahay MP, Kendall AW, Richardson WJ (eds) Ontogeny and systematics of fishes. Amer Soc Icht and Herp, Spec Pub, vol 1, pp 265–283
5. Fischer P (1999) Otolith microstructure during the pelagic, settlement and benthic phases in burbot. J Fish Biol 54:1231–1243
6. Fuiman LA (1983) Growth gradients in fish larvae. J Fish Biol 23:117–123
7. Geerinckx T, Verhaegen Y, Adriaens D (2008) Ontogenetic allometries and shape changes in the suckermouth armoured catfish *Ancistrus cf. triradiatus* Eigenmann (*Loricariidae, Siluriformes*), related to suckermouth attachment and yolk-sac size. J Fish Biol 72:803–814
8. Ghan D, Sprules WG (1993) Diet, prey selection, and growth of larval and juvenile burbot *Lota lota* (L.). J Fish Biol 42:47–64
9. Gisbert E (1999) Early development and allometric growth patterns in Siberian sturgeon and their ecological significance. J Fish Biol 54:852–862
10. Gisbert E, Merino G, Muguet JB, Bush D, Piedrahita RH, Conklin DE (2002) Morphological development and allometric growth patterns in hatchery-reared California halibut larvae. J Fish Biol 61:1217–1229
11. Hardy R, Paragamian VL, Neufeld MD (2008) Zooplankton communities and burbot relative abundance of some oligotrophic Lakes of Idaho, USA and British Columbia, Canada. Am Fish Soc Symp 59:79–89
12. Harzevili AS, De Chareloy D, Auwerx A, Vught I, Van Slycken J, Dhert P, Sorgeloos P (2003) Larval rearing of burbot (*Lota lota* L.) using *Brachionus calyciflorus* rotifer as started food. J Appl Ichthyol 19:84–87

13. Harzevili AS, Dooremont I, Vught I, Auwerx J, Quataert P, De Charleroy D (2004) First feeding of burbot, *Lota lota* (Gadidae, Teleostei) larvae under different temperature and light conditions. Aquac Res 35:49–55
14. Heming TA, Buddington RK (1988) Yolk sac absorption in embryonic and larval fishes. In: Hoar WS, Randall DJ (eds) Fish physiology, vol 11. Academic Press, New York, pp 407–446
15. Hofmann N, Fischer P (2001) Seasonal changes in abundance and age structure of burbot *Lota lota* (L.) and stone loach *Barbatula barbatula* (L.) in the littoral zone of a large pre-alpine lake. Ecol Freshw Fish 10:21–25
16. Kendall AW, Ahlstrom EH, Moser HG (1984) Early life history stages of fishes and their characters. In: Moser HG, Richards WJ, Cohen DM, Fahay MP, Kendall AW, Richardson SL (eds) Ontogeny and systematics of fishes: American Society of Ichthyologists and Herpetologists, Special Publication No. 1. Allen Press Inc, Lawrence, KS, pp 11–22
17. Koumoundouros G, Divanach P, Kentouri M (1999) Ontogeny and allometric plasticity of *Dentex dentex* (Osteichthyes: Sparidae) in rearing conditions. Mar Biol 135:561–572
18. Lahnsteiner F, Kletzl M, Weismann T (2012) The effect of temperature on embryonic and yolk-sac larval development in the burbot *Lota lota*. J Fish Biol 81:977–986. doi:10.1111/j.1095-8649.2012.03344.x
19. McPhail JD, Paragamian VL (2000) Burbot biology and life history. In: Paragamian VL, Willis DH (eds) Burbot biology, ecology, and management. American Fisheries Society, Fisheries Management Section Publication Number 1, Bethesda, Maryland, pp 11–23
20. Osse JW, van den Boogaart JGM (1995) Fish larvae, development, allometric growth, and the aquatic environment. Int Council Exp Sea Mar Sci Symp 201:21–34
21. Osse JW, van den Boogaart JGM (2004) Allometric growth in fish larvae: timing and function. In: Govoni JJ (ed) The development of form and function in fishes and the question of larval adaptation. American Fisheries Society, symposium 40, Bethesda, MD, pp 167–194
22. Osse JWM, van den Boogaart JGM, van Snik GMJ, van der Sluys L (1997) Priorities during early growth of fish larvae. Aquaculture 155:249–258
23. Paragamian VL, Pyper BJ, Daigneault MJ, Beamesderfer RCP, Ireland SC (2008) Population dynamics and extinction risk of burbot in the Kootenai River, Idaho, USA and British Columbia, Canada. Am Fish Soc Symp 59:213–234
24. Peña R, Dumas S (2009) Development and allometric growth patterns during early larval stages of the spotted sand bass *Paralabrax maculatofasciatus* (*Percoidei*: Serranidae). Sci Mar 73S1:183–189
25. Ryder RA, Pesendorfer J (1992) Food, growth, habitat, and community interactions of young-of-the-year burbot, *Lota lota* L., in a Precambrian shield lake. Hydrobiologia 243/244:211–227
26. van Houdt JK, Hellemans B, Volckaert FA (2003) Phylogenetic relationships among Palearctic and Nearctic burbot (*Lota lota*): Pleistocene extinctions and recolonisation. Mol Phylogenet Evol 29:599–612. doi:10.1016/S1055-7903(03)00133-7
27. van Houdt JK, de Cleyn L, Perretti A, Volckaert FAM (2005) A mitogenic view on the evolutionary history of the Holarctic freshwater gadoid, burbot (*Lota lota*). Mol Ecol 14:2445–2457. doi:10.1111/j.1365-294X.2005.02590.x
28. van Snik GMJ, van den Boogaart JGM, Osse WM (1997) Larval growth patterns in *Cyprinus carpio* and *Clarias gariepinus* with attention to the finfold. J Fish Biol 50:1339–1352
29. Wang N, Appenzeller A (1998) Abundance, depth distribution, diet composition and growth of perch (*Perca fluviatilis*) and burbot (*Lota lota*) larvae and juveniles in the pelagic zone of Lake Constance. Ecol Freshw Fish 7:176–183 Wocher H, Harsányi A, Schwarz JF (2013) Larviculture of burbot (*Lota lota* L.): larval rearing

using *Artemia* and weaning onto dry feed. Aquac Res 44:106–113. doi:10.1111/j.1365-2109.2011.03015.x
30. Wolnicki J, Kamiński R, Myszkowski L (2002) Temperature-influenced growth and survival of burbot *Lota lota* (L.) larvae fed live food under controlled conditions. Arch Pol Fish 10:109–113
31. Żarski D, Sasinowski W, Kucharczyk D, Kwiatkowski M, Krejszeff S, Targońska K (2009) Mass initial rearing of burbot *Lota lota* (L.) larvae under controlled conditions. Pol J Nat Sci 24:76–84

# CHAPTER 4

# Biofloc Technology: An Emerging Avenue in Aquatic Animal Healthcare and Nutrition

**Irshad Ahmad[1], A. M. Babitha Rani1, A. K. Verma[1], Mudasir Maqsood[2]**

*[1]ICAR-Central Institute of Fisheries Education, Panch Marg, Yari Road, Versova, Andheri (W), Mumbai 400061, India*

*[2]Sher-e-Kashmir University of Agricultural Sciences and Technology (Kashmir), Srinagar, Jammu & Kashmir 190006, India*

## ABSTRACT

Biofloc is a conglomeric aggregation of microbial communities such as phytoplankton, bacteria, and living and dead particulate organic matter. Biofloc technology involves manipulation of C/N ratio to convert toxic nitrogenous wastes into the useful microbial protein and helps in improving water quality under a zero water exchange system. It may act as a complete source of nutrition for aquatic organisms, along with some bioactive compounds that will enhance growth, survival, and defense mechanisms, and acts as a novel approach for health management in aquaculture by stimulating innate immune system of animals. Nutritionally, the floc biomass provides a complete source of nutrition as well as various bioactive compounds that are useful for improving the overall welfare indicators of aquatic organisms. Beneficial microbial bacterial floc and its derivative compounds such as organic acids, polyhydroxy acetate and polyhydroxy butyrate, could resist the growth of other pathogens, thus serves as a natural probiotic and immunostimulant. The technology is useful in maintaining optimum water quality parameters under a zero water exchange system, thus prevents eutrophication and effluent discharge into the surrounding environment. Moreover, the technology will be useful to ensure biosecurity, as there is no water exchange except sludge removal. The technology is economically viable, environmentally sustainable, and socially acceptable.

**Keywords:** Bioflocs, C/N ratio, Immunostimulation, Nutritional composition, Carbon sources, Biosecurity, Zero water exchange systems, Sustainability,

## INTRODUCTION

To support the rapidly growing human population, food production industries such as aquaculture also needs horizontal as well as vertical expansion. The rapid growth of global aquaculture industry cannot be over emphasized because environmental and economical limitations can hamper this growth. Intensification of the aquaculture activities generates an immense amount of excess organic pollutants that are likely to cause acute toxic effects and long-run environmental risks (Piedrahita 2003). The foremost common method of dealing with this problem has been the utilization of continuous replacement of the pond water through the exchange (Gutierrez-Wing and Malone 2006). The water volume required for even small to medium culture systems can reach up to several hundreds of cubic meters per day. Parenthetically 20 $m^3$ of water is required for the production of 1 kg shrimp (Wang 2003). A recirculating aquaculture system (RAS) is another approach for the removal of major toxic pollutants from the culture water without causing environmental concerns (Gutierrez-Wing and Malone 2006). The beneficial effect of this technology is that only 10% of the total water volume is needed to be replaced on a daily basis (Twarowska et al. 1997), but due to the high operational and maintenance cost, the adoption of RAS among the farming community especially in developing countries is low. Therefore, there was a wide search for years for a low cost, sustainable, and environment-friendly technology for large-scale adoption. Biofloc technology has gained attentions recently as a sustainable and eco-friendly method of aquaculture which controls water quality, along with the production of value added microbial proteinaceous feed for the aquatic organisms. The use of BFT systems in the marine shrimp aquaculture has been extensively studied (Emerenciano et al. 2011, 2012b, 2013b; Ray et al. 2011; Xu and Pan 2012; da Silva et al. 2013; Schveitzer et al. 2013; de Souza et al. 2014; Kumar et al. 2014). The technology is cost-effective and environment-friendly and supports sustainable aquaculture (Naylor et al. 2000;

Avnimelech and Kochba 2009). The different aspects of such a system can be summarized under the following headings.

i. Biofloc technology
ii. Carbon-nitrogen ratio
iii. Microbial community in bioflocs
iv. Nutritional composition of bioflocs
v. Biofloc as a dietary stimulant
vi. Immuno-physiological response by bioflocs
vii. Bioflocs in aquaculture

## Biofloc technology

Biofloc technology (BFT) is based on the maintenance of high levels of microbial bacterial floc in suspension using constant aeration and addition of carbohydrates to allow aerobic decomposition of the organic material (Avnimelech and Weber 1986). By adding carbohydrates, heterotrophic bacterial growth is stimulated and production of microbial proteins takes place through nitrogen uptake (Avnimelech 1999). Maintaining the C/N ratio in the aquaculture system, through the external addition of carbon source or elevated carbon level in the feed, water quality can be improved along with the production of high-quality single-cell microbial protein (Crab et al. 2012). Under such conditions, dense microorganisms develop, functioning both as bioreactor controlling water quality (Avnimelech et al. 1989) and also acts as protein food source for the fishes and shrimps (Avnimelech et al. 1994). Immobilization of toxic nitrogen species occurs much more rapidly in bioflocs because the growth rate and microbial biomass production per unit substrate of heterotrophs are 10 times higher than those of autotrophic nitrifying bacteria (Hargreaves 2006). The technology works on the basic principle of flocculation (co-culture of heterotrophic bacteria and algae) within the system (Avnimelech 2006). Biofloc technology (BFT) has been successfully implemented in aquaculture especially shrimp farming due to economical, environmental, and marketing advantages over a conventional culture system. Compared to conventional aquaculture techniques, biofloc technology provides more economical alternative and sustainable technique in terms of minimal water exchange and reduced feed input making it a low-cost sustainable technology for sustainable future aquaculture development (Avnimelech and Kochba 2009; De Schryver et al. 2008). As a startup for biofloc technology, it might be interesting to investigate the effect of adding nucleation sites to the lined ponds which include adding calculated amounts of pond bottom clay to the water at startup to stimulate floc formation. In addition, the inoculation with water from existing biofloc ponds or a prepared inoculum (20 g of pond bottom soil, 10 mg ammonium sulfate $(NH)_4SO_4$ and 200 mg of carbon sources in 1 l of water) might allow an accelerated startup. McAbee et al. (2003) and Gaona et al. (2011) suggested the use of nucleation sites such as soil or biofloc-rich water as inoculum from a prior production cycle accelerates the formation of microbial flocs in the new culture system. Compared to the productivity of various eco-friendly farming practices, the intensive farming methods with limited water exchange provide a sustainable alternative for farming intensification and biosecurity.

## Carbon-nitrogen ratio

Carbon-nitrogen ratio (C/N) in the aquatic environment plays an important role in the immobilization of toxic inorganic nitrogen compounds into useful bacterial cells (single-cell protein) that may act as a direct source of food for the cultured organisms (Avnimelech 1999). Immobilization of inorganic nitrogen takes place when the C/N ratio of the organic matter is higher than 10 (Lancelot and Billen 1985). Thus, alteration in the C/N ratio may result in a shift from an autotrophic to a heterotrophic system (Avnimelech 1999; Browdy and Bratvold 2001). Once a mature biofloc community is established, TAN and $NO_2$-N concentrations can be effectively controlled by either heterotrophic assimilation or autotrophic nitrification that helps to maintain their concentrations at acceptable ranges for the cultured organisms even at higher stocking densities (Xu et al. 2016). By adding a carbon source (direct or indirect C-sources) to the culture medium in limited-discharge systems (changing C/N ratio), it is possible to obtain a significant enhancement of useful microbial growth and the fixation of toxic nitrogen metabolites (Ebeling et al. 2006; Hari et al. 2006; Avnimelech and Kochba 2009; Crab et al. 2010). As the C/N ratio of bacterial cells is 5:1 (Rittmann and McCarty 2001) and the conversion efficiency of bacteria is 40–60%, C/N ratio of 10 or more in the feed is required for the growth of heterotrophic microorganisms (Avnimelech 1999). The bacterial process transforms the toxic form of nitrogen (ammonia and nitrite) to one that is toxic only at high concentrations (nitrate) by the process called nitrification. The BFT being zero water exchange system thus tends to accumulate the nitrate in the long run, and hence usually the nitrate level in biofloc systems increases as the culture progresses. Kuhn et al. (2009) observed that carbon supplementation enhanced the removal rates of TAN at 26% per hour compared to 1% per hour in a control system. The C/N ratio of around 10 is maintained in most of the feeds used in semi-intensive aquaculture ponds, but bacteria require about 20 units of carbon per unit of nitrogen assimilated (Avnimelech 1999). So, when C/N ratio is low in the feed, carbon becomes the limiting nutrient for the growth of heterotrophic bacterial populations in the aquaculture ponds (Asaduzzaman et al. 2009) and hence the heterotrophic bacterial population will not inflate beyond a certain point due to the limited availability of carbon in the system. The C/N ratio in an aquaculture system can be increased by adding different locally available cheap carbon sources (agricultural by-products) and also by the reduction of protein content in the feed (Avnimelech 1999; Hargreaves 2006). Different organic carbon sources (glucose, cassava, molasses, wheat, corn, sugar bagasse, sorghum meal, etc.) are used to enhance production and to improve the nutrient dynamics through altered C/N ratio in shrimp culture (Avnimelech 1999), and C/N ratio is also widely used as a guide for analyzing the decomposition of organic matter (Alexander and Ingram 1992). The reduction of toxic nitrogenous compounds from the intensive, well-aerated systems can be achieved by the application of organic carbon sources and by altering the C/N ratio in the feed (Avnimelech 1999; Browdy and Bratvold 2001). The biofloc system maintained with C/N ratio of higher than 15–20 will be developing sufficient microbial floc to assimilate toxic nitrogenous species under intensive farming with limited discharge. Recently, a lot of work has been published in biofloc technology regarding the manipulation of C/N ratio, and also *Biofloc Technology: A Practical Guide Book* that is directed to farmers and researchers is a tremendous step forward in providing information on this technology (Avnimelech 2015).

## Microbial community in bioflocs

Two functional categories of bacterial populations are primarily responsible for water quality maintenance in minimal or zero water exchange systems (intensive systems) viz., heterotrophic ammonia-assimilative and chemoautotrophic nitrifying bacteria (Ebeling et al. 2006; Hargreaves 2006). The color changes from green to brown which takes place as the culture progresses due to the transition from a mostly algal-dominant to a bacterial biofloc-dominant system. The number of bacteria in biofloc ponds can be between $10^6$ and $10^9$/ml of floc plug which contains between 10 and 30 mg dry matter making the pond a biotechnological industry (Avnimelech 2007). Microbial communities formed consist of phytoplankton, bacteria, and aggregates of living and dead particulate organic matter (Hargreaves 2006). According to Ju et al. (2008), bioflocs collected from *Litopenaeus vannamei* tanks contained 24.6% phytoplankton (dominated by diatoms like *Thalassiosira*, *Chaetoceros*, and *Navicula*), 3% bacterial biomass (two thirds was gram-negative and one third gram-positive ), a small amount of protozoan community (98% flagellates, 1.5% rotifers, and 0.5% amoeba), and 33.2% detritus, and the remaining quantity was ash (39.25%). Only 2–20% of the organic fraction of sludge flocs is believed to be living (microbial cells) while the rest is of total organic matter (60–70%) and total inorganic matter (30–40%) (Wilen et al. 2003). Dominant bacterial species that are present in the bioflocs include Proteobacterium, *Bacillus* species, and Actinobacterium. Besides this, there are some other minor bacterial species such as *Roseobacter* sp. and *cytophaga* sp. (Zhao et al. 2012). In conventional activated sludge systems, efficient aggregation is of principal importance, since their operational success depends heavily on good settling sludge (Bossier and Verstraete 1996). The heterotrophic bacterial population utilizes the ammonium in addition to the organic nitrogenous wastes to synthesize single-cell microbial protein (Schneider et al. 2006) which act as natural feed for shrimps (Burford et al. 2004). The sinking rate of floc aggregate will be slow at a velocity of 1 to 3 m/h when densities of the microbial biomass move slightly above 1 g wet weight/ml (Sears et al. 2006). As the bacteria move through the water column, efficient laminar regime (Reynolds envelope) which is always present around bacteria which are smaller than 100 μm interferes with the nutrient mass transfer. When the rate of substrate consumption exceeds the rate of substrate supply, mass transfer limitations take place (Simoni et al. 2001).

## Nutritional composition of bioflocs

Nutritionally, the floc biomass could provide a complete source of nutrition as well as various bioactive compounds (Akiyama et al. 1992). The nutritional value of bioflocs is dependent on several factors such as food preferences by the animal, their ability to ingest and digest microbial protein, and the floc density in the water (Hargreaves 2006). The single-cell protein formed by heterotrophic bacterial population through uptake of inorganic N can be utilized as a source of food for cultured animals like shrimps, tilapia, and carps (Rahmatullah and Beveridge 1993; Burford et al. 2004; Mahanand et al. 2013). In terms of quality, biofloc contains 38% protein, 3% lipid, 6% fiber, 12% ash, and 19 kJ/g energy (on dry matter basis) (Azim and Little 2008). Azim and Little 2008 observed 50% crude protein, 2.5% crude lipid, 4% fiber, 7% ash, and 22 kJ $g^{-1}$ energy and reported that the quality of biofloc is independent of the quality of feed used for biofloc production (35 and 22% crude protein). Ballester et al. (2010) reported that bioflocs contain 30.4% crude protein, 4.7% crude lipid, 8.3% fiber, 39.2% ash, and 29.1% nitrogen free extract on dry matter basis when wheat bran and molasses were used as carbohydrates sources. Thus the change in the carbon source changes the nutritional composition and quality index of the flocs. Besides these characteristics, the type of carbon source also influences the palatability and digestibility of the cultured organisms (Crab 2010a; Crab et al. 2009). Overall, bioflocs produced on glycerol gave the best results (Crab 2010a). Biofloc enhances ingestion rate, nutrient absorption, and assimilation, and provides a complete source of cellular nutrition (Tacon et al. 2002). Broodstock diets fortified with biofloc supplementation improve reproductive performance in terms of fecundity, spawning, and egg biochemical composition in *Farfantepenaeus duorarum* and *L. Vannamei* (Emerenciano et al. 2012a, 2014). Tilapia culture inactivated suspension ponds indicated that the fish grew well on low-protein feed (Avnimelech 1999; Milstein et al. 2001). Bioflocs are rich in proteins, vitamins, and minerals (Brown et al. 1997; Tacon et al. 2002). Ju et al. (2008) studied the amino acid profile of the biofloc and reported that bioflocs have a better essential amino acid index (0.92–0.93) with histidine and taurine as the most abundant amino acids. However, arginine and lysine were found to be the limiting amino acids in the biofloc (Avnimelech 1999).

Consumption and regeneration of bioflocs can increase feed utilization efficiency of the microbial population by recycling feed residues and/or recovery of some fraction of excreted nutrients (Hargreaves 2006). The microorganism not only removes excess nutrient but also improves growth rate, feed conversion ratio, and weight gain in shrimps and tilapia (Burford et al. 2004; Wasielesky et al. 2006). Although bioflocs meet nutritional standards, nutritional properties and the ability to maintain water quality in the BFT system depend on the carbon source used to produce the flocs. Different carbon sources are not only used to manipulate the C/N ratio and stimulate the specific bacteria, protozoa, and algae but also to influence the microbial composition and community organization of the bioflocs (Crab et al. 2009; Crab 2010a). *L. vannamei* production rates and water quality were maintained without water exchange using a biofloc system supplemented with dextrose or molasses (Antonio et al. 2015). According to Avnimelech (2007) and Emerenciano et al. (2013b), the bioflocs are rich in natural protein and lipid and hence serve as natural in situ food for culture organisms while at the same times act as bio-control to the system by treating the feeding waste and reducing ammonium concentrations (Crab et al. 2007; De Schryver et al. 2008; Hargreaves 2013) thereby maintaining the water quality. However, further research should focus on the use of cheap and fermented non-conventional agro-industrial residues as carbon sources to upgrade wastes as healthy feeds and to find means of fish meal replacement for the aquatic organisms.

## Bioflocs as dietary stimulant

Bioflocs or its attached microorganisms could exert a positive effect on the digestive enzyme activity of shrimp (Xu and Pan 2012). Inclusion of bioflocs in the diet at BFT 75% results in improved growth performances and digestive enzyme activity of the common carp ( Najdegerami et al. 2016); also biofloc as a dietary supplement at a 4% level in shrimp feed can enhance the growth and digestive enzyme activities in *P. monodon* (Anand et al. 2014). Bioflocs have been recently projected as a possible novel strategy for disease management with the "natural probiotic effect" in contrast to conventional approaches such as antibiotic, antifungal, and external probiotic and prebiotic application (Emerenciano et al. 2013a). According to the original definition, probiotics are "organisms and substances which contribute to intestinal microbial balance". Fuller (1989) revised the definition as "live microbial feed supplement which beneficially affects the host animal by improving its intestinal microbial balance". Therefore, several terms such as "friendly", "beneficial", or "healthy" bacteria are also commonly used to describe probiotics. Presently, research is going on in the field of probiotic biofloc (bio-engineered biofloc) in which beneficial probiotic bacterial strains are added to the biofloc to increase the bio-control over the pathogenic microbes and to improve the quorum-sensing ability of beneficial bacterial population present in the floc, as probiotic-based biofloc showed a bio-control effect towards the pathogenic *vibrio* species (Crab 2010a). Bioactive compounds present in the floc cultures were considered to be effective enhancers of growth and immunity in fishes and shrimps (Burford et al. 2004; Ju et al. 2008; Linan-Cabello et al. 2002). In biofloc-based rearing units, excess nitrogen and carbon sources will lead to the development of mixed cultures which accumulate polyhydroxy butyrate (PHB) (Salehizadeh and Van Loosdrecht 2004). When the bacterial cell death or lysis happens in the culture system, there will be degradation of PHB through the activity of extracellular PHB depolymerase enzymes which are widely found in the microbes (Jendrossek and Handrick 2002). The activity of extracellular PHB depolymerase enzymes results in the release of 3-hydroxyl butyrate into the surrounding environment (Trainer and Charles 2006). PHB can act as a prebiotic for the cultured organisms. Prebiotics also known as immunosaccharides are indigestible food ingredients that selectively and beneficially affect the host by stimulating the growth of one or a limited number of bacteria in the colon thus changing the composition of gut microflora. The carbohydrates added to the BFT can also act as a source of prebiotics to the cultured organisms and could modulate the gut microbial population which is beneficial to improve the health status of the cultured organisms. Further research should be directed to know the role and activity of immunostimulatory and bioactive compounds present in the biofloc, since the technology deals with bacteria and bacterial products.

## Immuno-physiological response by bioflocs

The knowledge of immunology has diversified greatly due to the variety of fish that are farmed and the inclusion of species such as zebra fish to the animal model repertoire in comparative immunological studies (Van Der Sar et al. 2004; Yoder et al. 2002). The immune system of fish acts as a crossroad between innate and adaptive immune responses and is hardened to the environment and the poikilothermic nature of the fish (Tort et al. 2003). The immune responses initiated by injury or by pathogenic invasion will entail phagocytosis and inflammatory processes (Corbel 1975) assisted by non-specific immune cells such as macrophages, neutrophils, and non-specific cytotoxic cells. When bacterial load increases in the adjacent water, they stimulate the release of high molecular weight glycoproteins from fish skin mucus (van der Marel et al. 2010). A number of humoral factors are released by the fishes, once they are exposed to the pathogens, such as cytokines, anti-proteases, peroxidases, lysozymes, etc. Among them, lysozyme is a preferred marker of the immune response. Lysozyme acts as an anti-inflammatory and antiviral agent, besides its high potential for bactericidal or bacteriolytic activity against pathogenic gram-positive and gram-negative bacteria (Saurabh and Sahoo 2008). Recently, researchers have hypothesized possibilities of immunostimulatory features of the bioflocs leading to enhancement of the immunity and antioxidant status of shrimps and fishes to provide broad-based resistance towards many infections (Crab et al. 2012; Xu and Pan 2013; Ahmad et al. 2016). The heterotrophic microbial biomass is suspected to have a controlling effect on pathogenic bacteria (Michaud et al. 2006). Ju et al. (2008) reported that floc carotenoids have been shown to provide essential nutritional and many bioactive physiological functions in animal tissue, including stimulating the animal immune system. The disruption of quorum-sensing, bacterial cell-to-cell communication system is a new strategy to control bacterial infection in aquaculture. Interestingly, a similar phenomenon was observed in bioflocs grown on glycerol against *Vibrio harveyi* in *Artemia franciscana* culture (Crab et al. 2010); it was also speculated that microbial flocs contain many strains of probiotics bacteria (Bairagi et al. 2002).

Most of the existing immunostimulants available are a group of live and synthetic compounds derived from bacteria and bacterial products and also extracts from plants and animals (Wang et al. 2008). Since biofloc technology deals with the bacterial environment, it might also contain some immunostimulatory compounds that are beneficial for the health of cultured organisms. Microorganisms and their cell components have been studied and applied as probiotics or immunostimulants in order to improve the innate immunity and antioxidant status of the shrimp, thereby enhancing their disease resistance (Ninawe and Selvin 2009; Smith et al. 2003; Vazquez et al. 2009). Even though bioflocs have been confirmed as being the richest source of natural microorganisms and bioactive compounds, little effort was made to study its effect on the physiological health of cultured shrimp, particularly concerning immune and antioxidant defense systems. Jang et al. (2011) found that the expression of a prophenol oxidase-activating enzyme (lvPPAE1) in hemocytes of *L. vannamei* was enhanced significantly when shrimp were reared in a biofloc for a long term. More recently, Becerra-Dorame et al. (2014) reported that *L. vannamei* reared in biofloc-based systems showed improved physiological performance as indicated by selected hemolymph parameters including superoxide dismutase activity. Most probably, some active microorganisms enter a shrimp body continuously along with the process of ingesting biofloc (Johnson et al. 2008) and then modulate the immune system of the host whether as viable microbes or microbial components (Jang et al. 2011). Therefore, in order to improve the welfare of fishes in aquaculture, further work needs to be carried out in the exact nature of humoral innate or cellular immune response and to determine the protective biofloc life of bacteria. Finally, this paper proposes a long-term perspective of bioflocs that can be used as a strategic control of diseases in aquaculture.

## Bioflocs in aquaculture: future prospectus

A number of beneficial features are associated with biofloc technology along with 10–20% potential feed gain as estimated by application of biofloc technology (Crab et al. 2007; De Schryver et al. 2008) to nitrogen recovery from the culture system. This increase was based on the internal recirculation of nutrients through the formation of new microbial biomass, which was subsequently grazed by the fish (Avnimelech 2006). The advantages of the technology in aquaculture has been well documented which includes low feed and water input (economical), less risk of pathogen introduction and diseases, more biosecurity, increased growth and survival, and hence increased crop yield (Otoshi et al. 2009; Crab et al. 2009; Samocha et al. 1998, 2007; Krummenauer et al. 2011; Perez-Fuentes et al. 2013). It also lowered the feed conversion rate by utilizing the in situ natural feed and has small footprints, hence reducing environmental impacts (Krummenauer et al. 2014). It is also robust, easy to operate, and economically feasible (Crab et al. 2012). The zero water exchange system has advantages of maintaining temperature and heat fluctuations (Crab et al. 2009). It supports nitrogen removal even when organic matter and biochemical oxygen demand of the system is high (Avnimelech 2015). It will be important for the future BFT to understand the microbial mechanisms involved in the process of flocculation viz., quorum sensing and controlling effect on pathogenic microbes. BFT also improves sustainability and biosecurity and development of high-intensity grow out systems with no water discharge over the entire crop cycle. It can also serve as a cheap and effective immunostimulant for the cultured organisms, so more research should focus on the optimal way to manage the BFT in the culture ponds.

## CONCLUSION

The scarcity of water, growing demand for protein food, and conflict for land usage for the expansion of aquacultural practices have become a major problem at the global level. To cater to the growing demand for animal protein, intensive aquaculture is one of the major options. But intensification of aquaculture practices will generate a lot of effluents which will damage the aquatic environment. Moreover, the intensification will lead to heavy dependence on fish meal which is a scarce commodity, disease outbreaks in the cultured organisms, environmental degradation, and socioeconomic conflicts. In an attempt to minimize the impact of the environmental, health, and economic problems associated with aquaculture, BFT has become increasingly popular as a sustainable alternative for intensification (Avnimelech 1999, 2006; Browdy et al. 2001; Crab et al. 2007; De Schryver et al. 2008). The requirements for sustainable and eco-friendly aquaculture development can be fulfilled by the use of biofloc technology.

## REFERENCES


1. Ahmad I, Verma AK, Babitha Rani AM, Rathore G, Saharan N, Gora AH (2016) Growth, non-specific immunity and disease resistance of *Labeo rohita* against *Aeromonas hydrophila* in biofloc systems using different carbon sources. Aquaculture 457:61–67
2. Akiyama D, Doming WG, Lawrence AL (1992) Penaeid shrimp nutrition. In: Fast AW, Lester LJ (eds) Marine shrimp culture: principles and practices. Elsevier, Amsterdam, pp 535–568
3. Alexander JB, Ingram GA (1992) Noncellular nonspecific defence mechanisms of fish. Ann Rev Fish Dis 2:249–279
4. Anand PS, Kohli MPS, Kumar S, Sundaray JK, Roy SD, Venkateshwarlu G, Sinha A, Pailan GH (2014) Effect of dietary supplementation of biofloc on growth performance and digestive enzyme activities in *Penaeus monodon*. Aquaculture 418:108–115
5. Antonio M, Lorenzo D, Schveitzer R, Manoel C, Legarda EC, Quadros W, Vieira N (2015) Intensive hatchery performance of the Pacific white shrimp in biofloc system. Aquacultural Engineering 67:53–58
6. Asaduzzaman M, Wahab MA, Verdegem MCJ, Benerjee S, Akter T, Hasan MM, Azim ME (2009) Effects of addition of tilapia *Oreochromis niloticus* and substrates for periphyton developments on pond ecology and production in C/N-controlled

freshwater prawn *Macrobrachium rosenbergii* farming systems. Aquaculture 287(3):371–380
7. Avnimelech Y, Weber B (1986) Studies in circulated fish ponds: organic matter recycling and nitrogen transformation. Aquaculture Fisheries Management 17:231–242
8. Avnimelech Y, Mokady S, Schroeder G (1989) Circulated ponds as efficient bioreactors for single cell protein production. Isr J Aquac Bamidgeh 41:58–66
9. Avnimelech Y, Kochva M, Diab S (1994) Development of controlled intensive aquaculture systems with a limited water exchange and adjusted carbon to nitrogen ratio. Isr J Aquac Bamidgeh 46:119–131
10. Avnimelech Y (1999) Carbon/nitrogen ratio as a control element in aquaculture systems. Aquaculture 176:227–235
11. Avnimelech Y (2006) Bio-filters: the need for an new comprehensive approach. Aquac Eng 34:172–178
12. Avnimelech Y (2007) Feeding with microbial flocs by tilapia in minimal discharge bio-flocs technology ponds. Aquaculture 264:140–147
13. Avnimelech Y, Kochba M (2009) Evalution of nitrogen uptake and excretion by tilapia in biofloc tanks, using $^{15}N$ tracing. Aquaculture 287:163–168
14. Avnimelech Y (2015) Biofloc Technology, a Practical Guidebook, 3d Edition, pp 258, World Aquaculture Soc
15. Azim ME, Little DC (2008) The biofloc technology (BFT) in indoor tanks: water quality, biofloc composition, and growth and welfare of Nile tilapia, *Oreochromis niloticus*. Aquaculture 283:29–35
16. Bairagi A, Ghosh KS, Sen SK, Ray AK (2002) Enzyme producing bacterial flora isolated from fish digestive tracts. Aquac Int 10:109–121
17. Ballester ELC, Abreu PC, Cavalli RO, Emerenciano M, de Abreu L, Wasielesky JW (2010) Effects of practical diets with different protein levels on the performance of *Farfantepeneau spaulensis* juveniles nursed in zero exchange suspended microbial flocs intensive system. Aquac Nutr 16:163–172
18. Jesús Becerra-Dorame M, Martinez-Cordova LR, Martínez-Porchas M, Hernández-López J, López– Elías JA, Mendoza–Cano F (2014) Effect of using autotrophic and heterotrophic microbial-based-systems for the pre-grown of *Litopenaeus vannamei*, on the production performance and selected haemolymph parameters. Aquaculture Research 45(5): 944-948
19. Bossier P, Verstraete W (1996) Triggers for microbial aggregation in activated sludge? Appl Microbiol Biotechnol 45(1–2):1–6
20. Browdy C, Bratvold D (2001) Perspectives on the application of closed shrimp culture systems In: Jory, E.D., Browdy, C.L. (Ed.), the new Wave, Proceedings of the Special Session on Sustainable Shrimp Culture, The World Aquaculture Society, Baton Rouge, LA, USA, pp. 20–34
21. Browdy CL, Bratvold D, Stokes AD, McIntosh RP (2001) Perspectives on the application of closed shrimp culture systems. In: The Rising Tide, Proceedings of the Special Session on Sustainable Shrimp Farming, The World Aquaculture Society, Baton Rouge, LA, USA, pp. 20–34
22. Brown MR, Jeffrey SW, Volkam JK, Dunstan GA (1997) Nutritional properties of microalgae for mariculture. Aquacult 151:315–331
23. Burford MA, Thompson PJ, McIntosh RP, Bauman RH, Pearson DC (2004) The contribution of flocculated material to shrimp, *Litopenaeus vannamei* nutrition in a high-intensity, zero-exchange system. Aquacult 232:525–537
24. Corbel M (1975) The immune response in fish: a review. J Fish Biol 4:539–563
25. Crab R, Lambert A, Defoirdt T (2010) The application of bioflocs technology to protect brine shrimp, *Artemia franciscana* from pathogenic *Vibrio harveyi*. J Appl Microbiol 5:1643–1649
26. Crab R (2010a) Bioflocs technology: an integrated system for the removal of nutrients and simultaneous production of feed in aquaculture. PhD thesis, Ghent University. pp. 178

27. Crab R, Avnimelech Y, Defoirdt T, Bossier P, Verstraete W (2007) Nitrogen removal techniques in aquaculture for a sustainable production. Aquaculture 270:1–14
28. Crab R, Chielens B, Wille M, Bossier P, Verstraete W (2009) The effect of different carbon sources on the nutritional value of bioflocs, a feed for *Macrobrachium rosenbergii* postlarvae. Aquac Res 41:559–567
29. Crab R, Defoirdt T, Bossier P, Verstraete W (2012) Biofloc technology in aquaculture: beneficial effects and future challenges. Aquaculture 356-357:351–356
30. da Silva KR, Wasielesky W, Abreu PC (2013) Nitrogen and phosphorus dynamics in the biofloc production of the Pacific white shrimp. Litopenaeus vannamei. J. World Aquac. Soc. 44:30–41
31. De Schryver P, Crab R, Defoirdt T, Boon N, Verstraete W (2008) The basics of bio-flocs technology: the added value for aquaculture. Aquaculture 277:125–137
32. de Souza DM, Suita SM, Romano LA, Wasielesky W, Ballester ELC (2014) Use of molasses as a carbon source during the nursery rearing of Farfantepenaeus brasiliensis (Latreille, 1817) in a Biofloc technology system. Aquac. Res. 45:270–277
33. Ebeling JM, Timmons MB, Bisogni JJJ (2006) Engineering analysis of the stoichiometry of photoautotrophic, autotrophic, and heterotrophic removal of ammonia–nitrogen in aquaculture systems. Aquaculture 257:346–358
34. Emerenciano M, Ballester ELC, Cavalli RO, Wasielesky W (2011) Effect ofbiofloc technology (BFT) on the early postlarval stage of pink shrimp *Farfantepenaeus paulensis*: growth performance, floc composition and salinity stress tolerance. Aquac Int 19:891–901
35. Emerenciano M, Ballester ELC, Cavall RO, Wasielesky W (2012a) Biofloc technology application as a food source in a limited water exchange nursery system for pink shrimp *Farfantepenaeus brasiliensis* (Latreille, 1817). Aquac Res 43:447–457
36. Emerenciano M, Cuzon G, Goguenheim J, Gaxiola G (2012b) Floc contribution on spawning performance of blue shrimp *Litopenaeus stylirostris*. Aquac Res 44(1):75–85
37. Emerenciano M, Cuzon G, Paredes A, Gaxiola G (2013a) Evaluation of biofloc technology in pink shrimp *Farfantepenaeus duorarum* culture: growth performance, water quality, microorganisms profile and proximate analysis of biofloc. Aquac Int 21:1381–1394
38. Emerenciano M, Gaxiola G, Cuzon G (2013b) Biofloc technology (BFT): a review for aquaculture application and animal food industry. Biomass now: cultivation and utilization. InTech Rijeka Croatia 301–328
39. Emerenciano M, Cuzon G, Arevalo M, Gaxiola G (2014) Biofloc technology in intensive broodstock farming of the pink shrimp *Farfantepenaeus duorarum*: spawning performance, biochemical composition and fatty acid profile of eggs. Aquac Res 45(10):1713–1726
40. Fuller R (1989) Probiotic in man and animals. J Appl Bacteriol 66:365–378
41. Gaona CAP, Poersch LH, Krummenauer D, Foes GK, Wasielesky W (2011) The effect of solids removal on water quality, growth and survival of *Litopenaeus vannamei* in a biofloc technology culture system. International Journal of Recirculating Aquacult 12:54–57
42. Gutierrez-Wing MT, Malone RF (2006) Biological filters in aquaculture: trends and research directions for freshwater and marine applications. Aquac Eng 34:163–171
43. Hargreaves JA (2013) Biofloc Production Systems for Aquaculture. SRAC 4503:1–12
44. Hargreaves JA (2006) Photosynthetic suspended-growth systems in aquaculture. Aquac Eng 34:344–363
45. Hari B, Kurup BM, Varghese JT, Schrama JW, Verdegem MCJ (2006) The effect of carbohydrate addition on water quality and the nitrogen budget in extensive shrimp culture systems. AJSS 252:248–263
46. Jang IK, Pang Z, Yu J, Kim SK, Seo HC, Cho YR (2011) Selectively enhanced expression of prophenoloxidase activating enzyme 1 (PPAE1) at a bacteria clearance site in the white shrimp, *Litopenaeus vannamei*. BMC Immuno 12:70–80

47. Jendrossek D, Handrick R (2002) Microbial degradation of polyhydroxyalkanoates. Annu Rev Microbiol 56:403–432
48. Johnson CN, Barnes S, Ogle J, Grimes DJ, Chang YJ, Peacock AD, Kline L (2008) Microbial community analysis of water, foregut, and hindgut during growth of Pacific white shrimp, *Litopenaeus vannamei*, in closed-system aquaculture. J World Aquacult Soc 39:251–258
49. Ju ZY, Forster I, Conquest L, Dominy W (2008) Enhanced growth effects on shrimp, *Litopenaeus vannamei* from inclusion of whole shrimp floc or floc fractions to a formulated diet. Aquac Nutr 14:533–543
50. Krummenauer D, Peixoto S, Cavalli RO, Poersch LH, Wasielesky W (2011) Superintensive culture of white shrimp, *Litopenaeus vannamei*, in a biofloc technology system in southern Brazil at different stocking densities. J World Aquacult Soc 42(5):726–733
51. Krummenauer D, Poersch L, Romano LA, Lara GR, Encarnacao P, Wasielesky JW (2014) The effect of probiotics in a *Litopenaeus vannamei* biofloc culture system infected with *Vibrio parahaemolyticus*. J Appl Aquac 26(4):370–379
52. Kuhn DD, Boardman GD, Lawrence AL, Marsh L, Flick GJ (2009) Microbial floc meal as a replacement ingredient for fish meal and soybean protein in shrimp feed. Aquacult 296:51–57
53. Kumar S, Anand PSS, De D, Sundaray JK, Raja RA, Biswas G, Ponniah AG, Ghoshal TK, Deo AD, Panigrahi A, Muralidhar M (2014) Effects ofcarbohydrate supplementation on water quality, microbial dynamics andgrowth performance of giant tiger prawn (*Penaeus monodon*). Aquac Int 22:901–912
54. Lancelot C, Billen G (1985) Carbon-nitrogen relationships in nutrient metabolism of coastal marine ecosystems. Advances in Aquatic Microbiology 3:263–321
55. Linan-Cabello M, Paniagua-Michel J, Hopkins P (2002) Bioactive roles of carotenoids and retinoids in crustaceans. Aquac Nutr 8:299–309
56. Mahanand SS, Moulick S, Rao PS (2013) Water quality and growth of Rohu, *Labeo rohita*, in a biofloc system. J Appl Aquac 25(2):121–131
57. McAbee BJ, Browdy CL, Rhodes RJ, Stokes AD (2003) The use of greenhouse-enclosedrace-way systems for the super intensive production of pacific white shrimp *Litopenaeus vannamei*in the United States. Global Aquaculture Advocate 6:40–43
58. Michaud L, Blancheton JP, Bruni V, Piedrahita R (2006) Effect of particulate organic carbon on heterotrophic bacterial populations and nitrification efficiency in biological filters. Aquac Eng 34:224–233
59. Milstein A, Avnimelech Y, Zoran M, Joseph D (2001) Growth performance of hybrid bass and hybrid tilapia in conventional and active suspension intensive ponds. Isr J Aquac Bamidgeh 53:147–157
60. Najdegerami EH, Bakhshi F, Lakani FB (2016) Effects of biofloc on growth performance, digestive enzyme activities and liver histology of common carp (*Cyprinus carpio L.*) fingerlings in zero-water exchange system. Fish Physiol Biochem 42:457–465
61. Naylor RL, Goldburg RJ, Primavera JH, Kautsky N, Beveridge MC, Clay J, Folke C, Lubchenco J, Mooney H, Troell M (2000) Effect of aquaculture on world fish supplies. Nature 405:1017–1024
62. Ninawe AS, Selvin J (2009) Probiotics in shrimp aquaculture: avenues and challenges. Crit Rev Microbiol 35:43–66
63. Otoshi CA, Tang LR, Moss DR, Arce SM, Holl CM, Moss S (2009) Performance of Pacific white shrimp, *Penaeus (litopenaeus)* vannamei, cultured in biosecure, super-intensive, recirculating aquaculture systems. *BROWDY, CL; JORY, DE The rising tide-proceedings of the special session on sustainable shrimp farming. Baton Rouge: World Aquaculture Society*
64. Perez-Fuentes JA, Perez-Rostro CI, Hernandez-Vergara MP (2013) Pond-reared Malaysian prawn *Macrobrachium rosenbergii* with the biofloc system. Aquaculture 400:105–110

65. Piedrahita RH (2003) Reducing the potential environmental impact of tank aquaculture effluents through intensification and recirculation. Aquaculture 226:35–44
66. Rahmatullah SM, Beveridge MCM (1993) Ingestion of bacteria in suspension Indian major carps (*Catla catla, Labeo rohita*) and Chinese carps (*Hypophthalmichthys molitrix, Aristichthys nobilis*). Hydrobiologia 264:279–284
67. Ray AJ, Dillon KS, Lotz JM (2011) Water quality dynamics and shrimp (Litopenaeus vannamei) production in intensive, mesohaline culture systems with two levels of biofloc management. Aquac. Eng. 45:127–136
68. Rittmann B, McCarty P (2001) Environmental biotechnology: principles and applications. McGraw-Hill, New York, p 229
69. Salehizadeh H, Van Loosdrecht MCM (2004) Production of polyhydroxyalkanoates by mixed culture: recent trends and biotechnological importance. Biotechnol Adv 22:261–279
70. Samocha TM, Guajardo H, Lawrence AL, Castille FL, Speed M, McKee DA, Page KI (1998) A simple stress test for Penaeus vannamei postlarvae. AJSS 165:233–242
71. Samocha TM, Patnaik S, Speed M, Ali AM, Burger JM, Almeida RV, Ayub Z, Harisanto M, Horowitz A, Brock DL (2007) Use of molasses as carbon source in limited discharge nursery and grow-out systems for Litopenaeus vannamei. Aquac. Eng. 36:184–191
72. Saurabh S, Sahoo PK (2008) Lysozyme: an important defence molecule of fish innate immune system. Aquac Res 39:223–239
73. Schneider O, Sereti V, Eding E, Verreth J (2006) Molasses as C source for heterotrophic bacteria production on solid fish waste. AJSS 261:1239–1248
74. Schveitzer R, Arantes R, Costodio PFS, Santo CMD, Arana LV, Seiffert WQ, Andreatta ER (2013) Effect of different biofloc levels on microbial activity, water quality and performance of *Litopenaeus vannamei* in a tank systemoperated with no water exchange. Aquac Eng 56:59–70
75. Sears K, Alleman JE, Barnard JL, Oleszkiewicz JA (2006) Density and activity characterization of activated sludge flocs. J Environ Eng 132:1235–1242
76. Simoni SF, Schafer A, Harms H, Zehnder AJB (2001) Factors affecting mass transfer limited biodegradation in saturated porous media. J Contam Hydrol 50:99–120
77. Smith VJ, Brown JH, Hauton C (2003) Immunostimulation in crustaceans: does it really protect against infection. Fish Shellfish Immunol 15:71–90
78. Tacon A, Cody JJ, Conquest LD, Divakaran S, Forster IP, Decamp OE (2002) Effect of culture system on the nutrition and growth performance of pacific white shrimp, *Litopeneaus vannamei* (bonne) fed different diets. Aquac Nutr 8:121–137
79. Tort L, Balasch JC, Mackenzie S (2003) Fish immune system. A crossroads between innate and adaptive responses. Immunologia 22:277–286
80. Trainer MA, Charles TC (2006) The role of PHB metabolism in the symbiosis of rhizobia with legumes. Appl Microbiol Biotechnol 71:377–386
81. Twarowska JG, Westerman PW, Losordo TM (1997) Water treatment and waste characterization evaluation of an intensive recirculating fish production system. Aquac Eng 16:133–147
82. Van der Marel M, Caspari N, Neuhaus H, Meyer W, Enss ML, Steinhagen D (2010) Changes in skin mucus of common carp *Cyprinus carpio* after exposure to water with a high bacterial load. J Fish Dis 33:431–439
83. Van Der Sar AM, Appelmelk BJ, Vandenbroucke-Grauls CMJE, Bitter W (2004) A star with stripes: zebrafish as an infection model. Trends Microbiol 12:451–457
84. Vazquez L, Alpuche J, Maldonado G, Agundis C, Pereyra-Morales A, Zenteno E (2009) Review: immunity mechanisms in crustaceans. Innate Immunity 15:179–188
85. Wang JK (2003) Conceptual design of a microalgae-based recirculating oyster and shrimp system. Aquac Eng 28:37–46
86. Wang YC, Chang PS, Chen HY (2008) Differential time-series expression of immune-related genes of Pacific white shrimp Litopenaeus vannamei in response to dietary inclusion of β-1, 3-glucan. Fish & Shellfish Immunology 24(1):113–121

87. Wasielesky W, Atwood H, Stokes A, Browdy CL (2006) Effect of natural production in a zero exchange suspended microbial floc based super-intensive culture system for white shrimp, *Litopenaeus vannamei*. Aquaculture 258:396–403

88. Wilen BM, Jin B, Lant P (2003) The influence of key chemical constituents in activated sludge on surface and flocculating properties. Water Res 37:2127–2139

89. Xu WJ, Morris TC, Samocha TM (2016) Effects of C/N ratio on biofloc development, water quality, and performance of *Litopenaeus vannamei* juveniles in a biofloc-based, high-density, zero-exchange, outdoor tank system. Aquaculture 453:169–175

90. Xu WJ, Pan LQ (2012) Effects of bioflocs on growth performance, digestive enzyme activity and body composition of juvenile *Litopenaeus vannamei* in zero-water exchange tanks manipulating C/N ratio in feed. Aquaculture 356-357:147–152

91. Xu WJ, Pan LQ (2013) Enhancement of immune response and antioxidant status of *Litopenaeus vannamei* juvenile in biofloc-based culture tanks manipulating high C/N ratio of feed input. Aquaculture 412:117–124

92. Yoder JA, Nielsen ME, Amemiya CT, Litman GW (2002) Zebrafish as an immunological model system. Microbes and Infect 4:1469–1478

93. Zhao P, Huang J, Wang XH, Song XL, Yang C, Zhang XG, Wang GC (2012) The application of bioflocs technology in high-intensive, zero exchange farming systems of *Marsupenaeus japonicus*. Aquaculture 354-355:97–106

# CHAPTER 5

# Seaweed Aquaculture in Norway: Recent Industrial Developments and Future Perspectives

**Pierrick Stévant[1,2] Céline Rebours[1,3]**

[1]*Møreforsking Ålesund AS, PO Box 5075, 6021 Ålesund, Norway*

[2]*Department of Biotechnology, Norwegian University of Science and Technology (NTNU), 7491 Trondheim, Norway*

[3]*Norwegian Institute for Bioeconomy Research (NIBIO), Frederik A. Dahlsvei 20, 1432 Ås, Norway*

## ABSTRACT

The use of cultivated seaweeds as a feedstock for multiple industrial applications has gained increasing interest in the Western World over the past decades. Norway has an extensive coastline and a well-established aquaculture sector offering suitable preconditions for developing large-scale cultivation of seaweed biomass both in monoculture and in Integrated Multi-Trophic Aquaculture (IMTA) systems. Recent efforts from research, industry and public authorities have been committed to develop a Norwegian bio-economy based on cultivated seaweed, focusing on cultivation and processing of the biomass. This review reports on the status of seaweed aquaculture in Norway, supported by production data collected since the delivery of the first commercial cultivation permits at sea in 2014. Although novel product developments are currently limited, future industrial perspectives based on cultivated biomass are being discussed.

Upscaling from experimental cultivation schemes to commercial production requires a thorough assessment of the risks and benefits associated with seaweed aquaculture, as well as the development of a regulative framework adapted to this industry. Issues associated with upscaling the macroalgal production that needs to be addressed includes (i) genetic interactions between cultivated and wild crops, (ii) impacts of seaweed cultivation on surrounding ecosystems, (iii) epiphytes and diseases, (iv) area utilization and (v) threats from climate change. Addressing these issues and adapting production practices will ensure the environmental and economic sustainability of an emerging industry based on cultivated seaweed biomass in Norway.

**Keywords**: Bioeconomy Integrated multitrophic aquaculture (IMTA) Regulation Macroalgae Seaweed Sustainability

## INTRODUCTION

Aquaculture has a great potential for meeting challenges of increasing demand for food. It is the fastest growing food production sector and accounts for almost 50% of aquatic resources globally (FAO 2016a). The world's total aquaculture production in 2014 was in excess of 100 million t, of which 27 million t were aquatic plants, which are mostly seaweeds with a commercial value of 5.6 billion US$ (FAO 2016b). Global seaweed cultivation is largely dominated by Asian countries, i.e. China (producing more than 13 million t), Indonesia, The Philippines, North and South Korea and Japan. In these countries, the biomass is either used for direct human consumption (both fresh and dried), or it is further processed for other food applications (mainly phycocolloids such as alginates, agar and carrageenan). In addition, non-food applications include fertilizers, pharmaceuticals and cosmetics. Aside from having multiple industrial applications, brown, red and green seaweeds have other assets which make them suitable alternatives to terrestrial biomass production—as autotrophic organisms, marine macroalgae utilize dissolved carbon dioxide and nutrients (e.g. dissolved inorganic nitrogen and phosphorus) in combination with light for growth and do not require feed nor fertilizer. They can be cultivated on large scales in coastal areas with little or no demand on fresh water resources in production cycles. Moreover, the fact that macroalgal growth rates exceed those of terrestrial plants (Brinkhuis et al. 1987), have spurred the interest for cultivation of seaweed biomass, illustrated by a doubling of global production during the past decade (FAO 2016b).

The rapid development of fed aquaculture (e.g. finfish or shrimps) is often associated with environmental concern regarding the discharge of nutrients to coastal areas, potentially leading to deterioration of local marine environments (Buschmann et al. 2006). Integrated multitrophic aquaculture (IMTA) systems represent a practical solution for mitigating the negative effects of fish farming wastes by utilising excess nutrients as a valuable resource for marine plant production (Chopin et al. 2008). Seaweeds in such integrated cultivation systems, e.g. in proximity to fish farms, function as extractive components within a cultivation food web. In addition to reducing the environmental impact of intensive fish aquaculture, IMTA systems add value to the investment in finfish aquaculture by increasing the yield of total biomass produced on a single site (Chopin et al. 2001; Neori et al. 2004; Troell et al. 2009; Chopin et al. 2012).

For more than five decades, seaweeds have been the subject of considerable industrial utilization in Norway based on harvesting of natural biomass. Harvested wild seaweed biomass of primarily *Laminaria hyperborea* (commonly known as 'forest kelp', Northern kelp' or 'cuvie') but also *Ascophyllum nodosum* (knotted wrack) amounted to over 154,000 t fresh weight in 2014 (FAO 2016b). Harvesting of wild *L. hyperborea*, using specifically designed trawling equipment, remains somewhat controversial as the removal of and interference with natural habitats (specifically kelp forests) has the potential to affect local biodiversity and ecosystem integrity, and may contribute to coastal erosion (Mork 1996; Christie et al. 1998; Christie et al. 2003; Lorentsen et al. 2010; Smale et al. 2013). Although Norwegian management regimes for seaweed harvesting are considered to be among the best in the world (Smale et al. 2013), regional conflicts persist between the seaweed trawling industry and other coastal zone users (representing for instance coastal fisheries, marine conservation groups and ornithologists). Even with positive developments in knowledge-based integrated coastal zone management with broad stakeholder involvement, and the consideration of a prospective geographic expansion of kelp trawling areas northward, it is not foreseen that future harvesting of wild kelp biomass will be able to meet the growing demands of an industry that is mainly producing alginate from *L. hyperborea*.

Norway's long and complex coastline extending over 100,000 km and characterized by fjords, islands and skerries is highly suited for aquaculture. The cold temperate waters of the Northeast Atlantic are home to more than 400 species of brown, red and green seaweeds (Rueness 1998), many of which have recognized commercial value. Norway is the world's second largest exporter of seafood and has natural advantages which create opportunities to increase seafood production through aquaculture. In 2015, 1.38 million t of salmon and trout were farmed in Norway (Statistisk Sentralbyrå 2016). As salmon and trout production in Norway is predicted to increase by approximately 3 to 5% per year (Olafsen et al. 2012) over the coming decades, measures must be taken to limit the negative effects of excess nutrient discharge, especially environmental eutrophication and resulting ecological degradation and biodiversity loss. In addition, finite global phosphorus resources (Cordell and White 2011) provide a powerful argument for recovering excess nutrients, specifically phosphorus, from all sources including aquaculture and in view of their high value. The benefit of employing IMTA in this context, i.e. recovering organic and inorganic nutrients discharged from fish farms, is emphasized in a study by Wang et al. (2012) which highlights the potential for seaweed production in Norway, based on the estimated total amount of dissolved inorganic nutrients released by salmon farms. In this context, the development of seaweed aquaculture in Norway is in line with the European strategy driving 'Blue Growth' initiatives (European Commission 2012).

## Status of seaweed aquaculture in Norway

Experimental cultivation of kelps (Laminariales) at sea in Norway started around 2005 following successful trials conducted in France, Germany, Ireland and Scotland. Joint efforts of key actors from both research and private sectors primarily focused on developing cultivation technology in order to upscale the biomass production. Current efforts regarding the industrial scale production of seaweeds, mainly sugar kelp (*Saccharina latissima*), are now being directed at developing efficient farming strategies that reduce the need for technical maintenance, along with marine technology, including mechanisation of seedling deployment, biomass harvest and crop handling logistics (MACROSEA, www.sintef.no/projectweb/macrosea/). Efficient cultivation technology along with access to sufficient area at sea is a prerequisite for the sustainable development of an industry based on the utilization of seaweed biomass. Although activities to date have been characterised by research and pilot scale production, the first commercial permits for cultivation of seaweeds were granted in 2014 following the creation of a specific–interim-licensing system for macroalgae by public authorities. The Ministry of Trade, Industry and Fisheries coordinates the processing of seaweed farming applications and considers them according to the Aquaculture Act. The evaluation of applications involves several authorities, including the Directorate of Fisheries, The Norwegian Coastal Administration, the Norwegian Food Safety Authority, The County Governor's Environmental Department and The Norwegian Water Resources and Energy Directorate, each one considering potential conflicts of the application within their area of responsibility. Whether the seaweed farm is in concordance with coastal zone spatial plans, is considered by the concerned municipal authorities, as coastal zone management lies within the municipalities' responsibility.

Since 2014, several permits for sea cultivation of seaweeds, including in IMTA systems, have been granted following the rapid increase in the number of private actors involved in this activity (Table 1). The surface allocated to seaweed cultivation has more than tripled between 2014 and 2016, currently reaching a total of ca. 277 ha along the entire Norwegian coast and corresponding to a production potential of approximately 16,000 t (as estimated from the total maximum production by applicants on each cultivation sites). However, this potential by far exceeds the real production output, given that most companies involved in this activity are at a start-up phase characterized by experimental production. In fact, only 10 of 16 companies holding a cultivation permit were operating in 2015 (Directorate of Fisheries 2016a, 2016b). In Norway as in the rest of Europe, efforts for large-scale cultivation of seaweed biomass have been focussing largely on kelp species, particularly *S. latissima*, because of the species' potential for high biomass yields (Kraan et al. 2000; Handå et al. 2013) and valuable nutritional content (Schiener et al. 2015). The cultivation of *S. latissima* represented 96% (49 t wet weight) of the total production output, while cultivation of winged kelp (*Alaria esculenta*) represented the remaining 4% (2 t wet weight). Although licences for cultivation of a wide range of species (e.g. *Palmaria palmata, Ulva* spp., *Laminaria digitata, Porphyra* spp.) were granted, none of these are currently cultivated at sea. Given the data reported by the Directorate of Fisheries (2016b) for the year 2015, the retail value of *S. latissima* and *A. esculenta* was 399 US\$ $t^{-1}$ and 1099 US\$ $t^{-1}$, respectively. Although no information is available regarding the retail markets of the produced biomass, *A. esculenta* is often sold in dried or fresh form as a high value food ingredient or sea vegetable whereas *S. latissima* has a broader range of market outputs, e.g. dried for human consumption or as meal in animal feed, which may explain the difference in market value. A higher value for *A. esculenta* may also simply result from a lower biomass availability on the market. As a comparison, retail value of the global seaweed production output is estimated to 206 US\$ $t^{-1}$ in 2014 based on figures mentioned earlier (FAO 2016b).

**Table 1** Statistics from seaweed cultivation in Norway between 2014 and 2016. Source: Directorate of Fisheries (2016a, 2016b). Production data for 2016 have not yet been registered (as per January 2017)

| | **2014** | **2015** | **2016** |
|---|---|---|---|
| Total seaweed cultivation sites at sea | 8 | 24 | 36 |
| Total surface area for seaweed cultivation at sea (ha) | 76 | 194 | 277 |
| Seaweed production output (t wet weight) | na | 51 | na |
| Commercial value (US$)[a] | na | 21,760 | na |
| Number of companies involved in seaweed cultivation | 6 | 16 | 20 |

[a]Values initially expressed in Norwegian krone (NOK). The exchange rate of 1NOK = 0.1221 US$ was used

Despite a long coastline and an increase in the number of permits at sea, seaweed cultivation sites are still scarce and not evenly distributed along the Norwegian coast (Fig. 1). A very large part of the allocated surface area (87%) shared by 28 cultivation sites is concentrated on ten municipalities, localized in few counties, namely Hordaland and Sogn and Fjordane in the west, Sør- and Nord-Trøndelag in the mid-part and Nordland in the northern part of the country. Hence, there is a large potential to establish seaweed cultivation in other areas, especially the Møre and Romsdal county in the west, as well as Troms and Finnmark counties in the north, both of which are characterized by extensive areas and the presence of salmon aquaculture (see possibilities for integration of seaweed and salmon farming above). At present, only a few of the existing cultivation sites are in the vicinity of fish farms, hence characterized as IMTA. Most of the so-called IMTA sites were developed as 'add-on' to a pre-existing fish farm and to date, few have been fully designed for the purpose of fish and seaweed co-culture. However, many sites of seaweed monoculture are located in areas of intense salmon production (Directorate of Fisheries' interactive map: http://kart.fiskeridir.no/default.aspx?gui=1&lang=2) thus, are likely to directly benefit from nutrient releases originating from fish farming activities.

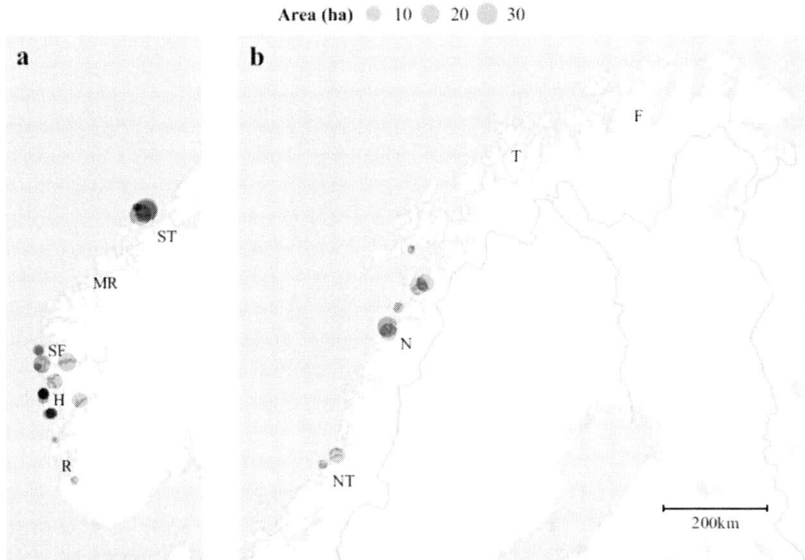


**Figure. 1** Surface area allocated to seaweed cultivation sites along the coast of southern Norway (**a**) and Northern Norway (**b**) as of January 2017. Counties are listed from south to north: Rogaland (*R*), Hordaland (*H*), Sogn and Fjordane (*SF*), Møre and Romsdal (*MR*), Sør-Trøndelag (*ST*), Nord-Trøndelag (*NT*), Nordland (*N*), Troms (*T*) and Finnmark (*F*). Darker circles indicate several cultivation sites close to each other. Data from the Directorate of Fisheries (2016a)

The cultivation cycle of the kelp species is based on their life cycle (see Bartsch et al. 2008 for more details on the life cycle of the Laminariales). Sporelings are grown from microscopic stages i.e. zoospores or gametophytes (derived from zoospores) under controlled conditions and directly seeded onto appropriate substrata (e.g. twine, nets, ribbons, sheets) which are later on deployed at sea for further growth under natural conditions. Although the production of spores in natural kelp populations is seasonal and species-specific (Bartsch et al. 2008), sorus (reproductive structure) formation can be induced artificially by, for example, controlling the light regimes provided to adult sporophytes. In this way, viable sporelings can be produced at all seasons (Forbord et al. 2012). However, biofouling of the biomass from epiphytic organisms limits the cultivation period at sea from autumn to spring (see Risk assessment 1—epiphytes and diseases).

## Product applications—industrial perspectives

Seaweed cultivation has been subject to increasing interest in Europe over the past decades, and activities have benefited from recent trends towards a bio-economy based on natural, and especially marine, biological resources. In Norway, research efforts to develop seaweed cultivation and processing technology were driven initially by the interest for producing bioenergy from seaweed's carbohydrates. However, the production of biofuel and biogas on commercial scales is impaired by the difficulties associated with producing large biomass volumes, since neither cultivation technology nor access to aquaculture sites at sea have yet been sufficiently developed for large-scale production. In addition, the market value for seaweed biomass as biofuel is low compared with other product types (Skjermo et al. 2014). Seaweed biomass is a rich source of a wide range of bioactive products of potentially high market value (Holdt and Kraan 2011). Relevant applications include (i) human food, (ii) domestic animal and fish feed products, (iii) fertilizer, (iv) prebiotics, (v) cosmetics, (vi) bioactive peptides and (vii) pharmaceuticals and nutraceuticals, and these are predicted to play an important role in the emerging Norwegian bio-economy based on seaweed cultivation (Skjermo et al. 2014).

In the past, mainly brown algae have been utilised in Norway. Primary applications have been alginates (used as food additives, health products and bio-medicinal ingredients, to name just a few) or agricultural products (i.e. animal feed supplement and soil enhancers/ fertilizers). In most cases, the seaweeds have been used for the functional properties of their polysaccharides or their mineral content. Although the seaweed chemical composition is variable depending on species, collection time and habitat, the nutritional value of several species is well characterized. Most species share a low lipid content on the one hand, while being rich in polysaccharides (both structural and storage, Dawczynski et al. 2007) and minerals on the other (Mabeau and Fleurence 1993; Rupérez 2002; MacArtain et al. 2007; Pereira 2011). Other nutritional properties, such as a relatively high protein content and high quality profiles of amino-acids (Fleurence 2004; Mæhre et al. 2014), lipids (Sánchez-Machado et al. 2004) and minerals (Cofrades et al. 2008; López-López et al. 2009) are highly relevant towards food and feed applications. Positive effects of including seaweeds in feed meals have been reported for both, marine and terrestrial organisms (Soler-Vila et al. 2009; Kandasamy et al. 2012; Evans and Critchley 2014; Wilke et al. 2015), and several feed products based on powdered dried seaweed are commercially available.

Seaweeds have traditionally been a major element of the diet in Asian countries and are used extensively as food in China, Japan, Korea and the Philippines where they are recognized for their nutritional value as well as for rich and unique flavours. In Europe, traditions of eating seaweeds have been limited to coastal communities, and over the past centuries, few regions have had an explicit culture for utilising seaweeds as a significant component of the people's diet. The use of seaweeds as food in Norway goes back to the Vikings who likely brought dried dulse on long expeditions at sea for its virtues in protecting against scurvy, following its high content of vitamins (Mouritsen et al. 2013). In recent decades, Asian dishes such as sushi, have enjoyed increasing popularity and contributed to introducing seaweeds to European palates. In Scandinavia, the renewal of the Nordic cuisine initiated by high-end restaurants and based on locally available natural ingredients has aroused interest for including seaweeds in the gastronomy (Mouritsen et al. 2012). Previous studies have revealed the potential of seaweed species native to the Atlantic coast of Europe (e.g. the red algae *P. palmata* and the kelps *S. latissima* and *A. esculenta*) to be used as food ingredients in a wide range of foodstuffs both as vegetables and flavour enhancers (Mouritsen et al. 2012; Chapman et al. 2015). In Western Europe, i.e. France, Ireland, Spain and UK, a number of small and medium-sized enterprises are using wild or cultivated seaweeds in food products. This new type of businesses has developed in recent years following the rising demand from European consumers (NETALGAE 2012). In France, edible seaweeds e.g. *S. latissima*, *P. palmata*, *Undaria pinnatifida* and *Himanthalia elongata* are sold in fresh or dried form, as "sea vegetables" or condiments, or are incorporated in various product types (e.g. spreads, dressings, soups) as identified by Le Bras et al. (2015a, 2015b) in a comprehensive study among a wide range of retailers across the country. The retail price of these products, mainly sold in organic or health food shops, varies greatly with product category, seaweed species and the amount of seaweed used, but the identified market value is generally high i.e. over 100 € $kg^{-1}$ (ca. 105 US$ $kg^{-1}$) on average for dried and flaked seaweeds (Le Bras et al. 2015b) . Although there is no data available regarding seaweed consumption in Norway, it is truly marginal, and few Nordic seaweed food products are currently available. The development of such 'niche' food products, based on market and consumer insight, can contribute to creating value from Norwegian produced seaweed biomass on a relative short-term. Thoroughly investigating the food safety of commercialized edible seaweeds (see Risk assessment 2—product related) is, however, essential.

Moreover, seaweeds contain a large variety of phytochemical constituents that can be used in the prevention and treatment of health diseases by various mechanisms (Holdt and Kraan 2011; Brown et al. 2014). Bioactive compounds of interest include polysaccharides (e.g. alginates, fucoidans), polyphenols and pigments, to name just a few. Besides being used mainly in the food industry as thickening or emulsifying agent, alginates are regarded as a source of dietary fibre with beneficial effects both on colonic and cardiovascular health (Brownlee et al. 2005). Alginate polymers can be artificially modified to achieve tailored properties suited in various medical applications including cell micro-encapsulation, tissue engineering and regeneration, and dental impressions (Ruvinov and Cohen 2016). The company leading the alginate production in Norway (FMC BioPolymer) manufactures ultrapure alginate from wild *Laminaria hyperborea* for research and high-end commercial applications. The same firm recently obtained the authorization to commercialize high purity extracts of fucoidan as a dietary ingredient. Fucoidan represents a group of polysaccharides from brown seaweeds, mainly composed of sulfated fucose with documented bioactivity (Berteau and Mulloy 2003). Other compounds which are currently not commercially exploited in Norway may offer industrial opportunities in the future. Carotenoid pigments from brown algae, namely fucoxanthin, are recognized for their antioxidant activity as well as positive health effects against obesity and type-2 diabetes (Maeda 2015), and can be found in dietary supplements and nutraceuticals. Phycoerythrin, a pigment from red algae, is used extensively in fluorescent applications in clinical and immunological analysis as well as natural dye in the food and cosmetic industry (Sekar and Chandramohan 2008). Extraction procedures of this pigment using common species along the Atlantic coast of Europe, e.g. *P. palmata,* have been suggested (Galland-Irmouli et al. 2000; Dumay et al. 2013) together with methodology for industrial upscaling (Denis et al. 2009). Polyphenols are secondary metabolites which are abundant in brown seaweeds such as *A. nodosum*, *Fucus* spp., and widely recognized for their antioxidant properties (Wang et al. 2009; Tierney et al. 2013). Polyphenol-rich extracts from brown algae have inhibitory effects against lipid oxidation in fish (Wang et al. 2010) and fish oil (Xiaojun et al. 1996). Hence, polyphenols may provide a safer alternative to synthetic commercial antioxidants such as butylated hydroxytoluene (BHT), butylated hydroxyanisole (BHA) and tert-butyl hydroxyquinone (TBHQ) which are currently used in the food and feed industry, and criticized for their potential toxicity and carcinogenic effects (Gupta and Abu-Ghannam 2011). The commercial exploitation of such high value-added products will require relatively low biomass volumes in comparison to bulk products (e.g. biofuels) and is predicted to generate economic value from cultivated macroalgal biomass (Holdt and Kraan 2011; Skjermo et al. 2014). However, the extraction of these substances is technology demanding thus, associated with high investment costs. More knowledge is also required regarding the market potential of seaweed bioactive compounds to identify commercial opportunities.

In contrast to current processes in the Norwegian alginate industry, in which only a fraction of the harvested biomass is utilized and by-products are discarded, recent trends towards bio-economy favour complete utilization of produced biomass, recycling of nutrients in complete cycles and hence, reducing negative environmental impacts. Integrated processes maximizing biomass utilization by recovering a stream of products from seaweed biomass have been studied with promising results (Hou et al. 2015; Baghel et al. 2016). An increasing number of initiatives involving both research and industrial partners are aiming to develop biorefinery processes of marine macroalgae in Norway, in order to optimize feedstock utilization towards a wide range of products. However, the high water content of seaweeds (70 to 90%, Jensen 1993) represents a challenge for conserving and transporting large amounts of biomass (as in industrial production) from harvesting to processing sites. Seaweeds are characterized by a rapid microbial decomposition once harvested (Enríquez et al. 1993), and thus require appropriate preservation methods to maintain biomass quality and ensure product safety. Primary processes such as drying can effectively stabilize the biomass but require technology and may be difficult to implement close to harvesting sites for large biomass volumes. Drying also affects the chemical content (Chan et al. 1997; Gupta et al. 2011) with consequences for the product's nutritional value as well as extraction yields of bioactive compounds during further processing. Post-harvest treatments that will contribute to maintaining biomass quality include seawater storage (Paull and Chen 2008), cold storage (Liot et al. 1993), silage (Herrmann et al. 2015) and freezing (Choi et al. 2012). A recent study by Broch et al. (2016) stressed that large-scale seaweed cultivation sets standards for efficient logistic strategies and suggests the use of energy surplus from existing industrial facilities along the coast for processing of large biomass volumes. Another national research project investigates the possibilities for using energy surplus from a waste incinerator located on the west coast of Norway for seaweed processing, including its technical, economic and logistic potential (PROMAC, www.promac.no/).

## Risk assessment 1—managerial tools associated with seaweed cultivation and IMTA

Despite a growing interest for seaweed cultivation in the western world and a wide range of industrial applications globally, the tools and methods for establishing a seaweed industry in Norway—whether they are technical, economic, environmental, social or regulatory—still need to be developed or adapted from Asian models to fit European frameworks. This will help aquaculture enterprises and policy makers gain a better understanding of the risks and benefits associated with seaweed cultivation in both monoculture and IMTA systems, and provide conditions suitable for the development of a sustainable bioeconomy based on macroalgal biomass.

By cultivating extractive species (e.g. seaweeds) near fish farms, IMTA systems increase the biomass production of an aquaculture site associated with a certain area license. Whereas seaweed cultivation integrated with fish farms is currently being viewed as an experimental 'add-on' to existing aquaculture sites of salmonid fishes, the potential of seaweed production for improving environmental and economic performance of fish aquaculture and consequently increasing biomass production, is significant. The regulatory frameworks largely differ among species (i.e. fish, seaweed, shellfish) and are non-existent for species combinations (particularly concerning the integration of finfish, shellfish and seaweed production). The current interim regulatory process in use includes seaweed cultivations and IMTA systems, but needs to be specifically adapted for the various seaweed farming scenarios. An adaptation of the legal framework needs to achieve the transition from fed-monoculture (fish or shellfish) to poly- and integrated aquaculture (Alexander et al. 2015). In addition, the adaptation of existing regulations and policies for IMTA will demand an assessment of the risks and benefits of developing new aquaculture systems or using new aquaculture species.

The "Strategy for an Environmentally Sustainable Norwegian Aquaculture Industry" (Ministry of Fisheries and Coastal affairs 2009) provides a guideline for current Norwegian aquaculture policies, focusing on five key problem areas, i.e. (i) genetic interaction and escape, (ii) pollution and emissions, (iii) disease, (iv) area utilization and (v) feed and feed resources. Whereas these policies were designed for fish aquaculture, they are equally relevant for seaweed cultivation, and are discussed below in this context.

## Genetic interactions

There are several infamous cases of accidental introductions of non-native seaweed species to new coastal environments, e.g. *Undaria pinnatifida* and *Sargassum muticum* to Western Europe and *Caulerpa taxifolia* to the Mediterranean Sea, which have resulted in massive population explosions of the non-native species and consequently significant ecosystem alterations. Because spreading of non-native species can have large environmental and economic impacts (Schaffelke et al. 2006), their intentional use in mariculture must be prevented.

Though seaweeds have been exploited in Asia for centuries, selective breeding of crops is a relatively recent practice, carried out for the purpose of increasing yields and/or value by optimizing seaweed growth, improving the biomass' general quality or increasing the content of one or several desirable compounds, as well as improving performance towards pathogens and competitors. Commercial varieties of *Saccharina japonica* (Zhang et al. 2007) as well as eucheumoid species like *Kappaphycus alvarezii* and *Eucheuma denticulatum* (Ask and Azanza 2002) illustrate successful efforts towards seaweed crop improvement. However, crop selection and the use of the selected varieties on a large scale may lead to gene swamping, where genetic material from cultivated crops is introduced to wild populations (Snow et al. 2010).

The genetic structure and degrees of isolation between populations of species that are relevant for aquaculture in Norway are still poorly understood. Recent studies conducted on kelp species showed increasing genetic isolation correlated with distance among populations. Water current patterns were identified as the primary force responsible for spreading of dispersal stages, resulting in gene flow among populations (Alberto et al. 2011; Robuchon et al. 2014). High extent of isolation may lead to phenotypic differences among populations, which can be affected by selection. Accordingly, Futsæter and Rueness (1985) reported morphological differences between sugar kelps from Northern and Southern Norway, as well as differences in growth under certain conditions. In this regard, the use and spreading of crop material along the coast of Norway must be regulated to limit the effects of "crop-to-wild" gene flows. Fredriksen and Sjøtun (2015) recommended using local ecotypes for crop breeding and restricting the produced crops within fjord systems and coastal areas defined by the eco-region based map catalogue from the Norwegian Environment Agency (Miljødirektoratet, https://kartkatalog.miljodirektoratet.no/Dataset/Details/501). Sterilization technologies for cultivated seaweeds have also been suggested as a measure to mitigate gene swamping (Loureiro et al. 2015), but examples of supposedly sterile aquaculture organisms adapting over time and eventually being able to reproduce after all (Jouaux et al. 2010) advise caution in such cases.

## Impact on ecosystems

As organisms which take up dissolved nutrients from surrounding seawater, seaweeds function as extractive components in culture systems. Thus, they assimilate dissolved inorganic nitrogen and phosphorus, which, in many coastal ecosystems, originate from terrestrial runoffs or from fish farming activities. In this way, seaweed cultivations can mitigate potential eutrophication of coastal environments (Kim et al. 2015). In addition to nutrients, seaweeds assimilate dissolved carbon in form of $CO_2$ as part of the photosynthetic processes. Pedersen et al. (2012) estimated the annual production of a Norwegian kelp forest (*L. hyperborea*) as between 14 and 25 t ha$^{-1}$ dry weight, corresponding to 5.6 to 10 t ha$^{-1}$ of carbon-bound biomass growth, ranking these kelp forests among the most productive vegetation systems on earth. Chung et al. (2013) estimated the potential of the brown alga *Ecklonia cava* cultivated on ropes to extract approximately 10 t $CO_2$ $ha^{-1}$ $year^{-1}$. Large-scale seaweed cultivation is justifiably regarded as a potential temporary $CO_2$ sink and may contribute substantially to the management of human impacts on atmospheric and oceanic carbon cycle interactions.

Seaweed cultivations create a temporary habitat for invertebrates and fishes, increasing habitat complexity, possibly functioning as artificial reefs and contributing to ecological interactions within nearby ecosystems. Although such an effect has not been systematically investigated, seaweed cultivation fields have been observed to house associated biota, i.e. mobile drifting or swimming organisms and attached fauna and flora (Stamenic 2015). These observations indicate an overall positive effect on local biodiversity, but environmental monitoring over time will provide evidence. Similarly, benthic environments associated with seaweed cultivations are bound to be altered, and the effects need to be documented systematically. An ongoing research project in collaboration with Norwegian fish farmers (TARELAKS, contact Céline Rebours) is addressing ecosystem effects of co-cultivation of seaweeds and salmonids at pilot IMTA-sites. In this context, the effects of seaweeds on water quality (e.g. current speed, dissolved nutrient content, chlorophyll *a* content) and fish health parameters, as well as on biological diversity near the seaweed cultivation sites are being assessed. Preliminary observations obtained from a Norwegian farm in a European project (IDREEM, www.idreem.eu/) indicate high abundances of juvenile fishes associated with macroalgae cultivations (Hughes et al. 2016). Especially lumpfish, *Cyclopterus lumpus,* appear to be common in seaweed farms (Stamenic 2015). This is a potentially valuable association, as lump suckers are being used as cleaning fish in salmon farms to remove parasites (salmon lice) from the farmed fish. Salmon lice infestations represent one of the major challenges facing the salmon aquaculture industry in Norway, and developing biological controls is a priority.

Nutrient transfers form seaweeds and kelp forests to other ecosystems are key processes in nature, increasing food availability to otherwise nutrient-poor environments (Vetter 1995; Krumhansl and Scheibling 2012). However, besides acting as a temporary nutrient sink and having potentially positive effects on surrounding pelagic environments, large-scale seaweed cultivations may also cause high material deposits on the seafloor which could have adverse consequences on benthic ecosystems and biodiversity through decomposition processes and associated oxygen depletion (Viaroli et al. 2008). In addition, physical shading of both pelagic and benthic ecosystem by large-scale seaweed cultivation may affect primary production in the case of shallow inshore areas, although these effects have not yet been studied.

## Epiphytes and diseases

Problems with pathogens and epiphytes of seaweeds may arise in large-scale and high-density cultivations. In Norway, the presence of epiphytic invertebrates, especially bryozoans (e.g. *Membranipora membranacea* and *Electra pilosa*), but also blue mussels (*Mytilus edulis*), hydroids and fouling macroalgae has been reported for sugar kelp cultivated at sea during summer months (Forbord et al. 2012; Handå et al. 2013; Lüning and Mortensen 2015; Førde et al. 2016). In worst case scenarios, such fouling may lead to extensive blade damage and loss of biomass or considerable quality deterioration. Whereas, the diversity of associated organisms in natural kelps beds carries important ecological functions in coastal areas (Christie et al. 2009), biofouling by epiphytes is a major constraint for the year-round cultivation of seaweed biomass, forcing producers to harvest in May–June, hence limiting the year-round seaweed cultivation at sea (Forbord et al. 2012; Peteiro and Freire 2013; Førde et al. 2016). Outbreaks of encrusting colonies of the bryozoan *M. membranacea* are the result of settlement of cyphonaut larvae, which are present in the water column of coastal area throughout the year (Ryland 1965; Saier and Chapman 2004) with a peak of abundance during the summer (Ryland 1965; Førde et al. 2016). Larval recruitment and the extent of biofouling which may vary among cultivation sites, are linked to zooplankton population dynamics, influenced by a wide range of both biotic and abiotic factors including the temperature history and wave action (Saunders and Metaxas 2008). Seawater temperature seems to play an important role since the onset of fouling is somewhat delayed at higher latitudes (Fieler and Matsson 2015). However, the same time delay is observed for growth of kelps, with growth seasons extended into late summer in Northern Norway, compared to more southern locations. Hence, the growing season for cultivated sugar kelp is limited from October to May in Southern Norway, and possibly to August in Northern Norway (Fieler and Matsson 2015), and the crops must be harvested before the onset of fouling. In the case of IMTA, the peak nutrient release from fish farms occurs during the summer months, due to high metabolic activity and growth of the fish (Wang et al. 2012). Developing crops resistant to fouling or alternative seaweed species as summer crops, is key to increasing the production efficiency in IMTA systems.

Seaweed production in IMTA systems will also alter the surrounding water quality and the fish-rearing environment in general, e.g. by changing the oxygen content resulting from photosynthesis and respiration, as well as hydrographic regimes to the fish cages. Large-scale seaweed cultivation may also influence fish health indirectly either by acting as a physical and/or biological barrier for pathogens, or in the opposite case, as a reservoir for fish disease agents.

Whereas, evidence from IMTA sites elsewhere indicates many benefits, both on conceptual, practical, environmental and economic levels (Barrington et al. 2008; Chopin et al. 2012), potential risks associated with integrated fish and seaweed production need to be evaluated and assessed in the context of specific geography, production systems (regarding both fish and seaweed), scales of production and associated technology. Such issues are also addressed in on-going research projects in Norway.

## Area utilization

A special commission mandated by the Norwegian Ministry of Trade, Industry and Fisheries stressed that a sufficient number of locations must be assigned to aquaculture with focus on efficient production and minimal environmental impact (Gullestad 2011). An economically viable seaweed aquaculture sector requires large areas for cultivation which may lead to conflicts with other users of the coastal zone. Integrated management tools applied across municipal boundaries are therefore necessary to avoid conflicts between seaweed cultivation on the one hand and other uses, including fisheries, fairways, nature protection, renewable energy production, kelp trawling areas as well as oil and gas installations on the other (Broch et al. 2016).

Although the current demand for seaweed biomass can be met by cultivation sites in the Norwegian coastal zone, model simulations including environmental variables critical to seaweed growth (e.g. light intensity, nutrient, temperature and current speed) showed better and more stable conditions off the continental shelf than closer to the coast (Broch et al. 2016). The modelling study by Broch et al. (2016) was carried out for the coast of Møre and Romsdal county with an overall conclusion of a prospective upscaling of seaweed biomass production placed offshore.

However, offshore working environments entail technical challenges including mooring of the cultivation structures as well as logistic aspects from deployment of seedlings at sea to harvesting and processing of the biomass. Knowledge and competence transfer from existing maritime industries (e.g. oil and gas) is pointed out by the authors as a key factor for the success of the future offshore seaweed production.

## Threats from climate change

Climate change is likely to impose threats to marine ecosystems and consequently aquaculture. Rising sea temperatures could eventually lead to significant changes both in terms of aquaculture species, optimal range for the production and localization patterns. Higher sea temperatures will involve a relocation of the boundaries of the distribution of organisms in the ocean, including seaweeds, and the general northern displacement of farmed organisms (NOU 2010). Higher summer sea surface temperatures may be problematic for farmed species that are adapted for life in cold water. As a result, the productivity of aquaculture species can be affected, and Southern Norway may be less suitable for species such as salmon, with socio-economic impacts (Hermansen and Heen 2012). Such climatic changes may eventually entail that northern areas will be better suited for mariculture than southern regions (Bergh et al. 2007). A study by Callaway et al. (2012) assessing the consequences of climate change on aquaculture species in the UK and Ireland, suggested that elevated sea temperatures and changes in hydrodynamic regimes will impact macroalgal cultivation; however, impacts will likely vary according to species and geographic locations. It is therefore particularly important to investigate the framework conditions for seaweed monoculture and IMTA along the entire Norwegian coast.

## Risk assessment 2—product related

In contrast to the many features of seaweeds beneficial for human health and nutrition, seaweeds may also accumulate undesired compounds such as heavy metals (lead, mercury, cadmium, arsenic) as well as minerals and trace elements which may be toxic above a certain limit (iodine, manganese, zinc) (Almela et al. 2006; Rose et al. 2007; Besada et al. 2009).

Variations in content and concentrations of undesirable and toxic compounds are dependent on species (Phaneuf et al. 1999), seasons and locations (Ar Gall et al. 2004), and in general, levels of toxic compounds are often related to the quality of the surrounding environment of harvested or cultivated seaweeds. However, few studies about negative consequences of consuming seaweeds have been carried out, but among the best documented are effects of high iodine content in some brown algae (Lüning and Mortensen 2015; Desideri et al. 2016). The potentially mitigating effects of primary processing of edible seaweeds on their content of toxins as well as antinutrients (i.e. compounds reducing digestibility and bio-availability of health-advancing ingredients) and on their microbial stability require further investigation.

Current research efforts in collaboration with the Norwegian Food Safety Authority (Duincker et al. 2016) are addressing these relevant questions in order to establish a regulatory framework for the use of macroalgae as human food and animal feed. Such a regulatory framework is a prerequisite for further industrial development of seaweeds as raw material for human consumption and animal production.

## CONCLUSIONS

The seaweed industry is a growing economic sector in Norway. Both research communities, public and private stakeholders are collaborating to further the sustainable development of the Norwegian bio-economy based on cultivated seaweed biomass. Although cultivation technology has come a long way, it still requires optimization to upscale the biomass production and reduce the need for technical operators as well as maintenance. Current needs from seaweed producers are directed towards energy efficient processing standards that will maintain or increase product quality, ensure consumer safety and maximize biomass utilization. Integrated models using industrial surplus heat as energy source for seaweed processing are suggested, and their feasibility are currently under study.

Despite a large body of scientific literature reporting on the quality of macroalgal phytochemical constituents, commercial outputs from Norwegian cultivated seaweeds are currently limited. High-value products to be used in food, cosmetics as well as medical and pharmaceutical applications, are predicted to play an important role in creating value from Norwegian cultivated seaweed biomass. Nevertheless, expert knowledge regarding markets and prices for different seaweed products is required in order to establish market strategies. Successful product developments along with efficient logistic solutions throughout the entire value chain will build competitiveness of the Norwegian seaweed sector. The future production of macroalgae on large-scale and at relatively low cost will open perspectives for using the biomass as feedstock for bulk products (high volume and low price) e.g. bio-fuels and -chemical, and animal feed.

Upscaling seaweed cultivation requires specific solutions regarding the technical, spatial and practical integration of the production and will be specific for each production site. Although the current demand for seaweed raw material can be met by cultivating biomass within the coastal zone, large-scale production is foreseen to occur outside the continental shelf of Norway, due to more stable growth conditions and avoidance of conflicts arising with other users of the coastal zone. The Norwegian regulations regarding seaweed cultivation (including monoculture and

IMTA) are not yet fully developed. Hence, the cooperation between public authorities, the industry and research institutions is needed in order to develop a knowledge-based legal framework for a sustainable industry based on cultivated seaweed biomass in Norway.

# REFERENCES


1. Alberto F, Raimondi PT, Reed DC et al (2011) Isolation by oceanographic distance explains genetic structure for *Macrocystis pyrifera* in the Santa Barbara Channel. Mol Ecol 20(12):2543–2554
2. Alexander KA, Potts TP, Freeman S et al (2015) The implications of aquaculture policy and regulation for the development of integrated multi-trophic aquaculture in Europe. Aquaculture 443:16–23
3. Almela C, Clemente MJ, Velez D et al (2006) Total arsenic, inorganic arsenic, lead and cadmium contents in edible seaweed sold in Spain. Food Chem Toxicol 44(11):1901–1908
4. Ar Gall E, Küpper FC, Kloareg B (2004) A survey of iodine content in *Laminaria digitata*. Bot Mar 47(1):30
5. Ask EI, Azanza RV (2002) Advances in cultivation technology of commercial eucheumatoid species: a review with suggestions for future research. Aquaculture 206(3–4):257–277
6. Baghel RS, Trivedi N, Reddy CRK (2016) A simple process for recovery of a stream of products from marine macroalgal biomass. Bioresource Technol 203:160–165
7. Barrington K, Ridler N, Chopin T et al (2008) Social aspects of the sustainability of integrated multi-trophic aquaculture. Aquacult Int 18(2):201–211
8. Bartsch I, Wiencke C, Bischof K et al (2008) The genus *Laminaria* Sensu Lato: recent insights and developments. Eur J Phycol 43(1):1–86
9. Bergh Ø, Asplin L, Boxaspen K et al. (2007). Climate change - its consequences for Norwegian aquaculture. Marine Research, Topics 2-2007. Institute of Marine Research. https://brage.bibsys.no/xmlui/bitstream/handle/11250/116175/No.2_2007.pdf?sequence=2
10. Berteau O, Mulloy B (2003) Sulfated fucans, fresh perspectives structures, functions and biological properties of sulfated fucans and an overview of enzymes active toward this class of polysaccharide. Glycobiology 13(6):29–40
11. Besada V, Andrade JM, Schultze F et al (2009) Heavy metals in edible seaweeds commercialised for human consumption. J Marine Syst 75(1–2):305–313
12. Brinkhuis BH, Levine HG, Schlenk CG et al. (1987). *Laminaria* cultivation in the far east and North America. In: Bird and Benson (ed) Seaweed cultivation for renewable resources. Development in aquaculture and fisheries science. Elsevier, New York: 107–146.
13. Broch OJ, Skjermo J and Handå A (2016). Potensialet for storskala dyrking av makroalger i Møre og Romsdal. SINTEF Fisheries and Aquaculture - Report A27869. https://www.sintef.no/publikasjoner/publikasjon/?pubid=SINTEF+A27869
14. Brown EM, Allsopp PJ, Magee PJ et al (2014) Seaweed and human health. Nutr Rev 72(3):205–216
15. Brownlee IA, Allen A, Pearson JP et al (2005) Alginate as a source of dietary fiber. Crit Rev Food Sci Nutr 45(6):497–510
16. Buschmann A, Riquelme V, Hernandez-Gonzalez M et al (2006) A review of the impacts of salmonid farming on marine coastal ecosystems in the Southeast Pacific. ICES J Mar Sci 63(7):1338–1345

17. Callaway R, Shinn AP, Grenfell SE et al (2012) Review of climate change impacts on marine aquaculture in the UK and Ireland. Aquat Conserv 22(3):389–421
18. Chan JCC, Cheung PCK, Ang PO (1997) Comparative studies on the effect of three drying methods on the nutritional composition of seaweed *Sargassum hemiphyllum* (turn.) C. Ag.†. J Agr Food Chem 45(8):3056–3059
19. Chapman AS, Stévant P, Emblem Larssen W (2015) Food or fad? Challenges and opportunities for including seaweeds in a Nordic diet. Bot Mar 58(6):423–433
20. Choi J-S, Lee B-B, An S et al (2012) Simple freezing and thawing protocol for long-term storage of harvested fresh *Undaria pinnatifida*. Fisheries Sci 78(5):1117–1123
21. Chopin T, Buschmann A, Halling C et al (2001) Integrating seaweeds into marine aquaculture systems: a key toward sustainability. J Appl Phycol 37:975–986
22. Chopin T, Robinson SMC, Troell M et al. (2008). Multitrophic Integration for Sustainable Marine Aquaculture. In: Jorgensen SE and Fath BD (ed) The Encyclopedia of Ecology, Ecological Engineering. Elsevier, Oxford: 2463–2475.
23. Chopin T, Cooper JA, Cross S et al (2012) Open-water integrated multi-trophic aquaculture: environmental biomitigation and economic diversification of fed aquaculture by extractive aquaculture. Rev Aquacult 4(4):209–220
24. Christie H, Fredriksen S, Rinde E (1998) Regrowth of kelp and colonization of epiphyte and fauna community after kelp trawling at the coast of Norway. Hydrobiologia 375(376):49–58
25. Christie H, Jørgensen NM, Norderhaug KM et al (2003) Species distribution and habitat exploitation of fauna associated with kelp (*Laminaria hyperborea*) along the Norwegian coast. J Mar Biol Assoc UK 83(4181):1–13
26. Christie H, Norderhaug KM, Fredriksen S (2009) Macrophytes as habitat for fauna. Mar Ecol-Prog Ser 396:221–233
27. Chung IK, Oak JH, Lee JA et al (2013) Installing kelp forests/seaweed beds for mitigation and adaptation against global warming: Korean project overview. ICES J Mar Sci 70(5):1038–1044
28. Cofrades S, López-López I, Solas MT et al (2008) Influence of different types and proportions of added edible seaweeds on characteristics of low-salt gel/emulsion meat systems. Meat Sci 79(4):767–776
29. Cordell D, White S (2011) Peak phosphorus: clarifying the key issues of a vigorous debate about long-term phosphorus security. Sustainability 3(10):2027
30. Dawczynski C, Schubert R, Jahreis G (2007) Amino acids, fatty acids, and dietary fibre in edible seaweed products. Food Chem 103(3):891–899
31. Denis C, Massé A, Fleurence J et al (2009) Concentration and pre-purification with ultrafiltration of a R-phycoerythrin solution extracted from macro-algae *Grateloupia turuturu*: process definition and up-scaling. Sep Purif Technol 69(1):37–42
32. Desideri D, Cantaluppi C, Ceccotto F et al (2016) Essential and toxic elements in seaweeds for human consumption. J Toxicol and Env Heal, Part A 79(3):112–122
33. Directorate of Fisheries (2016a). Akvakulturregisteret. Cited 11 Jan 2017. http://www.fiskeridir.no/Akvakultur/Registre-og-skjema/Akvakulturregisteret
34. Directorate of Fisheries (2016b). Akvakulturstatistikk (tidsserier)-Alger. Cited 02 Nov 2016. http://www.fiskeridir.no/Akvakultur/Statistikk-akvakultur/ Akvakultur statistikk-tidsserier/Alger
35. Duincker A, Sunde Roiha I, Amlund H et al. (2016). Potential risks posed by macroalgae for application as feed and food - a Norwegian perspective. National Institute of Nutrition and Seafood Research (NIFES). http://www. Mattilsynet .no/mat_og_vann/produksjon_av_mat/fisk_og_sjomat/rapport_makroalger_2016_nifes.23097/binary/Rapport%20makroalger%202016%20Nifes
36. Dumay J, Clément N, Morançais M et al (2013) Optimization of hydrolysis conditions of *Palmaria palmata* to enhance R-phycoerythrin extraction. Bioresource Technol 131(0):21–27
37. Enríquez S, Duarte CM, Sand-Jensen K (1993) Patterns in decomposition rates among photosynthetic organisms: the importance of detritus C:N:P content. Oecologia 94(4):457–471

38. European Commission, (2012). Communication from the commission to the european parliament, the council, the european economic and social committee and the committee of the regions: Blue Growth opportunities for marine and maritime sustainable growth, COM/2012/0494 final. http://eur-lex.europa.eu/legal-content/EN/ALL/?uri=CELEX:52012DC0494
39. Evans FD, Critchley AT (2014) Seaweeds for animal production use. J Appl Phycol 26:891–899
40. FAO, (2016a). The state of world fisheries and aquaculture: Food and Agriculture Organization of the United Nations (FAO), Fisheries and Aquaculture Department. Rome, 2012. http://www.fao.org/3/a-i5555e.pdf
41. FAO (2016b). FAO Yearbook - Fishery and Aquaculture Statistics Summary tables. Cited 01 Nov 2016. ftp://ftp.fao.org/FI/STAT/summary/default.htm
42. Fieler R and Matsson S, (2015). FoU aktiviteter ved Akvaplan-Niva. Results presented at the SIG (Special Interest Group) seaweed workshop - Cultivation technology. Industrial Biotech Network Norway, 28th of May, Trondheim, Norway. http://www.indbiotech.no/sites/default/files/Akvaplan_Niva.pdf
43. Fleurence J (2004). Seaweed proteins. In: Yada RY (ed) Proteins in food processing. Woodhead publishing, Cambridge: 197–213
44. Forbord S, Skjermo J, Arff J et al (2012) Development of *Saccharina latissima* (Phaeophyceae) kelp hatcheries with year-round production of zoospores and juvenile sporophytes on culture ropes for kelp aquaculture. J Appl Phycol 24(3):393–399
45. Førde H, Forbord S, Handå A et al (2016) Development of bryozoan fouling on cultivated kelp (*Saccharina latissima*) in Norway. J Appl Phycol 28(2):1225–1234
46. Fredriksen S and Sjøtun K (2015). Risikovurdering ved utsetting av ikke–stedegen tare. Miljødirektoratet (Norwegian Environment Agency). Rapport M-299. http://www.miljodirektoratet.no/Documents/publikasjoner/M299/M299.pdf
47. Futsæter G and Rueness J, (1985). Metode for kontroll av livssyklus, krysningsforsøk og masseproduksjon av sporofytter i kultur av *Laminaria saccharina* (L) Lamour. University of Oslo. 23 pp.
48. Galland-Irmouli AV, Pons L, Luçon M et al (2000) One-step purification of R-phycoerythrin from the red macroalga *Palmaria palmata* using preparative polyacrylamide gel electrophoresis. J Chromatogr B 739:117–123
49. Gullestad P, (2011). Effektiv og bærekraftig arealbruk i havbruksnæringen - areal til begjær. Rapport Fra Et Ekspertutvalg Oppnevnt Av Fiskeri- Og Kystdepartementet, Oslo, 4. feb 2011. https://www.regjeringen.no/globalassets/upload/fkd/ vedlegg/rapporter/2011/effektiv_og_baerekraftig_arealbruk_i_havbruksnaeringen.pdf
50. Gupta S, Abu-Ghannam N (2011) Recent developments in the application of seaweeds or seaweed extracts as a means for enhancing the safety and quality attributes of foods. Innov Food Sci Emerg 12(4):600–609
51. Gupta S, Cox S, Abu-Ghannam N (2011) Effect of different drying temperatures on the moisture and phytochemical constituents of edible Irish brown seaweed. Food Sci Technol-LEB 44(5):1266–1272
52. Handå A, Forbord S, Wang X et al (2013) Seasonal- and depth-dependent growth of cultivated kelp (*Saccharina latissima*) in close proximity to salmon (*Salmo salar*) aquaculture in Norway. Aquaculture 414-415:191–201
53. Hermansen Ø, Heen K (2012) Norwegian salmonid farming and global warming: socioeconomic impacts. Aquacult Econ Manage 16(3):202–221
54. Herrmann C, FitzGerald J, O'Shea R et al (2015) Ensiling of seaweed for a seaweed biofuel industry. Bioresource Technol 196:301–313
55. Holdt SL, Kraan S (2011) Bioactive compounds in seaweed: functional food applications and legislation. J Appl Phycol 23(3):543–597
56. Hou X, Hansen JH, Bjerre A-B (2015) Integrated bioethanol and protein production from brown seaweed *Laminaria digitata*. Bioresource Technol 197:310–317
57. Hughes AD, Corner RA, Cocchi M et al. (2016). Beyond fish monoculture: Developing Integrated Multi-trophic Aquaculture in Europe. IDREEM (Increasing Resource

Efficiency in European Mariculture) project report. Project funded by the European Union's Seventh Framework Programme for research, technological development and demonstration under grant agreement no. 308571. http: //www. idreem.eu/cms/wp-content/uploads/2016/10/IDREEM_FINALREPORT_PRINT_710_web_2.pdf
58. Jensen A (1993). Present and future needs for algae and algal products. In: Chapman ARO, Brown MT and Lahaye M (ed) Fourteenth International Seaweed Symposium. Springer Netherlands. 85: 15–23
59. Jouaux A, Heude-Berthelin C, Sourdaine P et al (2010) Gametogenic stages in triploid oysters *Crassostrea gigas*: irregular locking of gonial proliferation and subsequent reproductive effort. J Exp Mar Biol Ecol 395(1–2):162–170
60. Kandasamy S, Khan W, Evans F et al (2012) Tasco(R): a product of *Ascophyllum nodosum*enhances immune response of *Caenorhabditis elegans* against *Pseudomonas aeruginosa*infection. Mar Drugs 10(1):84–105
61. Kim JK, Kraemer GP, Yarish C (2015) Use of sugar kelp aquaculture in Long Island sound and the Bronx River estuary for nutrient extraction. Mar Ecol-Prog Ser 531:155–166
62. Kraan S, Verges Tramullas A, Guiry M (2000) The edible brown seaweed *Alaria esculenta*(Pheophyceae, Laminariales) hybridization growth and genetic comparisons of six Irish populations. J Appl Phycol 12:577–583
63. Krumhansl KA, Scheibling RE (2012) Production and fate of kelp detritus. Mar Ecol-Prog Ser 467:281–302
64. Le Bras Q, Lesueur M, Lucas S et al. (2015a). Etude du marché français des algues alimentaires. Catalogue et analyse des produits existants. Programme IDEALG Phase 2: Les publications du Pôle halieutique AGROCAMPUS OUEST, vol. 37. 41 pp. http://halieutique.agrocampus-ouest.fr/pdf/4661.pdf
65. Le Bras Q, Lesueur M, Lucas S et al. (2015b). Etude du marché français des algues alimentaires. Panorama de la distribution en magasins. Programme IDEALG Phase 2: Les publications du Pôle halieutique AGROCAMPUS OUEST, vol. 36. 42 pp. http://halieutique.agrocampus-ouest.fr/pdf/4662.pdf
66. Liot F, Colin A, Mabeau S (1993) Microbiology and storage life of fresh edible seaweeds. J Appl Phycol 5(2):243–247
67. López-López I, Cofrades S, Ruiz-Capillas C et al (2009) Design and nutritional properties of potential functional frankfurters based on lipid formulation, added seaweed and low salt content. Meat Sci 83(2):255–262
68. Lorentsen S-H, Sjøtun K, Grémillet D (2010) Multi-trophic consequences of kelp harvest. Biol Conserv 143(9):2054–2062
69. Loureiro R, Gachon CMM, Rebours C (2015) Seaweed cultivation: potential and challenges of crop domestication at an unprecedented pace. New Phytol 206(2):489–492
70. Lüning K, Mortensen LM (2015) European aquaculture of sugar kelp (*Saccharina latissima*) for food industries: iodine content and epiphytic animals as major problems. Bot Mar 58(6):449–455
71. Mabeau S, Fleurence J (1993) Seaweed in food products: biochemical and nutritional aspects. Trends Food Sci Tech 4:103–107
72. MacArtain P, Gill CIR, Brooks M et al (2007) Nutritional value of edible seaweeds. Nutr Rev 65(12):535–543
73. Maeda H (2015) Nutraceutical effects of Fucoxanthin for obesity and diabetes therapy: a review. J Oleo Sci 64(2):125–132
74. Mæhre HK, Malde MK, Eilertsen KE et al (2014) Characterization of protein, lipid and mineral contents in common Norwegian seaweeds and evaluation of their potential as food and feed. J Sci Food Agric 94(15):3281–3290
75. Ministry of Fisheries and Coastal affairs, (2009). Strategy for an Environmentally Sustainable Norwegian Aquaculture Industry. https://www. regjeringen.no /globalassets/upload/fkd/vedlegg/diverse/2009/strategy-for-an-sustainable-aquaculture.pdf

76. Mork M (1996). Wave Attenuation due to Bottom Vegetation. In: Grue J, Gjevik B and Weber JE (ed) Waves and Nonlinear Processes in Hydrodynamics. Springer Netherlands, Dordrecht: 371–382
77. Mouritsen OG, Williams L, Bjerregaard R et al (2012) Seaweeds for umami flavour in the new Nordic cuisine. Flavour 1(1):4
78. Mouritsen OG, Dawczynski C, Duelund L et al (2013) On the human consumption of the red seaweed dulse (*Palmaria palmata* (L.) Weber & Mohr). J Appl Phycol 25:1777–1791
79. Neori A, Chopin T, Troell M et al (2004) Integrated aquaculture: rationale, evolution and state of the art emphasizing seaweed biofiltration in modern mariculture. Aquaculture 231(1–4):361–391
80. NETALGAE, (2012). Overview of the seaweed industry by country: Ireland, France, Norway, Portugal, Spain, United Kingdom. A summary of the initial findings of the Netalgae project regarding the current status of the seaweed industries in the participant countries. Netalgae project financed by the Atlantic Area Operational Programme of the European Union. http://www. netalgae.eu/ uploadedfiles /posters_final_UK_1.pdf
81. NOU, (2010). Adapting to a changing climate. Norway's vulnerability and the need to adapt to the impacts of climate change. Recommendation by a committee appointed by Royal Decree of 5 December 2008, submitted to the Ministry of the Environment on 15 November 2010. Official Norwegian Reports, Noregs Offentlege Utgreiingar (NOU) https://www.regjeringen.no/contentassets/00f70698362f4f889cbe30c75bc a4a48/pdfs/nou201020100010000en_pdfs.pdf
82. Olafsen T, Winther U, Olsen Y et al. (2012). Value created from productive oceans in 2050. Report of a working group appointed by the Royal Norwegian Society of Sciences (DKNVS) and the Norwegian Academy of Technological Sciences (NTVA). https://www.sintef.no/contentassets/f025260af6b8435394eced5e03939e 11/value-created-from-productive-oceans-in-2050.pdf
83. Paull RE, Chen NJ (2008) Postharvest handling and storage of the edible red seaweed *Gracilaria*. Postharvest Biol Techn 48(2):302–308
84. Pedersen MF, Nejrup LB, Fredriksen S et al (2012) Effects of wave exposure on population structure, demography, biomass and productivity of the kelp *Laminaria hyperborea*. Mar Ecol-Prog Ser 451:45–60
85. Pereira L (2011). A review of the nutrient composition of selected edible seaweeds. In: Pomin VH (ed) Seaweed: Ecology, Nutrient Composition and Medicinal Uses. Nova Science Publishers, Inc.: 15–50.
86. Peteiro C, Freire Ó (2013) Epiphytism on blades of the edible kelps *Undaria pinnatifida* and *Saccharina latissima* farmed under different abiotic conditions. J World Aquacult Soc 44(5):706–715
87. Phaneuf D, Côté I, Dumas P et al (1999) Evaluation of the contamination of marine algae (seaweed) from the St. Lawrence River and likely to be consumed by humans. Environ Res 80(2):S175–S182
88. Robuchon M, Le Gall L, Mauger S et al (2014) Contrasting genetic diversity patterns in two sister kelp species co-distributed along the coast of Brittany, France. Mol Ecol 23(11):2669–2685
89. Rose M, Lewis J, Langford N et al (2007) Arsenic in seaweed—forms, concentration and dietary exposure. Food Chem Toxicol 45(7):1263–1267
90. Rueness J (1998). Alger i farger. En felthåndbok om kystensmakroalger. Almater Forlag, Oslo.
91. Rupérez P (2002) Mineral content of edible marine seaweeds. Food Chem 79:23–26
92. Ruvinov E, Cohen S (2016) Alginate biomaterial for the treatment of myocardial infarction: progress, translational strategies, and clinical outlook: from ocean algae to patient bedside. Adv Drug Deliver Rev 96:54–76
93. Ryland JS (1965). Polyzoa (Bryozoa). Order Cheilostomata, cyphonautes larvae. In: Fiches d'identification du zooplancton, vol. 107. Conseil International pour l'Exploration de la Mer, ICES Copenhagen, p 6.

94. Saier B, Chapman AS (2004) Crusts of the alien bryozoan *Membranipora membranacea* can negatively impact spore output from native kelps (*Laminaria longicruris*). Bot Mar 47:265
95. Sánchez-Machado DI, López-Cervantes J, López-Hernández J et al (2004) Fatty acids, total lipid, protein and ash contents of processed edible seaweeds. Food Chem 85(3):439–444
96. Saunders M, Metaxas A (2008) High recruitment of the introduced bryozoan *Membranipora membranacea* is associated with kelp bed defoliation in Nova Scotia, Canada. Mar Ecol-Prog Ser 369:139–151
97. Schaffelke B, Smith JE, Hewitt CL (2006) Introduced Macroalgae—a growing concern. J Appl Phycol 18(3):529–541
98. Schiener P, Black KD, Stanley MS (2015) The seasonal variation in the chemical composition of the kelp species *Laminaria digitata*, *Laminaria hyperborea*, *Saccharina latissima* and *Alaria esculenta*. J Appl Phycol 27(1):363–373
99. Sekar S, Chandramohan M (2008) Phycobiliproteins as a commodity: trends in applied research, patents and commercialization. J Appl Phycol 20(2):113–136
100. Statistisk Sentralbyrå (2016). Akvakultur, 2013, endelige tall. Cited 16 Dec 2016. http://www.ssb.no/jord-skog-jakt-og-fiskeri/statistikker/fiskeoppdrett
101. Skjermo J, Aasen IM, Arff J et al. (2014). A new Norwegian bioeconomy based on cultivation and processing of seaweeds: opportunities and R&D needs. SINTEF Fisheries and Aquaculture - Report A25981. https://www.sintef.no /contentassets/04aa6cd738b245979b334715163c0506/seaweed-bioeconomy-2014-revised-2.pdf
102. Smale DA, Burrows MT, Moore P et al (2013) Threats and knowledge gaps for ecosystem services provided by kelp forests: a Northeast Atlantic perspective. Ecol Evol 3(11):4016–4038
103. Snow AA, Culley TM, Campbell LG et al (2010) Long-term persistence of crop alleles in weedy populations of wild radish (*Raphanus raphanistrum*). New Phytol 186(2):537–548
104. Soler-Vila A, Coughlan S, Guiry M et al (2009) The red alga *Porphyra dioica* as a fish-feed ingredient for rainbow trout effects on growth. Feed efficiency and carcass composition. J Appl Phycol 21:617–624
105. Stamenic J (2015). From monoculture to Integrated Multi-Trophic Aquaculture (IMTA): mapping and valuation of ecosystem services after the integration of seaweed cultivation on a Norwegian salmon farm. Thesis submitted in partial fulfillment of the requirements for the Degree of Msc in Environmental Management for Business, Cranfield University.
106. Tierney MS, Smyth TJ, Rai DK et al (2013) Enrichment of polyphenol contents and antioxidant activities of Irish brown macroalgae using food-friendly techniques based on polarity and molecular size. Food Chem 139(1–4):753–761
107. Troell M, Joyce A, Chopin T et al (2009) Ecological engineering in aquaculture—potential for integrated multi-trophic aquaculture (IMTA) in marine offshore systems. Aquaculture 297(1–4):1–9
108. Vetter EW (1995) Detritus-based patches of high secondary production in the nearshore benthos. Mar Ecol-Prog Ser 120(1):251–262
109. Viaroli P, Bartoli M, Giordani G et al (2008) Community shifts, alternative stable states, biogeochemical controls and feedbacks in eutrophic coastal lagoons: a brief overview. Aquat Conserv 18(S1):S105–S117
110. Wang T, Jónsdóttir R, Ólafsdóttir G (2009) Total phenolic compounds, radical scavenging and metal chelation of extracts from Icelandic seaweeds. Food Chem 116(1):240–248
111. Wang T, Jónsdóttir R, Kristinsson HG et al (2010) Inhibition of haemoglobin-mediated lipid oxidation in washed cod muscle and cod protein isolates by *Fucus vesiculosus* extract and fractions. Food Chem 123(2):321–330

112. Wang X, Olsen LM, Reitan KI et al (2012) Discharge of nutrient wastes from salmon farms: environmental effects, and potential for integrated multi-trophic aquaculture. Aquac Environ Interact 2(3):267–283
113. Wilke T, Faulkner S, Murphy L et al (2015) Seaweed enrichment of feed supplied to farm-raised Atlantic salmon (*Salmo salar*) is associated with higher total fatty acid and LC n-3 PUFA concentrations in fish flesh. Eur J Lipid Sci Tech 117(6):767–772
114. Xiaojun Y, Xiancui L, Chengsu Z et al (1996) Prevention of fish oil rancidity by phlorotannins from *Sargassum kjellmanianum*. J Appl Phycol 8(3):201–203
115. Zhang Q-S, Tang X-X, Cong Y-Z et al (2007) Breeding of an elite *Laminaria* variety 90-1 through inter-specific gametophyte crossing. J Appl Phycol 19(4):303–311

# CHAPTER 6

# People Matter in Animal Disease Surveillance: Challenges and Opportunities for The Aquaculture Sector

**Cecile Brugere[a,*], Dennis Mark Onuigbo[b], Kenton Ll. Morgan[c]**

[a]*Soulfish Research and Consultancy/Stockholm Environment Institute, c/o Environment Building, Wentworth Way, University of York, Heslington, York YO10 5NG, United Kingdom*

[b]*School of Postgraduate Studies, Department of Agricultural Economics, Institut Pertanian Bogor, C/o Gedung Sekolah Pascasarjana Lt.1, Jl. Raya Darmaga Kampus IPB Darmaga, Bogor, 16680, West Java, Indonesia*

[c]*Institute of Ageing and Chronic Diseases and School of Veterinary Science, Leahurst Campus, Neston, Wirral, CH64 7TE & Apex Building, West Derby St., Liverpool, United Kingdom*

## ABSTRACT

Recurring epidemics and the emergence of new aquatic diseases are increasingly threatening the growth of aquaculture. The fast pace of aquaculture development and on-going global environmental, social and economic change are challenging epidemiologists in their capacity to surveil and control the spread of diseases and avert losses to farmers and impacts on their livelihoods and environment. By placing farmers as the starting point of disease surveillance, we contend that farmer-based syndromic disease surveillance holds potential to overcome the current limitations of conventional disease surveillance, and demonstrate its relevance for aquaculture, particularly in resource constrained environments. Drawing on the literature on aquaculture, epidemiology, farmers' decision-making, technology adoption in animal health management and participation in animal disease surveillance, we highlight the complex interplay of behavioural (economic and social) factors behind farmers' reporting of disease. To this we add insights from institutional economics to analyse the constraints and dilemmas disease surveillance poses to institutions. Whilst information technologies are playing a significant supporting role in disease surveillance, our central argument is that if data collection for epidemiological monitoring is about technology, surveillance itself is about people. Stakeholder involvement and perception of surveillance benefits, value of epidemiological data collected, farmers' knowledge, motivation and trust and institutions' functioning are key considerations in the design of successful syndromic disease surveillance programmes. These human dimensions constitute important knowledge gaps in animal disease surveillance in general, and in particular in aquaculture. Interdisciplinary collaboration in disease surveillance is essential. It is crucial in an environment where diseases are emerging and spreading in increasingly complex, interconnected and dynamic social-ecological systems and is the key to unlocking the numerous benefits of farmer-based syndromic aquatic disease surveillance.

"When major health problems arise someone must make decisions... Good surveillance does not ensure the making of the right decisions but it reduced the chances of wrong ones"

[Alexander Langmuir (1963: 191)]

# 1. INTRODUCTION

The impact of liberalisation in international trade of animals and animal products (Oidtmann et al., 2013a; Rogers et al., 2011; Thiermann, 2005), accidental translocation of non-native species (both host and vector) and mass global movement of people on the risk of spread of infectious agents (Fèvre et al., 2006; Tatem et al., 2006a,b) has changed the paradigm of infectious disease control. Driven by the Sanitary and Phytosanitary (SPS) Agreement (WTO, 1995), which raised infectious disease to one of the few remaining barriers to free trade, and the International Health Regulations (IHR) (WHO, 2005), which placed the onus of epidemic or emergent human and zoonotic disease detection, assessment, reporting and response on the country of origin, epidemiological intelligence, risk assessment and certification have moved centre stage.

In parallel with these developments, the increased demand for affordable food, housing and power have resulted in husbandry, climatic and ecological changes which have altered the dynamics of the relationship between individual animals and infectious agents, and between people and animals (domesticated and wild). The result has been the emergence of new diseases or the appearance of recognised diseases in epidemic form.

Nowhere are these changes more marked than in aquaculture. As the crop of farmed aquatic species is set to exceed the wild catch for the first time in history (reviewed by Ottinger et al., 2016), domestication of aquatic animals and our transition from hunter-gatherers to farmers of this blue planet is taking place in real time (Hedgecock, 2012; Teletchea, 2015). The global change in species distribution (Doupé and Lympery, 2000; Flegel, 2006), population density (Frazer et al., 2012; Krkosek, 2010), host-parasite-environment interaction (Ashander et al., 2012), fish and shellfish consumption (reviewed by Ottinger et al., 2016) and demand for fish as ornaments, companions or pets (Whittington and Chong, 2007) is proceeding at a pace unmatched by scientific investigation or epidemiological capacity (e.g. Jones et al., 2015). The last 20 years have witnessed three recognised aquatic pandemics (Kamilya and Baruah, 2014; Lightner, 2011) and the current human burden of zoonotic aquatic parasites is estimated at 100 million (Keiser and Utzinger, 2009; reviewed by Lima dos Santos and Howgate, 2011).

Epidemiological intelligence transforms the strategy of infectious disease control from relying on control by biosecurity, at the point of invasion into a new population or country, to targeting changes in the distribution, frequency and determinants of disease in the source population. It enables barriers and contingencies to be enhanced by early warning systems, risk estimates and preparedness. In defining epidemiological intelligence, Langmuir (1963)considered it synonymous with disease surveillance. He is credited with the first definition of modern disease surveillance as the "*Continued watchfulness over the distribution and trends of incidence through the systematic collection, consolidation and evaluation of morbidity and mortality reports and other relevant data. Intrinsic in this concept is the regular dissemination of the basic data and interpretations to all who have contributed and all others who need to know. The concept however does not encompass direct responsibility for control activities.*" (Langmuir, 1963). This definition still resonates today (OIE, 2014).

Conceptually simple, intuitive and logical, the implementation of effective and efficient disease surveillance systems has proven problematic. Disease surveillance challenges the epidemiological principles that the lower limits and uncertainties around disease detection should be within known bounds and that the reporting of disease frequency should be unbiased and representative of the population. Furthermore, epidemiological rigor requires that the probability of a disease being reported given that animals are truly diseased (sensitivity) and the probability of not reporting given that the animals are truly healthy (specificity) are known and are within acceptable limits. High specificity reduces the number of false positive reports and high sensitivity reduces the number of false negative reports.

In essence, the epidemiological challenges to an effective surveillance system relate to rapid detection, representative reporting and accurate diagnosis. These challenges have spawned a gamut of research which addresses these issues and focusses on methods for: combining multisource, non-representative data (Gubbins, 2008; Hay et al., 2013; Martin et al., 2011), capture-recapture techniques (Vergne et al., 2015), using high risk populations as a cost-effective proxy (Cameron, 2012; Diserens et al., 2013; Marques et al., 2015; Oidtmann et al., 2013b; Stärk et al., 2006), the application of molecular techniques to pen-side, pool side or point of care detection and diagnosis (reviewed by Teles and Fonseca, 2015), the use of mobile phone technologies to report data (Brinkel et al., 2014; Robertson et al., 2010a), natural language processing (Gerbier et al., 2011) and content analysis (Butler et al., 2007; Lam et al., 2007) to read and translate the richness of text based data and the harnessing of statistical methods (reviewed by Robertson et al., 2010a) and artificial intelligence to detect anomalies and classify, collate and transform these data to information (Dórea et al., 2015; Fanaee and Gama, 2014, Hepworth et al., 2012; Wong et al., 2005). None of these however address the need to understand and account for the human factors that underpin the implementation of an effective surveillance system.

The objective of this paper is to argue for farmer-based, syndromic surveillance as a way of overcoming current limitations of conventional disease surveillance and to demonstrate its relevance for aquaculture, particularly in resource limited environments. In doing so we will focus on the human aspects of disease detection and reporting. We suggest that the need to understand and harness these parameters is an essential component of effective disease surveillance systems and extends beyond the boundaries of epidemiology and veterinary science.

We start by reviewing the epidemiological challenges that disease surveillance raises, notably in terms of sampling, diagnosis, effectiveness and cost-benefits. We focus on establishing freedom from disease, estimating disease prevalence and perhaps most importantly for aquaculture, the detection of emerging diseases or shifts in the pattern of recognised disease. Certification of disease freedom is a passport to international trade; disease prevalence allows prioritisation and resource allocation for endemic diseases and emergence may require a global response.

We then turn to exploring the human dimensions that influence the effectiveness of syndromic disease surveillance. In particular, we examine the motivations behind farmer reporting of diseases to national authorities and competent authority reporting to international organisations such as the Office International des Epizooties (OIE). Knowledge gaps and challenges of farmer-based syndromic disease surveillance for aquaculture are discussed, and the challenges and opportunities for the sector are highlighted.

# 2. EPIDEMIOLOGICAL CHALLENGES OF DISEASE SURVEILLANCE

## 2.1. Sampling and confidence

One of the epidemiological challenges of establishing disease freedom is to decide at which level of prevalence the absence of detection means freedom from disease and the confidence with which this conclusion is reached, i.e. does the absence of detectable disease presence really mean the presence of disease absence? This involves consideration of statistical probability but also of the host population, its structure and the nature of the infectious agent.

Farmed populations are hierarchical: individual animals live in ponds, farms may have more than one pond and there may be more than one farm on a common water source or catchment. This lack of independence challenges one of the basic statistical tenets. Furthermore, ponds and farms may be multispecies or even multi-phyla with differences in susceptibility to infection or expression of disease. The numbers of animals per pond may vary from a few to thousands. Aquatic animals inhabit murky ponds and may be invisible or even, as in the early stages of shrimp grow-out, absent because of early mortality. This poses practical challenges to random sampling, aggravated by the stressful and potentially life-threatening removal and handling of animals outside their aquatic environment.

Transmission rates also influence sampling. Infectious diseases which have a high transmission rate may affect the majority of the population. In a fully susceptible population in which there are no births (e.g. a grow-out population) and no innate or adaptive immunity, the epidemic will die out; if infection is fatal, so will the hosts. In the presence of an immune response, or variably expressed innate immunity phenotypes, and a dynamic (regenerating) population, endemic equilibrium may be reached and the prevalence of disease or infection, although variable, will be maintained at a much lower level (Keeling and Rohani, 2007; Van den Driessche and Watmough, 2002).

The importance of prevalence and the confidence of detection is that it influences the sample size and cost of studies aimed at both establishing disease freedom and estimating disease prevalence. For example, in a country with a large aquaculture industry with tens of thousands of farms, a random sample of about 150 fish would be required to be 95% confident of detecting at least one diseased or infected farm if the minimum prevalence was thought to be 2%. This assumes that all ponds and all the fish on the farm are infected! If a minimum of 10% of ponds per farm were thought to be infected and the minimum prevalence of infected fish in each pond was considered to be 25%, numbers would escalate. If the number of ponds per farm was between 1 and 10 with a median of five and each pond containing 100 fish, then every pond per farm and 10 fish per pond would need to be included, totalling 7500 samples (150 × 5 × 10). This assumes a perfect diagnostic test – which of course they never are! A diagnostic test with a specificity of 99% and sensitivity of 95% would increase this to almost 70,000 (540 × 5 × 18). This is clearly impractical and has resulted in the development of techniques which target farms with a higher disease risk (Cameron, 2012; Diserens et al., 2013; Marques et al., 2015; Oidtmann et al., 2013b; Stärk et al., 2006). Risk-based surveillance is now required by the EU (EU Council directive 2006/88/EC on animal health, EU, 2006). In support of this, online calculators have been developed which incorporate risk-based surveillance and cost into their sample size estimates (Sergeant, 2016).

## 2.2. Diagnostic tests

The laboratory-based identification of necessary infectious agents or pathognomonic lesions, visible either macroscopically or microscopically, has until recently underpinned the diagnosis of infectious diseases. Diagnostic tests influence surveillance in a number of ways. Sometimes they are simply not available, not good enough or inappropriate for disease surveillance. Many diagnostic tests for aquatic diseases rely on detecting the infectious agent, which may only be present in an individual host for a matter of days. In contrast, many tests used for detecting infectious agents in terrestrial animals rely on historical markers of exposure such as serum antibodies. These can last for years. The use of antibody as a biomarker is inappropriate in crustaceans as they do not mount an adaptive immune response. Diagnostic tests may also be unacceptable to farmers because they involve sacrificing apparently healthy individual animals. Validated diagnostic tests for aquatic diseased are listed and updated by the OIE (2015).

Diagnostic tests influence the cost of surveillance through the unit cost of testing and by influencing the sample size, as illustrated in the previous section. This cost escalation has been addressed by decreasing the sophistication of laboratory equipment required, focusing on general markers of infection (Andre et al., 2004), increasing the range of infectious agents or diseases detected in each test and developing tests which can be used pond-side or at the "point of surveillance" (Teles and Fonseca, 2015).

### 2.2.1. "Point of care" (POC), "pond-side" and "point of surveillance" tests

Technological developments and the demand for rapid, high throughput, diagnosis in resource limited settings have driven the development of a cluster of diagnostic tests known synonymously as "rapid diagnostic (RDT)", "bed-side" ("pond-side" and "pen-side" in veterinary parlance), "near-patient", "field tests", "point of surveillance" or most commonly "point of care" (POC) tests (reviewed by Adams and Thompson, 2011; Teles and Fonseca, 2015). These aspire to the ASSURED criteria: A-Affordable, S-Sensitive, S-Specific, U-User-friendly (simple to perform in a few steps with minimal training), R-Robust and rapid (results available in < 30 min), E-Equipment-free, D-Deliverable to those who need the test (Kettler et al., 2004). Although collectively defined as "point of care", the precise meaning of this term is unclear. Pai et al. (2012) classified "point of care" into 5 hierarchical levels according to their deployment to different actors and institutions involved in diagnosis of human disease in resource-limited settings. These were: home, community health worker, clinic or health post, peripheral laboratory or hospital (Pai et al., 2012).

At the home level, dipstick or lateral flow devices (LFD), also called lateral flow assays (LFA) or lateral flow immunochromatographic assays (LFIA), are rapid, robust and reliable enough for use by untrained individuals. Pregnancy diagnosis kits, commercially available for 40 years are the classic example. Lateral flow refers to the detection, by specific antibody, of an infectious agent by its lateral diffusion through a porous substrate and the appearance of two coloured spots or lines. Although relatively inexpensive, the cost of these tests (< 1€) may be a limiting factor in the resource-constrained environments where much of the world's aquaculture occurs. Currently, the reaction takes place in a plastic cassette or frame. Future use of disposable paper assays may change this. Lateral flow devices have been reported for White Spot Syndrome Virus (WSSV) of shrimp (Sithigorngul et al., 2006), Infectious Salmon Anaemia Virus (ISAV) (Adams and Thompson, 2010), and Cyprinid Herpesvirus 3 (CyHV-3) (Vrancken et al., 2013).

Although LFDs for infectious disease diagnosis have the potential to be used by farmers, community animal health workers or in animal health posts, evidence from human medicine suggests that this does not happen; most are used in established laboratories, hospitals, or small stand-alone laboratories (Moore, 2013; Pai et al., 2012). Furthermore, the use of these tests by farmers has raised a number of concerns, particularly in relation to notifiable diseases where mandatory state reporting to OIE is required. Identification and culling of non-zoonotic diseases may occur without notification and if this takes place on a farm whose simulated position in a contact network indicates that the risk of disease spread is low (Jonkers et al., 2010), it may never be reported.

In addition to detection by specific antibody, infectious agents may be detected using nucleic acid based tests. These amplify and detect nucleotide sequences unique to specific infectious agents. In laboratories, Polymerase Chain Reaction (PCR) assays are used for DNA and Reverse Transcriptase PCR (RT-PCR) for RNA. These require two sophisticated pieces of equipment: one exposes the reaction mix to a fixed number of alternating cycles of high and low temperatures (thermal cycler) to enable amplification of the target sequence, and the other uses the migration of the product in a gel exposed to an electrical charge and to detect the product (electrophoresis). Traditionally, each PCR test detects a single specific organism. Multiplex assays reduce the cost by allowing detection of multiple species of organisms. These have been developed for some shrimp diseases (Xie et al., 2008) but they still require specialist laboratory equipment.

Nucleic acid tests have been revolutionised by the development of Loop Mediated Isothermal Amplification (LAMP) assays (Notomi et al., 2000). These take place at a fixed temperature and the product can be detected visually either by turbidity or using a dye. LAMP assays have been developed for WSSV (Kono et al., 2004), Koi Herpesvirus (Soliman and El-Matbouli, 2005) and *Perkinsus*spp. (Feng et al., 2013). LAMP assays have also been adapted for multiple infectious agents (Zhou et al., 2014). Use of these tests in aquaculture has recently been reported (Caipang et al., 2015). Although not suitable for most farmer or community animal health workers, these assays offer the potential for use in peripheral field laboratories.

The decentralisation of diagnostic tests promises to change the focus of information and with it the relationship between stakeholders involved in its generation, transmission and use. Although a number of issues concerning the use of these tests need to resolved (reviewed by Teles and Fonseca, 2015) before they are used as a management aid by individual farmers, they will not contribute to an understanding of the disease in the population in the absence of a formally-established surveillance system.

### 2.2.2. Emerging diseases and syndromic surveillance

Emerging diseases raise specific issues with regard to detection and reporting as by their nature there are no diagnostic tests or formal reporting systems. Thus, "it is not yet clear how best to build surveillance systems for unknown pathogens" (Halliday et al., 2012: 2877). Even the cheapest, ubiquitous pond-side diagnostic tests can only detect known diseases. Tests may also focus on those diseases which are considered important enough to warrant the financial investment needed to develop the assays.

One approach to the detection of emerging diseases is syndromic surveillance. It has a number of interpretations but involves the "use of health-related information that may be indicative of a probability of change in the health of a population that merits further research or enables a timely impact assessment and action requirement" (Rodríguez-Prieto et al., 2015: 6). In veterinary medicine, it is derived from the signs of disease that can be detected by the human senses. In human medicine, the physical or emotional symptoms reported by people are added. Syndromic surveillance is sometimes extended to include purchasing non-prescription medication, partial or complete carcase condemnations or submissions to laboratories. This interpretation is not used here.

Although syndromic surveillance is increasingly recognised as the most cost-effective approach to the detection of new and emerging diseases, detractors highlight the occurrence of infection in the absence of clinical signs. Here it is important to reiterate the difference between infection and disease. Disease is defined as the presence of clinical signs, whether these be as obvious as increased mortality or as subtle as decreased growth rate or fertility. If an infectious agent is a necessary cause of disease, then it will also be present around the time of clinical disease. If the agent is present in the absence of disease, then this is more accurately defined as infection (or subclinical infection) rather than disease. Infection cannot be detected until diagnostic tests are developed to identify the agent associated with disease. These diagnostic tests follow rather than precede disease detection. The process by which subclinical infection can be detected is: 1. identification of clinical signs; 2. identification of the agent; 3. development of laboratory diagnostic tests; 4. detection of subclinical disease. Only the first step of this process is syndromic surveillance. Syndromic surveillance by definition can only be used to detect clinical disease, not infection. However it might be argued that unless infection is associated with some physiological abnormality and is not proscribed as notifiable, it is irrelevant.

In contrast to "causative agent based surveillance" which, for cost and logistical reasons, can only be used on a limited proportion of the population, syndromic surveillance can be used repeatedly and regularly on a much larger proportion of the population. In addition, where the clinical signs of disease are pathognomonic, farmer diagnosis may be as good as laboratory tests (Morgan et al., 2014).

# 3. THE HUMAN DIMENSIONS OF SYNDROMIC DISEASE SURVEILLANCE

## 3.1. Detection of disease

The detection of overt clinical signs in animals requires no special training. Indeed there may be advantages to allowing untrained farmers to do this. They are not be constrained by the bias of prior disease knowledge, as many veterinarians or community animal health workers are. This offers the best chance of detecting new diseases.

In terrestrial animals, mortality is obvious, coughing and wheezing can be heard, diarrhoea can be seen and smelled, lumps, bumps and other skin abnormalities can be seen or felt, feeding or other behavioural change may be observed and growth measured. The regular appearance of one of these signs either alone, or in consistent combinations, constitutes a syndrome. 'Sign (or symptom) surveillance' would in fact be a better name. The importance and power of these sensory or organoleptic observations are reflected in the names of animal diseases (Table 1).

**Table 1.** Terrestrial and aquatic diseases named according to their clinical signs (physical description).

| Livestock[a] | Fish, shellfish, algae |
|---|---|
| Foot and mouth disease (artiodactyla) | White Spot Disease Syndrome (*Penaeus* shrimp) |
| Bluetongue (cattle/sheep) | Yellowhead disease (shrimp) |
| Lumpy skin disease (cattle) | White tail disease (*Macrobrachium* shrimp) |
| Foul in the foot (cattle) | Early Mortality Syndrome (shrimp) |
| Pink eye (cattle) | Epizootic Ulcerative Syndrome (freshwater fish) |
| Lumpy jaw (cattle) | Sleeping disease syndrome (trout) |
| Vomiting and wasting disease (pigs) | Whirling disease (fish) |
| Greasy pig disease | Orange sickness (mussels) |
| Strangles (horses) | Brown ring disease (clams) |
| Gapes (chickens) | Ice-ice disease (seaweed) |

[a] Reviewed in Radostits et al. (2007).

Observing abnormalities of appearance, behaviour, growth, feeding, reproduction and survival may be hampered in the aquatic environment by lack of visibility either because the water is murky or because the fish are at some depth. These disadvantages may be overcome by using divers, lift trays, cast nets or novel developments in behavioural assessment such as video or ultrasonic analysis. However, even in the absence of these technologies, the opportunity for the detection of abnormal signs still exists, e.g. the location and pattern of swimming, poor growth, lack of food consumption, mortality rate and physical appearance of dead animals (Bondad-Reantaso et al., 2001; Mohan et al., 2002). In finfish, abnormal swimming behaviours such as rubbing against solid objects or "flashing" can indicate surface irritation; cork-screwing may also indicate neurological problems and air gulping oxygen deprivation. General listlessness, belly-up or rolling motion may also be seen. The physical appearance of the fins, skin and eyes can also be observed. Fin damage may indicate mechanical damage or the result of infection. The occurrence and distribution of red spots or erosions on the skin and the presence of macroscopic ectoparasites can be noted. Similar observation may be made with shrimp. In addition to changes in feeding behaviour, colonisation, erosion of the cuticle, broken antennae, loss of limbs, white or black spots and general changes in colour may indicate disease. The detection of behavioural changes in molluscs is more difficult but failure of shell closure or "gaping" on removal from the water may indicate weakness. Abnormal smell may occur. Although changes to shell shape, and colonisation or damage of its external surface may be normal, these changes on the inside of the shell may indicate disease as may changes in appearance or colour of the soft tissue or the presence of water blisters or abscesses. An important component of these observations is knowledge of the normal, a state farmers become aware of from daily contact with these species. Indeed farmers may use a number of almost subliminal senses to conclude that "something is wrong." The importance of these methods of disease detection has been described in detail (Bondad-Reantaso et al., 2001) and is also reflected in the names of many of the important diseases of fish and shellfish (Table 1).

## 3.2. Reporting disease

The accurate reporting of disease is as important as its detection. The effect of under-reporting on the effectiveness of disease surveillance programmes is acknowledged (Halliday et al., 2012). However, over-reporting, mis-reporting and non-representative reporting, either purposely or by ignorance, are equally distorting.

The framework depicted in Fig. 1 allows the reasons for poor quality and unrepresentative reporting to be teased apart. These may be categorised broadly into technological and behavioural realms. These two main drivers influence the ability to detect and report disease and the willingness and motivation to do so. They affect individuals and institutions. The risk of losing trade, credibility and reputation, can affect the willingness of national institutions to report disease to international authorities. Such disincentives reflect interplay between external economic, social and institutional constraints and influences (blue boxed arrows in Fig. 1).

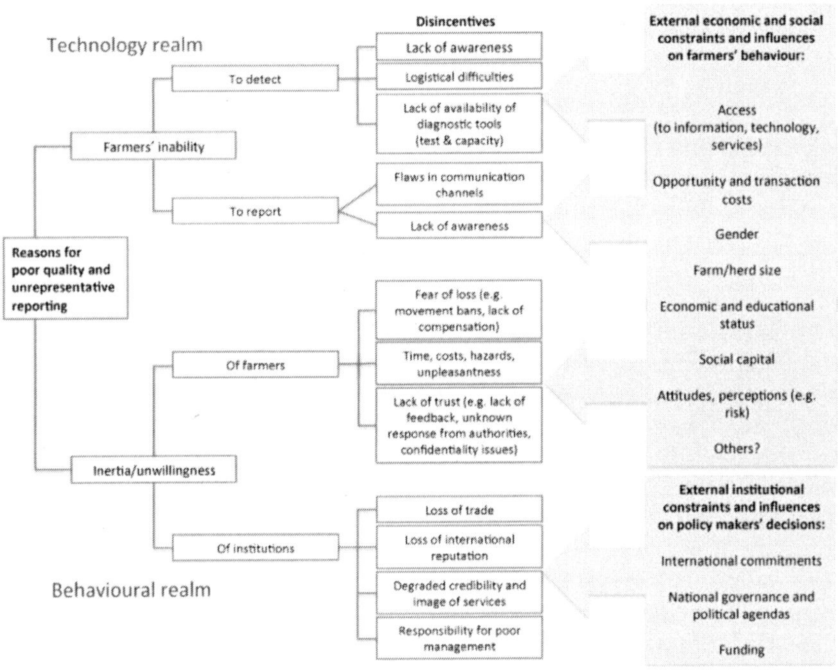


**Figure. 1.** Reasons and influences for poor quality and unrepresentative reporting in disease surveillance. Developed from World Bank (2010a); Halliday et al. (2012).

Because of the fundamental role of the participation of farmers and institutions in syndromic surveillance, the impact of the disincentives and external influences on their behaviour needs to be clearly understood. We explore these issues by considering farmers' adoption of technologies supporting syndromic surveillance, individual and collective influences underpinning farmers' reporting behaviour and participation, and the function of informal and formal institutions involved in disease surveillance.

## 3.3. Technology: a mixed blessing for farmer-based syndromic surveillance

Farmers' active participation in disease surveillance involves the adoption of facilitating technologies and good reporting practices. Disease detection and reporting, and syndromic surveillance more generally, rely largely on technical means.[1] Modern communication technologies, including mobile phones and internet-based mapping systems, have been suggested as powerful aids for disease surveillance (Chunara et al., 2012, Freifeld et al., 2010). They can however be a mixed blessing for farmer-based surveillance on both epidemiological and social grounds, the latter in relation to their acceptance and adoption (Renaud and Van Biljon, 2008).[2]

Despite their increasing availability worldwide, the use of modern information and communication technologies (ICT) for surveillance raises important epidemiological issues. These relate to:

(i) Sampling and targeting of populations, which are affected by the uneven distribution of mobile phones, internet access and network coverage, and by literacy issues, in particular in developing countries (Watkins et al., 2012). This can lead to a bias towards increased reporting from countries with greater electronic communications infrastructures and public health resources (Brownstein et al., 2008). Evidence from Sri Lanka and Kenya indicated that this can challenge the accuracy of disease surveillance (Robertson et al., 2010b; Walker et al., 2011).

(ii) Confidence in the data collected, which raises validation and quality control issues, in particular over data collected via 'crowdsourcing' (Freifeld et al., 2010).

(iii) Anonymisation of data and the need for privacy protection. This is an additional challenge for data crowdsourced using smartphone and internet-based tools (Chunara et al., 2012; Freifeld et al., 2010), although methods to overcome this are being developed (Clarke and Steele, 2014).

The adoption and diffusion of technologies supporting syndromic disease surveillance are just as critical. Adoption hinges on a mix of social and economic factors, individual attitudes, perceptions and status (Fig. 1) (Duncombe, 2015). The livestock health literature shows the influence of exogenous factors as diverse as social status (Heffernan et al., 2011), age, income and distance (Oladele and Jood, 2010), level of concern for animal welfare and understanding of the disease, economic environment, veterinary support and severity of the situation (Alarcon et al., 2014) on the adoption of technological advances for improved animal health management. Such detailed literature is not available for aquaculture, although research on the adoption of improved farmed management practices more generally reveals the influence of a range of similar exogenous factors, regardless of the production system (e.g. farm ownership and distance (Baticados et al., 2014); education level, contacts with extension workers, access to seed and markets (Adeogun et al., 2008); age, extension, perceived profitability, marketability and risk levels (Wetengere, 2011)). Little is known however about aquatic farmers' motivation to adopt technologies which support syndromic surveillance, opening a potentially a large field of enquiry.

Some useful insights into the adoption of technologies which support syndromic surveillance are starting to emerge from the use of ICT and mobile phones to provide farmers with information and services aimed at improving productivity (Duncombe, 2015). In spite of reducing information costs and asymmetries[3] and alleviating overstretched extension services (Aker, 2011, Mittal et al., 2010), the often assumed potential positive developmental outcomes of ICT-based innovations (Avgerou, 2010) remain to be verified. For example, ICT is not gender-neutral (Arun et al., 2004), and adequate consideration of gender differences in the access and use of mobile technologies, and the involvement of end-users in the design of ICT-driven services, is still largely lacking (Duncombe, 2015).

## 3.4. Motivation and other influences behind farmers' willingness or inertia to report disease

Farmer-centred, participatory approaches (Chambers, 1994; Pretty, 1994) have been promoted by some epidemiologists to generate farmer buy-in to surveillance programmes (Catley et al., 2001). These approaches have had some success in the control of terrestrial animal diseases (e.g. Ashley-Robinson et al., 2004; Catley et al., 2012; Jost et al., 2007; Mariner et al., 2014). They are congruent with the idea that syndromic surveillance should be placed in the hands of farmers because they are able to recognise signs of disease in the animals in their care.

In aquaculture however, to our knowledge, there are only two examples of participatory syndromic surveillance. In one, participatory rural appraisal was used to assess farmers' knowledge and observations of fish diseases in India (Sahu et al., 1999). This was however insufficient for epidemiological monitoring. In the other, French oyster farmers were surveyed to determine what to them constituted 'increased mortality', a trigger for notification to the EU. This was an attempt to understand the factors behind their reporting of disease (Lupo et al., 2014b). Results indicated that the degree of awareness of mortality, reporting requirements, individual perceptions about compensation, personal involvement and past experience in surveillance programmes were important in doing so (Lupo et al., 2014a). More studies are clearly needed to understand individual and collective behaviours and incentives which impact on farmer-based syndromic surveillance.

### 3.4.1. Individually

#### *3.4.1.1. Behavioural determinants*

According to the theory of planned behaviour (Ajzen, 1991), farmers' "*behaviour*" in reporting disease is a reflection of the level of their "*intention*" to carry out actions to reduce or manage disease risk, their "*attitudes*", i.e. values, (e.g. pursuit of profit), priorities (e.g. personal fulfilment) and personalities, and external expectations placed upon them (also called '*subjective norms*'), and of their "*perceived behavioural control*", i.e. their ability put practices of their choice into effect (Alarcon et al., 2014; Garforth et al., 2004; Garforth et al., 2013).

Studies of the impact of these drivers in terrestrial animal disease control have been undertaken (Alarcon et al., 2014) but are non-existent in aquaculture. Understanding their importance and interactions is essential as it will elucidate farmers' values, commitment, motivation and trust, and their perceptions of the benefits of surveillance. This knowledge would be instrumental in tailoring the design of reporting systems and would inform the *modus operandi* of syndromic surveillance programmes in aquaculture. Feedback, in the form of acknowledging reports, providing diagnostic test results and advice on disease management are non-monetary incentives that may enhance the willingness and overcome the inertia to report (Halliday et al., 2012). Confidence in anonymisation and identity protection may have a similar effect (Clarke and Steele, 2014). Anonymity may be important not only for epidemiological integrity, but also to protect farmers with emerging diseases from ostracization (Corsin et al., 2009; Mariner et al., 2014).

#### *3.4.1.2. Monetary incentives*

Financial compensation for animal losses incurred during disease outbreaks is meant to incentivise farmers to report early signs of disease. The extent to which this works is debatable. It appears to vary according to species, disease, farming context and history. Lupo et al. (2014a) reported that following the introduction of mandatory mortality reporting, oyster farmers who had received compensation before the law was passed were more likely to report mortalities than those who had not. They cite similar findings for the reporting of avian influenza by Dutch poultry farmers and scrapie by Norwegian sheep farmers but contrast these with the reporting of classical swine fever by Dutch pig farmers.

These variations are not surprising. The effectiveness of compensation in stimulating reporting assumes that farmers, as rational agents, will seek to maximise their utility; in other words, that profit maximisation from farming will be a key objective for farmers. Yet, studies in agriculture show time and again that farmers exhibit behaviour and decision-making patterns which do not follow this assumption. They are strongly influenced by other, non-pecuniary, motivations (Howley, 2015). These detract from compliance. If perceived as inadequate by farmers, compensation will not trigger the expected reporting response (Enticott and Lee, 2015; Lupo et al., 2014a). Compensation can be a deterrent to the implementation of improved health management procedures and give rise to free-riding behaviours. Inconsistencies in policies, whereby some notifiable diseases are compensated for and others not (e.g. Infectious Salmon Anaemia-ISA in the UK), and lack of transparency of their ultimate purpose (Enticott and Lee, 2015), raise questions over their potential impact on disease control. Further research is needed to understand how compensation and insurance may affect farmers' behaviour in aquaculture.

*3.4.1.3. Gender and reporting: a big unknown*

Women play a significant role in farming terrestrial and aquatic species (FAO, 2011a) but the way in which they manage disease in animals under their care is under-studied (e.g. FAO, 2012). To our knowledge, the extent to which the differential in opportunities and constraints observed between men and women farmers in livestock (e.g. Kristjanson et al., 2010; Miller, 2011) and in aquaculture (e.g. Brugere et al., 2001; Williams et al., 2012) would affect their engagement in syndromic disease surveillance and would influence epidemiological study results, has not been documented. As a consequence, we simply do not know whether men and women would notice and report disease differently, nor whether women's engagement in disease surveillance has the potential to trigger the transformational change that progressing towards greater gender equality in livestock farming and aquaculture requires.

### 3.4.2. Collectively: influence of social capital, networks and collective action

Social capital is a very important dimension of farming communities. Social capital refers to social resources, such as trust, reciprocity, norms, formal and informal membership of groups, collectives and networks. These shape the interactions of individuals or groups and can be used to facilitate their actions and achieve their objectives (Bourdieu, 1980; Coleman, 1990; Putnam et al., 1993; Woolcock and Narayan, 2000). The role of social capital in epidemiology has been mainly considered from a public health perspective (e.g. Kawachi and Berkman, 2000; Szreter and Woolcock, 2004) and in the context of farmers' perceptions of, and response to, disease risk (e.g. Naylor and Courtney, 2014). Parallels exist with animal disease surveillance: mutual trust and confidence must exist between the actors in the reporting chain to enable control and eradication (World Bank, 2010a).

Social networks are an integral part of social capital (Putnam, 2001; Sabatini, 2009). Social network analysis (SNA) can establish the position, embedment and influence of individuals within networks. It also provides an understanding of their structure, operation and effectiveness. Interestingly, whilst SNA is used extensively in epidemiology to study the spread of infectious agents (e.g. Brigas-Poulin et al., 2006; Jonkers et al., 2010; Martin et al., 2011; Ortiz-Pelaez et al., 2006), and in natural resources management to study environmental governance (Bodin et al., 2011), this approach has not been extrapolated to understanding how relationships between individual farmers and their social groups might influence their decision to report disease.

The concept of collective action, which also stems from the notion of social capital (Putnam et al., 1993; Woolcock and Narayan, 2000) and is mainly used in relation to natural resources management,[4] is also relevant to disease surveillance. Social capital is enabling when used as a productive asset facilitating transactions and generating cooperation. However it can have a potentially harmful effect on collective action when opportunistic behaviours, such as rent-seeking and free-riding, are displayed (Dasgupta and Serageldin, 2001). This is of particular importance for disease surveillance and in outbreaks, where a rapid, coordinated and collective response is required. In studies of oyster farming, collective intentions towards improved farming practices aimed at preventing disease emergence were overridden by individual profit-maximisation objectives (Carlier et al., 2013) and diverging individual interpretations of what constituted a case definition hampered the collective response necessary to halt disease spread (Lupo et al., 2014b).

Collective action also concerns institutions. It is manifest through the networks that have been created by international organisations and national institutions to strengthen the coordination of animal and human disease surveillance and cross border control (Wibulpolprasert et al., 2013).

## 3.5. Institutional influences on disease surveillance effectiveness

There is a need to integrate political economy and institutional perspectives in animal health research to understand the factors that influence the behaviour of institutions with regard to the implementation of disease surveillance (Rushton et al., 2007). Institutional behaviour may be influenced by legal requirements and by functional, reputational and economic incentives.

### 3.5.1. Legal requirements

The need to minimise the potential for disease spread and the economic impacts of restricted movement on trade requires complex interactions and commitments among national and international authorities. These interactions are regulated by the SPS Agreement (WTO, 1995) for animal health, and the updated IHR (WHO, 2005) for public health, including zoonoses. The IHR seeks to facilitate disease information sharing and surveillance across borders. The SPS agreement operates at the interface of health and trade. It uses disease freedom as leverage for trade transactions and access to markets. Both treaties provide legal frameworks which strengthen the role of the OIE and WHO. In compliance with the SPS, countries have to report the occurrence of diseases listed in the Aquatic Animal Health Code to the OIE. Countries are expected to take reasonable action to prevent the spread of disease, but the requirements of the SPS Agreement and OIE standards are not consistently translated, resourced or enforced in national legislations (Otte et al., 2004, Oidtmann et al., 2011). This weakens the process of disease reporting to international authorities, particularly if the diseases are emerging and not yet listed by the OIE. It also reduces the overall efficacy of national and international biosecurity frameworks. This is aggravated by the fact that the OIE does not have a policing authority (Oidtmann et al., 2011). Similar issues arise with regard to the compliance, implementation and effectiveness of the IHR, and the enforcement role of the WHO particularly in developing countries (Youde, 2011).

### 3.5.2. Institutional '(dis)functions'

The role of institutions is to create stable structures for human interactions (North, 1990). International, national and local institutions are critical in the surveillance, control and management of disease. Their role is however often hampered by principal-agent relationship breakdowns, i.e. when one 'agent' at a lower level has to respond to multiple 'principals' with whom authority rests and whose interests are not necessarily aligned (Dixit, 2003). In addition, information asymmetries give rise to uncertainty, conflict (Bardhan, 1989) and high transaction costs (North, 2000). For example, separate competent authorities with overlapping mandates, or the delegation of responsibilities for disease diagnosis and prevention to third, private, entities can create communication gaps, confuse reporting channels and increase the number and costs of transactions. These undermine the enforcement of regulations and effectiveness of disease detection and control (Halliday et al., 2012). Mixed messages from animal health institutions to farmers (Mariner et al., 2014), or lack of transparency and clarity with respect to the notification process between farmers and veterinary authorities (Elbers et al., 2010) are also typical cases of information asymmetries and principal-agent breakdowns.

In response to these limitations, integrated approaches such as "One Health", which aim to integrate disease prevention and control across multiple ministries, such as those responsible for human and animal health, tourism, environment, wildlife and trade, have been proposed to facilitate communication, coordination and responsiveness (World Bank, 2010a; Zinsstag et al., 2013). Similarly, improvements in the speed and transparency of disease notification and information dissemination have been achieved through the development of surveillance networks at regional and sub-regional levels. In human health, CORDS - Connecting Organizations for Regional Disease Surveillance (Wibulpolprasert et al., 2013) is one such example, as is the OIE's World Animal Health Information System (WAHIS) and Database (WAHID) (Corsin et al., 2009). However, only listed diseases are considered (Oidtmann et al., 2011) and false positive reports can perpetuate mistrust, and result in unnecessary actions and wasted resources (Halliday et al., 2012). The precise extent to which such network approaches will improve the governance of disease surveillance and control is therefore yet to be determined.

### 3.5.3. Reputational (dis)incentives

Veterinary and other government authorities can find themselves trapped between their obligation of disease notification and its potentially negative impact on how the status of animal health and performance of their veterinary services are viewed (Mariner et al., 2014). Potentially large economic and political risks may be associated with the damage done to a country's reputation by reporting disease and implementing corrective measures, e.g. threats to exports, effects on tourism, perceptions of under-performance and discredit of veterinary services (e.g. Mariner et al., 2014).

### 3.5.4. Economic (dis)incentives

Although schemes that compensate farmers financially for the slaughter of their stocks tend to increase the sensitivity of surveillance, they are not available everywhere and for every disease (Oidtmann et al., 2011). They are expensive (Inamura et al., 2015) and are not feasible in countries where resources are limited (Halliday et al., 2012). Returns on public investment in disease surveillance and control of notifiable diseases are also not always guaranteed and vary according to the type of disease. In farmed salmon for example, they are higher for ISA and viral haemorrhagic septicaemia (VHS) than for infectious haemorrhagic necrosis (IHN) (Moran and Fofana, 2007). The establishment of a global fund for the financial compensation of national governments who follow international regulations in reporting disease outbreaks and implementing control measures has been proposed (World Bank, 2010a). However, in the case of non-zoonotic diseases where public health risks are minimal, this begs the question of who should be responsible for the costs of surveillance.

For national governments, deciding whether to compensate or invest in prevention, through surveillance and intervention, is a complex policy decision. Economically optimizing resource allocation in the context of disease management is neither straightforward nor intuitive (Howe et al., 2013). Furthermore, deciding how much protection from disease is appropriate, and who should bear the costs of this, is particularly important in the context of transboundary diseases. It involves consideration of the economic notions of public good, externality and equity (Otte et al., 2004). Surveillance, epidemiological and veterinary research, information and service provision to farmers are typically considered as "public goods". When carried out unilaterally by all countries, such public goods generate global societal benefits. But they can also give rise to free-riding behaviour displayed at national level when uncertainties exist regarding the risks and impacts of existing and emerging transboundary diseases. Governments may relax their efforts and responsibilities for reporting and control in the belief that their benefits from global surveillance will remain unchanged. Whilst, on equity grounds, the cost of disease externalities, i.e. the uncompensated damage generated by the spread of disease on third parties, should be shared between those who impose a higher risk or spread disease to others, and those who benefit from protection from this risk, this is difficult in practice (Otte et al., 2004). Although not quantified, it is most likely that, by influencing the reporting behaviour of governments, these economic (dis)incentives are undermining the effectiveness of the SPS agreement, and ultimately the efforts of the international community towards improved disease communication and control.

# 4. DISCUSSION: CHALLENGES AND OPPORTUNITIES FOR SYNDROMIC DISEASE SURVEILLANCE IN AQUACULTURE

New and emerging diseases remain a major challenge to aquaculture development. The nature of the sector's production systems, on-going intensification, reliance on international trade and importance for livelihoods and food security in resource-limited countries, make it prone and extremely sensitive to the impact and spread of disease.

Could the proximity and observational skills of fish farmers be used as a basis for syndromic surveillance? Although fish farmers' ability to accurately detect and report signs of diseases has not been demonstrated, the most threatening diseases for aquaculture have obvious clinical signs (cf. Table 1). There is no reason to believe that fish farmers would not be able to notice these deviations from the norm and to identify existing and emerging diseases. This will only work if the factors that influence individual and institutional behaviours, and consequently the quality and representativeness of reporting, are understood and used to inform the design of surveillance and control programmes. By focussing on these human dimensions, the aquaculture sector could set an example in this area.

Technological developments such as pond-side tests, internet-based technologies and smart phone apps are becoming globally available to farmers and are likely to be pivotal in reporting disease and developing farmer-based syndromic surveillance. Aquaculture should not miss these opportunities. However, jumping on the "*m*-surveillance" bandwagon is not without dangers. Evidence from human medicine shows that the use of mobile phones and related ICT does not scale up beyond pilot studies (Andreassen et al., 2015). Overcoming these "plagues of pilots" (ibid) requires a focus on the interactions between technology and people. Designing new technologies aimed at farmer use is easy. But designing a technology-driven system for collecting epidemiological data that is meaningful to farmers, veterinarians, researchers, and policy-makers is complex. Human aspects need to be adequately accounted for. Epidemiological data collection is about technology, but effective disease surveillance is about people. This is a neglected aspect of epidemiology and disease surveillance (Lupo et al., 2014a; Mariner et al., 2014). Understanding what motivates fish farmers and aquaculture authorities and devising appropriate incentives to maintain their engagement to report disease over long periods is essential in developing sustainable, dynamic and adaptable surveillance systems.

Promoting farmer-based syndromic aquatic disease surveillance (FASADS) requires marshalling disciplines such as veterinary science, epidemiology, information technology, biology, economics, psychology and social science on an equal footing in order to design systems which are inclusive of people and technology. Calls for greater inclusion of social sciences and economics in epidemiology are not new (e.g. Perry et al., 2001; Rushton et al., 2007) but until recently these have largely been ignored. Animal disease surveillance has thus remained focused on what Halliday et al. (2012) describe as "tangible elements, such as laboratory diagnostic infrastructure and communications technology." These veterinary, epidemiological and technical components of surveillance are well established but social, economic and institutional elements are open research areas, particularly in syndromic surveillance and in the aquaculture sector. Considering these dimensions from the outset will strengthen the framework of aquatic disease surveillance systems by identifying their components and boundaries (Morgan et al., 2015).

Interdisciplinary collaboration in surveillance will be all the more crucial as diseases emerge, evolve and spread in increasingly complex, interconnected and dynamic social-ecological systems (Leung et al., 2012). Over the years, disease surveillance driven by veterinary and epidemiological sciences has progressively lost the human connotation that was evident in Langmuir's original conception. Yet, for it to become effective, understanding the human and institutional dimensions affecting decisions about disease reporting and control is paramount (Rich and Perry, 2011). This is particularly important in developing countries, where resources are limiting and rural livelihoods are dependent on healthy animals and successful rearing cycles. The importance of understanding and harnessing social capital in doing this has been demonstrated by use of cluster management in the form of farmer groups and aquaclubs in Vietnam and India to develop and promote best management practices to improve shrimp productivity (Corsin et al., 2008). Similarly, animal disease surveillance needs to be about people.

By embedding health and disease in broader social-ecological contexts, "One Health" (World Bank, 2010a) lends itself to interdisciplinary interactions and the incorporation of farmers and institutions behavioural influences in disease surveillance and control. It is contributing to moving disease surveillance beyond disciplinary silos and promoting the systemic study and understanding of interactions between people, their animals and the environment. This will increase the likelihood of positive impacts on livelihoods and the resilience of social-ecological systems more generally (Zinsstag et al., 2013). One Health still needs to better incorporate aquatic animal health however. If it does this, it offers a promising vehicle for the development of aquatic disease surveillance (MacKenzie et al., 2015).

Two major questions in developing disease surveillance systems are: Who benefits? and Who pays? Many stakeholders are affected, directly or indirectly, by animal disease surveillance. The aquaculture sector is no different in this regard: farmers, researchers, diagnosticians, national and international authorities, private sector organisations and consumers are linked by the epidemiological data collected. Private input providers (e.g. pharmaceutical companies) may be interested in paying for accessing the data generated by a disease surveillance programme run by a national authority. A farmers' collective may wish to set up its own surveillance system across large areas or clusters of privately owned farms, and generate the epidemiological data which national surveillance programmes will require access to and be prepared to pay for through cost sharing agreements. How the value of this data (and for whom) will be captured will be an essential component of the long-term sustainability of surveillance programmes. However, in addition to the issue of sharing the costs of surveillance, sharing the data it generates will be just as sensitive.

In a context of growing numbers of cases of aquatic zoonoses, seafood consumers may also drive demand and be willing to pay premiums for certified disease-free aquaculture commodities. The potential of certified disease-free products to command a price premium is an immediately tangible, incentive for surveillance (Halliday et al., 2012). This is potentially very relevant to aquaculture. Debates over aquaculture product certification are raging, but have so far paid little attention to the place of disease freedom in product certification (e.g. Bush et al., 2013), or relegated the issue to one of compliance with the provisions of the OIE Aquatic Animal Health Code (FAO, 2011b). The issue of certification is intimately linked to the implementation of the SPS Agreement. Failure to accede to certification of disease freedom resulted in a trade ban and dispute between Canada and Australia regarding the import of farmed salmon to protect Australia's recreational fisheries from exotic diseases (Taylor, 2000). The opportunity to trade is obviously important, but does not necessarily translate into an immediate, additional reward for farmers' efforts to surveil and maintain disease-free stocks. To be an incentive for all actors in the surveillance 'chain', the benefits of improved surveillance should be spread equitably among all the stakeholders of this chain. Although the role of the OIE has become more prominent since the SPS Agreement, one may question the value, as well as ethics, of an over-reliance on a trade-related agreement in governing animal health and disease control[5] without providing the OIE with resources to assist countries and producers in meeting, and benefiting from, disease freedom.

# 5. CONCLUDING COMMENTS

Farmer-based syndromic aquatic disease surveillance constitutes a real opportunity to overcome barriers inherent to traditional, laboratory-based surveillance in aquaculture. However, the long-term sustainability of surveillance will necessitate overcoming farmers' and institutional inertia, i.e. their reluctance to change. This will mean addressing the psychological, organizational, and political barriers that have become ingrained in the behaviour of farmers and institutions (World Bank, 2010b). Much however remains to be explored in this field in relation to its consequences on disease reporting and its combination with the atypical - but overall understudied - risk preferences fish farmers can display pre and post disease outbreaks (Castinel et al., 2015). The future global governance of disease surveillance may require a rethink. Currently the IHR is concerned chiefly with human health and represents the process through which human diseases can be controlled on a global scale; the SPS sanctions this in the context of animal health. The boundaries between the two treaties are however likely to become increasingly blurred with the increasing emergence of zoonoses. Aquatic diseases should not be exempt from these discussions.

Progressing farmer-based aquatic syndromic disease surveillance requires the creation of a new "culture of surveillance" (Halliday et al., 2012; Soto et al., 2008) that shifts the paradigm from surveillance as the unique prerogative of veterinarians and diagnostic laboratories, to one in which farmers, acknowledged as the starting point of disease surveillance, are given equal power and responsibility. Technological change will only facilitate this if the human components of the system are understood, respected and optimised.

## ACKNOWLEDGEMENTS

We acknowledge the Newton Fund Global Research Partnership in Aquaculture of the UK's Biotechnology and Biological Sciences Research Council (BBSRC), the UK Department for International Development (DFID) and the Indian Department of Biotechnology (DBT) for supporting our attendance at a workshop in Kerala, India, in February 2015, during which the ideas presented in this paper were first conjectured. No financial support was further received from these organisations or any other source towards its preparation. We also acknowledge the valuable suggestions of anonymous referees to improve our manuscript.

## REFERENCES

1. Adams, A., Thompson, K.D., 2010. Development and Validation of a Lateral Flow Kit to Detect ISAV (Workshop on use of diagnostic kits for the detection of fish diseases. Aarhus, Denmark May 26–28), pp. 12–13.
2. Adams, A., Thompson, K.D., 2011. Development of diagnostics for aquaculture :challenges and opportunities. Aquac. Res. 42 (Suppl. 1), 93–102.
3. Adeogun, O.A., Ajana, A.M., Ayinla, O.A., Yarhere, M.T., Adeogun,M.O., 2008. Application of logit model in adoption decision: a study of hybrid clarias in Lagos State, Nigeria. Am. Eurasian J. Agric. Environ. Sci. 4 (4), 468–472.

4. Ajzen, I., 1991. The theory of planned behavior. Organ. Behav. Hum. Decis. 50, 179–211.
5. Aker, J.C., 2011. Dial "A" for agriculture: a review of information and communication technologies for agricultural extension in developing countries. Agric. Econ. 42, 631–647.
6. Alarcon, P., Wieland, B., Mateus, A.L., Dewberry, C., 2014. Pig farmers' perceptions, attitudes, influences and management of information in the decision-making process for disease control. Prev. Vet. Med. 116, 223–242.
7. Andre, M., Schwan, A., Odenholt, I., 2004. The use of CRP tests in patients with respiratory tract infections in primary care in Sweden can be questioned. Scand. J. Infect. Dis. 36, 192–197.
8. Andreassen, H.K., Kjekshus, L.E., Tjora, A., 2015. Survival of the project: a case study of ICT innovation in health care. Soc. Sci. Med. 12, 62–69.
9. Arun, S., Heeks, R.,Morgan, S., 2004. ICT Initiatives, Women and Work in Developing Countries: Reinforcing or Changing Gender Inequalities in South India? Institute for Development Policy and Management. University of Manchester, Manchester.
10. Ashander, J., Krkošek, M., Lewis, M.A., 2012. Aquaculture-induced changes to dynamics of a migratory host and specialist parasite: a case study of pink salmon and sea lice. Theor. Ecol. 5, 231–252.
11. Ashley-Robinson, R., Catley, A., Hird, D., 2004. Significance of participatory epidemiology in veterinary public health community-based systems. Expert consultation on community-based veterinary public health (VPH) Systems, Rome, 27–28 October 2003. FAO Animal and Health Proceedings 2. FAO, Rome.
12. Avgerou, C., 2010. Discourses on ICT and development. Inf. Technol. Int. Dev. 6, 1–18.
13. Baland, J.-M., Platteau, J.-P., 1996. Halting Degradation of Natural Resources: Is There a
14. Role for Rural Communities? Clarendon Press, Oxford
15. Bardhan, P., 1989. The new institutional economics and development theory: a brief critical assessment. World Dev. 17, 1389–1395.
16. Baticados, D.B., Agbayani, R.F., Quinitio, E.T., 2014. Community-based technology transfer in rural aquaculture: the case of mudcrab Scylla serrata nursery in ponds in northern Samar, Central Philippines. Ambio 43, 1047–1058.
17. Bodin, Ö., Ramirez-Sanchez, S., Ernstson, H., Prell, C., 2011. A social relational approach to natural resource governance. In: Bodin, Ö., Prell, C. (Eds.), Social Networks and Natural Resource Management: Uncovering the Social Fabric of Environmental Governance. Cambridge University Press, Cambridge, pp. 3–28.
18. Bondad-Reantaso, M.G., McGladdery, S.E., East, I., Subasinghe, R.P., 2001. Asia Diagnostic Guide to Aquatic Animal Diseases. FAO Fisheries Technical Paper No. 402, Supplement 2. Rome, FAO 240 p.
19. Bourdieu, P., 1980. Le capital social. Act. de la Rech. en Sci. Social. 31, 2–3.
20. Brigas-Poulin, M., Thompson, R.A., Chriel, M., Mortensen, S., Greiner, M., 2006. Network analysis of Danish cattle industry trade patterns as an evaluation of risk potential for disease spread. Prev. Vet. Med. 76, 11–39.
21. Brinkel, J., Krämer, A., Krumkamp, R., May, J., Fobil, J., 2014. Mobile phone-based mHealth approaches for public health surveillance in sub-Saharan Africa: a systematic review. Int. J. Environ. Res. Public Health. 11, 11559–11582.
22. Brownstein, J.S., Freifeld, C.C., Reis, B.Y., Mandl, K.D., 2008. Surveillance sans frontiers: internet-based emerging infectious disease intelligence and the HealthMap Project. PLoS Med. 5 (7), e151.
23. Brugere, C., McAndrew, K., Bulcock, P., 2001. Does cage aquaculture address gender goals in development? Results of a case study in Bangladesh. Aquac. Econ. Manag. 5, 179–189.
24. Bush, S.R., Belton, B., Hall, D., Vandergeest, P., Murray, F.J., Ponte, S., Oosterveer, P., Islam, M.S., Mol, A.P.J., Hatanaka, M., Kruijssen, D., Ha, T.T.T., Little, D.C., Kusumawati, R., 2013. Certify sustainable aquaculture? Science 341, 1067–1068.

25. Butler, S.F., Venuti, S.W., Benoit, C., Beaulaurier, R.L., Houle, B., Katz, N., 2007. Internet surveillance: content analysis and monitoring of product-specific internet prescription
26. opioid abuse-related postings. Clin. J. Pain 23, 619–628.
27. Lima dos Santos, C.A.M., Howgate, P., 2011. Fishborne zoonotic parasites and aquaculture: a review. Aquaculture 318, 253–261.
28. Caipang, C.M.A., Muegue,M.F.S., Geduspan, J.S., 2015. Utilization of the loop-mediated isothermal amplification (LAMP) for the detection of pathogens in shrimp and fish aquaculture in the Philippines. Biotechnological Advances in Shrimp Health Management in the Philippines. 661, pp. 149–163.
29. Cameron, A.R., 2012. The consequences of risk-based surveillance: developing outputbased standards for surveillance to demonstrate freedom from disease. Prev. Vet. Med. 105, 280–286.
30. Carlier, M., Prou, J., Mille, D., Lupo, C., 2013. Oyster farmers' perception of spat mortality outbreaks: more a firm than a farm issue. Society for Veterinary Epidemiology and Preventive Medicine. Proceedings of a meeting held in Madrid, Spain, 20–22 March 2013, pp. 226–237.
31. Castinel, A., Fletcher, L., Dhand, N., Rubio, A., Whittington, R., Taylor, M., 2015. OsHV-1 mortalities in Pacific oysters in Australia and New Zealand: the farmer's story. Report Prepared for the Ministry of Business, Innovation and Employment (MBIE). Cawthron Report No. 2567. Cawthron Institute, Nelson.
32. Catley, A., Okoth, S., Osman, J., Fison, T., Njiru, Z.,Mwangi, J., Jones, B.A., Leyland, T.J., 2001. Participatory diagnosis of a chronic wasting disease in cattle in southern Sudan. Prev. Vet. Med. 51, 161–181.
33. Catley, A., Alders, R.G., Wood, J.L.N., 2012. Participatory epidemiology: approaches, methods, experiences. Vet. J. 191, 151–160.
34. Chambers, R., 1994. Challenging the Professions. Intermediate Technology Publications, London, Frontiers for Rural Development.
35. Chunara, R., Freifeld, C.C., Brownstein, J.S., 2012. New technologies for reporting real-time emergent infections. Parasitology 139 (14), 1843–1851.
36. Clarke, A., Steele, R., 2014. Local processing to achieve anonymity in a participatory health e-research system. Procedia – Soc. Behav. Sci. 147, 284–2292.
37. Coleman, J.S., 1990. Foundations of Social Theory. Harvard University Press, Cambridge.
38. Corsin, F., Mohan, C.V., Padiyar, A., Yamamoto, K., Chanratchakool, P., Phillips, M.J., 2008. Codes of practice and better management: a solution for shrimp health management? In: Bondad-Reantaso, M.G., Mohan, C.V., Crumlish, M., Subasinghe, R.P. (Eds.), Diseases in Asian Aquaculture VI. Fish Health Section, Asian Fisheries Society, Manila, Philippines, pp. 419–432
39. Corsin, F., Georgiadis, M., Hammell, K.L., Hill, B., 2009. Guide for Aquatic Animal Health Surveillance. World Organisation for Animal Health (OIE), Paris.
40. Dasgupta, P., Serageldin, I. (Eds.), 2001. Social Capital: A Multifaceted Perspective. World Bank Publications, World Bank, Washington D.C.
41. Diserens, N., Presi, P., Bernet, D., Health, W., 2013. Risk Assessment for the Design of a Risk-Based Surveillance Programme for Fish Farms in Switzerland (in Accordance With Council Directive 2006/88/EC of the European Union). 32 pp. 1–29.
42. Dixit, A., 2003. Some lessons from transaction-cost politics for less-developed countries. Econ. Polit. 15, 107–133.
43. Dórea, F.C.,Widgren, S., Lindberg, A., 2015. Vetsyn: an R package for veterinary syndromic surveillance. Prev. Vet. Med. 122, 21–32.
44. Doupé, R.G., Lympery, A.J., 2000. Managing translocations of aquatic species. Aquac. Res. 31 (151–156), 15.
45. Duncombe, R., 2015. Mobile phones for agricultural and rural development: a literature review and suggestions for future research. Eur. J. Dev. Res. (5 March 2015) http:// dx.doi.org/10.1057/ejdr.2014.60.

46. Elbers, A.R.W., Gorgievski-Duijvesteijn, M.J., Zarafshani, K., Koch, G., 2010. To report or not to report: a psychosocial investigation aimed at improving early detection of avian influenza outbreaks. OIE Rev. Sci. Tech. 29, 435–449.
47. Enticott, G.P., Lee, R., 2015. Buying biosecurity: UK compensation for animal diseases. In: Havinga, T., van Waarden, F., Casey, D. (Eds.), The Changing Landscape of Food Governance. Edward Elgar Publishing, Cheltenham, pp. 57–78.
48. EU, 2006. Council Directive 2006/88/EC of 24 October 2006 on animal health requirements for aquaculture animals and products thereof, and on the prevention and control of certain diseases in aquatic animals. http://eur-lex.europa.eu/legal-content/EN/ALL/?uri=CELEX%3A32006L0088 (Accessed 22nd Jan 2016).
49. Fanaee, T.H., Gama, J., 2014. EigenEvent: An Algorithm for Event Detection From Complex Data Streams in Syndromic Surveillance. arXiv Preprint arXiv: 1406.3496. pp. 1–20.
50. FAO, 2011a.Women in agriculture: closing the gender gap for development. The State of Food and Agriculture 2010–11. Rome, FAO.
51. FAO, 2011b. Technical Guidelines on Aquaculture Certification. FAO, Rome.
52. FAO, 2012. Invisible Guardians -Women Manage Livestock Diversity. FAO Animal Production and Health Paper No. 174. FAO, Rome.
53. Feng, C., Wang, C., Lin, X., Zhang, Y., Lv, J., Deng, J., Wu, S., 2013. Development of a loopmediated isothermal amplification method for detection of Perkinsus spp. in mollusks. Dis. Aquat. Org. 104, 141–148.
54. Fèvre, E.M., Bronsvoort, B.M.D.C., Hamilton, K.A., Cleaveland, S., 2006. Animal movements and the spread of infectious diseases. Trends Microbiol. 14, 125–131.
55. Flegel, T.W., 2006. The special danger of viral pathogens in shrimp translocated. ScienceAsia 32 (3), 215.
56. Frazer, L.N., Morton, A., Krkosek, M., 2012. Critical thresholds in sea lice epidemics: evidence, sensitivity and subcritical estimation. Proc. R. Soc. B Biol. Sci. 279, 1950–1958.
57. Freifeld, C.C., Chunara, R., Mekaru, S.R., Chan, E.H., Kass-Hout, T., Ayala Iacucci, A., Brownstein, J.S., 2010. Participatory epidemiology: use of mobile phones for community-based health reporting. PLoS Med. 7 (12), e1000376.
58. Garforth, C., Rehman, T., McKemey, K., Tranter, R., Cooke, R., Yates, C., Park, J., Dorward, P., 2004. Improving the design of knowledge transfer strategies by understanding farmer attitudes and behaviour. J. Farm. Manag. 12, 17–32.
59. Garforth, C.J., Bailey, A.P., Tranter, R.B., 2013. Farmers' attitudes to disease risk management in England: a comparative analysis of sheep and pig farmers. Prev. Vet. Med. 110, 456–466.
60. Gerbier, S., Yarovaya, O., Gicquel, Q., Millet, A.-L., Smaldore, V., Pagliaroli, V., Metzger, M.-H., 2011. Evaluation of natural language processing from emergency department computerized medical records for intra-hospital syndromic surveillance. BMC Med. Inform. Decis. Mak. 11, 50.
61. Gubbins, S., 2008. Prevalence of sheep infected with classical scrapie in Great Britain: integrating multiple sources of surveillance data for 2002. J. Roy. Soc. Interface 5, 1343–1351.
62. Halliday, J., Daborn, C., Auty, H., Mtema, Z., Lembo, T., Bronsvoort, B.M., Handel, I., Knobel, D., Hampson, K., Cleaveland, S., 2012. Bringing together emerging and endemic zoonoses surveillance: shared challenges and a common solution. Philos. Trans. R. Soc. Lond. Ser. B Biol. Sci. 367, 2872–2880.
63. Hay, S.I., George, D.B., Moyes, C.L., Brownstein, J.S., 2013. Big data opportunities for global infectious disease surveillance. PLoS Med. 10 (4), e1001413.
64. Hedgecock, D., 2012. Aquaculture, the next wave of domestication. In: Gepts, P., Famula, T.R., Bettinger, R.L., et al. (Eds.), Biodiversity in Agriculture Domestication. Publ CUP,
65. Evolution and Sustainability, pp. 538–548.

66. Heffernan, C., Thomson, K., Nielsen, L., 2011. Caste, livelihoods and livestock: an exploration of the uptake of livestock vaccination adoption among poor farmers in India. J. Int. Dev. 23, 103–118.
67. Hepworth, P.J., Nefedov, A.V., Muchnik, I.B., Morgan, K.L., 2012. Broiler chickens can benefit from machine learning: support vector machine analysis of observational epidemiological data. J. R. Soc. Interface 9, 1934–1942.
68. Howe, K.S., Häsler, B., Stärk, K.D.C., 2013. Economic principles for resource allocation decisions at national level to mitigate the effects of disease in farm animal populations. Epidemiol. Infect. 141, 91–101.
69. Howley, P., 2015. The Happy Farmer: The Effect of Non-pecuniary Benefits on Farmers' Behaviour. Contributed Paper Prepared for Presentation at the 89th Annual Conference of the Agricultural Economics Society, University of Warwick, England 13–15 April 2015.
70. Inamura, M., Rushton, J., Antón, J., 2015. Risk Management of Outbreaks of Livestock Diseases. OECD Food, Agriculture and Fisheries Papers, No. 91. OECD Publishing, Paris.
71. Jones, A.C., Mead, A., Kaiser, M.J., Austen, M.C.V, Adrian, A.W., Auchterlonie, N.A. Sutherland, W.J., 2015. Prioritization of knowledge needs for sustainable aquaculture: a national and global perspective. Fish Fish. 16, 668–683.
72. Jonkers, A.R.T., Sharkey, K.J., Thrush, M.A., Turnbull, J.F.,Morgan, K.L., 2010. Epidemics and control strategies for diseases of farmed salmonids: a parameter study. Epidemics 2, 195–206.
73. Jost, C.C., Mariner, J.C., Roeder, P.L., Sawitri, E., Macgregor-Skinner, G.J., 2007. Participatory epidemiology in disease surveillance and research. Rev. Sci. Tech. (OIE) 26, 537–549.
74. Kamilya, D., Baruah, A., 2014. Epizootic ulcerative syndrome (EUS) in fish: history and current status of understanding. Rev. Fish Biol. Fish. 24, 369–380.
75. Kawachi, I., Berkman, L., 2000. Social cohesion, social capital, and health. In: Kawashi, I., Berkman, L. (Eds.), Social Epidemiology. Oxford University Press, Oxford, pp. 174–190.
76. Keeling, M.J., Rohani, P., 2007. Introduction to simple epidemic models. In: Keeling, M.J., Rohani, P. (Eds.), Modeling Infectious Diseases in Humans and Animals. Princeton University Press, Princeton, pp. 15–52.
77. Keiser, J., Utzinger, J., 2009. Food-borne trematodiases. Clin. Microbiol. Rev. 22, 466–483. Kettler, H.,White, K., Hawkes, S., 2004. Mapping the landscape of diagnostics for sexually transmitted infections. WHO/TDR publication. http://www.who.int/tdr /publications/ documents/mapping-landscape-sti.pdf (Accessed 28/1/16).
78. Kono, T., Savan, R., Sakai, M., Itami, T., 2004. Detection of white spot syndrome virus in shrimp by loop-mediated isothermal amplification. J. Virol. Methods 115, 59–65.
79. Kristjanson, P.,Waters-Bayer, A., Johnson, N., Tipilda, A., Njuki, J., Baltenweck, I., Grace, D., MacMillan, S., 2010. Livestock and Women's Livelihoods: A Review of the Recent Evidence. Discussion Paper No. 20. ILRI, Nairobi.
80. Krkosek, M., 2010. Host density thresholds and disease control for fisheries and aquaculture. Aquac. Environ. Interact. 1, 21–32.
81. Lam, K., Parkin, T., Riggs, C., Morgan, K., 2007. Use of free text clinical records in identifying syndromes and analysing health data. Vet. Rec. 161, 547–551.
82. Langmuir, A.D., 1963. The surveillance of communicable diseases of national importance. N. Engl. J. Med. 268, 182–192.
83. Leung, Z., Middleton, D., Morrison, K., 2012. One Health and EcoHealth in Ontario: a qualitative study exploring how holistic and integrative approaches are shaping public health practice in Ontario. BMC Public Health 12, 358.
84. Lightner, D.V., 2011. Virus diseases of farmed shrimp in the Western Hemisphere (the Americas): a review. J. Invertebr. Pathol. 106, 110–130.
85. Lupo, C., Osta Amigo, A., Mandard, Y.V., Peroz, C., Renault, T., 2014a. Improving the early detection of exotic or emergent oyster diseases in France: identifying factors

associated with shellfish farmer reporting behavior of oyster mortality. Prev. Vet. Med. 116, 168–182.

86. Lupo, C., Osta Amigo, A., Marcé, C., Prou, J., 2014b. Participatory case definition between stakeholders and implications for early detection: a case study of the oyster production in France. Paper Presented at the 2nd International Conference on Animal Health Surveillance, 7–9 May 2014. Cuba, La Havana.

87. MacKenzie, S., Turnbull, J., Weidmann, M., 2015. Toward a 'One' Health Concept in Finfish Aquaculture. Keynote Presentation at the "Aquaculture 2015: Cutting Edge Science in Aquaculture" International Conference, 23–26 August 2015, Montpellier, France.

88. Mariner, J.C., Jones, B.A., Hendrickx, S., El Masry, I., Jobre, Y., Jost, C.C., 2014. Experiences in participatory surveillance and community-based reporting systems for H5N1 highly pathogenic avian influenza: a case study approach. EcoHealth 11, 22–35.

89. Marques, A.R., Pereira, M., Ferreira Neto, J.S., Ferreira, F., 2015. Design and prospective evaluation of a risk-based surveillance system for shrimp grow-out farms in northeast Brazil. Prev. Vet. Med. 122, 355–362.

90. Martin, V., Zhou, X., Marshall, E., Jia, B., Fusheng, G., FrancoDixon, M.A., DeHaan, N., Pfeiffer, D.U., Soares Magalhães, R.J., Gilbert, M., 2011. Risk-based surveillance for avian influenza control along poultry market chains in South China: The value of social network analysis. Prev. Vet. Med. 102, 196–2005.

91. Miller, B., 2011. The Gender and Social Dimensions to Livestock Keeping in South Asia: Implications for Animal Health Interventions (Report prepared for GALVmed).

92. Mittal, S., Gandhi, S., Tripathi, G., 2010. Socio-economic impact of mobile phones on Indian agriculture. Working Paper No. 246. Indian Council for Research on International Economic Relations, New Delhi. http://www. eaber.org /sites/default/files/documents/Socio-economic%20Impact%20of%20Mobile%20Phones%20on%20Indian%20Agriculture.pdf (Accessed 17 November 2015).

93. Mohan, C.V., Corsin, F., Thakur, P.C., Padiyar, P.A.,Madhusudan, M., Turnbull, J.F., Hao, N.V., Morgan, K.L., 2002. Usefulness of dead shrimp specimens in studying the epidemiology of white spot syndrome virus (WSSV) and chronic bacterial infection. Dis. Aquat. Org. 50, 1–8.

94. Moore, C., 2013. Point-of-care tests for infection control: should rapid testing be in the laboratory or at the front line? J. Hosp. Infect. 85, 1–7.

95. Moran, D., Fofana, A., 2007. An economic evaluation of the control of three notifiable fish diseases in the United Kingdom. Prev. Vet. Med. 80, 193–208.

96. Morgan, K.L., Handel, I.G., Tanya, V.N., Hamman, S.M., Nfon, C., Bergman, I.E., Malirat, V., Sorensen, K.J., de C. Bronsvoort, B.M., 2014. Accuracy of herdsmen reporting versus serologic testing for estimating foot-and-mouth disease prevalence. Emerg. Infect. Dis. 20, 2048–2054.

97. Morgan, K.L., Cameron, A., Gustafson, L., 2015. Making surveillance happen: a practical approach to determining disease status and freedom from farm to country. J. Appl. Aquac. 27, 263–278.

98. Naylor, R., Courtney, P., 2014. Exploring the social context of risk perception and behaviour: Farmers' response to bovine tuberculosis. Geoforum 57, 48–56.

99. North, D.C., 1990. Institutions, Institutional Change and Economic Performance. Cambridge University Press, New York.

100. North, D.C., 2000. A revolution in economics. In: Menard, C. (Ed.), Institutions, Contracts and Organisations: A New Institutional Economics Perspective. Edward Elgar Publishing Ltd, Cheltenham, UK, pp. 37–41.

101. Notomi, T., Okayama, H., Masubuchi, H., Yonekawa, T., Watanabe, K., Amino, N., Hase, T., 2000. Loop-mediated isothermal amplification of DNA. Nucleic Acids Res. 28, E63.

102. Oidtmann, B.C., Thrush, M.A., Denham, K.L., Peeler, E.J., 2011. International and national biosecurity strategies in aquatic animal health. Aquaculture 320, 22–33.

103. Oidtmann, B., Johnston, C., Klotins, K., Mylrea, G., Van, P.T., Cabot, S., Berthe, F., 2013a. Assessment of the safety of aquatic animal commodities for international trade: the OIE aquatic animal health code. Transbound. Emerg. Dis. 60, 27–38.
104. Oidtmann, B., Peeler, E., Lyngstad, T., Brun, E., Bang Jensen, B., Stärk, K.D.C., 2013b. Riskbased methods for fish and terrestrial animal disease surveillance. Prev. Vet. Med. 112, 13–26.
105. OIE, 2015. Manual of diagnostic tests for aquatic animals. OIE, Paris. http:// www.oie.int/ international-standard -setting /aquatic-manual/access-online/ (Accessed Jan 22nd 2015).
106. OIE (World Organisation for Animal Health), 2014. Appendix 3.8.1. General guidelines for animal health surveillance. Terrestrial Animal Health Code. 23rd Ed. OIE, Paris.
107. Oladele, O.I., Jood, M., 2010. Factors affecting adoption of livestock identification and traceback system among cattle farmers in Kgalagadi district, Botswana. Livest. Res. Rural. Dev. 22 (Article #147).
108. Ortiz-Pelaez, A., Pfeiffer, D.U., Soares-Magalhães, R.J., Guitian, F.J., 2006. Use of social network analysis to characterise the pattern of animalmovements in the initial phases of the 2001 foot and mouth disease (FMD) epidemic in the UK. Prev. Vet. Med. 76, 40–55.
109. Ostrom, E., 1990. Governing the Commons: The Evolution of Institutions for Collective Action. Cambridge University Press, Cambridge.
110. Otte, M.J., Nugent, R., McLeod, A., 2004. Transboundary Animal Diseases: Assessment of Socio-economic Impacts and Institutional Responses. Livestock Policy Discussion Paper No. 9, Livestock Information and Policy Branch, FAO, Rome.
111. Ottinger, M., Clauss, K., Kuenzer, C., 2016. Aquaculture: relevance, distribution, impacts and spatial assessments – a review. Ocean Coast. Manag. 119, 244–266.
112. Pai, N.P., Vadnais, C., Denkinger, C., Engel, N., Pai, M., 2012. Point-of-care testing for infectious diseases: diversity, complexity, and barriers in low- and middle-income countries. PLoS Med. 9, e1001306.
113. Perry, B.D., McDermott, J., Randolph, T., 2001. Can epidemiology and economics make a meaningful contribution to national animal-disease control? Prev. Vet. Med. 48, 231–260.
114. Pretty, J.N., 1994. Alternative systems of inquiry for a sustainable agriculture. IDS Bull. 25,
115. 37–47.
116. Prévot, D., 2009. Balancing Trade and Health in the SPS Agreement: The Development Dimension. Wolf Legal Publishers, Nijmegen, The Netherlands.
117. Putnam, R.D., 2001. Social capital. Measurement and consequences. Can. J. Policy Res. 2, 41–51.
118. Putnam, R.D., Leonardi, R., Nanetti, R., 1993.Making DemocracyWork: Civic Traditions in Modern Italy. Princeton University Press, Princeton.
119. Radostits, O.M., Gay, G.C., Hinchcliff, K.W., Peter, D., Constable, P.D., 2007. VeterinaryMedicine: A Textbook of the Diseases of Cattle, Horses, Sheep, Pigs and Goats. 10th ed. Saunders (W.B.) Co Ltd, London.
120. Renaud, K., Van Biljon, J., 2008. Predicting Technology Acceptance and Adoption by the Elderly: A Qualitative Study. Annual Research Conference of the South African Institute of Computer Scientists and Information Technologists, Wilderness, South Africa, 6–8 Oct 2008. pp. 210–219.
121. Rich, K.M., Perry, B.D., 2011. The economic and poverty impacts of animal diseases in developing countries: new roles, new demands for economics and epidemiology. Prev. Vet. Med. 101, 133–147.
122. Robertson, C., Nelson, T.A., MacNab, Y.C., Lawson, A.B., 2010a. Review of methods for space-time disease surveillance. Spat. Spatiotemporal Epidemiol. 1, 105–116.
123. Robertson, C., Sawford, K., Daniel, S.L.A., Nelson, T.A., Stephen, C., 2010b. Mobile phonebased infectious disease surveillance systems, Sri Lanka. Emerg. Infect. Dis. 16, 1524–1531.

124. Rodríguez-Prieto, V., Vicente-Rubiano, M., Sánchez-Matamoros, A., Rubio-Guerri, C., Melero, M., Martínez-López, B., Martínez-Avilés, M., Hoinville, L., Vergne, T., Comin, A., Schauer, B., Dórea, F., Pfeiffer, D.U., Sánchez-Vizcaíno, J.M., 2015. Systematic review of surveillance systems and methods for early detection of exotic, new and reemerging diseases in animal populations. Epidemiol. Infect. 143, 2018–2042.
125. Rogers, C.J., Mohan, C.V., Peeler, E.J., 2011. The spread of pathogens through trade in aquatic animals and their products. Rev. Sci. Tech. 30, 241–256.
126. Rushton, J., Viscarra, R., Otte, J., McLeod, A., Taylor, N., 2007. Animal health economics—where have we come from and where do we go next? CAB Rev.: Perspect. Agric. Vet. Sci. Nutr. Nat. Res. 1, 1–10.
127. Sabatini, F., 2009. Social capital as social networks: a new framework for measurement and an empirical analysis of its determinants and consequences. J. Socio-Econ. 38, 429–442.
128. Sahu, B.B., Radheyshy, A.M., Kumar, K., Mukherjee, S.C., Ayyappan, S., 1999. Farmer participatory fish disease surveillance and monitoring: use of PRA tools. Trop. Agric. Res. Ext. 2, 118–123.
129. Sergeant, E.S.G., 2016. Epitools epidemiological calculators. AusVet Animal Health Services and Australian Biosecurity Cooperative Research Centre for Emerging Infectious Disease (Available at: http://epitools.ausvet.com.au.).
130. Sithigorngul, W., Rukpratanporn, S., Pecharaburanin, N., Longyant, S., Chaivisuthangkura, P., Sithigorngul, P., 2006. A simple and rapid immunochromatographic test strip for detection of white spot syndrome virus (WSSV) of shrimp. Dis. Aquat. Org. 72, 101–106.
131. Soliman, H., El-Matbouli, M., 2005. An inexpensive and rapid diagnostic method of Koi Herpesvirus (KHV) infection by loop-mediated isothermal amplication. Virol. J. 2, 83–88.
132. Soto, G., Araujo-Castillo, R.V., Neyra, J., Fernandez, M., Leturia, C., Mundaca, C.C., Blazes, D.L.L., 2008. Challenges in the implementation of an electronic surveillance system in a resource-limited setting: Alerta, in Peru. BMC Proc. 2 (Suppl. 3), S4.
133. Stärk, K.D.C., Regula, G., Hernandez, J., Knopf, L., Fuchs, K., Morris, R.S., Davies, P., 2006. Concepts for risk-based surveillance in the field of veterinary medicine and veterinary public health: review of current approaches. BMC Health Serv. Res. 6, 20.
134. Sunding, D.L., Zilberman, D., 2001. The agricultural innovation process: research and technology adoption in a changing agricultural sector. In: Gardner, B.I., Rausser, G.C. (Eds.), Handbook of Agricultural EconomicsAgricultural Production Vol. 1A. Elsevier Science, Amsterdam, pp. 207–261.
135. Szreter, S.,Woolcock,M., 2004. Health by association? Social capital, social theory, and the political economy of public health. Int. J. Epidemiol. 33, 650–667.
136. Tatem, A.J., Rogers, D.J., Hay, S.I., 2006a. Global transport networks and infectious disease spread. GlobalMapping of Infectious Diseases: Methods, Examples and Emerging Applications.\ 62, pp. 293–343.
137. Tatem, A.J., Hay, S.I., Rogers, D.J., 2006b. Global traffic and disease vector dispersal. PNAS 103, 6242–6247.
138. Taylor, M.D., 2000. The WTO panel decision on Australia's salmon import guidelines: evidence that the SPS agreement can effectively protect human health interests. Pac. Rim. Law Pol. J. 9, 473–506.
139. Teles, F., Fonseca, L., 2015. Nucleic-acid testing, new platforms and nanotechnology for\ point-of-decision diagnosis of animal pathogens. In: veterinary infection biology: molecular diagnostics and high-throughput strategies. Methods Mol. Biol. 1247, 253–283.
140. Teletchea, F., 2015. Domestication of marine fish species: update and perspectives. J. Mar. Sci. Eng. 3, 1227–1243.
141. Thiermann, A.B., 2005. Globalization, international trade and animal health: the newroles of OIE. Prev. Vet. Med. 67, 101–108.

142. Van den Driessche, P.,Watmough, J., 2002. Reproduction numbers and sub-threshold endemic equilibria for compartmental models of disease transmission. Math. Biosci. 180, 29–48.
143. Vergne, T., Vilas, V.J.D.R., Cameron, A., Dufour, B., Grosbois, V., 2015. Capture-recapture approaches and the surveillance of livestock diseases: a review. Prev. Vet. Med. 120, 253–264.
144. Vrancken, R., Boutier, M., Ronsmans, M., Reschner, A., Leclipteux, T., Lieffrig, F., Vanderplasschen, A., 2013. Laboratory validation of a lateral flow device for the detection of CyHV-3 antigens in gill swabs. J. Virol. Methods 193 (2), 679–682.
145. Walker, J.G., Ogola, E., Knobel, D., 2011. Piloting mobile phone-based syndromic surveillance of livestock diseases in Kenya. Epidémiologie et Santé Animale 59 (60), 19–21.
146. Watkins, J., Kitner, K.R., Mehta, D., 2012. Mobile and smartphone use in urban and rural India. Continuum J. Media Cul. Studies 26 (5), 685–697.
147. Wetengere, K., 2011. Socio-economic factors critical for intensification of fish farming technology. A case of selected villages in Morogoro and Dar es Salaam regions. Tanzania. Aquacult. Int. 19, 33–49.
148. Whittington, R.J., Chong, R., 2007. Global trade in ornamental fish from an Australian perspective: the case for revised import risk analysis and management strategies. Prev. Vet. Med. 81, 92–116.
149. WHO, 2005. International Health Regulations (2005). second ed.World Health Organization, Geneva.
150. Wibulpolprasert, S., Gresham, L.S., Smolinski,M.S. (Eds.), 2013. A world united against infectious diseases: Connecting organisations for regional disease surveillanceEmerg. Health Threats J. (Supplement 1).
151. Williams, M.J., Agbayani, R., Bhujel, R., Bondad-Reantaso, M.G., Brugere, C., Choo, P.S., Dhont, J., Galmiche-Tejeda, A., Ghulam, K., Kusakabe, K., Little, D., Nandeesha, M.C., Sorgeloos, P.,Weeratunge, N., Williams, S., Xu, P., 2012. Sustaining aquaculture by developing human capacity and enhancing opportunities for women. In: Subasinghe, R.P., Arthur, J.R., Bartley, D.M., De Silva, S.S., Halwart, M., Hishamunda, N., Mohan, C.V., Sorgeloos, P. (Eds.), Farming the Waters for People and Food. Proceedings of the Global Conference on Aquaculture 2010, Phuket, Thailand22–25 September 2010. FAO, Rome and NACA, Bangkok, pp. 785–874.
152. Wong, W.K.,Moore, A., Cooper, G.,Wagner, M., 2005. What's Strange About Recent Events (WSARE): An Algorithm for the Early Detection of Disease Outbreaks. J. Mach. Learn. Res. 6, 1961–1998.
153. Woolcock, M., Narayan, D., 2000. Social capital: implications for development theory, research, and policy. World Bank Res. Obs. 15, 225–249.
154. World Bank, 2010a. People, pathogens and our planet. Volume 1: Towards a One Health Approach for Controlling Zoonotic Diseases. D.C, The World Bank, Washington.
155. World Bank, 2010b. Chapter 8: Overcoming behavioral and institutional inertia. World Development Report 2010a: Development and Climate Change. The World Bank, Washington, D.C.
156. WTO, 1995. Agreement on the Application of Sanitary and Phytosanitary Measures. World Trade Organisation, Geneva, p. 21.
157. Xie, Z., Xie, L., Pang, Y., Lu, Z., Xie, Z., Sun, J., Deng, X., Liu, J., Tang, X., Khan, M., 2008. Development of a real-time multiplex PCR assay for detection of viral pathogens of Penaeid shrimp. Arch. Virol. 153, 2245–2251.
158. Youde, J., 2011. Mediating risk through the International Health Regulations and biopolitical surveillance. Polit. Stud. 59, 813–830.
159. Zhou, Q.-J., Wang, L., Chen, J., Wang, R.-N., Shi, Y.-H., Li, C.-H., ... Zhang, Y.-J., 2014. Development and evaluation of a real-time fluorogenic loop-mediated isothermal amplification assay integrated on a microfluidic disc chip (on-chip LAMP) for rapid and simultaneous detection of ten pathogenic bacteria in aquatic animals. J. Microbiol. Methods 104, 26–35.

160. Zinsstag, J., Schelling, E., Waltner-Toews, D., Tanner, M., 2013. From "one medicine" to "one health" and systemic approaches to health and well-being. Prev. Vet. Med. 101, 148–156.

# END NOTES

[1]We assume here, and in the rest of this section, that the potential demand, appropriate knowledge base and right institutional setup that underpin the emergence of technological innovation (Sunding and Zilberman, 2001) for new surveillance technologies, already exist.

[2]According to these authors, technology acceptance and adoption are linked but different dimensions. Whereas adoption is the process through which one first becomes aware of a technology and ends up embracing it and making full use of it, acceptance is the attitude towards this technology. Accepting a technology will ensure that it is adopted (for example, if a technology becomes dysfunctional and is not well accepted, it is unlikely to be replaced - and therefore adopted - by the user/owner).

[3]Information asymmetry is said to occur in transactions where one party has more or better information than the other, creating an imbalance in decision-making power and a skewed decision outcome.

[4]The majority of the literature on management of common-pool natural resources tends to be grounded in the works of Ostrom (1990) and Baland and Platteau (1996) on informal solutions and collective action.

[5]The interface and duality between globalised trade and globalised health in the SPS Agreement is extensively discussed in Prévot (2009).

# CHAPTER 7

# Growth Phase Significantly Decreases The DHA-To-EPA Ratio in Marine Microalgae

**Peter Boelen[1,3] Audrey van Mastrigt[1] Henk H. van de Bovenkamp Hero J. Heeres[2] Anita G. J. Buma[1]**

[1]Department of Ocean Ecosystems, Energy and Sustainability Research Institute, University of Groningen, Nijenborgh 7, 9747 AG Groningen, The Netherlands

[2]Green Chemical Reaction Engineering, Engineering and Technology Institute Groningen,

University of Groningen, Nijenborgh 4, 9747 AG Groningen, The Netherlands

[3]BioSolar Cells, P.O. Box 98, 6700 AB Wageningen, The Netherlands

## ABSTRACT

Microalgae are the principal producers of long-chain polyunsaturated fatty acids (LC-PUFAs) such as eicosapentaenoic acid (EPA) and docosahexaenoic acid (DHA) in marine ecosystems. Algae are used in aquaculture systems as direct or indirect feed for zooplankton, filter-feeding mollusks and larval stages of crustaceans and fish. Therefore, it is necessary to select nutrient-rich strains, with high levels of EPA and/or DHA, preferably during the stage of rapid growth. During the course of algal growth (exponential to stationary phase), many microalgal species accumulate lipids, especially triacylglycerols. However, relatively little is known about the effect of growth phase on LC-PUFA accumulation. In the present study, absolute and relative EPA and DHA levels of seven representative species of marine microalgae were determined during different growth phases in batch culture. Four species (*Phaeodactylum tricornutum, Thalassiosira weissflogii, Thalassiosira pseudonana* and *Rhodomonas salina*) accumulated fatty acids during growth. In all these species, intracellular EPA levels were higher during the late stationary growth phase than during exponential growth. In contrast, an increase in DHA content was not observed and therefore the DHA-to-EPA ratio was significantly lower in late stationary phase cultures. These results can be used to improve the nutritional value of microalgae cultivated for application in marine aquaculture systems.

**Keywords:** Aquaculture Growth phase Fatty acids LC-PUFAs Phytoplankton

# INTRODUCTION

Due to their high nutritional value, microalgae are important feed sources in aquaculture systems (Patil et al. 2005). Cultivated marine algae can be used directly, as live feed for (larval stages of) bivalves and crustaceans, or indirectly, as food for zooplankton such as rotifers which, in turn, are used to feed crustaceans or small fish larvae (Muller-Feuga 2000; Chauton et al. 2014). Besides this, several microalgal species are currently being investigated by the aquaculture industry as fish meal or fish oil replacement in the diet of commercially farmed fish (Kousoulaki et al. 2015; Sprague et al. 2015; Sørensen et al. 2016). Marine algae are key organisms in the production of essential long-chain polyunsaturated fatty acids (LC-PUFAs) such as eicosapentaenoic acid (EPA) and docosahexaenoic acid (DHA). These *n*-3 PUFAs are necessary for optimal nutrition and stress tolerance of marine fish, especially at the larval and juvenile stages (Khozin-Goldberg et al. 2011, and references therein). Most fatty acids in algal cells are present as part of membrane lipids (e.g., phospho- or glycosyl-glycerides) or as part of storage lipids, mainly triacylglycerols (TAGs), in the cytosol (Thompson 1996). LC-PUFA content and composition may show significant differences between and within algal classes. Each algal class has roughly its own fatty acid composition, and the EPA–DHA content between algal classes is highly variable (Guschina and Harwood 2006; Lv et al. 2010; Boelen et al. 2013). For a cost-effective production of nutrient-rich marine microalgae, with high levels of EPA and/or DHA, it is necessary to select highly productive strains while optimizing cultivation conditions. Besides taxon-specific EPA and DHA variability, intracellular LC-PUFA content may be influenced by growth conditions, such as temperature, irradiance and $CO_2$concentration (e.g., Jiang and Gao 2004; Pal et al. 2011; Boelen et al. 2013). During the course of batch growth or when grown under stress, many microalgal species accumulate lipids, especially TAGs (Hu et al. 2008). For example, when cultured under nitrate or silicate starvation, TAGs accumulated up to 14–18 % of total dry weight in the marine diatoms *Thalassiosira pseudonana* and *Phaeodactylum tricornutum* (Yu et al. 2009). Although most of the LC-PUFAs can be found in structural membrane lipids (Roessler 1990), LC-PUFAs could be partitioned to or deposited in TAGs during the stationary phase of growth (Khozin-Goldberg et al. 2002; Tonon et al. 2002; Guihéneuf and Stengel 2013). In three of the four species investigated by Tonon et al. (2002) (*Nannochloropsis oculata*, *T. pseudonana* and *Pavlova lutheri*), this resulted in a higher cellular EPA content. Tonon et al. (2002) also reported accumulation of DHA in *T. pseudonana* and *P. lutheri*, although the incorporation into TAGs was significantly lower and accumulation started later during the course of batch growth. In other studies, EPA or DHA content per algal biomass was reduced or not significantly affected by culture age or nutrient limitation (Klein Breteler et al. 2005; Hsiao and Blanch 2006; Gong et al. 2013; Nalder et al. 2015).

The aim of this study was to investigate intracellular accumulation of the essential *n*-3 LC-PUFAs EPA and DHA during different growth phases. Therefore, absolute and relative EPA and DHA levels of seven representative species of marine microalgae were determined during growth in batch culture. The implications of our results for the nutritional value of marine microalgae and thereby their application in aquaculture systems will be discussed.

# MATERIALS AND METHODS

## Experimental setup

Seven species of marine algae were selected, representing different taxonomic groups (Table 1). The selected species were known to have a high EPA and/or DHA content and were often already used in aquaculture. The cultures were obtained from the Roscoff Culture Collection (RCC, Roscoff, France) and the National Center for Marine Algae and Microbiota (NCMA, formerly CCMP, Maine, USA). The cultures were grown in a climate room at 20 °C in *f*/2 enriched (Guillard and Ryther 1962) filter-sterilized seawater adjusted to a salinity of 35 ‰. The species were subjected to a light–dark cycle of 16:8 h (L:D), with a photon flux density of 90 μmol photons $m^{-2}$ $s^{-1}$. Before the start of the experiment, stock cultures were acclimated for at least 2 weeks to the culture conditions in a semi-continuous mode. The cultures were cultured in triplicate in Fernbach flasks with a working volume of approximately 0.5 L. Cell counts and photosystem II (PSII) fluorescence parameters were determined regularly to define growth phase and maximum quantum efficiency of PSII. PSII fluorescence measurements are widely used to recognize stress conditions of the photosynthetic apparatus. A decline in maximum quantum efficiency of PSII often occurs after the exponential phase of growth, when nutrients become limited. Samples for fatty acid analysis and algal biovolume were collected during exponential growth (EXP), at the end of the exponential growth phase or the beginning of the stationary phase (STAT) and after at least 5 days after the beginning of the stationary phase, usually after 14 days of batch growth (LATE STAT). To be able to measure true inter- and intraspecific differences, taking into account differences in total fatty acid content relative to algal biomass, the absolute amounts of EPA and DHA were determined, using algal biovolume as biomass unit. Algal biovolume was calculated from cell counts and cell size measurements. Cell numbers were determined using a Coulter XL-MCL flow cytometer (Beckman Coulter, Miami, FL, USA) as described by van de Poll et al. (2005). To calculate the cell volume of the algae, samples of about 1.5 mL of culture were analyzed using an inverted microscope. The sizes of 50 cells were measured, and cell volume was calculated according to stereometric formulas as given in Hillebrand et al. (1999).

**Table** 1 Details of the investigated species

| Strain | Class | Strain# | Abbreviation |
|---|---|---|---|
| *Phaeodactylum tricornutum* | Bacillariophyceae | CCMP2558 | PT |
| *Thalassiosira weissflogii* | Bacillariophyceae | CCMP1049 | TW |
| *Thalassiosira pseudonana* | Bacillariophyceae | RCC950 | TP |
| *Skeletonema costatum* | Bacillariophyceae | RCC70 | SC |
| *Emiliania huxleyi* | Prymnesiophyceae | CCMP2112 | EH |
| *Isochrysis galbana* | Prymnesiophyceae | RCC179 | IG |
| *Rhodomonas salina* | Cryptophyceae | CCMP1319 | RS |

## PAM fluorometry

PSII fluorescence parameters were measured with a pulse-amplitude-modulated chlorophyll fluorometer (Water-PAM; Heinz Walz GmbH, Germany). Samples (5 mL) were dark-adapted for 15 min and placed in a custom-made cuvette (2 × 2 cm). Basal fluorescence ($F_0$) was measured under weak measuring light, and maximal fluorescence ($F_m$) was determined after a saturating light pulse (0.8 s, 4000 µmol photons $m^{-2}$ $s^{-1}$). The maximum quantum efficiency of PSII was calculated as $F_v/F_m = (F_m - F_0)/F_m$ (Maxwell and Johnson 2000).

## Fatty acid analysis

Lipids were extracted using a modified Bligh and Dyer extraction as described by Guckert et al. (1988). Subsequently, the lipids were subjected to a mild alkaline methanolic transesterification (White et al. 1979) to produce fatty acid methyl esters (FAMEs). Samples of 100 mL of algal culture were centrifuged, and the remaining pellets were freeze-dried for 48 h. The pellets were extracted for 18 h in a single-phase solvent system of 3.5 mL chloroform, 7.5 mL methanol and 2.8 mL 50 mM phosphate buffer (pH 7.4), and a known amount of nonadecanoic methyl ester (Sigma) was added as an internal standard. The samples were transferred to 50-mL centrifuge tubes, and 3.5 mL of chloroform and 3.5 mL of ultrapure Milli-Q water (Millipore) were added. After centrifugation for 5 min, the lower chloroform phase was dried under $N_2$ and the remaining lipids were resuspended in 1 mL of a methanol/toluene mixture (1:1, v/v). One milliliter of 0.2 N MeOH–KOH was added, and the mixture was incubated for 15 min at 37 °C. After neutralizing to pH 6 with 1 M acetic acid, 2 mL of chloroform and 2 mL of Milli-Q water were added. After centrifugation, the chloroform phase with the FAMEs was recovered and the nonpolar FAMEs were separated from the polar compounds over a small $Al_2O_3$ column using dichloromethane (DCM) as eluent. The FAMEs were analyzed on a Hewlett-Packard 5890 gas chromatograph equipped with both a flame ionization detector (FID) and a HP 5972 mass spectrometer detector (MSD). Samples of 5 μL were injected using an HP 6890 autosampler at 60 °C on a Restek RTX 1701 (60 m × 0.25 mm, film thickness 0.25 μm) column. The temperature was held at 60 °C for 1 min and then increased at a rate of 10 °C $min^{-1}$ until 180 °C and then to a final temperature of 250 °C at 5 °C $min^{-1}$, which was maintained for 15 min. The injection temperature was 250 °C, and the detector temperatures were both 280 °C. Long-chain fatty acids (number of carbons ≥14) were identified from mass spectra and retention times by comparison with those of PUFA No. 1 standard mixture (Matreya LLC, USA). Quantification of fatty acids was done by integration of appropriate peak areas calibrated with the known concentration of the added internal standard.

## Statistical analysis

Differences in $F_v/F_m$, fatty acid content or DHA-to-EPA ratio between growth phases and species were analyzed by a two-way analysis of variance (ANOVA) followed by a simple main effects analysis when there was a significant interaction. Significant ($p < 0.05$) differences among growth phases were further analyzed by pairwise comparisons using the

Šidak adjustment for multiple comparisons. Values of relative EPA and DHA content (% of total fatty acids) were square-root-transformed prior to analysis. All statistical analyses were performed using IBM SPSS Statistics (version 22) software (IBM Corporation, Armonk, NY, USA).

## Results

Relative (% of total fatty acid) and absolute (normalized to biovolume) EPA and DHA content showed large variability between the tested species (Table 2; Fig. 1). The diatoms (Bacillariophyceae) showed high EPA and a relatively low DHA percentages. In contrast, the two prymnesiophytes were characterized by a high DHA and a low EPA content. *Rhodomonas salina*, a representative cryptophyte species, showed both high EPA and DHA percentages compared to the other species. Although EPA percentages in *S. costatum* were comparable with the other diatoms (9 % of total fatty acids on average), absolute levels of EPA normalized to biovolume were low (<1 fg $\mu m^{-3}$). DHA percentages were highest in *E. huxleyi* (up to 20 % of total fatty acids), while absolute levels were comparable with *I. galbana* and *R. salina* (between 3 and 6 fg $\mu m^{-3}$). Also, the total fatty acid content varied significantly between species (Table 2; Fig. 2a). The lowest amounts (between 1.6 and 8.2 fg $\mu m^{-3}$) were found in *S. costatum*, while the highest total fatty acid content (122.6 ± 29.8 fg $\mu m^{-3}$) was measured in *T. pseudonana*during the late stationary phase of growth. For all species, the maximum quantum efficiency of PSII ($F_v/F_m$) decreased during growth (Fig. 2b), indicating that the cultures were subjected to stress during the late stationary phase. The interactive effects of species × growth phase were highly significant (Table 2), indicating that the effect of growth phase on total fatty acid content was not the same for all species. Out of the seven species tested, four species (*P. tricornutum*, *T. weissflogii*, *T. pseudonana* and *R. salina*) showed a significantly enhanced total fatty acid content in the late stationary phase as compared to the exponential growth phase (Fig. 2a). Also, the absolute amount of EPA in these species was higher during the late stationary growth phase than during exponential growth, although this effect was not statistically significant for *P. tricornutum* (Fig. 1b). In *P. tricornutum*, absolute EPA and total fatty acid content initially decreased with growth, but both levels were significantly higher at the late stationary growth phase as compared to the stationary growth phase. Highest absolute EPA levels were found in *T. pseudonana* and *R. salina* (8.9 and 7.5 fg $\mu m^{-3}$, respectively). The highest increase in EPA during batch growth was found in *T. weissflogii*. In this species, the absolute EPA content was 3.4 times higher during the late stationary growth phase than during exponential growth. On the other hand, absolute levels of DHA during the late stationary phase were not significantly higher than levels in exponentially growing cultures (Table 2; Fig. 1d). DHA-to-EPA ratios were calculated for species with initial EPA and DHA levels both higher than 1 % (Table 3). In all these species, the DHA-to-EPA ratio decreased significantly during growth. The highest DHA-to-EPA ratio was measured in *R. salina* during exponential growth (0.72 ± 0.02). The highest decrease was observed in *P. tricornutum*. In this species, the DHA-to-EPA ratio was almost four times lower during the late stationary growth phase than during exponential growth.

**Table** 2 Statistical results of two-way ANOVA for effects of growth phase and algal species on relative (% of total fatty acids), absolute (normalized to biovolume) EPA and DHA content, total fatty acid content (fg μm$^{-3}$), maximum quantum efficiency of PSII ($F_v/F_m$) and DHA-to-EPA ratio

| Effect on | Source of variation | *df* | MS | *F* | *p* |
|---|---|---|---|---|---|
| EPA % | Species | 6 | 12.221 | 190.2 | **<0.001** |
| | Growth phase | 2 | 0.444 | 6.9 | **0.003** |
| | Species × growth phase | 12 | 0.577 | 9.0 | **<0.001** |
| | Residual | 42 | 0.064 | | |
| DHA % | Species | 6 | 18.798 | 97.9 | **<0.001** |
| | Growth phase | 2 | 0.180 | 0.9 | 0.400 |
| | Species × growth phase | 12 | 0.542 | 2.8 | **0.006** |
| | Residual | 42 | 0.192 | | |
| EPA | Species | 6 | 61.606 | 69.3 | **<0.001** |
| | Growth phase | 2 | 19.174 | 21.6 | **<0.001** |
| | Species × growth phase | 12 | 4.245 | 4.8 | **<0.001** |
| | Residual | 42 | 0.89 | | |
| DHA | Species | 6 | 30.479 | 58.8 | **<0.001** |
| | Growth phase | 2 | 1.158 | 2.2 | 0.120 |
| | Species × growth phase | 12 | 0.799 | 1.5 | 0.148 |
| | Residual | 42 | 0.519 | | |
| Total fatty acids | Species | 6 | 6924.594 | 67.9 | **<0.001** |
| | Growth phase | 2 | 5849.409 | 57.3 | **<0.001** |
| | Species × growth phase | 12 | 840.981 | 8.2 | **<0.001** |
| | Residual | 42 | 101.932 | | |
| $F_v/F_m$ | Species | 6 | 0.026 | 15.5 | **<0.001** |
| | Growth phase | 2 | 0.230 | 138.6 | **<0.001** |
| | Species × growth phase | 12 | 0.018 | 10.6 | **<0.001** |
| | Residual | 42 | 0.002 | | |
| DHA-to-EPA ratio | Species | 2 | 0.321 | 67.7 | **<0.001** |
| | Growth phase | 2 | 0.158 | 33.3 | **<0.001** |
| | Species × growth phase | 4 | 0.014 | 3.0 | 0.052 |
| | Residual | 16 | 0.005 | | |

Statistically significant $p$ values ($p < 0.05$) are printed in bold

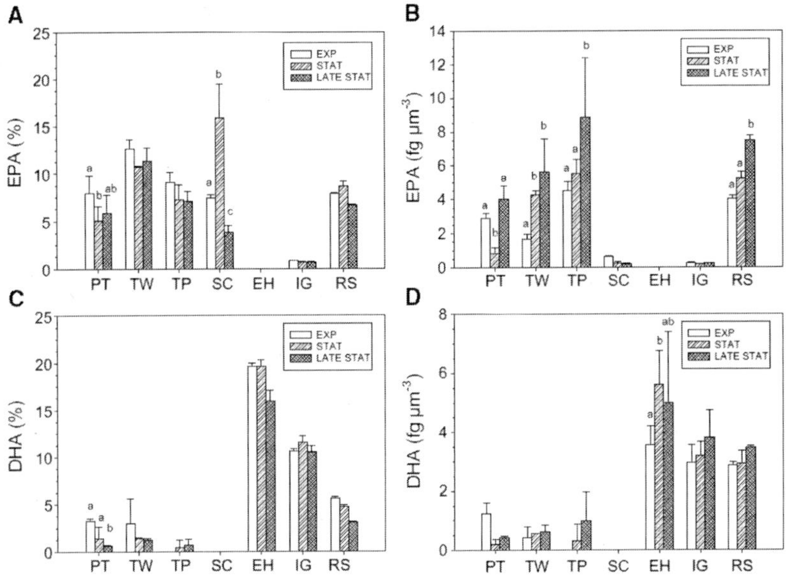


**Figure**. 1 Relative (% of total fatty acids) (**A, C**) and absolute (normalized to biovolume) (**B, D**) EPA and DHA content in seven species of marine algae during different growth phases. Values represent averages (±SD) of three replicate cultures. *Different letters* (per species) represent significant differences at $p < 0.05$ among growth phases according to the analysis of simple main effects. EXP = exponential phase, STAT = stationary phase, LATE STAT = late stationary phase. PT = *Phaeodactylum tricornutum*. TW = *Thalassiosira weissflogii*. TP = *Thalassiosira pseudonana*. SC = *Skeletonema costatum*. EH = *Emiliania huxleyi*. IG = *Isochrysis galbana*. RS = *Rhodomonas salina*

**Table** 3 Effect of growth phase on DHA-to-EPA ratio in three species of marine microalgae. DHA-to-EPA ratios were calculated for species with initial EPA and DHA levels both higher than 1 % (relative to total fatty acids)

| Strain | EXP | STAT | LATE STAT |
|---|---|---|---|
| *Phaeodactylum tricornutum* | $0.43 \pm 0.14^{a}$ | $0.41 \pm 0.17^{a}$ | $0.11 \pm 0.01^{b}$ |
| *Thalassiosira weissflogii* | $0.35 \pm 0.00^{a}$ | $0.13 \pm 0.01^{b}$ | $0.11 \pm 0.01^{b}$ |
| *Rhodomonas salina* | $0.72 \pm 0.02^{a}$ | $0.56 \pm 0.05^{b}$ | $0.46 \pm 0.02^{b}$ |

Values represent averages (±SD) of three replicate cultures. Different letters per row (species) represent significant differences at $p < 0.05$ among growth phases according to the analysis of simple main effects. EXP = exponential phase, STAT = stationary phase, LATE STAT = late stationary phase

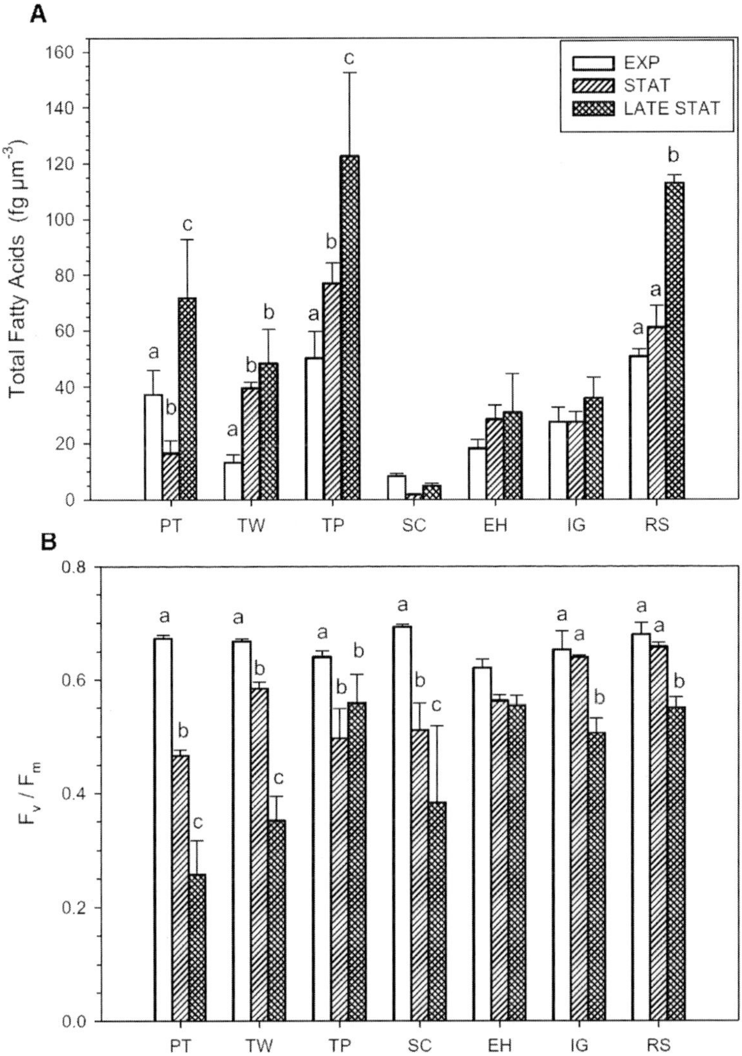


**Figure**. 2 Total fatty acid content (fg $\mu m^{-3}$) (**A**) and maximum quantum efficiency of PSII ($F_v/F_m$) (**B**) of seven species of marine microalgae during different growth phases. Values represent averages (±SD) of three replicate cultures. *Different letters* (per species) represent significant differences at $p < 0.05$ among growth phases according to the analysis of simple main effects. EXP = exponential phase, STAT = stationary phase, LATE STAT = late stationary phase. PT = *Phaeodactylum tricornutum*. TW = *Thalassiosira weissflogii*. TP = *Thalassiosira pseudonana*. SC = *Skeletonema costatum*. EH = *Emiliania huxleyi*. IG = *Isochrysis galbana*. RS = *Rhodomonas salina*

# DISCUSSION

In this study, we focused on the accumulation of the essential *n*-3 LC-PUFAs EPA and DHA in batch cultures of marine algae. Although many studies have demonstrated accumulation of (neutral) lipids in starved cultures of microalgae, relatively little was known about the effect of growth phase on EPA and DHA accumulation. Our study confirms previous results that the absolute amounts of EPA and DHA, but also the total fatty acid content varies considerably between species (Volkman et al. 1989; Viso and Marty 1993; Tonon et al. 2002; Mansour et al. 2005; Patil et al. 2007; Boelen et al. 2013). Consequently, high interspecific variability in their nutritional value exists.

Four out of the seven investigated microalgal species accumulated fatty acids during batch growth. For *E. huxleyi*, *I.* galbana and *S. costatum*, no significant accumulation of lipid compounds could be demonstrated. Possibly, other photosynthetic products (e.g., carbohydrates) accumulated in these species when entering the nutrient-depleted stage (Harrison et al. 1990; Granum et al. 2002).

In all lipid-accumulating species, absolute cellular EPA levels were higher during the late stationary growth phase than during exponential growth. The relative abundances (% of total fatty acids) of EPA in these algae remained relatively constant during the stage of growth, while total cellular fatty acid content was significantly increased. This could be an indication that at least part of the EPA is transferred into or present in TAGs, since in most algae stress-induced lipids are predominantly in the form of TAGs (Hu et al. 2008). It is hypothesized that microalgal TAG can serve as a depot for LC-PUFAs, which could be mobilized for the construction of chloroplastic membranes under sudden changes in environmental conditions (Cohen et al. 2000; Khozin-Goldberg et al. 2005). High proportions of LC-PUFAs present in TAGs could be an advantage when culturing microalgae to produce an alternative for fish oil (Tonon et al. 2002; Leu and Boussiba 2014).

The absolute amount of DHA, however, was not affected, and therefore, the DHA-to-EPA ratio in species containing DHA as well as EPA decreased significantly during growth. This implies that in these species, the accumulation of EPA and DHA seems to be regulated by two different mechanisms. This effect could have consequences for the nutritional quality, since the nutritional value of algae is dependent not only on the total amount of EPA and/or DHA but also on the relative proportion of these LC-PUFAs (Brown 2002; Spolaore et al. 2006). For fish larval nutrition, DHA-to-EPA ratios of between 1 and 2 are advised (Rodríguez et al. 1998). For *R. salina*, which contains significant percentages of both

DHA and EPA and is used as aquaculture feed (e.g., Gagné et al. 2010), this would imply that although the total cellular amount of fatty acids is higher, the nutritional value of this species is lower at the end of the growth phase.

Currently, several microalgal species are being investigated by the aquaculture industry as fish meal or fish oil replacement in the diet of commercially farmed fish (Kousoulaki et al. 2015; Sprague et al. 2015; Sørensen et al. 2016). Replacing the traditional EPA- and DHA-rich, but finite, marine ingredients, fishmeal and fish oil present a significant challenge for the aquaculture industry (Sprague et al. 2016). Promising fish oil alternatives are lipids from microalgae which can potentially grow at high growth rates and which can reach high maximum cell densities with high levels of EPA and/or DHA. Much research has been focused on heterotrophic DHA-producing species such as *Schizochytrium* (Kousoulaki et al. 2015; Sprague et al. 2015). However, Sprague et al. (2015) showed that the absence of EPA in *Schizochytrium*-based fish diet significantly impairs the overall nutritional value. Sørensen et al. (2016) investigated the potential use of the phototrophic EPA-rich *P. tricornutum* as a fish meal replacement in diet for Atlantic salmon. They showed that *P. tricornutum* can replace up to 6 % of the fish meal without adverse effects on nutrient digestibility, utilization of the feed and growth performance. In future studies, it would be useful to test mixtures of DHA-rich *Schizochytrium* and EPA-rich phototrophic algal species as alternative sources of EPA and DHA in aquafeeds. In our study, we showed that intracellular EPA levels in *P. tricornutum* and three other marine algal species can be further enhanced by optimizing the harvest time.

We conclude that in some marine algal species EPA accumulates during batch growth. The highest increase in EPA was measured in *T. weissflogii*, while highest levels were measured in *T. pseudonana* and *R. salina*. Absolute levels of DHA were not affected, which can have consequences for the nutritional value of these algae. This effect has to be taken into account when batch culturing microalgae as a food source in aquaculture systems.

## ACKNOWLEDGMENTS

We thank Willem van de Poll, Patrick Rozema and Ronald Visser for constructive suggestions and technical support. This project was carried out within the research program of BioSolar Cells, co-financed by the Dutch Ministry of Economic Affairs.

# REFERENCES


1. Boelen P, van Dijk R, Sinninghe Damsté JS, Rijpstra WIC, Buma AGJ (2013) On the potential application of polar and temperate marine microalgae for EPA and DHA production. AMB Express 3:26
2. Brown MR (2002) Nutritional value and use of microalgae in aquaculture. In: Cruz-Suárez LE, Ricque-Marie D, Tapia-Salazar M, Gaxiola-Cortés MG, Simoes N (eds) Avances en Nutrición Acuícola VI. Memorias del VI Simposium Internacional de Nutrición Acuícola. 3 al 6 de Septiembre del 2002. Cancún, Quintana Roo, México
3. Chauton MS, Reitan KI, Norsker NH, Tveterås R, Kleivdal HT (2014) A techno-economic analysis of industrial production of marine microalgae as a source of EPA and DHA-rich raw material for aquafeed: research challenges and possibilities. Aquaculture 436:95–103
4. Cohen Z, Khozin-Goldberg I, Adlerstein D, Bigogno C (2000) The role of triacylglycerol as a reservoir of polyunsaturated fatty acids for the rapid production of chloroplastic lipids in certain microalgae. Biochem Soc Trans 28:740–743
5. Gagné R, Tremblay R, Pernet F, Miner P, Samain JF, Olivier F (2010) Lipid requirements of the scallop *Pecten maximus* (L.) during larval and post-larval development in relation to addition of *Rhodomonas salina* in diet. Aquaculture 309:212–221
6. Gong Y, Guo X, Wan X, Liang Z, Jiang M (2013) Triacylglycerol accumulation and change in fatty acid content of four marine oleaginous microalgae under nutrient limitation and at different culture ages. J Basic Microb 53:29–36
7. Granum E, Kirkvold S, Myklestad SM (2002) Cellular and extracellular production of carbohydrates and amino acids by the marine diatom *Skeletonema costatum*: diel variations and effects of N depletion. Mar Ecol Prog Ser 242:83–94
8. Guckert J, Cooksey K, Jackson L (1988) Lipid solvent systems are not equivalent for analysis of lipid classes in the microeukaryotic green alga, *Chlorella*. J Microbiol Meth 8:139–149
9. Guihéneuf F, Stengel DB (2013) LC-PUFA-enriched oil production by microalgae: accumulation of lipid and triacylglycerols containing *n*-3 LC-PUFA is triggered by nitrogen limitation and inorganic carbon availability in the marine haptophyte *Pavlova lutheri*. Mar Drugs 11:4246–4266
10. Guillard RRL, Ryther JH (1962) Studies of marine planktonic diatoms. I. *Cyclotella nana*hustedt, and *Detonula confervacea* (Cleve) Gran. Can J Microbiol 8:229–239
11. Guschina IA, Harwood JL (2006) Lipids and lipid metabolism in eukaryotic algae. Prog Lipid Res 45:160–186
12. Harrison P, Thompson P, Calderwood G (1990) Effects of nutrient and light limitation on the biochemical composition of phytoplankton. J Appl Phycol 2:45–56
13. Hillebrand H, Durselen C, Kirschtel D (1999) Biovolume calculation for pelagic and benthic microalgae. J Phycol 35:403–424
14. Hsiao TY, Blanch HW (2006) Physiological studies of eicosapentaenoic acid production in the marine microalga *Glossomastix chrysoplasta*. Biotechnol Bioeng 93:465–475
15. Hu Q, Sommerfeld M, Jarvis E, Ghirardi M, Posewitz M, Seibert M, Darzins A (2008) Microalgal triacylglycerols as feedstocks for biofuel production: perspectives and advances. Plant J 54:621–639
16. Jiang H, Gao K (2004) Effects of lowering temperature during culture on the production of polyunsaturated fatty acids in the marine diatom *Phaeodactylum tricornutum*(Bacillariophyceae). J Phycol 40:651–654
17. Khozin-Goldberg I, Bigogno C, Shrestha P, Cohen Z (2002) Nitrogen starvation induces the accumulation of arachidonic acid in the freshwater green alga *Parietochloris incisa*(Trebuxiophyceae). J Phycol 38:991–994

18. Khozin-Goldberg I, Shrestha P, Cohen Z (2005) Mobilization of arachidonyl moieties from triacylglycerols into chloroplastic lipids following recovery from nitrogen starvation of the microalga *Parietochloris incisa*. Biochim Biophys Acta 1738:63–71
19. Khozin-Goldberg I, Iskandarov U, Cohen Z (2011) LC-PUFA from photosynthetic microalgae: occurrence, biosynthesis, and prospects in biotechnology. Appl Microbiol Biotechnol 91:905–915
20. Klein Breteler W, Schogt N, Rampen S (2005) Effect of diatom nutrient limitation on copepod development: the role of essential lipids. Mar Ecol Prog Ser 291:125–133
21. Kousoulaki K, Østbye T-KK, Krasnov A, Torgersen JS, Mørkøre T, Sweetman J (2015) Metabolism, health and fillet nutritional quality in Atlantic salmon (*Salmo salar*) fed diets containing *n*-3-rich microalgae. J Nutr Sci 4:e24
22. Leu S, Boussiba S (2014) Advances in the production of high-value products by microalgae. Ind Biotechnol 10:169–183
23. Lv X, Zou L, Sun B, Wang J, Sun M-Y (2010) Variations in lipid yields and compositions of marine microalgae during cell growth and respiration, and within intracellular structures. J Exp Mar Biol Ecol 391:73–83
24. Mansour MP, Frampton DMF, Nichols PD, Volkman JK, Blackburn SI (2005) Lipid and fatty acid yield of nine stationary-phase microalgae: applications and unusual C24–C28 polyunsaturated fatty acids. J Appl Phycol 17:287–300
25. Maxwell K, Johnson GN (2000) Chlorophyll fluorescence—a practical guide. J Exp Bot 51:659–668
26. Muller-Feuga A (2000) The role of microalgae in aquaculture: situation and trends. J Appl Phycol 12:527–534
27. Nalder TD, Miller MR, Packer MA (2015) Changes in lipid class content and composition of *Isochrysis* sp. (T-Iso) grown in batch culture. Aquacult Int 23:1293–131
28. Pal D, Khozin-Goldberg I, Cohen Z, Boussiba S (2011) The effect of light, salinity, and nitrogen availability on lipid production by *Nannochloropsis* sp. Appl Microbiol Biotechnol 90:1429–1441
29. Patil V, Reitan KI, Knutsen G, Mortensen LM, Källqvist T, Olsen E, Vogt G, Gislerød HR (2005) Microalgae as source of polyunsaturated fatty acids for aquaculture. Curr Top Plant Biol 6:57–65
30. Patil V, Källqvist T, Olsen E, Vogt G, Gislerød HR (2007) Fatty acid composition of 12 microalgae for possible use in aquaculture feed. Aquacult Int 15:1–9
31. Rodríguez C, Pérez J, Badía P, Izquierdo M, Fernández-Palacios H, Hernández AL (1998) The *n*-3 highly unsaturated fatty acids requirements of gilthead seabream (*Sparus aurata* L.) larvae when using an appropriate DHA/EPA ratio in the diet. Aquaculture 169:9–23
32. Roessler PG (1990) Environmental control of glycerolipid metabolism in microalgae: commercial implications and future research directions. J Phycol 26:393–399
33. Sørensen M, Berge GM, Reitan KI, Ruyter B (2016) Microalga *Phaeodactylum tricornutum* in feed for Atlantic salmon (*Salmo salar*) -effect on nutrient digestibility, growth and utilization of feed. Aquaculture 460:116–123
34. Spolaore P, Joannis-Cassan C, Duran E, Isambert A (2006) Commercial applications of microalgae. J Biosci Bioeng 101:87–96
35. Sprague M, Walton J, Campbell PJ, Strachan F, Dick JR, Bell JG (2015) Replacement of fish oil with a DHA-rich algal meal derived from *Schizochytrium* sp. on the fatty acid and persistent organic pollutant levels in diets and flesh of Atlantic salmon (*Salmo salar*, L.) post-smolts. Food Chem 185:413–421
36. Sprague M, Dick JR, Tocher DR (2016) Impact of sustainable feeds on omega-3 long-chain fatty acid levels in farmed Atlantic salmon, 2006–2015. Sci Rep 6:21892
37. Thompson GA (1996) Lipids and membrane function in green algae. Biochim Biophys Acta 1302:17–45

38. Tonon T, Harvey D, Larson TR, Graham IA (2002) Long chain polyunsaturated fatty acid production and partitioning to triacylglycerols in four microalgae. Phytochemistry 61:15–24
39. van de Poll WH, van Leeuwe MA, Roggeveld J, Buma AGJ (2005) Nutrient limitation and high irradiance acclimation reduce PAR and UV-induced viability loss in the Antarctic diatom *Chaetoceros Brevis* (Bacillariophyceae). J Phycol 41:840–850
40. Viso A, Marty J (1993) Fatty acids from 28 marine microalgae. Phytochemistry 34:1521–1533
41. Volkman JK, Jeffrey SW, Nichols PD, Rogers GI, Garland CD (1989) Fatty acid and lipid composition of 10 species of microalgae used in mariculture. J Exp Mar Biol Ecol 128:219–240
42. White DC, Davis WM, Nickels JS, King JD, Bobbie RJ (1979) Determination of the sedimentary microbial biomass by extractible lipid phosphate. Oecologia 40:51–62
43. Yu ET, Zendejas FJ, Lane PD, Gaucher S, Simmons BA, Lane TW (2009) Triacylglycerol accumulation and profiling in the model diatoms *Thalassiosira pseudonana* and *Phaeodactylum tricornutum* (Baccilariophyceae) during starvation. J Appl Phycol 21:669–681

# CHAPTER 8

# Impacts of Aquaculture on Habitats and Best Management Practices (BMPs)

**Gulnihal Ozbay[1], Grant Blank[1] and Taworn Thunjai[2]**

[1] *Department of Agriculture and Natural Resources, Delaware State University, Dover, USA*

[2] *Department of Fisheries, Kasetsart University, Ladyao, Jatujak, Bangkok, Thailand*

## 1. INTRODUCTION

The demand for food must be met as the human population reaches an estimated nine billion people by the year 2050. This means we must increase overall food production by 70% and this increase must be sustainable and food price affordable (United Nations FAO 2009). Most of the population growth is expected to continue in underdeveloped countries with limited technologies and venues (United Nations FAO 2009). As a popular high protein food source, seafood contains omega-3 fatty acids that are required for healthy human development (UMD Medical Center 2013). Seafood is low in calories, total fat, and saturated fat, while high in vitamins and minerals including vitamins A and D, phosphorus, magnesium, selenium, and iodine (FAO FOCUS 2013). Fish have been shown to have numerous health benefits (Table 1). Seafood is a healthy, low-fat alternative to beef, poultry, and pork and significant omega-3 fatty acids much higher than vegetable-based diets (FAO FOCUS 2013). Specifically, omega-3 fatty acids contained within fish oil are critically important for infants and babies to develop a normal brain (FAO FOCUS 2013).

Population growth and economic development trends are the most important drivers for the demand for high quality and nutritional seafood products (Ewart 2013). With wild capture fisheries exceeding the maximum sustainable harvest capacity, aquaculture has become a bridge in closing the gap between rising demand and traditional seafood sources (Figure 1). Today, farmed seafood accounts for about 50% of overall production in the global marketplace (Bush et al. 2013). The United States aquaculture industry, valued at over $1.1 billion, produces a variety of fish and shellfish species for food, recreation, and industrial needs (Ewart 2013). However the United States is in a seafood deficit, importing more seafood to meet the demands for seafood consumption than it can produce (NOAA Office of Aquaculture 2013).

Commercial aquaculture is a young and rapidly expanding industry in the United States and the need for information on sustainable growth and development has increased dramatically during the past few decades (Wilson et al. 2002; FAO FOCUS 2013). Aquaculture in the simplest terms is the farming of aquatic plants and animals. Furthermore, the National Oceanic and Atmospheric Administration (NOAA) Office of Aquaculture (2013) describes aquaculture on a broader scale as the breeding, rearing, and harvesting of plants and animals in all types of water environments, including ponds, rivers, lakes, and the ocean. Similar to agriculture, aquaculture can take place in the natural environment or in a manmade environment where controlled cultivation and husbandry of aquatic plants and animals are achieved. Using aquaculture techniques and technologies, researchers, aquaculturists and the aquaculture industry are "growing," "producing," "culturing," and "farming" all types of freshwater and marine species (NOAA Office of Aquaculture 2013). According to Ewart (2013), aquaculture has a long history dating back a few thousands of years in China and Egypt. Aquaculture within the United States dates back to the late 1800s, when hatchery technologies were utilized to cultivate fish for restoration of depleted inland freshwater fishes (Ewart 2013). Ewart (2013) stated with a short commercial history (about 50 years), the United States aquaculture industry has a current annual farm gate value of $1.9 billion. Included in the domestic aquaculture production are variety of fish and shellfish species for food, recreation (stock enhancement, restoration, ornamental fish, aquatic plants, live bait), and industrial applications (food additives).

**TABLE 1.** Nutrition facts on various seafood species (The United States Food and Drug Administration 2008).

# Seafood

## Nutrition Facts

Cooked (by moist or dry heat with no added ingredients), edible weight portion. Percent Daily Values (%DV) are based on a 2,000 calorie diet.

| Seafood Serving Size (84 g/3 oz) | Calories | Calories from Fat | Total Fat g | Total Fat %DV | Saturated Fat g | Saturated Fat %DV | Cholesterol mg | Cholesterol %DV | Sodium mg | Sodium %DV | Potassium mg | Potassium %DV | Total Carbohydrate g | Total Carbohydrate %DV | Protein g | Vitamin A %DV | Vitamin C %DV | Calcium %DV | Iron %DV |
|---|---|---|---|---|---|---|---|---|---|---|---|---|---|---|---|---|---|---|---|
| Blue Crab | 100 | 10 | 1 | 2 | 0 | 0 | 95 | 32 | 330 | 14 | 300 | 9 | 0 | 0 | 20g | 0% | 4% | 10% | 4% |
| Catfish | 130 | 60 | 6 | 9 | 2 | 10 | 50 | 17 | 40 | 2 | 230 | 7 | 0 | 0 | 17g | 0% | 0% | 0% | 0% |
| Clams, about 12 small | 110 | 15 | 1.5 | 2 | 0 | 0 | 80 | 27 | 95 | 4 | 470 | 13 | 6 | 2 | 17g | 10% | 0% | 8% | 30% |
| Cod | 90 | 5 | 1 | 2 | 0 | 0 | 50 | 17 | 65 | 3 | 460 | 13 | 0 | 0 | 20g | 0% | 2% | 2% | 2% |
| Flounder/Sole | 100 | 15 | 1.5 | 2 | 0 | 0 | 55 | 18 | 100 | 4 | 390 | 11 | 0 | 0 | 19g | 0% | 0% | 2% | 0% |
| Haddock | 100 | 10 | 1 | 2 | 0 | 0 | 70 | 23 | 85 | 4 | 340 | 10 | 0 | 0 | 21g | 2% | 0% | 2% | 6% |
| Halibut | 120 | 15 | 2 | 3 | 0 | 0 | 40 | 13 | 60 | 3 | 500 | 14 | 0 | 0 | 23g | 4% | 0% | 2% | 6% |
| Lobster | 80 | 0 | 0.5 | 1 | 0 | 0 | 60 | 20 | 320 | 13 | 300 | 9 | 1 | 0 | 17g | 2% | 0% | 6% | 2% |
| Ocean Perch | 110 | 20 | 2 | 3 | 0.5 | 3 | 45 | 15 | 95 | 4 | 290 | 8 | 0 | 0 | 21g | 0% | 2% | 10% | 4% |
| Orange Roughy | 80 | 5 | 1 | 2 | 0 | 0 | 20 | 7 | 70 | 3 | 340 | 10 | 0 | 0 | 16g | 2% | 0% | 4% | 2% |
| Oysters, about 12 medium | 100 | 35 | 4 | 6 | 1 | 5 | 80 | 27 | 300 | 13 | 220 | 6 | 6 | 2 | 10g | 0% | 6% | 6% | 45% |
| Pollock | 90 | 10 | 1 | 2 | 0 | 0 | 80 | 27 | 110 | 5 | 370 | 11 | 0 | 0 | 20g | 2% | 0% | 0% | 2% |
| Rainbow Trout | 140 | 50 | 6 | 9 | 2 | 10 | 55 | 18 | 35 | 1 | 370 | 11 | 0 | 0 | 20g | 4% | 4% | 8% | 2% |
| Rockfish | 110 | 15 | 2 | 3 | 0 | 0 | 40 | 13 | 70 | 3 | 440 | 13 | 0 | 0 | 21g | 4% | 0% | 2% | 2% |
| Salmon, Atlantic/Coho/Sockeye/Chinook | 200 | 90 | 10 | 15 | 2 | 10 | 70 | 23 | 55 | 2 | 430 | 12 | 0 | 0 | 24g | 4% | 4% | 2% | 2% |
| Salmon, Chum/Pink | 130 | 40 | 4 | 6 | 1 | 5 | 70 | 23 | 65 | 3 | 420 | 12 | 0 | 0 | 22g | 2% | 0% | 2% | 4% |
| Scallops, about 6 large or 14 small | 140 | 10 | 1 | 2 | 0 | 0 | 65 | 22 | 310 | 13 | 430 | 12 | 5 | 2 | 27g | 2% | 0% | 4% | 14% |
| Shrimp | 100 | 10 | 1.5 | 2 | 0 | 0 | 170 | 57 | 240 | 10 | 220 | 6 | 0 | 0 | 21g | 4% | 4% | 6% | 10% |
| Swordfish | 120 | 50 | 6 | 9 | 1.5 | 8 | 40 | 13 | 100 | 4 | 310 | 9 | 0 | 0 | 16g | 2% | 2% | 0% | 6% |
| Tilapia | 110 | 20 | 2.5 | 4 | 1 | 5 | 75 | 25 | 30 | 1 | 360 | 10 | 0 | 0 | 22g | 0% | 2% | 0% | 2% |
| Tuna | 130 | 15 | 1.5 | 2 | 0 | 0 | 50 | 17 | 40 | 2 | 480 | 14 | 0 | 0 | 26g | 2% | 2% | 2% | 4% |

Seafood provides negligible amounts of trans fat, dietary fiber, and sugars.

U.S. Food and Drug Administration
(January 1, 2008)

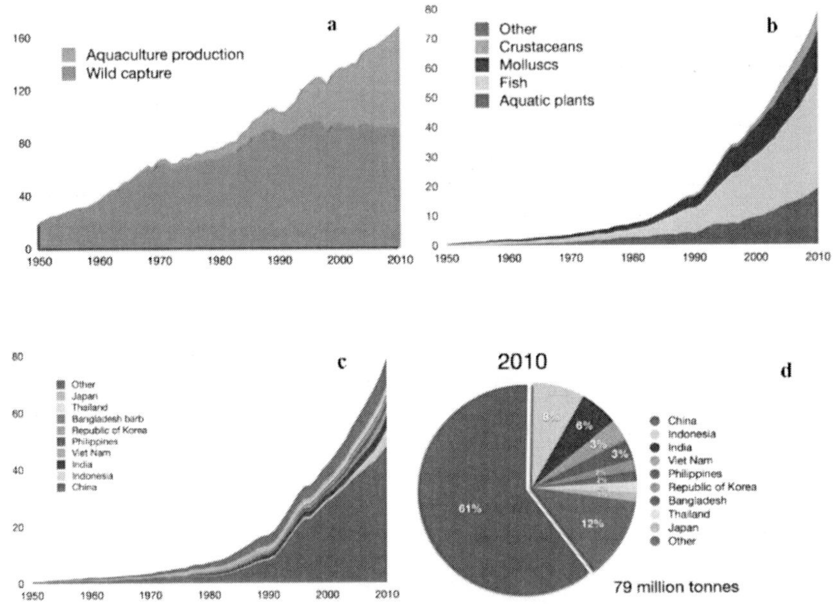


**FIGURE 1.** Global harvest of aquatic organisms in millions of tons between 1950-2010 (FAO 2011); b. Global aquaculture production in millions of tons between 1950-2010 (FAO 2011); c. Main aquaculture countries between 1950-2010 (FAO 2011); and d. Main aquaculture countries in 2010 (FAO 2011).

Aquaculture can benefit more than human economies and diets. Oyster shellfish aquaculture provides many of the same ecosystem services as natural oyster reefs (Dealteris et al. 2004; Erbland and Ozbay 2008). Unlike some finfish farming practices, rearing shellfish in high densities in shallow water with abundant phytoplankton concentrations can have positive effects on the environment and may promote biodiversity (Shumway et al. 2003; Dealteris et al. 2004; O'Beirn et al. 2004; Tallman and Forrester 2007; D'Amours et al. 2008; Erbland and Ozbay 2008; Taylor and Bushek 2008).

As stated by Emerson (1999), the process of aquaculture has been under increasing scrutiny as the world tries to supply food for a population which is currently over seven billion. This criticism is happening regardless of how aquaculture is perceived as an economic windfall for developing countries or potential food industries. Aquaculture is the fastest growing food production sector in the world but its sustainability is not fully satisfied (FAO 2013). This chapter will reassure the ultimate question we ask ourselves: is sustainable aquaculture our solution?

Emerson (1999) discussed how pollution, destruction of sensitive coastal habitats, threats to aquatic biodiversity and significant socio-economic costs must be balanced against the substantial benefits and how aquaculture has great potential for food production and the alleviation of poverty for people living in coastal areas where most of the poorest in the world live. He also stated a delicate balance between food security and the environmental costs of production must be achieved. This leads us to our second question: how do we make the world's fastest growing food sector environmentally and socially responsible?

As we search for answers to these questions, the World Wildlife Federation (WWF 2013) gives additional reasons for why aquaculture must become more responsible. According to the WWF (2013), over 53% of the fisheries worldwide are exploited when over 32% are either depleted, overexploited or recovering including are top ten marine fisheries and as much as 30% of all capture fisheries production are either fully exploited or overexploited. Over 90% of large fish were overfished including are several important commercial fishes (i.e. tuna, skipjack, cod, sturgeon) to the point their survival is threated. Whether it is fully exploited or overexploited, by 2048 fish species harvested for food will be collapsed unless urgent management practices are taken to improve the current conditions (WWF 2013). Unwanted fish (by-catch), like many other animals, die due to inefficient, illegal, and destructive fishing practices every year. This destructive fishing practice along with overfishing largely results in poor fisheries management, pirate fishes, subsidies, and unfair fisheries partnerships (WWF 2013).

Over the past 50 years in the Unites States, the demand for seafood has increased as the population reached over 300 million people (NOAA 2011). Seafood import is over 86% of total seafood demand in the United States (NOAA 2010). Unfortunately, many economically and ecologically important fish species are disappearing from our oceans through over-harvest, loss of habitat, and pollution as we stated earlier. As our most important fisheries are collapsing, fishermen and seafood processors are losing their businesses. A solution to this issue lies in aquaculture, particularly marine aquaculture. Although technological advances enable safe, profitable, and environmentally sustainable culturing of aquatic organisms, sustainability seems to be the key to long lasting aquaculture practices that are profitable and environmentally sound.

Considering the majority of fish we consume are farm-raised fish with over 100 species cultured globally, various culturing technique have been used including traditional earthen ponds to high-tech tank systems, each culture method/technique yields its own environmental foot print (Monterey Bay Aquarium Foundation 2013). Although most aquaculture facilities manage with the best intent for stress reduction, beneficial health and fast growth, many larger intensive aquaculture systems are managed where the stock is raised under stressful environmental conditions where there is little ecological balance. In compensating for poor environmental and health condition, managers have often relied heavily upon chemical, antibiotic and water treatments to get their fish to harvest before the system becomes too stressful for optimum growth. Managers risk rising production costs, stock mortalities, and the degradation of habitats that receive liquid waste discharges (Briggs and Funge-Smith 1994). Dalsgaard et al. (1995) explained the concept of ecological sustainability as maximization of internal feedback within a culture system. They refer this maximization as the integrated resources management practice similar to the agro-ecological engineering approach to integrated agriculture-aquaculture farming used in China. Such a system would minimize inputs and wasted outflows of resources and maximize profits.

As the aquaculture industry grows, the use of treatments unapproved by the Food and Drug Administration (FDA) and/or the misuse of chemicals and treatment strategies administered to culture seafood also grows (FDA 2008). To protect consumers, it is important to ensure that both imported and domestic aquaculture seafood products are free from potentially harmful drug, microbial, and heavy metal residues. These residues in food can cause acute, chronic or microbial effects on people. An acute response from hypersensitivity or allergenicity may occur (FDA 2008). Chronic effects can be long term and they are difficult to detect because these events are typically underreported. Cancer is a potential chronic long term effect (Virtanen et al. 2008). Microbial effects caused by drug residues have an effect on human intestinal flora which limits the activity of intestinal bacteria (FDA 2008). Moreover, antibiotic drug residues can affect the development of resistant bacterial populations. FDA (2008) provides one such example "the unapproved use of fluoroquinolones, such as ciprofloxacin, poses the risk of increasing antibiotic resistant bacteria with the potential for serious human health consequences from untreatable infections. In addition, chronic dietary exposure to high concentrations of fluoroquinolone residues, particularly during early growth, may result in a number of toxicities including joint and testicular lesions." The use of unapproved compounds or misuse of FDA approved new animal drugs, will impact the safety of aquaculture products for the consumers in the United States.

In order to identify potential ways to decrease unnecessary outflows from aquaculture systems in the United States, the National Pollutant Discharge Elimination System (NPDES) permitting is utilized by the United States Environmental Protection Agency (EPA) on a case-by-case basis, typically in larger aquaculture operations. It is difficult to make correlations between aquaculture effluents and environmental impacts without accurate records from each facility. It is crucial to examine the aquaculture practices not only in the United States, but also practices world-wide that can minimize our impacts on aquatic ecosystems and while simultaneously increasing food production.

Some other areas of concern regarding aquaculture include, but are not limited to: eutrophication, benthic enrichment, habitat alteration, erosion, disease, water quality, and effective implementation of best management practices (Coastal Habitat Protection Plan 2005). With the addition of nutrients and phytoplankton, bacteria and viruses, become even more important in regards to aquaculture water quality concerns. There is a direct correlation between bacterial diversity and nutrient content. Naturally occurring bacteria from the environment and the guts of cultured fish stocks thrive in nutrient-rich waters and the surface layers of sediments (Garland Science 2011). There are various pathways humans can be infected by the zoonotic pathogens including food and contact with contaminated environments (Friend 2006). Viruses are a special concern in non-native stocks, where introduced species and hybrids may bring new viral strains into an area. Even if the potential for introduced viruses is reduced, periodic outbreaks of viruses are not uncommon (Yanong and Erlacher-Reid 2012)Y. *Vibrio* bacteria are major fish pathogens that are particularly problematic in aquaculture settings (Chatterjee and Haldar 2012). Uncontrolled proliferation within farm operations appears to have made a direct contribution in the dispersion of *Vibrio* pathogens in receiving water bodies (Yanong and Erlacher-Reid 2012).

Integrated aquaculture may provide solutions to many of aquaculture's problems. Since no organism lives naturally in a vacuum, stocking production facilities with complementary species is a logical way to integrate multiple species while simultaneously increasing production for a given area. For instance, to control algae and plant growth, grass carp or other herbivores may be raised along with primary stock. Suspension feeding bivalves are useful organisms in filtering phytoplankton. Mori (1979) found that phytoplankton concentrations decreased by 94% after water was passed through eleven oyster rafts. Not only are the secondary stocks beneficial in controlling water quality, they often are valuable food products as well. Integrated multi-trophic aquaculture systems yield not only greater profit and lower cost but also enhance economic stability and provide more acceptable management practices (Bastin 2013).

Various fish farming techniques have been used depending on the species and their growth stage. Some of these include but are not limited to: ponds, open net pens or cages, hatchery, bag and rack, raceways, recirculating systems, shellfish culture, submersible net pens, suspended culture, tuna ranching, and aquaponics. Although our discussion is limited primarily to inland aquaculture practices with particular emphasis on pond aquaculture in this chapter, recirculating aquaculture and aquaponics systems are also discussed as popular aquaculture practices that are frequently employed to eliminate potential nutrient loads to the surrounding environment. More specifically, our discussion on recirculating aquaculture and aquaponics systems is due to use of recirculating aquaculture systems for commercial aquaculture species of high market value or application of aquaponics for their roles in minimizing nutrient loads from aquaculture water discharge and increase farm profits by growing alternate crops.

In order to address Best Management Practices (BMPs) in this chapter, we will explore studies from various countries including Thailand, South Africa, United States, Canada, and Australia. These studies address issues regarding how to best manage an aquaculture operation, while minimizing environmental effects and maintain profitable output. BMPs reflect the most technically practical and economically feasible methods which reduce environmental impacts and limit operation costs at aquaculture facilities. One primary goal is to discuss effluent treatment systems that reduce loads of organic matter, suspended solids, and nutrients to prevent polluting receiving waters. The best method to prevent soil and water quality problems includes selecting a site with appropriate soils and an adequate water supply, and maintaining moderate organism densities and feeding rates (Boyd 1989). Secondary management techniques to prevent soil and water quality imbalances include liming, fertilization, and aeration (Boyd 1989). Agricultural irrigation, created wetlands, settling basins, and biological filters also are practical methods for improving the quality of effluents from ponds that will be discussed within this chapter.

In the challenging area of integrating aquaculture Best Management Practices (BMPs), it is imperative that older, proven methods be incorporated with new and innovative ideas. Nearly 40 years ago at Woods Hole, MA and Fort Pierce, FL, Ryther et al. (1975) developed working integrated waste recycling systems utilizing commercially valuable mariculture stocks. Their systems proved so efficient that the final effluent of their system was incapable of supporting further growth or contributing to eutrophication. They suggested that similar systems can be developed for other aquaculture operations to desired needs and purposes. At the Eilat Laboratory in Israel, Neori et al. (1998) established a land-based integrated system that attempted to eliminate external food sources and water exchanges. Avnimelech (2012) provided detailed information on biofloc technology and how this technology can be used to increase farm profit and reduce the nutrient loads of the system. This manual discussed super-intensive biofloc shrimp farming and effects of biofloc technology on the sexual development of shrimp broodstock and other practices. The University of Virgin Islands Aquaculture Program (2013) established a biofloc system that produces tilapia every six months by using biologically active and suspended solids serving as the primary waste treatment in the farm. Additional management practices include good aeration, settleable solid removal, pH adjustments and anaerobic denitrification in this system. Even if this type of system may not be as profitable as growers would like, it is easy to see how the basic principles may be applied to a wide range of aquaculture systems. Unfortunately, there is little impetus to develop such systems unless discharge regulations are increased or the systems are shown to be profitable. Coupled with recirculating systems, aquaponics is described as a synergistic growing technique by the Aquaponics Association (2013) by growing fish and plants together in the same systems. The logic behind this system is that nitrate-nitrogen in fish waste serve as a fertilizer to grow the plants. Once the plants such as lettuce, basil, parsley remove nitrate-nitrogen, this water returns to the fish environment so no water has to be discharged to the environment (The Aquaponics Association 2013).

Integrated aquaculture is nothing new and has been used for thousands of years although their uses commercially are most recent (Bennett et al. 2012). With the demand for high protein food diet, limited resources and environmental concerns, integration provides a solution to maximize profits and reduce potential impacts on the surrounding habitat. By culturing multiple species, farmers can offset the negative impacts in the environment. In China, farmers have been using integrated farming practices for years, although not at the commercial scale at the present time, and have maximized the resource uses to feed their growing population (Bennett et al. 2012).

In this chapter, we will further discuss water quality, eutrophication and disease causing organisms concerns along with effective treatment methods for aquaculture effluents and best farm management strategies in the interest of giving aquaculture professionals, educational professionals, students, and decision-makers a better perspective on how to move forward in a rapidly-changing global market.

Aquaculture will play very important role in feeding about nine billion people by the year 2050 (Nutreco 2011). Meeting this demand can only be possible if seafood is farmed in a sustainable way, both environmentally and economically. As we work together we will find better ways to improve quantity, quality, and sustainability of food supply within the aquaculture sector.

# 2. ISSUES OF SPECIAL CONCERN IN AQUACULTURE

## 2.1. Water Quality and Eutrophication

The highly variable nature of any aquatic environment is often held in a delicate balance by several mechanisms which are common in undisturbed habitats. When anthropogenic stressors (e.g. discharge from aquaculture, farming practices) are introduced into the environment this delicate balance can be disturbed. As a result of an increased aquaculture activity and related farming practices, the effects of seepage and discharge off farms can disturb the healthy conditions of aquatic ecosystems within entire watersheds As described by SAMS (2013), high concentrations of nutrients may lead to deleterious effects, especially in receiving water bodies with the limited water exchange such as lochs. The harmful effects that occur come about as a result of changes in microbial growth and community composition. These changes often result in toxic conditions arising from harmful algae blooms, de-oxygenation of water and sediments from excessive microbial growth, and the transfer/concentration of toxic compounds through the food web. Dissolved and particulate materials in estuaries and coastal environment increase from both natural and anthropogenic sources such as rivers, sewage outfalls, agriculture, and fish farms. These dissolved and particulate materials provide nutrients for phytoplankton and bacteria because they are sources rich in carbon, nitrogen, and phosphorus. Particulate and dissolved materials can also be carriers of heavy metals and drug residues harmful to aquatic life (SAMS 2013). Science Daily (2013) describes eutrophication as the enrichment of an ecosystem with nitrogen or phosphorus, or a mixture of both chemicals. Regarding eutrophication and healthy aquatic system, the water quality variables with the highest concentrations in pond effluents relative to the normal criteria allowed by National Pollution Discharge Elimination System (NPDES) permits are our major concern and discussed in detail throughout the chapter. This includes total dissolved solids, total phosphorus, and biochemical oxygen demand. Eutrophication is the leading problem associated with nutrient runoff of phosphorus (Boyd 2001). Resulting phytoplankton blooms often create an increase in organic matter by two to four times the original amount of metabolic wastes, multiplying the negative effects (Boyd and Queiroz 1997).

As we previously stated, total dissolved solids, total phosphorus, and biochemical oxygen demand are the water quality variables that have the highest concentrations in pond effluents relative to NPDES permits for standard water quality for effluents (Shireman and Cichra 1991; Schwartz and Boyd 1994a). These variables have especially high concentrations in the final 25% of effluent when ponds are completely drained (Boyd

1978; Schwartz and Boyd 1994b; Seok et al. 1995). According to Boyd et al. (2000), total suspended solids and total phosphorus are water quality variables consistently higher in concentration in aquaculture effluents than the typical concentration in effluents of other industries in the southern United States. In comparable studies of the effects of aquaculture effluents on water quality from catfish facilities between Alabama and Mississippi, Hariyadi et al. (1994) found greater concentrations of suspended clay, turbidity, dissolved inorganic phosphorus, total ammonia, and nitrite concentrations. Although effluents from aquaculture facilities with less commonly cultured species have not been studied as thoroughly as channel catfish pond effluents, it is reasonable to assume discharge off aquaculture ponds with other benthic species will have similar nutrient concentrations because of feeding and intensive culture methods. However, the methods of management will vary depending on the species cultured and life stage being cultured. Methods need to be developed for reducing effluent volume and improving the quality of aquaculture effluents in general. Developing specific procedures for removing or reducing suspended solids, total phosphorus and biochemical oxygen demand from pond effluent are especially important. The goal is to develop methods to treat aquaculture discharge so that the materials meet NPDES water quality criteria for effluents (Boyd et al. 1998).

Many techniques have been developed that can be effective in reducing the volume and enhancing the quality of aquaculture pond effluent. These methods include but are not limited to the use of proper site evaluation and design procedures, good construction practices, use of high quality feeds and good feed management, attention to erosion control, moderate stocking densities, reduction in water exchange, seine harvest, and the use of settling basins (Boyd and Tucker 1998). Suitable methods for removing aquaculture waste within effluents include sedimentation, filtration, and mechanical separation using screens, chemical and biological amendments, and using high quality fish-meal (Wheaton 1977, Boyd et al. 1998, Coloso et al. 2001). Boyd and Tucker (1998) summarized methods for using and improving effluents from ponds. These methods have advanced over the years and include hydroponics, irrigation, the development of culture medium for other aquatic organisms, constructed wetlands, settling basins, biological filters, nutrient removal by water hyacinths or other floating macrophytes, and fluidized-bed filters. Queiroz et al. (1998) tested the effectiveness of various bioorganic catalysts[1] - on water quality, soil organic carbon, and channel catfish production and recorded higher concentrations of dissolved oxygen and a slight increase in phytoplankton productivity.

According Boyd and Tucker (1998), the most efficient procedures for treating effluents appear to be irrigation, settling basins, and wetlands.

Filter-feeding fish, mollusks and certain plants have been successfully cultured in aquaculture pond effluents to reduce nutrient and organic matter loadings. Tucker et al. (1996) reported that harvesting fish without draining ponds between fish crops maintained water storage potential and reduced average annual nutrient and organic matter discharge by over 60% relative to annually drained ponds.

Unlike nitrogen or carbon, phosphorus can only enter the watershed via land-use runoff and coastal areas (Thompson and Polz 2006). Release of phosphorus into the aquatic environment is dependent on soil type, landscape slope, rainfall intensity, and the particle trapping capabilities of the watershed in question because phosphorus is considered a particle bound nutrient. Soluble Reactive Phosphorus (SRP) is a biologically available inorganic form of phosphorus often measured in estuarine systems to better assess the available phosphorus used by the aquatic organisms (Mitsch and Gosselink 2007). Through their bioactivity, oysters transport more phosphorus to sediments than they re-mineralize through metabolism (Dame et al. 1989). Mitsch and Gosselink (2007) stated that phosphorus removal within a system occurs through algal cell absorption and co-precipitation of phosphates in high pH waters. Therefore oysters and algae, both of which have been raised in an aquaculture setting, may provide an economical solution to improving the condition of certain effluents.

One of the most efficient methods for removing excess nutrients in water is seaweed culture. Seaweeds absorb the dissolved nutrients, nitrate and phosphate through their whole plant body. The nutrient absorption is very efficient as seaweeds are immersed and waste no energy for uptake and transport of either water or nutrients (SES 2013).

The most severe consequence of eutrophication on estuarine ecosystems is the depletion of dissolved oxygen (Becker et al. 2008) (Figure 2). Oxygen is consumed during the decomposition of organic matter, resulting in hypoxic and/or anoxic conditions unless dissolved oxygen is replaced. Excess organic matter increases microbial populations which utilize the available dissolved oxygen in order to break down the organic matter. Along with the increase in microbial populations, increases in nutrients from organic matter result in phytoplankton blooms. Phytoplankton cannot produce oxygen at night but instead uses up dissolved oxygen in the system that might otherwise be needed by various other resident organisms. If these conditions are sustained over time this can lead to low levels of dissolved oxygen referred to as a hypoxic condition (NOAA 1998). Low dissolved oxygen levels, specifically less than 5mg/L, can result in large fish kills in estuarine waters with limited tidal or water exchange activities and can have a detrimental impact on various commercially important species (Becker et al. 2008). Boesch et al. (2001) stated "in addition to the obvious requirements for fish and shellfish growth, lack of oxygen also limits nitrification and subsequent denitrification, compounding the effects of eutrophication. Rivers, lakes, estuaries, and coastal areas receiving the nutrient rich water with low dissolved oxygen become impaired and ecosystem health is compromised more often." More often dissolved oxygen is the limiting condition in waters of intensive pond aquaculture facilities and this condition is mostly as a result of poor management and bad planning (Boyd 1998).

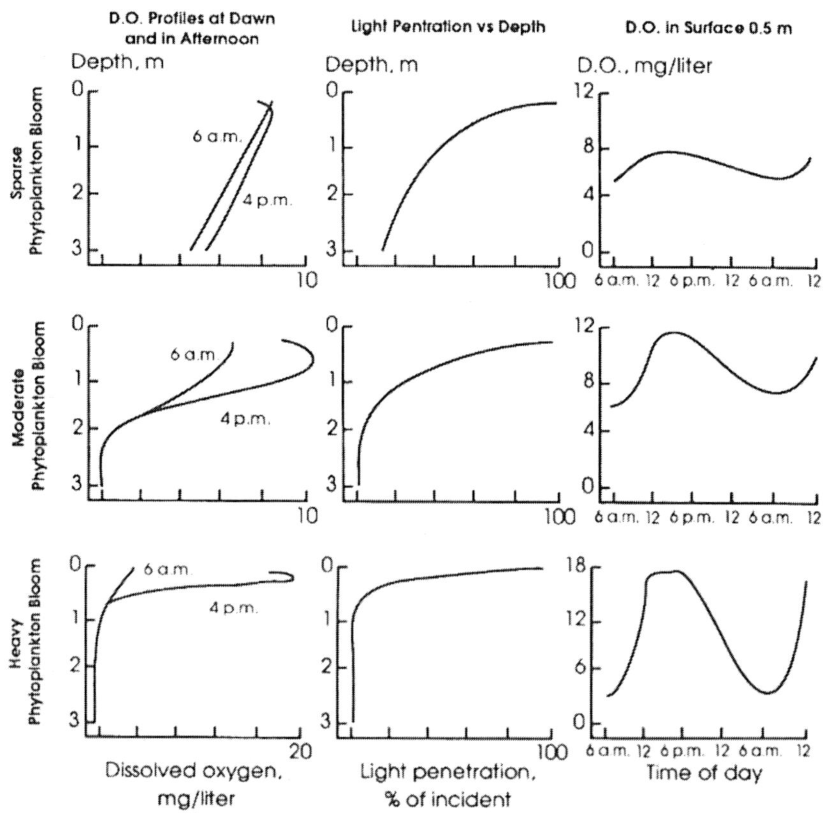


**FIGURE 2.** Relationships among phytoplankton density, dissolved oxygen, and light penetration in fish ponds (Boyd 1990).

Boyd and Musig (1992) summarized that the discharge of effluents below the permitted limits are very important. Effluent discharges by one farm may contaminate the water source of another farm downstream. Therefore, if intake water used for filling pond and for water exchanges are highly polluted, water quality problems can occur at even very low feeding rates.

Poor water quality in incoming water may increase the risk of disease transfer and intensity of any potential diseases. Pollution load created by aquaculture should not exceed the assimilative capacity of the ponds and water supply of that area. Boyd and Queiroz (1997) stated that receiving stream waters assimilate pollutants through various physical, chemical, and biological processes. As long as the pollution load in the pond effluents does not exceed the assimilative capacity of a water body, adverse environmental changes should not occur.

Boyd (1995a) suggests that the best method to prevent soil and water quality problems is by selecting a site with good soils, an adequate supply of high quality water and to maintain moderate levels of fish densities and feeding rates. Secondary management techniques to prevent soil and water quality imbalances include liming, fertilization and aeration. Sedimentation basins may still needed to be considered to prevent ponds from discharging excess sediments.

Similar to nitrogen, phosphorus and dissolved oxygen, pH, alkalinity, hardness, salinity and ammonia are a few other water quality variables that require constant monitoring in modern aquaculture systems because these variables may become a threat to the habitat in receiving waters (Ozbay 2002). The Best Management Practices section at the end of this chapter describes in detail how the impacts of aquaculture farming are minimized and how striving for sustainability is the key for the long term profitable and environmental friendly farming practices.

## 2.2. Pathogens and Disease Risks

As we stated previously, aquaculture refers to culture of organisms (animals or plants) under controlled or semi-controlled conditions. In order to be commercially successful, aquaculture establishments generally have to operate at high density and under conditions which facilitate fast growth. Whatever the species or the type of aquaculture operation (i.e. pond, recirculation, aquaponics, and raceways) in question, maintaining good stock health is the key to successfully operating a profitable aquaculture facility (Bowser 2012). Even when present in low numbers, most disease-causing agents including bacteria, viruses, parasites, and fungi can cause problems and have significant impacts on the fish and associated habitat (Bowser 2012).

The presence of bacteria or viruses in the aquaculture system can be detrimental to the overall operation and surrounding environment. As Pietrak et al. (2010) stated, infection and disease can invade from multiple sources of water, wild fish or shellfish, newly-introduced farmed fish or shellfish, contaminated equipment, predators (i.e. birds, turtles), and human visitors. Newly introduced disease-causing pathogens can lead to production loss from mortality, lost marketability of products, and an inability to transport the product to other locations and farms (Pietrak et al. 2010).

Most diseases and related issues can be prevented by using proper management techniques. It is easier and more cost-effective to prevent disease-causing pathogens from entering the systems than it is treating the pathogens after they have already been introduced into the facility (Bowser 2012). As Bowser (2012) stated, maintaining optimum water quality conditions and keeping the facility clean and well organized are some of the key factors to reduce various stressors which fish are exposed to and will reduce the likelihood of a disease problem. Water quality problems listed as critical are: temperature, dissolved oxygen, pH, alkalinity, hardness, un-ionized ammonia-$NH_3$, nitrite, and potentially toxic substances including heavy metals, drug residues, pesticides, and $CO_2$ (Bowser 2012).

Limited to the few intensively studied commercial aquaculture species, there is currently a large gap in our knowledge concerning diseases associated with other species with potential commercial and ecological importance. Included in this group are Enterobacteriaceae and fecal *Streptococci*, which threaten swimming beaches as well as wild fauna and can easily spread and persist in natural environment (Figueras et al. 2000). Viruses are a special concern in non-native stocks, where introduced species and hybrids may bring new viral strains into an area. Pathogenic bacteria such as ones belonging to the genus *Vibrio* have caused devastating disease outbreaks in shellfish larviculture (Thompson et al. 2004). These outbreaks resulted in substantial financial losses for commercial hatcheries and culture facilities (Austin 2010).

Numerous *Vibrio* species present in the aquatic environment are also common human pathogens including *V. cholerae*, *V.vulnificus*, and *V. parahaemolyticus* and can cause wound infections and gastro-intestinal disease (Austin 2010). Urakawa and Rivera (2006) and Austin (2006) reported other species such as *V.anguillarum*, *V. logei*, and *V. tapetis* as finfish and bivalve pathogens known to cause vibriosis, disease, and in some cases mortality in aquaculture facilities and hatcheries.

Similar to fish and shellfish pathogens, *V. shilonii* and *V. coralliilyticus* are a few *Vibrio* species linked to coral reef bleaching events, having detrimental impacts on the health and biodiversity of highly productive ecosystems (Thompson et al. 2004). Thompson and Polz (2006) reported that *Vibrio* play an important role in nutrient cycling in the aquatic environment by excreting different chitin-degrading enzymes when attached to zooplankton. This could also be devastating to commercial

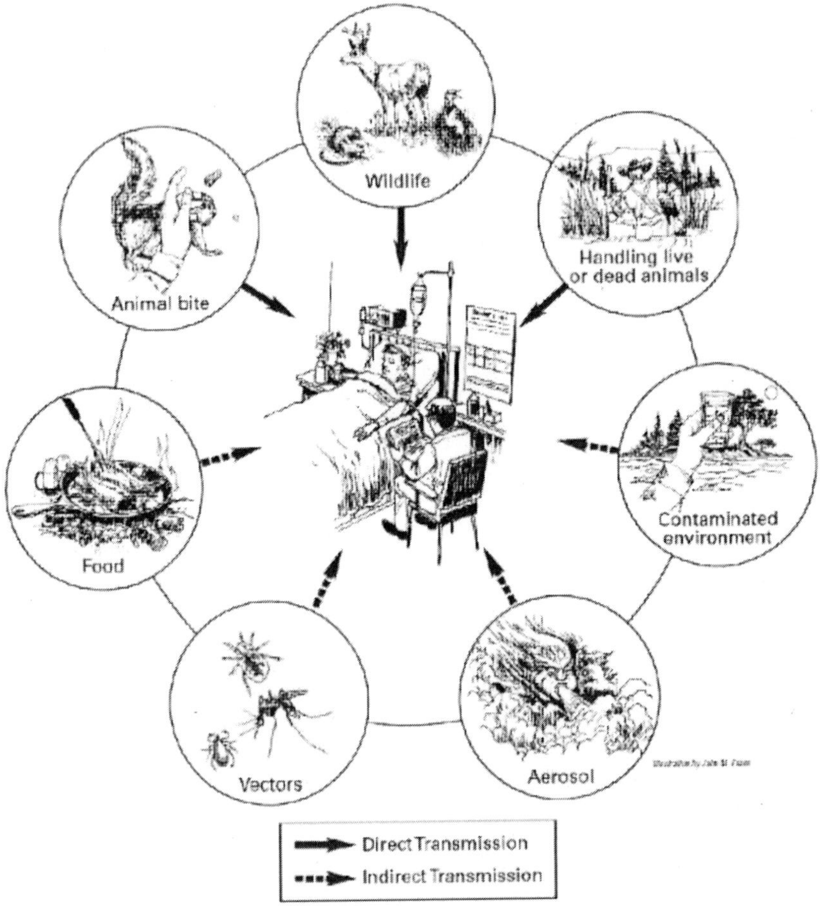


**FIGURE 3.** Common routes for potential transmission of infectious diseases and how they are transferred from animals to human and water to human and animal and others (Friend 2006).

species such as crabs and lobster which rely on chitin exoskeletons. Thompson and Polz (2006) reported *V. cholerae* to occur as a free living form in the water column and attached to zooplankton. The direct relationship between nutrient enrichment via eutrophication and the occurrence of *V. cholerae* in estuarine and coastal environments calls for further investigations (Grimes 1991). Threats of shellfish-borne disease from *V. parahaemolyticus* (Vp) and *V. vulnificus* (Vv) are of significant public health concern in the United States (Baker-Austin et al. 2010). Increased *Vibrio*-related disease incidence and changes in *Vibrio* populations are a likely consequence of changing environmental conditions (Lipp et al. 2002). As Friend (2006) show in Figure 3, humans can be contaminated by eating seafood grown in infected waters. Escaped *Vibrio* from aquaculture can disrupt natural systems and can be a potential threat to wildlife or livestock. Infected wildlife or livestock entering into systems can be a threat the aquatic health as well (Friend 2006)Although our discussion in this chapter is limited to bacterial pathogens, other pathogens causing diseases in fish includes viral infections, fungal infections, water mold infections, such as *Saprolegnia* sp., metazoan parasites, such as copepods, unicellular parasites, such as *Ichthyophthirius multifiliis* (Moyle and Cech 2004).

Preventive measures are the most cost effective and practical ways to minimize disease outbreaks in these types of establishments. The common problems can be avoided by strictly adhering to the following practices: avoiding the movement of animals and farm traffic, having a good background check of the stocks brought into the farm, or certified stocks "pathogen free", utilizing good quality pasteurized feeds and tools to monitor water quality, and keeping good farm records (Pietrak et al. 2010).

Water quality has a direct and vital impact on the transmission of pathogens. Good water quality reduces the risks of transmission and mortality rates. Regardless of outbreak history at a farm, each farm should develop a biosecurity plan and the plan must be adapted to the specific farm and operation, location and culture method, consider existing threats in the area and avoid environmental contaminant risks (Pietrak et al. 2010).

Yanong and Erlacher-Reid (2012) stated biosecurity in aquaculture as the best practice to minimize the risk of introducing an infectious disease into a facility. Likewise biosecurity minimizes the risks where a diseased fish or infectious agents leaves the facility and is able to spread to other facilities and infect other susceptible species. The biosecurity goals they discussed include: animal management, pathogen management, and people management. —According to Yanong and Erlacher-Reid (2012), the main management practice is to obtain healthy animals (eggs, fry, juveniles, brood stocks) and optimize their health and immunity through good husbandry practices. Pathogen management primarily includes prevention, reduction and elimination of pathogens. While preventative practices can be cost-effective and easy to follow through, pathogen reduction and elimination can be very expensive and may cause further environmental and economic damages if the methods fail.

People management practices include educating everyone involved including visitors and suppliers. Well planned and coordinated facility work schedules and periodic worker trainings are the keys to ensure that people follow tight biosecurity plans and keep it in their minds as they complete daily tasks (Yanong and Erlacher-Reid 2012). There are various factors which play important roles for facilitating pathogen entries into a facility, spreading from unit to another, from one species to another in the facility and finally infect the whole facility. These factors depend on the species of concern, their immune status, life stages and susceptibility to pathogens, husbandry practices, and water quality conditions. In addition, understanding further characteristics of a particular pathogen (i.e. biology and life cycle of pathogen, reservoir potential), its survivorship in the facility, on the tools and equipment, application of the approved treatment options, understanding regulatory status and compliance with biosecurity protocols are additional biosecurity measures to minimize disease outbreak risks in a facility.

One application particularly useful for treating and disinfecting pond bottoms is to dry out ponds for one or two weeks, or longer if necessary (Boyd et al. 2012). As Boyd et al. (2012) states, parasites and disease organisms and their vectors survive in areas where puddles and wet areas remain or when the area has constant rain. When those areas cannot be dried they can be treated with burnt lime, calcium oxide, hydrated lime or calcium hydroxide. The purpose of these various chemical applications is to raise the pH above 10 to kill potentially harmful organisms. Boyd et al. (2012) suggested until natural food organisms have re-established, stocking shrimp or fingerling fish in the ponds should be avoided due to the toxicity risks of lime residues. Coagulation with alum, limestone or polyelectrolytes is effective in reducing virus counts (Boyd and Tucker 1998).

A well thought out biosecurity plan is necessary to minimize the potential for catastrophic losses from infectious disease in the facility. Knowledge is the key to understanding the risks associated with disease outbreaks. Knowing your animal, where your fish comes from, water source of the facility, how pathogens may potentially enter, live and persist in the facility, good husbandry practices, diagnostic tools and legal treatment options (Yanong and Erlacher-Reid 2012). Further practices include: having experts aid in the development of a biosecurity plan (production specialist, animal health professional, engineer, scientists...etc.), planning the facility sanitation, disinfection and system management schemes, good water quality monitoring, separating populations by their life stages, good planning of disposal and facility re-arrangement if necessary. Keeping good records of every operation in the farm (i.e. hazardous waste disposal, chemical use, water quality, fish growth and survivorship, feeding, vaccine application) is also critical in maintaining a productive and healthy facility (Yanong and Erlacher-Reid 2012).

Depending on the type of aquaculture operation (ponds, raceways, cages, or recirculating systems) specific biosecurity measures and management practices should be used. Although fundamental practices are generally the same for many biosecurity plans, practices may vary depending on the species cultured, life stage of the animals, pathogen, type of operation and many others listed earlier. The most important aspect of this plan is to prevent disease outbreaks so that economic and environmental risks are reduced. A biosecurity plan, along with the facility sanitation and disinfection practices, is part of the best management practices used in various successful modern aquaculture settings.

# 3. METHODS OF MINIMIZING ENVIRONMENTAL IMPACTS

## 3.1. Wetlands

There are various definitions on what wetlands are and what best describes wetlands. Kalff (2002) described wetlands as the transition zones between terrestrial and aquatic systems where the soils are waterlogged for at least part of the year or covered by shallow water, and which are typically occupied by rooted aquatic vegetation (*macrophytes*); not all wetlands are physically connected to lakes or lotic systems. Occupying three times the surface area of lakes, wetlands cover about 8.6 million $km^2$, or 6.4%, of Earth's land area (Shine and de Klemm 1999). There are tremendous benefits associated with the presence of wetlands (USEPA 2006).Figure 4a shows a healthy wetland and Figure 4b demonstrate the schematic representation of nutrient cycling in the soil-water column of a wetland. Many biogeochemical transformations occur in wetlands and mostly anaerobic conditions exist at the soil water interface. The plants also create an aerobic zone near the roots and different oxidation reduction mechanisms occur in the soil leading to nutrient cycling (USEPA 2008). Within the aerobic zone surrounding plant roots, ammonia is oxidized to nitrate by a process called nitrification; nitrate is then readily diffused into adjacent anaerobic soil. Nitrate is reduced to molecular nitrogen through denitrification, or may be reduced to ammonium under certain conditions through the dissimilatory nitrate reduction process (Ruckauf et al. 2004; Reddy and Delaune 2008). The nitrogen cycle is shown in Figure 4c.

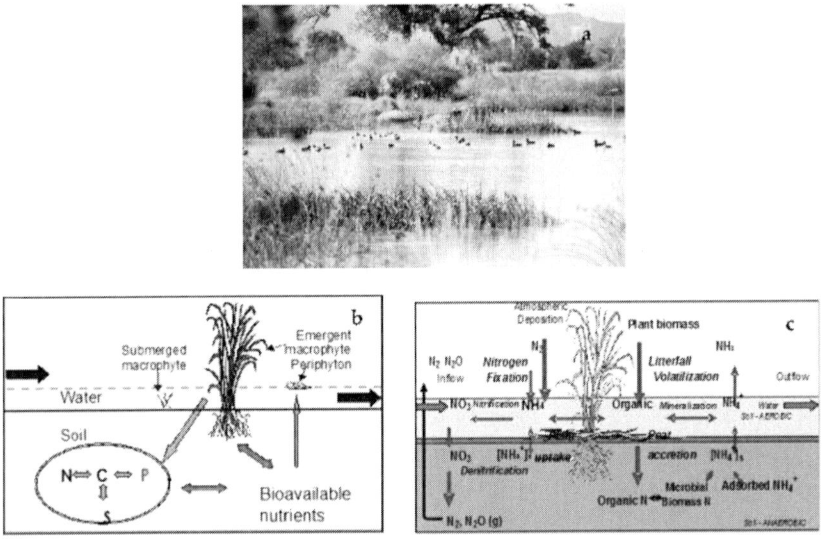


**FIGURE 4.** a. Healthy Wetland. Courtesy of http://www. New bedfordguide.com/wp-content/uploads/2011/02/wetlands1.jpg; b. Basics nutrient cycles in soil-water column of a wetland (USEPA 2008); and c. Nitrogen cycle in wetlands (USEPA 2008).

Phosphorous enters wetlands in different forms ($PO_4^-$, $PO_3^-$...etc.); both biotic and abiotic mechanisms regulate accumulation and transformation of phosphorous compounds within the water column and soil. Biotic processes include assimilation by vegetation, plankton, periphyton, and microorganisms and abiotic processes include sedimentation, adsorption by soils, precipitation, and exchange processes between soil and the overlying water column (Reddy and Delaune 2008). Transformations of nitrogen, phosphorous, sulfur, iron, manganese, and carbon occur in the anaerobic environment and are mostly microbial mediated. Transportation and translocation of transformed constituents occur in the oxidized layer, providing a barrier to translocate some reduced constituents (USEPA 2008). The value of wetlands for flood control, water storage, and water purification are estimated to be $15,000 ha/year (Kalff 2002). Their value as for fish and wildlife habitat, recreation, or maintaining biodiversity must also be considered (Mitch and Goselink 2007).

Unfortunately, half of these wetlands are disappearing and being converted for agricultural uses such as rice monoculture and aquaculture (Kalff 2002; Figure 5). Many nations restrict development in the wetlands because of the ecological value placed on the wetlands (Boyd and Tucker 1998). Considering their significant roles in removing excess nutrients, breaking down harmful metals and toxic substances via microorganisms living in soil, preventing soil erosion controls, capturing solids in flowing waters, providing habitat for many wildlife species, many countries insist restoration of degraded wetlands or the mitigation the creation of the new ones (Mitsch and Gosselink 1993). Mitsch and Gosselink (1993) stated that natural wetlands and constructed wetlands are very effective in reducing nutrient and organic matter concentrations in wastewater. Wetlands act as biological filters by removing suspended minerals and organic matters from water. Natural and constructed wetlands can be used for treating agricultural, municipal, and industrial wastewaters (Moshiri 1995). Boyd and Tucker (1998) describe the removal processes of suspended minerals and organic matter from water by a wetland as: sedimentation of suspended particles, filtration of suspended particles by plant materials, uptake of nutrients by plants and bacteria, decomposition of organic matter, denitrification, nitrification, and adsorption of ions by the soil. Macrophytes in wetland systems play a key role as substrate for periphyton and actively transport oxygen to the rhizosphere, which serves to facilitate chemical transformations in the sediment (Schwartz and Boyd 1995).

**FIGURE 5.** a. Degraded mangroves in Vietnam, courtesy © EJF/Thornton; and b. Shrimp Farm in South America courtesy WWF (http://www.worldwildlife.org/cci/aquaculture_photos.cfm)

Schwartz and Boyd (1995) passed pond effluent through constructed wetlands which drastically reduced concentrations of potential pollutants in channel catfish effluents. Concentrations of total settleable solids, total suspended solids, and total phosphorus were reduced 50% or more by the constructed wetlands except for total ammonia nitrogen due to low hydraulic residence time (HRT) of the wetlands in their study. The greatest removal of total phosphorus (TP), 84% and nitrate-nitrogen ($NO_3$-N), 75% were obtained in wetlands with a four-day HRT (Schwartz and Boyd 1995). Passing water through wetlands was more effective in removing pollutants than simply holding water in the ponds in their study.

There are two basic types of ponds used to raise channel catfish; levee ponds and watershed ponds (Boyd 1985). Levee ponds discharge little water following rains because of their limited watershed area. However, watershed ponds discharge larger water volume following heavy rains due to their larger watershed areas (Schwartz and Boyd 1994a). Most channel catfish farming is conducted in levee ponds where ponds consist of the inside slopes and tops of levees, resulting in high seepage rates especially from rain during the winter. Watershed ponds are usually located much farther apart than levee ponds, so it is typically not feasible to transfer water between ponds for reuse. Boyd and Tucker (1998) suggested that large wetlands could be used to treat effluent when ponds are drained. A smaller wetland could be used for treating the last 25% of highly concentrated effluent from watershed ponds. In a study by Shpigel et al. (2013), the authors demonstrated that nitrogen, phosphorous, and total suspended solids were efficiently removed using *Salicorna* as a biofilter within a constructed wetland. In another study by Lymbery et al. (2013), wetlands removed 60-90% of total nitrogen loads and at least 85% of total phosphorus, and orthophosphate loads from the aquaculture effluent.

Some advantages of utilizing wetlands in the wastewater treatment process include the elimination of chemical treatments, an inexpensive construction process, and wetlands contribute to wildlife habitat and plant communities and to local hydrologic processes. Therefore, using natural wetlands for aquaculture should be minimized to prevent them from disappearing (Schwartz and Boyd 1995; Kalff 2002). Because of the need for large areas, concern arises over the feasibility of using wetlands for treating aquaculture effluents (Schwartz and Boyd 1995; Boyd and Tucker 1998). Integration with pond effluent management procedures might reduce the area of wetland needed for treating fish farm effluents (Schwartz and Boyd 1995). One of best management options allows for the maximization of fish production while maintaining a good pond environment with minimal impacts on the adjacent coastal system including maintaining good stocking densities to improve food assimilation efficiency in a biogeochemical energy model (Serpa et al. 2013).

## 3.2. Settling Basins and Retention Ponds

Settling basins can be built to remove turbidity and suspended solids from pond water supplies. Sediment ponds should be fairly deep to minimize land requirements and to provide hydraulic residence time. In general, a hydraulic residence time of at least 6-8 h is necessary but 2-3 days of retention is preferred (Boyd 1995b). Preliminary sedimentation studies on catfish pond effluents suggested that settleable solids and total phosphorus could be removed as effectively in settling basins as in wetlands (Seok et al. 1995; Boyd et al. 1998). Sedimentation can reduce biochemical oxygen demand by 40 to 50% (Boyd et al. 1998). Schwartz and Boyd (1994a) obtained information on the quality of effluent released from channel catfish ponds during pond draining and fish harvest in watershed ponds. The concentrations of total nitrogen, ammonia nitrogen, soluble reactive phosphorus, total phosphorus, and biochemical oxygen demand started increasing as early as the seining phase (Boyd et al. 2000). Schwartz and Boyd (1994a) suggested that the best way to minimize the pollution potential of aquaculture pond effluents is to harvest ponds as quickly as possible and not to discharge water during seining or to discharge fairly contaminated water into a settling basin or retention pond (Figure 6). Cathcart et al. (1999) suggested that harvesting catfish during late summer/ early fall can significantly decrease effluent discharge from the production ponds due to low water level. This may apply to other species cultured in the ponds such as shrimp and tilapia; however this practice may not be the right fit for other culture systems. Boyd and Musig (1992) found that settleable solids were seldom present in measurable quantities in effluents discharged at shrimp harvest, as seines are not used.

**FIGURE 6.** a. Settling basin. Picture courtesy of Auburn University; b. Constructed wetlands in Mississippi (Tucker 2009) (BMP No. 6 2002).

The maximum instantaneous settleable solids rate allowed by the EPA (1979) is 1 ml/L for 30-day average and 2ml/L for daily maximum (USEPA NPDES 2010). Boyd and Tucker (1998) found that the effluents from catfish ponds might contain settleable matter higher than the allowable limit, and have a moderate oxygen demand at harvest. Boyd et al. (1998) demonstrated that a settling time of 8 hours sufficiently reduce total suspended solids and total phosphorus by 75%, and turbidity and 5-day Biochemical Oxygen Demand ($BOD_5$) by more than 40% in catfish pond effluents. Teichert-Coddington et al. (1999) studied the final effluent from draining shrimp ponds to settling ponds, and obtained near maximum sedimentation of most variables within 6 hours residence, with a removal of 88% of total suspended solids, 100% of settleable solids, 63% of $BOD_5$ and 55% of total phosphorus.

Using separate settling ponds to treat aquaculture effluent can be a problem on commercial fish farms because of the land requirements and construction costs. Production ponds can be utilized as settling ponds, but this would result in a loss of production capacity. Seok et al. (1995) suggested holding the last fraction of pond water for several days in production ponds before discharging it to the environment as a practical way to allay effluent impact.

The characteristics of solids in pond effluents have been studied to provide information for designing and operating pilot sedimentation basins to test their efficiency for treating pond effluents (Ozbay and Boyd 2004; Ozbay and Boyd 2003a; Ozbay and Boyd 2003b; Boyd 1999). Recommendation is made to lower water 25% of its full volume and settle pond effluents for minimum of 2 to 4 hours and more if necessary to remove over 90% of settleable matters, 75% of total suspended solids and over 50% phosphorus loads in the nearby pond used as a settling basin. Cathcart et al. (1999) studied the reduction of effluent discharge and groundwater use in catfish ponds in Mississippi. Deepening the settling ponds receiving overflows from adjacent production ponds reduced the effluent discharges of ponds by 40 – 90%. Hargreaves et al. (2003) summarized in a SRAC Report that over 50% of total suspended solids, total nitrogen, total phosphorus, and biochemical oxygen demand are related to particles less than 5 micrometers in diameter. Boyd (2000) suggested from the estimate of runoff from watershed studies that settling basins used to treat storm runoff from typical watershed type catfish ponds would need to have volumes of 30 to 40% of pond volume in order to provide a retention time of 8 hours. Thus, because of the large volume required, settling basins do not appear to be a feasible solution for treating storm runoff. Settling basins for treating intentional discharge for partial or complete draining would need to be around 10 to 20% of the volume of the largest pond on the farm (Boyd 2000).

Among the frequently applied practices for treating pond effluents such as coagulant application, water exchanges, and settling basins, many farmers advocate settling ponds for effluents. Even though settling has certain benefits in removing solids stirred into water during catfish harvesting, at other times, nutrients and organic matter in effluents are likely to be phytoplankton or dissolved substances, which do not settle easily (Boyd 2000). Schwartz and Boyd (1996) suggested that after seining, the last 25% of effluent water can be held in the pond for two to three days (depending on the farm operation and timing) to allow solids to settle before they are drained completely. This reduced the discharge of settleable solids by 96%, total nitrogen by 74%, and total phosphorus by 69% and organic matter by 59%. This level reduction is very effective but may not be feasible considering limited space availability in most aquaculture farms. Settling basins are not recommended to treat storm runoff of watershed type ponds because of the large volume of pond water required to reach desirable effluent qualification with a retention time of 8 hours (Hargreaves et al. 2003). Lutz et al. (2011) suggest additional buffer strips to allow plants to pick up excess nutrients and allow water to further slow down before it reaches any downstream creeks. Table 2 provides application discharge data on the wastewater treatment plant and recommended maximum daily loads on water quality parameters as main concerns to EPA (USEPA NPDES 2010). Depending on the type of operation and inflow or existing water conditions, outflow water quality parameters are recommended not to exceed the concentrations provided in the table 2 for NPDES Permit.

According to Tucker and Hargreaves (2008), uneaten feed and fecal wastes are the primary producer of solids that potentially degrade environmental conditions at a farm. Solid accumulation can deteriorate the conditions in a facility and create a threat to the aquatic species cultured. Solids can damage fish gills or block shellfish from filtering and increase dissolved oxygen demand due to increased microbial activity in the accumulated organic materials (Tucker and Hargreaves 2008). Excess phosphorus and nitrogen in the sediment with high solids accumulation has a drastic effect in receiving water bodies causing eutrophic conditions (Tucker and Hargreaves 2009). As the most frequently applied tool for solids removal of pond effluents, settling basins or retention ponds are used to mitigate aquaculture effluent or overflows (depending on size and availability) where the water from ponds or other types of aquaculture facilities are treated by the natural processes to minimize or eliminate pollutants (Setty 2013). Particles settle if given enough time by gravity and microbial community which break down excess nutrients and other pollutants into a less harmful or harmless form (Setty 2013). Coupled with other practices suggested for the best management practices, settling basins and retention ponds remove a significant portion of contaminants and excess nutrients as we discussed in this section and have been recommended for pond aquaculture facilities (Boyd and Tucker 1995).

**TABLE 2.** Application discharge data (EPA NPDES 2010).

| Parameter | Units | Discharge Data[(1),(2)] | |
|---|---|---|---|
| | | Maximum Daily Discharge | Average Daily Discharge |
| Flow | MGD | -- | 0.06 |
| pH | Standard Units | 7.7-8.2 (min-max) | |
| Biochemical Oxygen Demand, 5-day ($BOD_5$) | mg/L | 4.9 | 0.6 |
| Total Suspended Solids (TSS) | mg/L | 25 | 7 |
| Ammonia (as N) | mg/L | 6.76 | 4.26 |
| Total Residual Chlorine | mg/L | 0.03 | 0.01 |
| Nitrate and Nitrite N | mg/L as N | 0.68 | 0.35 |
| Total Dissolved Solids (TDS) | mg/L | 553.00 | 496.00 |

## 3.3. Physical Amendments

Sedimentation and filtration are two of the most commonly used particle removal techniques in aquaculture. The applications of this technology have become priceless because untreated effluents or discharges may pose a threat to the environment by carrying various materials in excess quantities including soils, nutrients, and minerals (Ozbay 2002). According to Ebeling and Vinci (2013), total suspended solids (TSS), settleable solids, 5-day biochemical oxygen demand ($BOD_5$), and total phosphorus (TP) are the four major pollutants found in aquaculture effluents/discharge. These pollutants are regulated to ensure that their concentrations can be minimized through the removal of solids containing feces and uneaten feed.

It is important that pollutant concentrations associated with specific particle size ranges in the effluents are considered during the physical removal stage and knowledge on the characteristics of these particles make the removal process more successful (Ozbay and Boyd 2003a). Water quality requirements are frequently discussed but physical characteristics and particle size distribution of the pollutants in the water are not known well. Analytical technology including size fractionation using sieves, laser diffraction, size fractionation using membranes, and characterization using the Coulter registered method have all improved and have been applied in different industries depending on the effluent and discharge characteristics of the particles in question (Cripps 1993). Boyd (2000) reported that about 40% of total suspended solids (TSS), total phosphorus (TP), total nitrogen (TN), and biochemical oxygen demand were associated with particles 51 µm or larger in sizes in catfish ponds. Table 3 shows differences in concentrations of water quality parameters of the pond effluents before and after filtration (Ozbay and Boyd 2003a).Cripps (1995) found that aquaculture wastewaters typically have low TSS concentrations, compared to various industrial and municipal wastewaters, and numerous small particles which clog the 45 µm membranes used to filter the solids out of suspension in aquaculture waters. Suitable treatment techniques should separate particles from the primary effluent flow. Cripps (1995) indicated that by using filters with pore sizes ranging between 200 – 5 µm, increased treatment effects were achieved through sequential decrease in pore size, which removed more nutrients. It is important for a treatment process to remove relatively larger particles, resulting in a reduction in nutrient loading of the effluent.

Cripps (1995) found lower concentration of TN than TP after filtration, and an increase in the filtration effort reduced both nutrients. Most of the plankton/ solids of eutrophic pond waters, with over 50% of the TSS, are found in particles smaller than 10 µm. Ozbay and Boyd (2003a) found that removing particles down to a very small size provided the required benefits targeted for the pond effluents. They recorded percentage removal of total phosphorus (TP), total nitrogen (TN), 5-day biochemical oxygen demand ($BOD_5$), and particulate organic matters (POM) associated with total suspended particles (TSS) removal using filters of different pore sizes (41, 30, 20, 10, 8 and 5µm). Most water quality parameters except total nitrogen were substantially reduced except nitrogen after the effluent water was passed through the filter with 41µm pore size. Further removal was achieved with the consecutive filtration using filters with smaller pore sizes. They only monitored a noticeable reduction of total nitrogen with successively finer filter sizes.

Time required to remove different size particles was also studied by Ozbay and Boyd (2003a) and they suggested a settling time of 24 hours to remove about 30% of TSS and TP and 35% of POM, 25% $BOD_5$ and 20% of TN. Considering the length of time, they do not recommend using settling basins for treating storm overflow and pond draining effluent. In their later study, Ozbay and Boyd (2003b) used turbidity and found it was strongly correlated with total suspended solids and inorganic suspended solids. The relationship was stronger between total suspended solids and inorganic suspended solids as compared to total suspended solids and particulate suspended solids due to the fact that fluctuations in phytoplankton concentration over time changes particulate organic matter concentration.

**TABLE 3.** Percentage of total suspended particles removed by filters of different pore sizes (41, 30, 20, 10, 8 and 5μm). Percentages of total phosphorus, total nitrogen, 5-day biochemical oxygen demand, and particulate organic matters are removed with the removal of total suspended solids (Ozbay and Boyd 2003a).

| **Filter Pore Size (μm)** | **Average Percentage Removal (cumulative)** | | | | |
|---|---|---|---|---|---|
| | **TSS** | **POM** | **TP** | **TN** | **$BOD_5$** |
| **41** | 22.5 | 28.8 | 21.5 | 12.9 | 22.9 |
| **30** | 28.0 | 30.7 | 27.2 | 14.0 | 24.3 |
| **20** | 32.0 | 34.9 | 28.7 | 17.8 | 25.4 |
| **10** | 34.5 | 35.4 | 31.0 | 18.5 | 28.1 |
| **8** | 38.7 | 40.1 | 36.0 | 22.4 | 33.8 |
| **5** | 47.9 | 51.6 | 37.9 | 23.5 | 34.0 |

Ackefors and Enell (1994) found the majority of the phosphorus from fish farms is bound to the particles while nitrogen is not bound to the particles but more in a dissolved form in water. Similar to the phosphorus most of the biodegradable organic matter producing biochemical oxygen demand was in the particles in their study. Cripps (1992) studied the distribution of total phosphorus and total nitrogen in six serially filtered aquaculture effluent samples. Only the fraction containing particles smaller than 5 μm pore size added a disproportionately high nutrient load to the effluent in his study. Reduction rates of 60% for total phosphorus and 34% for total nitrogen were achieved using 5 μm pore size filter. Total nitrogen concentrations were greater in smaller particles than the large particles in his study. However, Cripps (1992) found 69% of the total phosphorus was associated with particles larger than 45 μm diameter. In the effluent the majority of suspended particles produced by the farms were within the size range of 4 – 120 μm diameters. Depending on the particle characteristics and farm effluent, excess nutrient removal can be achieved using the correct diameter in filtration. Cripps (1995) summarized the distribution of the particles for each successive size group and found phosphorus levels depend on the particle size distribution in the pond effluent. Removal of the relatively larger particles separated by the filters had little effect on the size distribution; hence changes in mean diameter were small but the effects were consequential. The phosphorus content in both suspended solids and particles increased significantly with decreased particle sizes (Cripps 1995). Bergheim et al. (1991) used screens with 200 μm or less pore sizes to remove particles. He found further reduction of phosphorus by using the screen with the filter pore size smaller than 5 μm produced negligible results, and in practice he found it difficult to implement. However, the phosphorus content of smaller particles (based on the total phosphorus concentration of water before the effluent was filtered) was significantly greater than larger particles. This difference may appear small but actually represents a large difference in filtration effort (5-200 μm pore size).

Particle size analysis, if used in conjunction with other techniques such as fractionation and nutrient analyses, can be used for the characterization of aquaculture wastes and for monitoring the improvement in wastewater treatment efficiency (Cripps 1994). Membranes can be used for fractionation; however these techniques on their own are limited in practical application. But when combined with other forms of analyses, such as nutrient concentration studies in a known volume of water, the determination of proper treatment techniques is simplified (Cripps 1996).

Although we provided a detailed overview on the feasibility of using filters with various pore sizes, specifically the effectiveness of filters with 5 μm or smaller pore sizes, for the purpose of removing phosphorus and nitrogen, sedimentation is probably the most practical application to remove the large particles in the effluent before further filtration is applied to remove the particles bound to smaller particles which cannot be effectively removed via sedimentation. Sedimentation is discussed in detail in the settling basin section of this chapter. Commonly, screens are placed in front of pond discharge areas to prevent fish, leaves, twigs, or other large debris from escaping in the pond effluent.

## 3.4. Chemical Amendments

The smaller particles of colloidal clay settle slowly and they may impart unwanted turbidity to pond water (Boyd 1998). Some of the chemicals applied to aquaculture ponds to remove this undesired turbidity in pond waters include coagulants like alum (aluminum sulfate), ferric chloride, gypsum, lime and polymers. Although using organic matter to reduce turbidity has advantages from an environmental standpoint, this is difficult to obtain and apply to ponds (Boyd 1990). Coagulants are most often added to alter the physical state of dissolved and suspended solids, thereby facilitating their removal by filtration and sedimentation (Boyd and Tucker

1998; Pepper et al. 1996) (Figure 7). Coagulants destabilize colloids, thereby permitting suspended particles to form aggregates that can settle out of solution. Coagulation with alum, limestone or polyelectrolytes is very effective in removing suspended matter and phosphorus from water (Boyd and Tucker 1998).

**FIGURE 7.** Application of the chemical amendment, lime, to an aquaculture pond in Auburn, Alabama (Photo courtesy of Ozbay).

Boyd (1995) demonstrated that phosphorus precipitates from pond water as insoluble iron, aluminum, or calcium phosphates. Alum and ferric chloride are commercially available sources of aluminum and iron. Aluminum, calcium or iron based coagulants added to poultry litter reduced soluble phosphorus concentrations (Moore and Miller 1994). Gypsum (calcium sulfate) is a source of calcium, however, it is only suggested for use in low alkalinity waters (Boyd 1990). Boyd (1990) observed that alum treatment caused almost immediate flocculation of clay particles, and a great reduction in turbidity within 2 hours in all the treated ponds. However, application of alum produces a strong acidic reaction in water and its use should be limited. Boyd (1998 and 1995) suggested alum for pond treatment if alkalinity is 50 mg/L or above in the water. Alum can remove organic particles and clay colloids in association with phosphorus in water through coagulation and entrapment. Generally, aluminum precipitates with inorganic phosphorus as aluminum phosphate compounds (Gensemer and Playle 1999). Welch and Cooke (1999) applied alum to the surface of eutrophic lakes at rates ranging from 5.5 to 10.9 g $Al/m^3$. They found a 50% decrease in total phosphorus and chlorophyll a concentrations. Cooke et al. (1993) investigated phosphorus removal in order to control algae blooms by using salts of iron, aluminum or calcium (ferric chloride, aluminum sulfate or calcium hydroxide). They reported aluminum salts as being most frequently used in lake restoration. Welch and Cooke (1999) observed decreased in cyanobacteria bio-volume after treatment with alum in unstratified lakes. Jacoby et al. (1994) found the magnitude and blooms of cyanobacteria reduced after 2 consecutive years of alum and sodium aluminate treatments in a polymictic lake. Phosphorus, total phytoplankton, and chlorophyll a concentrations in hypereutrophic lakes were reduced in 3 years following liquid alum, 10 mg Al/L treatment. However, Rowan (2001) found that application of alum at the rate of 50 mg/L immediately after seining resulted in a somewhat greater removal rate of some pollutants during the first hours of settling, but did not result in significantly improved water quality. She suggested using alum after the first hour of settling from seining, and higher application rates of alum would have been necessary to precipitate significant amounts of phytoplankton. Masuda and Boyd (1994) used alum as low as 20 mg/L concentration for catfish pond water, and found significant removal of Soluble Reactive Phosphorus (SRP) in the pond water. They reported no residual effects of alum treatment if used at low concentration (20 mg/L). Boyd (1995a) noted that to increase the amounts of solids removed from water for shrimp ponds utilizing alum would necessitate alum treatment of water in a settling basin. In most situations, settling ponds may be adequate to remove suspended solids (Boyd 1995a).

In ponds if acidity results from increased carbon dioxide and exchangeable aluminum in soil after chemical treatment, total alkalinity and total hardness concentrations are buffered by the applications of agricultural limestone, burnt lime, and hydrated lime (Boyd 1995a). Liming is applied simultaneously to neutralize $H^+$ ions, and eliminate or reduce the risks associated with alum toxicity. Masuda and Boyd (1994) used agricultural limestone and alum in catfish pond water in order to reduce nutrient concentrations. Twenty mg/L alum treatment reduced soluble reactive phosphorus, 80%; total phosphorus, 50%; and turbidity, 45% in their study. Precipitation of phosphorus after calcium hydroxide was rapid and higher than agricultural limestone. Calcium carbonate or calcium hydroxide treatments were also applied to hard water lakes by Prepas et al. (1990) and they reported significant decreases in the concentrations of total phosphorus and chlorophyll-a, resulting increased calcite precipitation and higher phosphorus binding affinity to the sediments. On the other hand, Salonen and Varjo (2000) applied gypsum to a hypereutrophic lake and observed that the treatment reduced the chlorophyll-a concentration. Masuda and Boyd (1994) suggest agricultural limestone or burnt lime in removal of Soluble Reactive Phosphorus (SRP) in the ponds. Schwartz and Boyd (1996) suggested that application of hydrated lime or quick lime to pond bottoms enhances ammonia volatilization, and kills pathogens, and they should not be used very frequently because of their inhibition on microbial activity.

Gypsum has a neutral reaction in water but it has been the least effective of the three (alum, ferric chloride, and gypsum) coagulants used in removing clay turbidity (Boyd 2000; 1990). However, Boyd (1995) suggested that gypsum treatment is better for use in low alkalinity waters because gypsum is a good source of calcium, and is more soluble than liming materials (agricultural limestone, burnt lime, and hydrated lime). Masuda and Boyd (1994) reported drastic decrease in SRP concentration, and lower phytoplankton concentration when calcium concentration was elevated with gypsum application. The effects of gypsum treatment on water quality in sunfish ponds with high alkalinity and low hardness conditions were studied and the gypsum treatment reduced phosphorus concentrations and phytoplankton abundance (Wu and Boyd 1990).

The effectiveness of several different compounds to immobilizing soluble reactive phosphorus found in soil from constructed wetlands was studied by Ann et al. (2000). They found that ferric chloride had immobilized the highest percentage of phosphorus in comparison to other amendments; alum, $Ca(OH)_2$, calcite, and dolomite. These amendments were only effective if applied at higher rates in their study. Cooke et al. (1993) reported that phosphorus inactivation with iron salts has shown only short term effectiveness, and subsequent failure was attributed to sediment anoxia because phosphorus precipitation with iron salts is possible if the water - sediment column is aerobic. However, Boyd (1995) reported that alum generally is cheaper than ferric chloride for pond treatment, and commercially available.

Ferric chloride is not suggested for frequent use in lake restoration because of the potential effects to redox reactions and relevant changes of pH on the solubility of iron-phosphate compounds. Under anaerobic condition, phosphorus bonded to the hydroxyl complexes of ferric iron is solubilized and released to the solution. Under anaerobic conditions, phosphorus from the sediments will be released to the water column therefore ferric chloride treatment for phosphorus precipitation can only be possible in aerobic condition (Rowan 2001).

Gutcho (1977) summarized the uses of polyelectrolytes and concluded that anionic, cationic, and nonionic polyelectrolytes are practical flocculating and clarifying agents in the clarification of water and sewage treatment. They are used in the removal of solids from various industrial wastes (mining, papermaking industries). Gutcho (1977) stated that

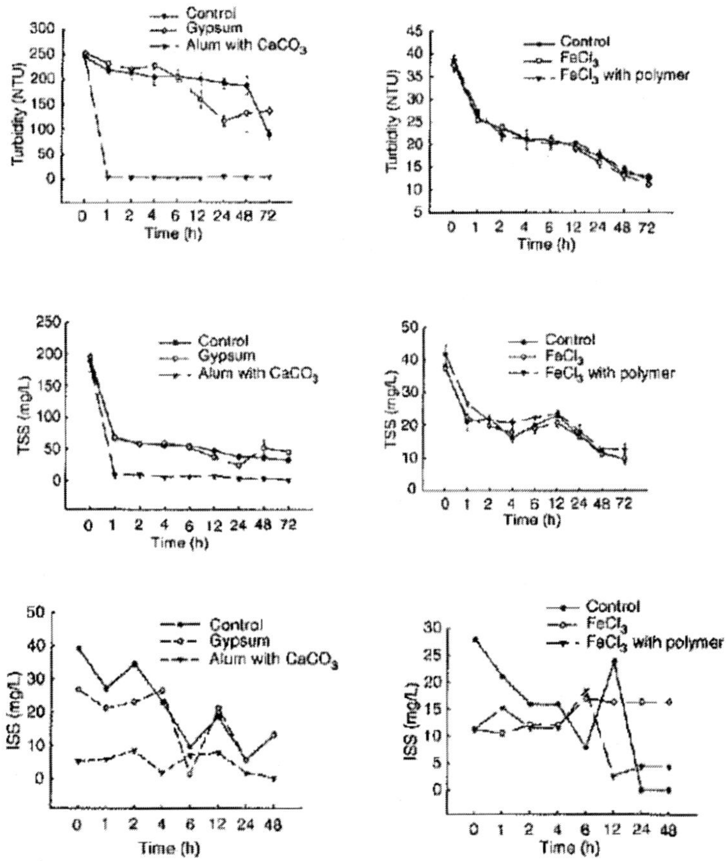


**FIGURE 8.** Means (±SE) of turbidity, total suspended solids (TSS), and inorganic suspended solids (ISS) concentrations in alum with agricultural limestone, gypsum, ferric chloride, and ferric chloride with polymer ((polyacrylamide) treated and control over a 72-72 hour sedimentation period, N=3 (Ozbay 2005).

polyelectrolytes if applied with ferric chloride are more effective in removal of phosphate and organic solids from municipal and wastewaters. Non-ionic polymer is generally applied to remove algae, diatoms, and organic contaminants in lakes and pond waters. Ozbay (2005) studied the effectiveness of gypsum, alum with agricultural limestone, ferric chloride, and ferric chloride with non-ionic polymer (polyacrylamide) removing excess nutrients and solids in the pond waters. She found alum with agricultural limestone treatment removed turbidity, suspended solids, and phosphorus during the sedimentation of pond effluents used in a laboratory set-up. Her research outcome was confirmed during her field application of alum with agricultural limestone and 1 hour was sufficient to remove most of the pollutants in the ponds. Figure 8 below shows the significant reduction in turbidity, TSS and ISS after alum with agriculture limestone application.

Regardless of the chemicals selected for the aquaculture pond treatments, chemical treatment options should be minimized and carefully selected considering their effects on pond sediment pH and potential to increase solubility of various harmful metals.

## 3.5. Biological Amendments

Increased awareness concerning environmental issues has been intensified by the constant pressure propagating from agricultural activities and more specifically from aquaculture farming. Many aquaculture farms have invested in alternative methods to minimize their environmental foot prints and keep their operations profitable. Phytoplankton is an important component of estuaries and coastal waters, reaching high population densities and accounting for a large fraction of the particulate matter in these systems (Wright et al. 1982). Considering high organic loading and detritus inputs from aquaculture effluents, further increase in phytoplankton abundance may become detrimental to estuarine health. Fish reared semi-intensively and intensively on formulated diets generate wastes containing organic particles and soluble nutrients. As a result, un-utilized feed and feces generate additional sources of nutrients, which result in higher abundances of phytoplankton (Lin et al. 1998). Phytoplankton blooms have a drastic effect on the water quality in receiving waters of estuaries and rivers.

During the last few decades, many studies have focused on reducing the secondary effects of poor water quality by means of introducing chemicals (copper sulfate, herbicides...etc.) or introducing herbivorous fish species (tilapia, carp, etc.), or introducing filter-feeding bivalves in order to eliminate the problems associated with heavy phytoplankton blooms (Ozbay 2002).

An alternative method which consisted of rearing manila clams to treat the marine fish pond effluents was attempted in Israel by Shpigel and Fridman (1990). The effluents from gilthead sea bream ponds which contained potential edible organic loading were then passed through manila clams in the effluent pond. They improved water quality by using manila clam as the filter-feeding bivalve in their study, and were able to simultaneously produce a high value product, the manila clam itself. Using manila clams to remove the organic loading, primarily phytoplankton, was an effective method to minimize the nutrient loads in pond effluents, and also produce an additional product for the market. Shpigel and Fridman (1990) also found manila clam to be very adaptable to changes in temperature, salinity, and high organic loading (i.e. phytoplankton) making them an ideal candidate for treating intensive aquaculture effluents. They suggested that this type of operation might have the potential for improving water quality depending on the pond conditions and species cultured.

Newell et al. (1999) found that the eastern oyster can exert a top-down control on phytoplankton stocks and also reduce turbidity, thereby increasing light available to benthic communities. Rehabilitation of natural oyster stocks has the beneficial effect of removing phytoplankton from the water column without stimulating further phytoplankton production. Rensel et al. (2011) investigated nutrient and phytoplankton removal and shellfish growth near the salmon pens. They monitored the highest oyster growth near the salmon pens due to food availability caused by the nutrients in fish feces. Although they did not find substantial differences in water quality parameters, phytoplankton was constantly available and removed by the Pacific oysters (*Crassostrea gigas*). Chrzanowski et al. (1986) investigated the ability of an oyster reef community to remove suspended microbial biomass and observed significant reduction in the suspended microbial biomass. Toro et al. (1999) found a significant negative relationship between oyster growth and amount of particulate inorganic and organic matter in water. Higher organic matters increase oyster growth via their filtration of phytoplankton in the organic matter. Ulanowicz and Tuttle (1992) observed that oyster abundance decreased phytoplankton productivity as well as stocks of pelagic microbes, ctenophores, medusae, and particulate organic carbon. Reduction in turbidity resulted from the removal of suspended solids including inorganic particles and phytoplankton by the oysters that oyster filtration plays an important role for increasing light penetration in the water column (Leffler 2001). Miura and Yamshiro (1990) recorded that phytoplanktivorous freshwater bivalves reduce phytoplankton blooms in the water outflow from fish tanks. Lowe et al. (1991) used mussels to increase water transparency in a lake and also observed a shift toward increased densities of benthic algae and recorded an increase in the visibility of water. Senichieva (1990) observed that actively filtrating mussels transform algae and microorganisms into feces and pseudo feces. Santelices and Martines (1986) found that the production of fecal material by filter feeders function as a fertilizer, and stimulated macroalgae growth that provides a venue to the farmers to integrate filter feeders and macroalgae.

Filter feeding bivalves provide a strong venue for the marine finfish farmers to cope with excess nutrient issues that are a result of un-eaten fish feeds and feces. Through their filtration activities, those filter feeders remove phytoplankton which results from additional nutrients introduced to the system. Various commercially and ecologically important species are dependent on oyster reefs for feeding, reproduction, and shelter from predators, including the naked goby (*Gobiosoma bosc*), skilletfish (*Gobiesox strumosus*), striped blenny, (*Chasmodes bosquianus*), and oyster toadfish (*Opsanus tau*) (Marenghi and Ozbay 2010a,b). There is a unique feeding cycle as these resident fishes feed primarily on benthic invertebrates and fish eggs and they also prey on other benthic fishes and will also eat each other while mud crabs (*Panopeus herbstii*) prey upon their eggs (Harding and Mann 2000). Although not oyster reef obligate, there are many other species that utilize oyster reefs including: black sea bass (*Centropristis striata*), northern pipefish (*Syngathus fuscus*), and Atlantic spadefish (*Chaetodipterus faber*) (Harding and Mann 2000). Oyster shells create habitat and serve as spawning substrate for the Florida blenny (*Chasmodes saburrae*), feather blenny (*Hypsoblennius hentz*) and the frillfin goby (*Bathygobius soporator*) (Tolley and Volety 2005). The larger, more transient, bottom-feeding species such as striped bass (*Morone saxatilis*), juvenile summer flounder (*Paralichthys dentatus*), juvenile winter flounder (*Pleuronectes americanus*), spot (*Leiostomus xanthurus*), black drum (*Pogonias cromis*), American eel (*Anguilla rostrata*), and Atlantic silverside (*Menidia menidia*) also utilize oyster reefs for feeding (Breitburg 1999).

Oyster reefs provide nursery habitat for many economically important juvenile species. Posey (1999) discussed why these reefs become important habitat for those species when natural seagrass beds are limited or absent because of environmental degradation. It is important to mention that 10 $m^2$ of restored oyster reef in the southeast United States is estimated to yield an additional 2.6 kg per year of production of fish and large mobile crustaceans (Peterson et al. 2003). Various ways reefs enhance fish production include increased recruitment, providing refuge from predation, and providing reef-associated prey (Peterson et al. 2003). Because an average size oyster filters 76 liters of water per day, they play a significant role in maintaining natural habitats (The Nature Conservancy 2013). Although aquacultured oysters provide limited but similar services as the natural oyster reefs, they can still effectively remove nutrients and control phytoplankton as they do in nature (Ozbay et al. 2013).

Cultured oysters can serve as broodstock, contributing to enhance and promote naturally occurring populations in the bays. Consecutive research by Marenghi and Ozbay (2010a) and Reckenbeil (2013) found newly settled juvenile oysters within floating oyster gear in man-made, residential canal systems, and on riprap shorelines for the first time around the Delaware

Inland Bays (DIB). It appears that the small scale oyster aquaculture for restoration yields hopeful results within the impaired estuarine conditions as more signs of natural recruitment were observed at several locations within the DIB (Marenghi and Ozbay 2010b).

Rice (2008) discussed how biodeposition of filter feeders, such as bivalves, transfer organic nitrogen in phytoplankton and suspended particles in the water to the sediment. Filter feeding bivalves cycle nitrogen and phosphorus which play an important role in maintaining aquatic productivity (Rice 2008). Similar to Rice (2008), Lin et al. (1993) stated that shrimp-bivalve integrated culture systems in Thailand served as a biological control on phytoplankton populations, thus relieving the nighttime BOD stress. Wright et al. (1982) observed bivalve filtration of natural marine bacterioplankton and their reduction in the presence of bivalves.

Boyd and Queiroz (1997) investigated the feasibility of using salt-tolerant plants (halophytes) that were used as crop plants to remove nutrients from the effluent wastewater stream. The plant-soil system sequestered inorganic nitrite and phosphorus, and removed over 94% and 97% of the applied inorganic nitrogen and phosphorus. Wilson et al. (2002) summarized the use of plankton-feeding fish threadfin shad with channel catfish and monitored improved water quality conditions and enhanced catfish survival in the ponds with threadfin fish. Improvement in catfish production through the use of the Partitioned Aquaculture System (PAS) indicated that PAS offers the potential to eliminate blue-green algal dominance and associated off-flavor problems, while recovering wasted nitrogen and phosphorus discharges, which pose the threat of eutrophication to surface and groundwater supplies (Wilson et al. 2002).

In Yingbin Bay, China, the farmers set a large integrated aquaculture system that is capable of removing excess nutrients. By integrating seaweed and abalone into their main operation for shrimp culture, they were able to improve water quality. Pond bottoms are passed through seaweed and abalone to allow nutrients to be removed before using for shrimp grow-out ponds (Bennett et al. 2012). The authors found that farmers prefer seaweed farming because it reduces financial risks and leads to more consistent profits than shrimp farming however shrimp farming is more profitable for them.

Boyd and Tucker (1998) stated that the grass carp, the common carp, and certain species of tilapia have been evaluated for control of larger plant forms, including filamentous macroalgae. Plankton-feeding fish such as silver carp, bighead carp, tilapias, and gizzard shad are frequently employed in the ponds. Figure 9 shows the pictures of integrated aquaculture farm practices around the world and last picture is the illustration of the multi-trophic aquaculture system.

**FIGURE 9.** Pictures of various integrated farm practices around the world; a. tilapia culture with hydroponics herbs culture (http://land.allears.net/blogs/jackspence); b. TamilNadu Agricultural University horticulture fish farming integrated system (agritech.tnau.ac.in); c. Malawi fish farm and fruit trees along the edges (http://www.afap.org/africa_masasa); d. shellfish macroalgae culture (E & T Magazine, eandt.theiet.org); and e. Integrated Multi-trophic Aquaculture Schema (Government of Canada, dfo-mpo.gc.ca).

Tseng et al. (1991) reported that low concentrations of ammonia nitrogen and optimum algal density are better for controlling dissolved oxygen levels in tilapia ponds. Generally, microalgae stabilize pond water quality via either ammonia uptake or oxygen production. Burke and Bayne (1986) studied the effects of paddlefish on zooplankton, chlorophyll *a*, total ammonia nitrogen, and nitrite in a yearling paddlefish-catfish polyculture system. Higher seasonal mean chlorophyll *a* concentrations associated with lower zooplankton densities were measured in paddlefish treatment ponds. Smith (1985) found that filter feeders reduced algal biomass as much as 99%, increased phytoplankton diversity, and improved silver carp growth compared with other ponds without filter feeders, because filter feeders allowed high densities of zooplankters to remain and be available for fish. Fott et al. (1979) introduced carp in whitefish ponds and observed an increase in light penetration while primary production of phytoplankton and small zooplankton concentrations decrease substantially in the ponds.

During the last few decades integrated aquaculture practices have become a popular method to reduce the nutrient loads and pollutants entering natural waterways, and also increase profits by culturing more than one species of animal and/or plant. Canadian Aquaculture (2012) describes integrated multi-trophic aquaculture systems as the farming of various aquaculture species together where feces of one species serve as the feed to another, as demonstrated in fish /bivalve relationships. This system also increases profits for the farm and decreases its negative environmental footprint. An aquaculture operation consisting of blue mussels and kelps located near pre-established Atlantic salmon aquaculture sites could substantially increase water quality and profits for the farmers in question. We provided a detailed overview on recirculating aquaculture systems and associated aquaponics systems, and their applications for integrated farming practices which ideally will result in economic and ecological benefits in our next sections.

## 3.6. Feeding and Diet Manipulation

Discharge of unused nutrients in effluents impacts eutrophication and different ecological measures. Impacts from aquaculture feed derived wastes have been observed on the natural environments (Gowen 1991). Boyd and Queiroz (1997) reported that in channel fish ponds in the United States, pond water quality was correlated to stocking and feeding rates. Water quality rapidly deteriorates at feeding rates above 100 - 120 kg/ha per day. In ponds utilized for fertilization and feeding, water quality is related primarily to nutrient input rates. Boyd and Queiroz (1997) stated that part of the nutrients in feeds and fertilizers is recovered in the harvested product, but the remaining nutrients enter the pond ecosystem as inorganic nitrogen, phosphorus, and carbon, dissolved and particulate organic matter. Therefore, relatively small percentages of nitrogen, phosphorus, and organic carbon are recovered in the harvested product. Consequently, the concentrations of nutrients and organic matter in the pond waters and the amount of organic matter settling onto bottoms increase as fertilization and feeding rates increase.

High quality feeds improve feed conversion ratios and reduce quantities of metabolic wastes and uneaten feed (Schwartz and Boyd 1996). Conservative feeding practices, and lower stocking rates also reduces feed inputs and improves feed conversion ratios (Boyd et al. 2000). Feeds are the ultimate source of organic matter pollution in catfish pond effluents (Boyd et al. 2000). The main types of wastes in aquaculture are residual feed particles, fecal matter, and metabolic by-products. Inefficient feed conversion results in poorer quality effluents and also decreases the concentrations of dissolved oxygen as shown in Figure 10. In a study by Filbrun and Culver (2013), dissolved oxygen levels in the ponds were increased by decreasing the feeding rates. The nitrogen in uneaten feed is transformed to ammonia by bacteria. Ammonia nitrogen tends to increase as feed application to a pond increases, and concentrations above 2 mg/L can be very harmful to aquaculture species at high pH (Gross et al. 1999). Ammonia is also added to ponds through fish excretions.

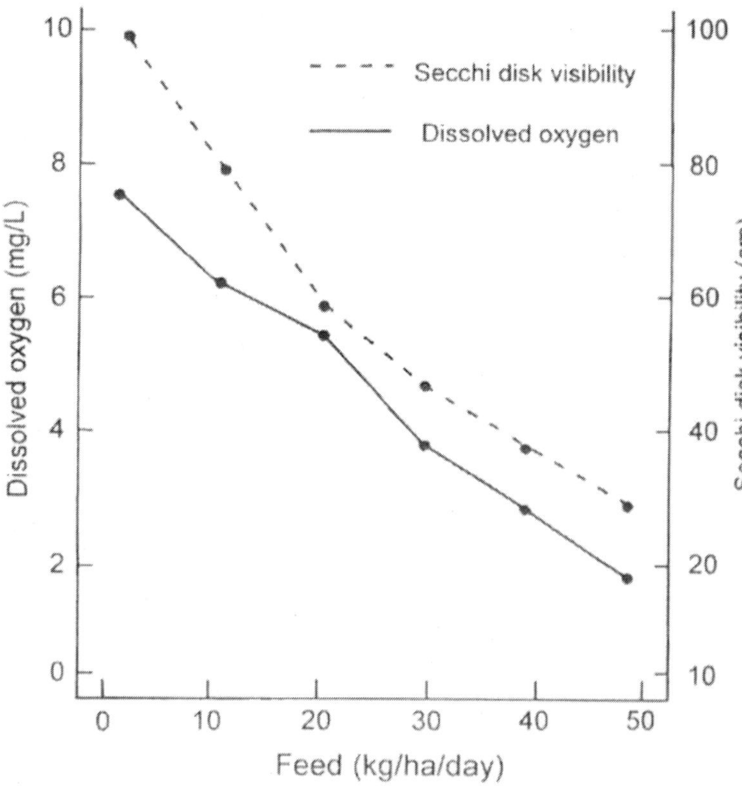


**FIGURE 10.** Effect of feeding rate on dissolved oxygen concentrations and Secchi disc visibilities at dawn (Boyd 2001).

Cripps (1995) stated that it is likely that ponds containing specific sized particles would have elevated nutrient concentrations, resulting from their origin in the diet. Abou et al. (2012) demonstrated that using fern (*Azolla spp.*) as a fish meal substitute for Nile Tilapia had tremendously limited phosphorous loss in the effluent and is considered environment friendly. Coloso et al. (2001) found that soluble phosphorus discharge in effluent water can be reduced in fish fed diets that contained little or no fishmeal, or in diets that were supplemented with a low level of dietary phosphorus. In their rainbow trout study, the dietary combination of low phosphorus and high vitamin $D_3$ decreased soluble and fecal phosphorus levels in the effluents, indicating a strategy whereby effluent phosphorus concentrations can be reduced by regulation of phosphorus metabolism. Increasing bioavailability of phosphorus will eliminate excess phosphorus in the effluent water.

According to Boyd and Queiroz (1997), increasing stocking density and feeding rates above assimilation capacity of pond water can harm the aquaculture pond in the long run. Heavy circulation, aeration, fish respiration and activities, plant abundance, feeding, fertilization, and stocking density in the ponds induced increases in concentrations of potential pollutants, which then require increased treatment efforts to reduce.

Nutrient manipulations were evaluated to promote more desirable phytoplankton communities by eliminating blue-green algae (Wilson et al. 2002). These methods include manipulating the ratio of nitrogen to phosphorus in the water, reducing the availability of phosphorus in bottom muds, enhancing the availability of inorganic carbon, increasing levels of salinity and potassium, and manipulating trace metal availability. Studies showed that various manipulations of waterborne plant nutrients have little promise for controlling phytoplankton community composition in catfish ponds with high feeding rates. Gross et al. (1999) found that the differences in phosphorus input among three feeds, containing 28, 32, and 36% crude protein, did not affect phosphorus concentrations in the effluents because most of the phosphorus from feed and fish excrement is absorbed by the soil. Gross et al. (1998) studied the phosphorus budgets for channel catfish ponds receiving one of five diets ranging from 0.60 to 1.03% phosphorus. They observed that low phosphorus diets did not decrease phytoplankton productivity or improve effluent quality. However, they suggested the use of low phosphorus diets in order to reduce the phosphorus load to bottom soils and to conserve the soils' ability to adsorb phosphorus.

Rangen Inc. (2013, Buhl, ID) provides important feeding tips of some of the commercially important aquaculture species. The tips that are species-specific would also minimize environmental impacts. Although recommended feeding practices differ by species, there are several common tips relevant to all species, including: 1) feeds should be broadcast well to allow all the fish to feed in the pond and minimize feed waste, 2) overfeeding should be avoided all costs to prevent from effluent pollution and gill damages, 3) feeding should be adjusted based on the percent body weight, 4) feed sizes should be adjusted as fish grows, 5) select the right feeding method for the species of interest, 6) feeding should be ceased before handling and shipping, 7) good record keeping is necessary to monitor fish growth and make the necessary adjustments, and 8) good storage and feeding management and feeding technique sanitation should be followed to avoid cross contamination and feed quality issues. In the last few decades, most fish farming has advanced from extensive rearing with few fish, to intensive rearing of high density populations in ponds and raceways. Cost effective good quality fish meals, proper feeding protocol, optimum growth and survival rates are the important goals of any fish farm operation.

## 3.7. Recirculating Aquaculture and Aquaponics

While many practices in aquaculture/mariculture (e.g., destruction of mangroves) have been criticized in years past for potential deleterious effects on the environment, the extent of any long term destruction due to aquaculture still remains debatable (Boyd and Schmittou 1990). Nevertheless in the United States and internationally, the most important environmental concern facing the aquaculture industry is the disposal of nutrient rich effluent water produced during the culture of aquatic animals (Goldburg and Triplett 1997; Frankic and Hershner 2003). Therefore, as aquaculture moves toward trying to feed the world, there is an inherent need to be stewards of the land, to protect, preserve and maintain conditions favorable to sustainability (Costa-Pierce 2002; United Nations FAO 2009). Recirculating aquaculture and its associated technology has largely developed out of concerns over water conservation and reducing environmental impacts (Martins et al. 2010). Besides growing fish, the purpose of a recirculating aquaculture system is to collect, concentrate, and process animal wastes rather than discharging wastes directly to the environment. Interestingly, an efficient recirculating aquaculture system is designed to reuse 90-99% of the water initially put into the system, while producing only a very small amount of waste or effluent (Chen et al. 2002, Hollingsworth et al. 2006; Badiola et al. 2012).

In their simplest form recirculating aquaculture systems (RAS) are similar to a home aquarium. Both a home aquarium and RAS have many of the same components including a tank or tanks, a pump to move water, some sort of filtration system, lighting, a heater or chiller, and fish. Also like a home aquarium, the RAS environment is very controlled to include lighting and room temperature good for the species of interest being cultured and other conditions in the facility. Fish are fed, water is added or taken out periodically, and water quality is monitored constantly and is often controlled through the addition of certain chemicals such as sodium bicarbonate. Unlike a home aquarium and relative to other types of aquaculture, RAS is very capital and energy intensive. It must rely on economic productivity for profitability, may require several additional components for processing water, and requires above average experienced personnel for successful management (Timmons et al. 2001; Timmons and Ebeling 2007; Ebeling and Timmons 2012).

With proper site selection, an advanced filtration capability, reduced water use and their small footprint, a recirculating aquaculture system lends itself to being a relatively environmentally friendly (Summerfelt and Vinci 2008; Ebeling and Timmons 2012, Losordo et al. 2013). Recirculating aquaculture systems do not rely on surface water for replenishment and with their ability to be located in close proximity to markets, they may be advantageous over other aquaculture systems (i.e., ponds, net pens, open ocean aquaculture) especially when comparing carbon footprint associated with food transport emissions (Martins et al. 2010). However, even with the positive attributes of a RAS, there is potential for it to be harmful to the environment and be considered unsustainable. Recirculating aquaculture systems are often described as the most effective way to grow large quantities of fish in a limited space. Furthermore, with their ability to control the environment they have the ability to grow fish year round (Hollingsworth et al. 2006; Ebeling and Timmons, 2012).

On a large production scale, recirculating aquaculture systems tend to be energy intensive and could be considered similar to other confined animal feeding operations, or CAFO's. In fact, under the 2004 United States federal aquaculture effluent limitation guidelines, recirculating aquaculture systems with an annual production exceeding 45,454 kg (100,000 lbs.) are classified as a concentrated aquatic animal production (CAAP) facility. Operations this large in scale are required to obtain a National Pollution Discharge Elimination System (NPDES) permit before discharging wastes into waters of the United States. Fortunately, the majority of recirculating aquaculture system operations in the United States choose alternatives to discharging effluent into natural waters, and instead either discharge into public municipal treatment works, collect the waste and apply it to nutrient deficient crops on land, or utilize treatment wetlands for processing effluent (Miller and Semmens 2002; Hollingsworth 2006; Summerfelt and Vinci 2008). A NPDES permit can be granted when the development of a facility specific Best Management Practices (BMP) plan specifies how discharges will be reduced of potential pollutants (Summerfelt and Vinci 2008).

The United States has a great deal of infrastructure that allows for regulation of discharge and more specifically, discharge into municipal treatment works, unfortunately the remainder of the world does not have this benefit. If recirculating aquaculture is to be adopted worldwide to raise fish in an environmentally and sustainable fashion, specific infrastructure is required. As previously mentioned, there is a plethora of literature available that describes RAS components and their efficiency at waste removal for large-scale fish culture. However, there is little information on dealing with the actual collected and concentrated solids that are generated from a large scale RAS. This is especially true when looking at recirculating aquaculture on an international capacity. Wetland ponds are often used in the United States and have been suggested on an international level, but wetland ponds have a limited lifetime and this is often a costly option. Another option for RAS effluent management that has been explored in other geographical areas is land application. Valencia et al. (2001) conducted a study to determine if effluent from a tilapia tank system could be used to replace nitrogen on guineagrass (*Panicum maximum*) managed as hay in a water limited area of the United States Virgin Islands. Interestingly, their results indicated that the tilapia tank system effluent could in fact serve as an adequate nitrogen and water replacement for guineagrass pasture, and hay production without changes in soil pH and phosphorous. Moreover, because this study used grass rather than row crops, it acts as a sink (similar to a wetland) with less risk of nutrient loss or leaching to the environment. The use of grass crops for assessing environmentally friendly ways to manage RAS effluent is but one step in the many ways research can explore repurposing and/or disposal of RAS effluent on and international level.

Overwhelmingly due to environmental concerns, but also to increase production efficiency, in recent years Best Management Practices have evolved across a number of industries from car manufacturing to food processing. Plain and simple, BMP's make sense and are a way of reducing multiple levels of risk. Within aquaculture, several entities including the Global Aquaculture Alliance (GAA) have created their own version of BMP's or Best Aquaculture Practices (BAP's). According to the GAA (2011), BAP's address environmental and social responsibility, animal welfare, food safety and traceability all on a national and international level. Through their BAP's the GAA further provides a certification program where they define elements of responsible aquaculture and evaluate adherence to these practices for each type of facility whether it be a hatchery, feed mill, farm or processing plant (www.gaalliance.org).

There are a number of BMP's that recirculating aquaculture system managers can use to make their facilities more environmentally friendly. Best management practices for recirculating aquaculture systems range in scope from choosing the right manager for the facility to using the most efficient types of filtration. Ensuring the use of high quality feeds with fewer fines will reduce nutrient levels and feed conversion ratios. Incorporating hybrid technology such as bioflocs which help to reduce feed costs and enhancing energy efficiency by using less and reusing energy where possible will all help the economic and environmental sustainability of recirculating aquaculture systems (Miller and Semmens 2002; Summerfelt and Vinci 2008; Hanna et al. 2010; Martins et al. 2010; Badiola et al. 2012). Summerfelt and Vinci (2008) have presented a thorough explanation of RAS BMP's beginning with site selection, working through solids storage, treatment and disposal, and complete facility operation and maintenance. Interestingly, Summerfelt and Vinci (2008) consider the point source waste stream to be the major facility level environmental issue (see also Hollingworth 2006).

Ultimately, for RAS to truly be environmentally friendly BMP's must be incorporated into their everyday function. Agriculture and its water counterpart aquaculture have been scrutinized due to various practices that have been employed over the many years of operation. In recent years, the colloquial buzzword has been "sustainable". You can't speak to anyone, go anywhere, or do anything anymore in any area of agriculture and natural resources without the mention of "sustainable". But with regard to agriculture what does sustainable mean? According to the United States Department of Agriculture, National Institute of Food and Agriculture (USDA-NIFA 2013), sustainable agriculture is an integrated system of plant and animal production practices having site specific application that over the long term will be able to (1) satisfy human food and fiber needs, (2) enhance environmental quality and the natural resource base upon which the agriculture economy depends, (3) it should further make the most efficient use of nonrenewable resources and on-farm resources and integrate, where appropriate, natural biological cycles and controls, (4) sustain the economic viability of farm operations, and (5) enhance the quality of life for farmers and society as a whole.

Aquaponics developed from hydroponics, or the culture of plants with little to no soil. In hydroponics, plants are raised in some sort of structure where the roots are submerged in either water or some type of media base where they are fed via a solution containing all the nutrients (fertilizer, trace minerals, etc.) that they need (McMurtry et al. 1990). Aquaponics, however are virtually the same as any other RAS, except that they use the metabolic byproducts of one crop (i.e., finfish) to produce a secondary crop (plants), thereby adding value by producing two crops instead of one (Rakocy 2012). In turn, the plants in an aquaponics system filter potentially lethal nutrients (nitrite, nitrate) from the water and return it back to the fish culture tank (Losordo et al. 2013; Rakocy et al. 1992). With the mention of sustainability, increasing pressure from environmental groups, governmental regulations and the fact that aquaculture continues to play an ever increasing role in supplying the worlds' food supply it is not surprising that interest in aquaponics has begun to take a foothold with regard to RAS, especially in urban area of the US.

A typical aquaponics system is set up so that water flows from the fish culture tanks (Figure 12a) to a settling chamber, or clarifier where solids are removed from the waste stream, water then enters a biofilter where ammonia in the water from the fish and excess feed is converted to nitrite nitrogen and then nitrate nitrogen. Water then exits the biofilter and proceeds toward the plant component where there may or may not be several other components included (i.e., base addition, degasser). The plant component is either a raceway with floating rafts, or could be what is called an NFT (nutrient film technique). In general, this is where the plants feed off of the nitrate nitrogen before the water returns to the fish culture tank relatively free of nitrogen. In this system the plants receive trace minerals via the fish food; however, there is often the need to supplement with things like iron, calcium, and potassium. (For a complete description of an aquaponics system, refer to Rakocy 2012).

Aquaponics is beneficial for a number of other reasons including that the cycling of the fish water to the plant component in an aquaponics system reduces the amount of concentrated discharge coming from this system relative to other RAS. Also important is that while fertilizers, herbicides, and pesticides may be utilized in or around greenhouses housing hydroponics, these are highly discouraged around aquaponics systems because of the deleterious effects they would have on the fish. Similarly, because an aquaponics system is a form of RAS, the use of antibiotics is discouraged within the system so as not to kill beneficial bacteria that are involved in the natural nitrification process. For these and other reasons, aquaponics systems are considered to be broaching the realm of organic. Organic farming is often considered to be environmentally friendly and sustainable. Unfortunately, the United States Department of Agriculture (USDA) has yet to provide aquaculture with an organic certification. While there are other private agencies worldwide that provide an organic certification for fish, the stringent guidelines provided by the USDA elevate this title to a higher level. Fortunately, in recent years there has been an ever increasing attempt at creating sustainable 'fishmeal reduced' and 'fishmeal free' diets for a growing number of fish species, especially with regard to highly prized carnivorous, saltwater species (see also Rhodes et al. 2013; Watson et al. 2013). The continued development of these diets may ultimately lead to a USDA organic certification for United States aquaculture.

While aquaponics systems are perhaps the most environmentally sustainable form of RAS to date, it does have drawbacks. Like any aquaculture venture, costs associated with initial investment, system components, their availability, construction and operation can have a significant impact on the economic sustainability of a system (Rackocy 2012). Hanna et al. (2010), for example have shown how different managers' management practices can affect the operation of identical RAS. It therefore becomes extremely important for aquaponics/RAS managers to follow Best management practices that will allow for a system to be profitable and sustainable. Best management practices for recirculating aquaculture and aquaponics systems have been described extensively in the literature (Chen et al. 2002; Summerfelt and Vinci 2008,). However, as we move toward trying to feed the world and keep RAS as environmentally friendly as possible, there are many important factors in operating a RAS. With regard to profitability and sustainability, perhaps offsetting initial investment, component and/or construction costs can be achieved by targeting highly sought after plant and fish species. Again, as fish meal free fish diets are developed for highly sought after marine species this reality becomes closer. One farmer at a recent national aquaponics conference suggested dealing directly with restaurants and "setting your price" rather than letting someone tell you how much something is worth (personal communication, National Aquaponics Conference, Tucson, AZ 2013). Similarly, it may also be advantageous for the owner/operator of a small scale RAS or aquaponics system to maximize profit and sustainability through raising high dollar plant and fish species as long as they have an established market (Frankic and Hershner 2003). With the recent surge in "farm to table" interest, it is very apparent that this concept can be profitable and environmentally friendly in the United States.

Ultimately, in trying to keep up with the worlds' population growth and food needs, RAS and aquaponics will continue to play a major role. Costa-Pierce (2002) and others suggest there has to be a behavioral shift in humans rather than technology in order for aquaculture to become truly sustainable. Many individuals only seem to see aquaculture in the sense that we need to produce as much fish as possible in as small an area as possible, however perhaps instead of trying to create RAS that are on the same level as a CAFO we instead look to systems that are sized according to the supporting the local community (Frankic and Hershner 2003). Again, one of the major advantages to a RAS is its' small footprint. By building a RAS with community size in mind it can be sized to feed the community on an ongoing basis. Having the RAS near or within a community would also reduce the carbon footprint by a reduction in fossil fuels needed for shipping etc. Figures 11 and 12 show a recirculating aquaculture system and an airlift aquaponics raft system in the Aquaculture Research and Demonstration Facility at Delaware State University, Dover, DE, USA.

**FIGURE 11.** a. An indoor multi-tank recirculating aquaculture system (RAS); b. A parabolic screen filter in a RAS (Photo courtesy of Blank).

**FIGURE 12.** a. Construction of an airlift aquaponics system; b. Aquaponics raft raceway showing aeration for plant roots; c. Planting herbs in a floating Styrofoam sheet (Photos courtesy of Blank).

## 4. BEST MANAGEMENT PRACTICES (BMPS) AND SUSTAINABLE AQUACULTURE

Best Management Practices (BMPs) have been used as an important management tool for various aquaculture practices and management may vary based on the species cultured, type of aquaculture practice, location, surrounding habitat, economy and policy conditions of the area. Although there are differences in the application and level of engagement with BMPs, there are common issues through which the application of BMPs can be applied to all. While there are many types of different aquaculture operations, because worldwide ponds are the most prevalent the recommendation and management practices in this section will focus on the pond management.

BMPs reflect the most technically practical and economically feasible methods to reduce environmental impacts and limit costs at aquaculture facilities. One primary goal is to develop simple effluent treatment systems that reduce organic matter loads, suspended solids, and nutrients in effluents to prevent polluting receiving waters. The best methods to prevent soil and water quality problems include selecting a site with appropriate soils and an adequate water supply, and maintaining moderate fish densities and feeding rates (Hargreaves et al. 2003). Secondary management techniques to prevent soil and water quality imbalances include liming, fertilization, and aeration (Boyd 1998). Agricultural irrigation, created wetlands, settling basins, and biological filters are also practical methods for reducing and improving the quality of effluent from ponds (Setty 2013; Tucker 2009; Ozbay and Jackson 2006).

Countries with regulations on farm practices and effluent standards follow guidelines and permit processes. However, countries without aquaculture regulations can also apply BMPs to minimize off-site water pollution and related environmental impacts. The code of conduct and codes of practice become useful tools for the farmers to adopt and serve as the guiding tool. As described by Boyd (2003) the code of conduct is a system of rules on how aquaculture should be conducted. The guiding principles for responsible aquaculture by the Global Aquaculture Alliance (GAA) serve as the code of conduct. In order to avoid social and environmental problems, codes of practice are used to solve the problems through management activities. Overall the goal of these codes is to minimize, or remove the negative impacts of an aquaculture operation (Boyd 2003). Boyd (2003) describes BMPs as management practices that are the most effective at reducing pollution levels and other environmental impacts which meet water quality or resource management goals.

Although not comprehensive, Boyd (2003) highlights some BMPs for pond aquaculture: a. use stocking and feeding rates that do not exceed the assimilative capacity of ponds; b. avoid overfeeding and apply a strict feeding plan; c. do not use fertilizers unless it is absolutely necessary to promote healthy phytoplankton growth; d. reduce water exchange; e. reuse water as much as possible; f. use a settling basin if available to treat pond effluents before final discharge. Application of these recommendations is based on farm operation and design, species cultured, and culture methods used (Boyd 2003). As Boyd states, selection of BMPs is case specific. One such example includes 10 codes of practice established for responsible shrimp farming by the GAA including mangroves, site evaluation, design and construction, feeds and feed use, shrimp health management, effluents, solids waste and few others. Although one BMP may be sufficient for one small farm, multiple BMPs may be necessary for others. ALEARN (www.alearn.info) listed over 20 best management practices for ponds, raceways, cages, effluents, safety, and others.

Shrimp aquaculture is a rapidly expanding field and is being closely scrutinized by environmentalists and government agencies. Due to the need for saltwater, discharge from shrimp facilities often flows into fragile coastal ecosystems. Problems associated with discharge include eutrophication due to nitrogen loading and detritus, low dissolved oxygen levels, sedimentation, along with other problems (Villalon 1991). These problems however are not without solutions; treatment of effluent should be regarded as an opportunity rather than just an obligation. Our goal is to provide broader perspectives on how basic principles and natural solutions can make shrimp aquaculture longer lived and be more sustainable. We discuss some of the management strategies current shrimp aquaculture operations along with best management practices for reducing potential impacts of shrimp aquaculture.

Although this may not be the ultimate solution, one particular recommendation discussed in this section would be to improve shrimp management practices. The mariculture of shrimp may provide one of the best opportunities for polyculture and integrated systems. Shrimp require higher water quality standards than many other cultured species and, thus, would benefit from a more stable ecosystem. The ability of shrimp to utilize a broad spectrum of the food web would allow them to be cultured with a number of other species. Feed and fertilization management can be geared toward supporting the food web to produce food items which shrimp prefer, rather than relying on the direct consumption of pelleted feed. In addition, Hopkins et al. (1991) discussed how dissolved oxygen levels were higher in polyculture ponds presumably due to a healthier phytoplankton community. Thus, polyculture may actually reduce aeration costs.

Hollingsworth et al. (2006) suggests growers may develop a farm-specific Standard Operating Procedure (SOP) manual and apply the code of conduct for significant farm practices in their SOP manuals. Although not required for all farms, the development of farm specific procedures will promote efficient management decisions including trouble shooting problems, training employees, planning future expansions or developing biosecurity and emergency procedure plans.

Many states in the United States have adopted BMPs over the years and some states and countries have implemented further policies and regulations based on scientific knowledge to sustain the environment and aquaculture industry. One specific example is that of Louisiana's aquaculture producers (Lutz et al. 2011). By implementation and application of best management practices, producers minimize potential pollutants (i.e., mainly excess nutrients) to the state's water resources and by doing so they reduce the cost that would be incurred to treat water quality problems, potential disease outbreaks and wild fish stock mortality related costs. Lutz et al. (2011) suggested that sediment runoff reduction should be one of the most important practices a pond aquaculture farmer must adopt to save money and reduce the environmental foot print of their operation. As an example, in Thailand shrimp aquaculture, scientists and policy-makers have developed new ways to improve the quality of the culture system, ecosystem, as well as the efficiency of regulations. It is critical that advances such as this and many other practices are discussed and maintained with integrity and strong regulations to improve the quality of our shared water resources for future generations. The key is to make aquaculture an asset to the environment while continuing to food production simultaneously.

Initial efforts and guidance on BMPs have been developed by Hargreaves et al. (2003). He has provided guide sheets on various topics and issues of concerns including reducing storm runoff into ponds, managing ponds to reduce effluent volume, erosion control on watersheds and embankments, pond management to minimize erosion, control of erosion by effluents, settling basins and wetlands, feed management, fertilization of catfish ponds, water quality protection to improve effluents, water quality enhancers, therapeutic agents, fish carcasses, general operations and worker safety, emergency response and management, and a few others added as the technology advanced in recirculating, bioflocs and aquaponics systems and other aquaculture operations.

The Best Aquaculture Practices (BAP) standards developed by the Global Aquaculture Alliance (GAA 2011) "address environmental and social responsibility, animal welfare, food safety and traceability in a certification program voluntary for aquaculture facilities." Certification for BAP ensures aquaculture operation is responsible and operates by the quantitative guidelines by which the farm operation is evaluated based on those practices. There are various standards developed in aquaculture sector including fish farm, hatchery, feed mill, and processing plant. The standard for the multi-species farming opens whole new area of attention with the new aquatic species used in integrated culture condition. Species BAP Standards used include channel catfish, shrimp, tilapia, and *Pangasius* initially and seabass, sea bream, cobia, seriola, trout, grouper, barramundi, perch, carp, flounder, turbot, striped bass, crabs, freshwater prawns, mussels, crawfish recent. According to GAA (2011), the new multi-species farm standards apply to all types of culture systems for finfish and crustaceans but not including cage-raised salmonids since this operation requires separate BAP standards. Seven of the most recent BAP standards listed in the GAA website (http://www.gaalliance. org/bap /standards.php) include Seafood processing/repacking plant, seafood processing plant, finfish and crustacean farm, salmon farm, mussel farm, shrimp hatchery, and feed mill.

The United States Environmental Protection Agency (USEPA 2004) initiated a new rule called the "effluent limitations guidelines (ELGs)' for concentrated aquatic animal production facilities including aquaculture facilities. This rule is applied to all commercial aquaculture facilities, with the below mentioned specifications, that discharge their wastewater from their farms directly into waters of the United States. According to the final rule, aquaculture facilities that "produce at least 45,360kg a year in flow-through and recirculating systems that discharge wastewater at least 30 days a year (used primarily to raise trout, salmon, hybrid striped bass and tilapia); at least 45,360 kg a year in net pens or submerged cage systems (used primarily to raise salmon)." The whole expectation with implementation of this rule is that the ELGs will help reduce discharges of conventional pollutants, primarily total suspended solids. As the solids are removed, it is expected that non-conventional pollutants such as nutrients will also be reduced. Other contaminants not discussed in this chapter include heavy metals, drug residues and other hormonal chemicals used in facilities to manage fish health and chemicals and better growth and this regulation is expected to be effective for reducing those contaminants in discharges of the facilities. With the implementation of this rule through National Pollution Discharge Elimination System (NPDES) permits the discharge of total suspended solids are expected to be reduced more than 226,796 kg per year, and the biochemical oxygen demand and nutrients in discharge is to be reduced by about 136,078 kg per year. With the application of this rule it is expected that water quality conditions will be improved and provide increased opportunity for other users, swimmers, fisherman and environmentalists concerned about keeping biodiversity in the streams, rivers and estuaries.

There are many definitions for sustainability and sustainability with regard to a catfish farm may not be sustainable for a shrimp farm. For aquaculture, sustainable aquaculture is an ultimate goal with the application of all the best aquaculture standards and management practices. Sustainability is described by the Northwest Earth Institute (2012) simply as meeting the needs of the present without compromising the ability of future generations to meet their own needs" (taken from UN World Commission on Environment and Development, Our Common Future). According to the Monterey Bay Aquarium (2013), environmental impact of fish farming varies depending on the species cultured, location of the farm, life stage of the organism, methods of culture and culture technique. Creating a sustainable farm should ensure species cultured will last long and habitat damage be minimal. The key factor with aquaculture sustainability is to operate with sound environmental management practices in place (FAO

FOCUS 2013). There are tremendous efforts being made to use integrated aquaculture-agriculture farming systems to sustain both aquaculture operations and maintain the healthy environmental conditions for aquatic life in rivers, streams and estuaries. Environmentally friendly methods are also beneficial to the species cultured and farm operation (FAO FOCUS 2013). Sustainable aquaculture should utilize the most readily available technology to produce high protein food diets while applying the same exact principle to reduce its environmental impact using similar technology. Sustainability is not a practice, it is a life style and condition we must grasp.

Although it is beyond the scope of this chapter, the application of innovative technology in sustainable aquaculture such as Geographic Information Systems (GIS) would be an effective tool for selecting sites for bivalve culture and farm management. Coupled with ecosystem models, this technology can assist in predicting the carrying capacity of estuaries (Newell et al. 2013). Similar to shellfish site selection and farm management, Clearwater Seafoods has utilized GIS to take an informed approach to harvesting which minimizes the impact of fishing activities and promotes sustainability both at sea and on land. By investing in GIS, this company saved and minimizes their impacts on ocean ecosystems and promotes a sustainable approach to fishing (ESRI News 2013).

## 5. CASE STUDY: SUSTAINABLE AQUACULTURE CULTURE IN THAILAND

Fisheries have long been integral to the Thai way of life. Management of fisheries in Thailand began in 1901 with the establishment of the Thailand Department of Fisheries (DOF). In 1901, the ministry of interior issued 3 guidelines to manage fisheries resources: 1) produce fisheries production for population in country, 2) produce fisheries production for country income, and 3) taxation for capture fisheries. However, only taxation was implemented because there was no fisheries biologist at that moment. In 1923, Dr. Hugh M. Smith, MD., LL.D (Commissioner of Fisheries USA) was invited as an advisor in fisheries to His Siamese Majesty's Government. After finishing his survey research, he published a book called "A Review of the Aquatic Resources and Fisheries of Siam, with Plans and Recommendation for the Administration, Conservation and Development". In 1926, the Department of Fisheries was established and Dr. Smith was appointed to be head official. Under his guidance, management systems were implemented and continue to be conducted. Fisheries production in the beginning relied mainly on wild capture since Thailand has many natural freshwater resources scattered all over the country, such as rivers, swamps, and reservoirs. Thailand also has a 3,500 kilometer-long coast line including both the Gulf of Thailand and Andaman Sea, including more than 900 islands. However, drastic changes to these habitats and overfishing have negatively impacted the wild capture fisheries. Therefore, production from aquaculture has gradually begun to play a more important role in maintaining total fisheries production (Figure 13).

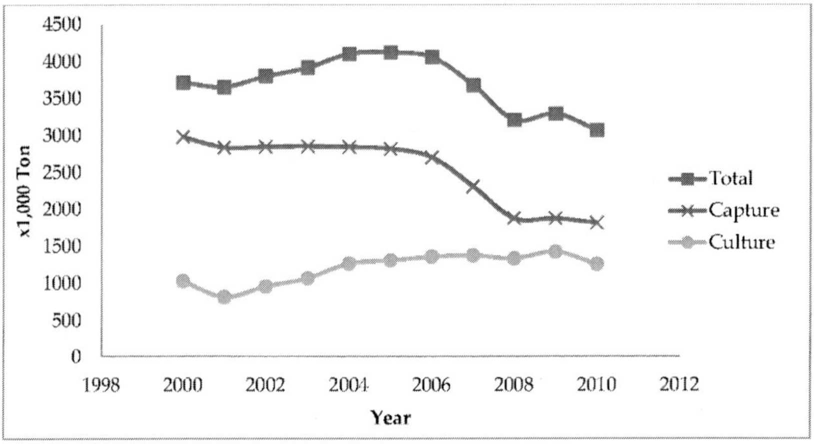


**FIGURE 13.** Thailand fisheries production from 2000-2010 (Thailand DOF Information System Center 2011).

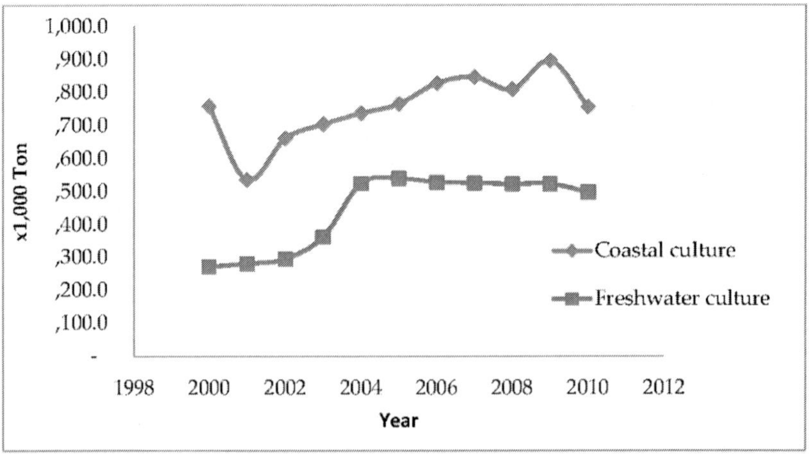


**FIGURE 14.** Freshwater and coastal aquaculture production from 2000-2010 (Thailand DOF 2011).

Aquaculture in Thailand evolved from traditional practice to modern science-based practices as aquaculture technology and innovations developed. Therefore, aquaculture in coastal areas, which contribute more national economics and provide more benefit to farmers, tended to increase and contribute more when compared to freshwater species, especially brackish water-cultured shrimp and prawn. However, aquaculture in freshwater areas has also increased due to population growth and market demands in the country (Figure 14) (Thailand DOF Information System Center 2011).

Of the three main groups of brackish water aquaculture – fish, shrimp, and shellfish – shrimp culture has increased dramatically while fish culture has decreased and become steady from 2000 – 2010, while shellfish culture production dropped because of shellfish diseases and natural changes (Thailand DOF Information System Center 2011) (Figure 15). However, shrimp culture is the most cultured species within brackish water aquaculture, globally.

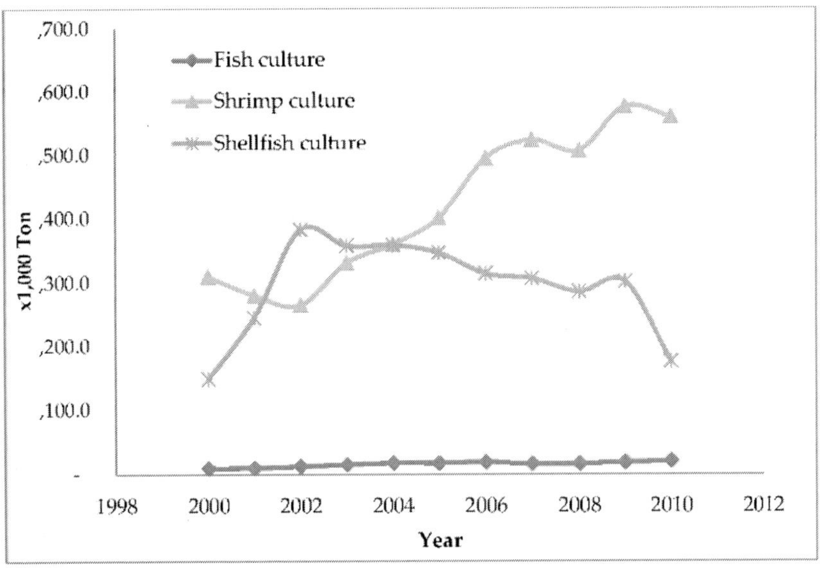


**FIGURE 15.** Brackish water aquaculture production by group from 2000-2010 (Thailand DOF 2011).

Shrimp culture in Thailand started more than 50 years ago. Production has been greatly increased within the last three decades as intensive farming techniques were developed and applied. In the beginning of brackish water shrimp culture the major shrimp species used was tiger prawn (*Peneaeus monodon*), which was substituted by white shrimp (*Peneaeus vannamei*) by more than 95%, by the year 2000 (Bureau of Agricultural Economic Research, 2011). Thailand now exports shrimp products at about 500,000 tons annually in the year 2010 (Center for Agricultural Information 2011), making Thailand one of the largest shrimp exporters in the world.

Cultured shrimp, in particular, has come under threat in key export markets due to adverse publicity concerning the environmental and social impact of some aquaculture activities (Kongkeo, 2001). Some of this publicity has served to highlight some of the negative environmental and social impacts that have occurred in the development of aquaculture. These include the destruction of wetlands and mangrove forests, water pollution, reduction of biodiversity, waste of natural resources, and loss of access to fishing grounds by artisanal fisherman (Boyd and Tucker, 1998). Therefore, management inputs that remedy these problems that occur between culturing period and after harvesting such as chemicals used, water management, water discharge, etc. must be considered in order to encourage sustainable growth and practices within the shrimp aquaculture sector.

Shrimp culture in Thailand is performed under intensive culture conditions which consumes heavy feed, water supply, and aeration; therefore, management must play a key role in helping to reduce problems and the impacts of effluent from culturing systems. Sustainable shrimp culture, economically speaking, is less about increasing production but more about the ability to maintain steady production, customer satisfaction and reliability, and mitigate social and environmental impact concerns.

Long before Good Aquaculture Practices (GAPs) and Food and Agriculture Organization (FAO) Code of Conduct for Responsible Fisheries (CoC) were integrated in Thailand, best management practices were implemented by the Department of Fisheries to increase farm productivity while minimizing the environmental impacts. The environmental issues caused by the effluent discharges of the shrimp culture facilities have been one major concern of the aquaculture operation. Some of these management practices facilitated by the Department of Fisheries at the National and local level have focused on increasing feed conversion ratios, better water exchange, aeration, and pond management, and, if available, applying integrated multi-trophic aquaculture technology to reduce nutrient loads from farm effluents and increase profits. Specifically, aquaponics using commercial crop systems become popular and practical to the farmers while the operation is sustained and become environmental friendly.

The Thailand DOF integrated GAP and CoC in 1998 with support from the World Bank. Under GAP standard, the requirement are farm registration;, farm management; use of veterinary drugs chemical, hazardous substances, and probiotics used in aquaculture; farm sanitation; harvest and postharvest prior to distribution; effluent and sediment management; energy source and fuel use; social and environmental responsibility; and record keeping. Code of Conduct for shrimp culture has two components: operational guidelines for hatcheries and farms, and guidelines for harvesting and transport. These practices - GAP and CoC - are standard for shrimp culture to ensure that shrimp culture has minimal to no chemical residues which protects consumers and applies environmental friendly practices. Three year GAP or CoC certificates are issued by the Thailand DOF to shrimp farms after they meet qualifications and comply with annual surveillance.

Shrimp culture systems in Thailand also have a traceability system from farm to product which initially started as a form of hatchery management. Some hatcheries complied with bio-secure systems to ensure that larvae produced are healthy and viable before selling to grow-out farmers. Hatcheries must provide Fry Movement Documents (FMD) to their customers to indicate the number of fry that a farmer purchases in addition to other hatchery information since this document is checked if any problems occur during grow-out. Shrimp farmers must also provide Movement Documents (MD) which indicate the weight of shrimp in the shipment in addition to other farm information to processing plants or their customers. A DOF officer checks MD, Hazard Analysis and Critical Control Point (HACCP), GMP, and product quality at the processing plant before issuing product health certificates.

Moreover, many policies and projects are established for sustainable aquaculture support. The agencies involved at the national level include, the Department of Fisheries, Pollution Control Department and Department of Marine and Costal Resources, while there are Provincial government and Local Administrative Organization involve at local level. Water quality testing program pond and discharge water responsible by DOF and water quality in natural water responsible by Pollution Control Department, etc. Mangrove re-habitation projects are established by the national and local government sectors, private sector, and Non-Governmental Organizations (NGOs) to increase mangrove forest area along the costal zones. Thailand also supports Non Illegal Unreported Unregulated Fishing (Non-IUU) and issued several programs for fish resource conservation, which include combatting IUU fishing, prohibiting certain fishing gears within spawning season, and expanding fish conservation areas. Although wild fisheries are not a part of the discussion in this chapter, it does play a part serving as a source of fish meal industry which is used in aquaculture feed. Therefore, control of IUU and certain fishing gear, will support sustainable aquaculture.

In conclusion, fisheries production in Thailand has decreased while aquaculture production has increased and plays a vital role for providing a

**FIGURE 16.** a. Clean and dried pond before start; b. Water quality monitoring; c. Water drained into treatment ponds; d. Aeration in culturing period using paddle wheel aerators; e. Shrimp health sampling with a cast net; f. Shrimp monitoring with a lift net; and g. Collected specimens are assessed for growth and survival rate.

high protein food source for economic development in the future. Aquaculture in Thailand evolved from traditional practice to science-based due to a number of policies and regulations put in the place to sustain both the aquaculture industry and the environment. Shrimp aquaculture in Thailand is an excellent example of why sustainable aquaculture practices are necessary and how they have become implemented. With the establishment of GAP and CoC, Thailand has ensured that shrimp farming results in production of a high quality product, safe from chemical residues, that is environmental friendly. However, truly sustainable aquaculture will only be attainable when the balance between food security, economic benefit, social benefit, and a reduction of environmental impact is achieved. Figure 16 below shows the steps involved in best management practices in shrimp aquaculture farm.

## 6. FINAL REMARKS

In recent years aquaculture has gone through the "blue revolution" in which there has been rapid growth worldwide in aquaculture production of both fresh and saltwater fish and shellfish species. In part this is due to the fact that the natural fisheries are close to their maximum sustainable yield. However, this rapid growth in aquaculture may also be attributed to the ever increasing world population and an increase in demand for high protein sources of seafood. In the past, aquaculture has been demonized for destruction of mangroves worldwide for shrimp production as well as causing potential eutrophication through unwelcomed discharges of nutrient rich effluents.

With increased environmental awareness and the general populations increased concern over its food sourcing, aquaculture has stepped up to the proverbial plate to try and fulfill the worlds seafood demands through increased production while trying to maintain more environmentally friendly practices of culturing fish through many technological advancements. Unfortunately, aquaculture has not yet truly reached its sustainability goal. However, in addition to much technological advancement, aquaculture has begun to incorporate best management practices to create a more environmentally friendly way of producing fish. In this chapter we have gone over several areas associated with BMPs and described them with regard to how their incorporation can impact or reduce the impacts of aquaculture on the environment. Many of the BMPs discussed are simple and rely on common sense approaches to nutrient problems. Others are more technologically advanced and require additional components and or descriptions that are beyond the scope of this chapter.

In the end we are all trying to get to a point at which aquaculture can be considered a sustainable farming entity so that its impacts are minimal at best to the surrounding environment, there is a continuous supply of food, and it is profitable for all of those who are involved. As we consider moving ahead we must continue to remember that the world's resources are there for all and we want to maintain them for future generations to come. Again, we quote Costa-Pierce (2002) in suggesting there has to be a behavioral shift in humans rather than technology in order for aquaculture to become truly sustainable.

# 7. TECHNICAL SUMMARY

Tremendous efforts have been made to improve aquaculture farm practices through disease prevention and treatment, planning and management of facilities, feeding, and advances in aquaculture technology and sustainable practices. However, the industry is not without its issues and faults. Although significant accomplishments have been made in minimizing the negative impacts of aquaculture operations on the environment, it is not reached sustainability worldwide. Further efforts ranging from husbandry practices to policy and regulations are essential to ensure the sustainability of aquaculture on a global scale. As aquaculture moves from feeding millions to billions of people in the last century, intensive culture practices have become common and require better management and monitoring efforts. Intensive production of fish farming requires significant inputs of nutrients in the form of inorganic fertilizers or feeds. Of these inputs, typically only 25% of the chemical constituents of the feed are assimilated into fish biomass while the rest is released into the water as metabolic wastes. In pond culture, fish are usually harvested after draining the pond partially or fully. The waste water expelled from these ponds into watershed, laden with organic matter and nutrients, concerns regulatory agencies as a point-source of pollution. In addition, in most countries, including the United States, statistics are lacking on the amounts of chemicals used, and as a result, regulations cannot truly be effective. Furthermore, it is difficult to make correlations of aquaculture effluents to environmental impacts without accurate records. The main goals of effluent management or, more often referred to as best management practices, are to minimize impacts to the environment while maintaining productivity. Fortunately, most of these strategies are as beneficial to the aquaculturists/ farmers as they are to the environment. Both production costs and effluent can be reduced by using stock-specific feeds applied in smaller quantities several times a day, good aeration, improved husbandry practices, and paying good attention not to exceed to the carrying capacity of the system. By lowering concentrations of phytoplankton, savings on herbicides and aeration are inevitable. Limited water exchange, integrated aquaculture, and good monitoring are further best management practices measures.

# 8. ACKNOWLEDGEMENTS

The authors would like to thank Dr. Karuna Chintapenta, Brian Reckenbeil, Kris Roeske, and Cory Janiak for their editorial assistance and review of the chapter. The authors would also like to thank Dr. Claude Boyd for the resources and guidance he provided both authors for their research and professional development. Some of the projects discussed in this chapter were funded by SRAC, USDA-NIFA, USDA Evans-Allen, NSF-EPSCOR, AU, DSU, and Government of Thailand.

## NOTES

[1] - Bioorganic catalysts are catalytic compounds that enhance the biological conversion abilities that would otherwise occur naturally (ICAP Bio-Organic 2013).

## REFERENCES

1. United Nations Food and Agriculture Organization. UNFAO: 2009. How to Feed the World 2050. High-Level Expert Forum. Global agriculture towards 2050. http://www.fao.org/fileadmin/templates/wsfs/docs/Issues (accessed 10 September 2013).
2. University of Maryland Medical Center. UMD: 2013. Omega-3 fatty acids. Overview. http://umm.edu/health/medical/altmed/supplement/omega3-fatty-acids (accessed 3 October 2013).
3. United Nations Food and Agriculture Organization. UNFAO: 2013. FOCUS. Fisheries and food security. Sustainable aquaculture development. http://www.fao.org/focus/e/fisheries/sustaq.htm (accessed 5 September 2013).
4. Ewart JW. 2013. Shellfish Aquaculture in Delaware's Inland Bays: Status, Opportunities, and Constraints.In Delaware Sea Grant Program; 2013. p 43.Available from http://darc.cms.udel.edu/ibsa/Inland (accessed 7 November 2013).
5. Bush SR, Belton B, Hall D, Vandergeest P, Murray FJ, Ponte S, Oosterveer P, Islam MS, Mol APJ, Hatanaka M, Kruijssen F, Ha TTT, Little DC, and Kusumawati R. Global Food Supply. Certify Sustainable Aquaculture? Science 2013;341(6150) 1067-1068.
6. National Oceanic and Atmospheric Administration. NOAA: 2013. 10 Myths about Marine Aquaculture. http://www.nmfs.noaa.gov/aquaculture (accessed 7 November 2013).
7. United States Food and Drug Administration. USFDA: 2008. Taken from Seafood Health Facts: Making Smart Choices. Balancing the Benefits and Risks of Seafood Consumption. http://seafoodhealthfacts.org/seafood_nutrition/practitioners/composition_chart.php (accessed 3 September 2013).
8. Wilson JL, Stone NM, Bayne D, Popma TJ, Boyd C, Brune DA, Collier JA, Schwedler TE, Burtle GJ, Lewis GW, Styer EL, Romaire RR, Hargeaves JA, Kingsbury SK, Nagle DG, Daniels HV, Hodson RG, Shrader KK, Zimba PV, Grimm CC. Control of Blue-green Algae in Aquaculture Ponds. SRAC Fifteenth Annual Progress Report 2002;30.
9. United Nations Food and Agriculture Organization. FAO: 2011. FAO STAT. http://faostat.fao.org/site/629/default.aspx (accessed 11 December 2013).
10. Dealteris JT, Kilpatrick BD, Rheault RB. A comparative evaluation of the habitat value of shellfish aquaculture gear, submerged aquatic vegetation and a non-vegetated seabed. Journal of Shellfish Research 2004;23(3) 867-874.
11. Erbland PJ, Ozbay G. A Comparison of the macrofaunal communities inhabiting a Crassostrea virginica oyster reef and oyster aquaculture gear in the Indian River Bay, Delaware. Journal of Shellfish Research 2008;27 757-768.

12. Shumway SE, Davis CR, Downey R, Karney J, Kraeuter J, Parsons R, Rheault, Wikfors G. Shellfish aquaculture: in praise of sustainable economies and environments. World Aquaculture 2003;34 15-18.
13. O'Beirn F X, Ross PG, Luckenbach MW. Organisms associated with oysters cultured in floating systems in Virginia, USA. Journal of Shellfish Research 2004;23(3) 825-829.
14. Tallman JC, Forrester GE. Oyster grow-out cages function as artificial reefs for temperate fishes. Transactions of the American Fisheries Society 2007;136(3) 790-799.
15. D'Amours O, Archambault P, McKindsey CW, Johnson LE. Local enhancement of epibenthic macrofauna by aquaculture activities. Marine Ecological Progress Series 2008; 371 73-84.
16. Taylor J, Bushek D. An Assessment of Habitat Value of Constructed Intertidal Oyster Reefs and Oyster Aquaculture Systems in Delaware Bay, USA. Journal of Shellfish Research 2008;27(4) 1056-1056.
17. Emerson C. Aquaculture impacts on the environment. Cambridge Scientific Abstracts 1999.
18. World Wildlife Foundation. WWF: Unsustainable fishing. Our oceans are being plundered. http://wwf.panda.org/about_our_earth/blue_planet/problems/problems _fishing/ (accessed 2 November 2013).
19. National Oceanic and Atmospheric Administration. NOAA: Fish watch U.S. Seafood Facts: Outside the U.S. 2011 http://www.fishwatch.gov/farmed_ seafood/outside_the_us.htm (accessed 6 October 2013).
20. National Oceanic and Atmospheric Administration. NOAA: 2010. U.S. Seafood Consumption Declines Slightly in 2009. http://www.noaanews.noaa.gov/stories2010/20100909_consumption.html (accessed 8 December 2013).
21. Monterey Bay Aquarium Foundation. MBAF: Fishing and Farming Methods. http://www.montereybayaquarium.org/cr/cr_seafoodwatch/sfw_gear.aspx (accessed 2 December 2013).
22. Briggs MRP, Funge-Smith S. Unsustainable shrimp culture- causes and potential solutions from experience in Thailand. In: Development of Strategies Shrimp Farming. In report to the Overseas Development Administration, Research Project R4751;1994. p31.
23. Dalsgaard JPT, Lightfoot C, Christensen V. Towards quantification of ecological sustainability in farming systems analysis. Ecological Engineering 1995;4 181-189.
24. Food and Drug Administration. FDA: Food and Drug Administration Compliance Program Guidance Manual, 2008. http://www.fda.gov/ downloads /Food/ComplianceEnforcement/ucm073192.pdf (accesses 10 October 2013).
25. Virtanen JK, Mozaffarian D, Chiuve SE, Rimm EB. Fish consumption and risk of major chronic disease in men. *American Journal of Clinical Nutrition* 2008;88(6) 1618–1625.
26. Soft Bottom. In Coastal Habitat Protection Plan; 2005. p365-448. Available from http://ncfisheries.net/habitat/chppdocs/G_Soft%20Bottom.pdf (accessed 3 August 2013).
27. Garland Science. In: Microbes in the Marine Environment; 2011. p23. Available from http://www.fisheries.go.th/library/app/ (accessed 3 November 2013).
28. Friend M. Disease Emergence and Resurgence: The Wildlife-Human Connection: Reston, Va., U.S. Geological Survey. 2006;400.
29. Yanong RPE, Erlacher-Reid C. Biosecurity in Aquaculture, Part 1: An Overview. Southern Regional Aquaculture Center 2012;(4707) 16.
30. Chatterjee S, Haldar S. Vibrio Related Diseases in Aquaculture and Development of Rapid Accurate Identification Methods. Journal of Marine Science 2012; 7.
31. Mori K. Effects of artificial eutrophication on the metabolism of the Japanese oyster Crassostrea gigas. Marine Biology 1979;53(4) 361-369.
32. Bastin L. International Aquafeed. Incorporating Fish Farming Technology 2013;16(1) 26.

33. Boyd CE. Water quality management and aeration in shrimp farming 1989: Auburn University, Department of Fisheries and Allied Aquacultures, Auburn, AL, Series No. 2. p183.
34. Ryther JH, Goldman JC, Gifford JE, Huguenin AS, Wing JP, Clarner LD, Williams, Lapontie BE. Physical models of integrated waste recycling of "Marine Polyculture Systems." Aquaculture 1975;5 163-177.
35. Neori A, Ragg NLC, Shpigel M. The integrated culture of seaweed, abalone, fish and clams in modular intensive land-based systems: II. Performance and nitrogen partitioning within an abalone (Haliotis tuberculata) and macroalgae culture system. Aquacultural Engineering 1998;17(4) 215-239.
36. Avnimelech Y., editor. Biofloc Technology - A Practical Guide Book. The World Aquaculture Society, Baton Rouge, Louisiana; 2012. p272.
37. The University of Virgin Islands Aquaculture Program. 2013. Biofloc Systems. http://www.uvi.edu/research/agricultural-experiment-station/aquaculture-home/biofloc-systems/default.aspx.
38. The Aquaponics Association. AA: What is aquaponics? http://www. aquaponicsassociation.org/ (accessed 8 November 2013).
39. Bennett L, Diana J, Qiuming L. Learning From Tradition. In: Integrated Aquaculture Practices in Yingbin Bay;2012. http://pdf.gaalliance.org/pdf/GAA-Bennett-Nov12.pdf (accessed 10 October 2013).
40. Nutreco. Feeding the Future? How can aquaculture contribute to feeding 9 billion people in 2050 in a Sustainable way? In Nutreco Prins Frederiklaan; 2011. http:// www. nutreco.com. (accessed 8 November 2013).
41. Scottish Association for Marine Science. SAMS: Aquaculture Impacts and Coastal Eutrophication. http://www.sams.ac.uk/microbial/aquaculture-impacts-and-coastal-eutrophication#sthash.pZzJ4ask.dpf (accessed 9 December 2013)
42. Science Daily. In Science Reference; 2013. Eutrophication. http://www. sciencedaily.com/articles/e/eutrophication.htm (accessed 8 November 2013).
43. Boyd CE. Soil and Water Quality Considerations in Shrimp Farming. In: Haws MC. and Boyd CE. (ed.) Methods for Improving Shrimp Farming in Central America. In Central American University Press; 2001. p17-20.
44. Boyd CE, Queiroz J. Aquaculture pond effluent management. Aquaculture Asia 1997;2 (2) 43-46.
45. Shireman JV, Cichra CE. Water quality from Florida catfish and fingerling culture ponds. In: Joseph St. (eds) 1996: national livestock, poultry, and aquaculture waste management, Proceedings of the National Workshop, American Society of Agricultural Engineers, St. Joseph, MI. p206-210.
46. Schwartz MF, Boyd CE.Channel catfish pond effluents. The Progressive Fish-Culturist 1994a;56(4) 273-281.
47. Boyd CE. Effluents from catfish ponds during fish harvest. Journal of Environmental Quality 1978; 7 59-63.
48. Schwartz MF, Boyd CE. Effluent quality during harvest of channel catfish from watershed ponds. The Progressive Fish-Culturist 1994b;56(1) 25-32.
49. Seok K, Leonard S, Boyd CE, Schwartz ME. Communications: Water Quality in Annually Drained and Undrained Channel Catfish Ponds over a Three-Year Period. The Progressive Fish-Culturist 1995;57(1) 52-58.
50. Boyd CE, Queiroz J, Lee J, Rowan M, Whitis GN, Gross A. Environmental assessment of channel catfish, Ictalurus punctatus, farming in Alabama. Journal of the World Aquaculture Society 2000;31(4) 511-544.
51. Hariyadi S, Tucker C, Steeby J, Van der Ploeg M, Boyd CE. Environmental conditions and channel catfish (Ictalurus punctatus) production under similar pond management regimes in Alabama and Mississippi. Journal of the World Aquaculture Society 1994;25 236-249.
52. Boyd CE, Gross A, Rowan M. Laboratory study of sedimentation for improving quality of pond effluents. Journal of Applied Aquaculture 1998;839-48.

53. Boyd CE, Tucker CS. Pond Aquaculture Water Quality Management. In Kluwer Academic Publishers; 1998. p541-575.
54. Wheaton FW. Aquaculture Engineering. In Robert E. Kreiger; 1977. Malabar, Florida, USA.1977. p414-462.
55. Coloso RM, Basantes SP, King K, Hendrix MA, Fletcher JW, Weis P, and Ferraris RP. Effect of dietary phosphorus and vitamin D3 on phosphorus levels in effluent from the experimental culture of rainbow trout (*Oncorhynchus mykiss*). Aquaculture 2001;202(1) 145-161.
56. Queiroz JF, Boyd CE, Gross A. Evaluation of a bio-organic catalyst in channel catfish, Ictalurus punctatus, ponds. Journal of Applied Aquaculture 1998;8(2) 49-61.
57. Tucker CS, Kingsbury SK, Pole JW, Wax CL. Effects of water management practices on discharge of nutrients and organic matter from channel catfish (Ictalurus punctatus) ponds. Aquaculture 1996;147(1) 57-69.
58. Thompson JR, Polz MF. Dynamics of Vibrio populations and their role in environmental nutrient cycling. The biology of Vibrios. In ASM; 2006. p90-203.
59. ICAP Bio-Organic. 2013. Technology. http://www.icapbio-organic.com /technology.html (accessed 8 November 2013).
60. Dame RF, Spurrier JD, Wolaver TG. Carbon, nitrogen and phosphorus processing by an oyster reef. Marine ecology progress series 1989;54(3) 249-256.
61. Seaweed Energy Solution. SES: Seaweed is simple. http://www. Seaweedenergy solutions.com/seaweed/ (accessed 8 November 2013).
62. Becker V, Huszar VLM, Crossetti LO. Responses of phytoplankton functional groups to the mixing regime in a deep subtropical reservoir. *Hydrobiologia*; 2008; 628(1) 137-151.
63. National Oceanic and Atmospheric Administration. NOAA: State of the Coastal Environment. Oxygen Depletion in Coastal Waters. http://oceanservice.noaa.gov /websites/retiredsites/sotc_pdf/HYP.PDF (accessed 8 December 2013).
64. Boesch D, Brinsfield R, Magnien R. Chesapeake Bay eutrophication: Scientific understanding, ecosystem restoration, and challenges for agriculture. Journal of Environmental Quality 2001;30 303-320.
65. Boyd CE. Water quality for pond aquaculture. Alabama Research and Development Series 1998; 43 37.
66. Boyd CE. Water Quality in Ponds for Aquaculture: Auburn University; 1990; p482.
67. Boyd CE, Musig Y. Shrimp pond effluents: observations of the nature of the problem on commercial farms. In: Proceedings of the Special Session on Shrimp Farming. World Aquaculture Society 1992, Baton Rouge, Louisiana 2006; 195-197.
68. Boyd CE., editor. Bottom Soils, Sediment, and Pond Aquaculture. Chapman and Hall; 1995a. p348.
69. Ozbay G. Studies of Channel Catfish *Ictalurus punctatus* Pond Effluent Characteristics and Treatment Options. PhD thesis. Auburn University Auburn; 2002; p109.
70. Ebeling JM, Vinci B. Solids Capture. Recirculating Aquaculture Systems Short Course; 2013. http://ag.arizona.edu/azaqua/ista/ISTA7/RecircWorkshop/Workshop %20PP%20%20&%20Misc%20Papers%20Adobe%202006/5%20Solids%20Captu re/Solids%20Control.pdf (accessed 15 November 2013).
71. Bowser PR. General Fish Health Management. Northeastern Regional Aquaculture Center 2012;111 p16.
72. Pietrak M, Leavitt D, Walsh M. Biosecurity on the Farm-Guidelines & Resources for Developing a Biosecurity Plan. Northeastern Regional Aquaculture Center 2010;208 p6.
73. Figueras MJ, Borrego JJ, Pike EB, Robertson W, Ashbolt N. Sanitary Inspection and Microbiological Water Quality. In: Bartram J. and Rees G. (eds.) Monitoring Bathing Waters-A Practical Guide to the Design and Implementation of Assessments and Monitoring Programmes. In E and FN Spon; 2000. p311.
74. Thompson FL, Iida T, Swings J. Biodiversity of Vibrios. Microbiology and Molecular Biology Review 2004;68(3) 403-431.

75. Austin. Vibrios as causal agents of zoonoses. Veterinary Microbiology 2010;140(3-4) 310-317.
76. Urakawa H., Rivera ING. Aquatic Environment. In: Thompson FL, Austin B, Swings J. (eds.) *The Biology of Vibrios*, D.C. ASM Press; 2006. p175–189.
77. Austin, B. Miscellaneous animal pathogens. In: Thompson, F.B., Austin, B. Swings, J. (eds.) The Biology of Vibrios In ASM Press; 2006. p423.
78. Grimes DJ. Ecology of estuarine bacteria capable of causing human disease: a review. Estuaries 1991;14(4) 345-360.
79. Baker-Austin CL, Stockely R, Rangdale, Martinez-Urtaza J. Environmental occurrence and clinical impact of *Vibrio vulnificus* and *Vibrio parahaemolyticus*: A European perspective. Environmental Microbiology Reports 2010;2 7-18.
80. Lipp EK, Huq A, Colwell RC. Effects of Global Climate on Infectious Disease: the Cholera Model. Clinical Microbiology Reviews 2002;15(4) 757-770.
81. Moyle PB, Cech JJ. 2004. Fishes: an introduction to ichthyology. Prentice Hall; 2004. p 465, 615.
82. Boyd CE, Boyd C, Chainark S. From The Shrimp Book: Shrimp Pond Soil and Water Quality Management. Global Aquaculture Advocate 2012;15(1):28-31.
83. Kalff, J. Limnology: inland water ecosystems. Prentice Hall; 2002. p592.
84. Shine C, de Klemm C. 1999. Wetlands, Water and the Law. Using law to advance wetland conservation and wise use. IUCN Environmental Policy and Law Paper 1999; p330.
85. United States Environmental Protection Agency. USEPA: Economic Benefits of wetlands. http://www.epa.gov/owow/wetlands (accessed 8 October 2013).
86. United States Environmental Protection Agency. USEPA: Nutrient Criteria Technical Guidance Manual- Wetlands. http://www2.epa.gov/sites/production /files/ documents/wetlands_full.pdf (accessed 6 November 2013).
87. Rückauf U, Augustin J, Russow R, Merbach, W. Nitrate removal from drained and reflooded fen soils affected by soil N transformation processes and plant uptake. Soil Biology and Biochemistry 2004;36(1) 77-90.
88. Reddy KR, Delaune RD. Biogeochemistry of Wetlands: Science and Applications. CRC Press; 2008. http://www.crcpress.com. (accessed 15 October 2013).
89. Mitsch WJ, Gosselink JG. Editors. Wetlands. In John Wiley and Sons, Inc., New York, NY. USA. 2007. p600.
90. Moshiri S. Identification of resistance source in barley to certain fungal pathogen in Khorasan 1995: conference proceedings, September 2-7, 1995, Karadj.
91. Schwartz MF, Boyd CE. Constructed wetlands for treatment of channel catfish pond effluents. The progressive fish-culturist 1995;57(4) 255-266.
92. Boyd CE. Hydrology and pond construction, In: Tucker CS. (ed.) Channel Catfish Culture. Elsevier; 1985. p107-134.
93. Shpigel M, Ben-Ezra D, Shauli L, Sagi M, Ventura Y, Samocha T, Lee JJ. Constructed wetland with Salicornia as a biofilter for mariculture effluents. Aquaculture 2013;412 52-63.
94. Lymbery AJ, Kay GD, Doupé RG, Partridge GJ, Norman HC. The potential of a salt-tolerant plant (Distichlis spicata) to treat effluent from inland saline aquaculture and provide livestock feed on salt-affected farmland. Science of The Total Environment 2013;445 192-201.
95. Serpa D, Pousão-Ferreira P, Caetano M, Cancela da Fonseca L, Dinis MT, Duarte P. A coupled biogeochemical-Dynamic Energy Budget model as a tool for managing fish production ponds. Science of the Total Environment 2013;463 861-874.
96. Boyd CE. Potential of sodium nitrate to improve environmental conditions in aquaculture ponds. Journal of the World Aquaculture Society 1995b;26 38-40.
97. Cathcart TP, Pote JW, Rutherford DW. Reduction of effluent discharge and groundwater use in catfish ponds. Aquacultural Engineering 1999;20(3) 163-174.
98. USDA/Natural Resources Conservation Service. [DVD]: Alabama Aquaculture Best Management Practices (BMP). Settling Basins and Wetlands 2002;BMP No. 6.

99. Tucker CS. Aquaculture Effluents and the Environment. Aquaculture Meeting; 2009 NC USA p61.
100. United States Environmental Protection Agency. USEPA: NPDES. Final Permit Fact Sheet. http://www.epa.gov/region9/water/npdes/pdf/ca/GoletaSdNpdes Permit FinalOrder.pdf (accessed 11 October 2013).
101. Teichert-Coddington DR, Rouse DB, Potts A and Boyd CE. Treatment of harvest discharge from intensive shrimp ponds by settling. Aquacultural Engineering 1999;19 147-161.
102. Ozbay G, Boyd CE. Treatment of Channel Catfish Pond Effluents in Sedimentation Basins. The World Aquaculture Society Magazine 2004;35(3)10-13.
103. Ozbay G, Boyd CE. Particle Size Fractions in Pond Effluents. The World Aquaculture Society Magazine 2003a;34(4) 56-59.
104. Ozbay G, Boyd CE. U.S. Study Finds Turbidity Useful in TSS, ISS Estimates. Global Aquaculture Advocate 2003b;6(4) 58-60.
105. Boyd CE. Aquaculture sustainability and environmental issues. World Aquaculture 1999;30(June) 10-15.
106. Hargreaves J, Cathcart T, Boyd C, Engle C, Romaire R, McClain R, Boardman G, Browdy C. Management of Aquacultural Effluents from Ponds. SRAC Sixteenth Annual Progress Report 2003; p18.
107. Boyd CE. Farm effluent during draining for harvest. Global Aquaculture Advocate 2000;3(4) 26-27.
108. Schwartz MF, Boyd CE. Suggested management to improve quality and reduce quantity of channel catfish pond effluents. Alabama Agriculture Experiment Station; 1996. Auburn University, Alabama, USA. Leaflet 108.
109. Lutz G, LeBlanc BD, Sheffield R, Nix KE. Aquaculture Environmental Best Management Practices. BMPs. LSU Ag Center Communications; 2011. Louisiana State University, LA, USA. p28.
110. Tucker CS, Hargreaves JA. (eds.). Environmental Best Management practices for aquaculture. John Wiley and Sons: Blackwell Publishing; 2008. p582.
111. Setty K. Design Manual: Retention Basin. Bren School of Environmental Science and Management; 2013. University of California, Santa Barbara, CA USA. p7.
112. Boyd CE, Tucker CS. Sustainability of channel catfish farming. World Aquaculture 1995;2645-53.
113. Cripps SJ. The application of suspended particle characterization techniques to aquaculture systems. In: Wang, J. (ed.) Techniques for Modern Aquaculture; 1993. p26–34.
114. Cripps SJ. Serial particle size fractionation and characterization of an aquacultural effluent. Aquaculture 1995;133 323–339.
115. Ackefors H, Enell M. The release of nutrients and organic matter from aquaculture systems in Nordic countries. Journal Applied Ichthyology 1994;10(4) 225-241.
116. Cripps SJ. 1992. Characterization of an aquaculture effluent based on water quality and particle size distribution data 1992: conference proceedings, May 21-25, 1992. Orlando, FL, USA.
117. Bergheim A, Tyvold T, Seymour EA. 1991. Effluent loadings and sludge removal from land based salmon farming tanks. In: Bredene EAS (ed.) Europe'91: Aquaculture and the Environment International Aquaculture Conference: proceedings of Aquaculture, Dublin, Ireland, 10–12 June, 1991. Belgium: EAS Special Publication.
118. Cripps SJ. Minimizing outputs: treatment. Journal of Applied Ichthyology 1994;10 284-294.
119. Cripps SJ., Kelly L. Reduction in wastes from aquaculture. In: Baird DG, Beveridge M, Kelly GA, Muir BF. (eds.) Aquaculture and Water Resource Management. Blackwell Scientific; 1996. p219.
120. Pepper VA, Sutterlin AM, Nicholls T, Collier C. Newfoundland experience with development of all-female and non-reproductive Atlantic salmon for marine aquaculture. Bulletin-Aquaculture Association of Canada 1996;14-23.

121. Moore PA, Miller DM. Decreasing phosphorus solubility in poultry litter with aluminum, calcium, and iron amendments. Journal of Environmental Quality 1994;23(2) 325-330.
122. Gensemer RW, Playle RC. The bioavailability and toxicity of aluminum in aquatic environments. Critical reviews in environmental science and technology 1999;29(4) 315-450.
123. Welch EB, Cooke GD. Effectiveness and longevity of phosphorus inactivation with alum. Lake and Reservoir Management 1999;15(1) 5-27.
124. Cooke GD., Welch EB., Peterson SA., Newroth PR. editors. Restoration and Management of Lakes and reservoirs. Lewis Publishers: CRC Press; 1993. p548.
125. Jacoby JM, Gibbons HL, Hanowell R and Bouchard DD. Wintertime blue-green algal toxicity in a mesotrophic lake. Journal of Freshwater Ecology 1994;9(3) 241-251.
126. Rowan M. Chemical phosphorus removal from aquaculture pond water and effluent. PhD thesis. Auburn University Auburn; 2001. p180.
127. Masuda K, Boyd CE. Effects of aeration, alum treatment, liming, and organic matter application on phosphorus exchange between soil and water in aquaculture ponds at Auburn, Alabama. Journal of the World Aquaculture Society 1994;25(3) 405-416.
128. Prepas EE, Murphy TP, Crosby JM, Walty DT, Lim JT, Babin J, Chambers PA. Reduction of phosphorus and chlorophyll a concentrations following calcium carbonate and calcium hydroxide additions to hypereutrophic Figure Eight Lake, Alberta. Environmental Science and Technology 1990;24(8) 1252-1258.
129. Salonen VP, Varjo E. Gypsum treatment as a restoration method for sediments of eutrophied lakes–experiments from southern Finland. Environmental Geology 2000;39(3-4) 353-359.
130. Wu R, Boyd CE. Evaluation of calcium sulfate for use in aquaculture ponds. The Progressive Fish-Culturist 1990;52(1) 26-31.
131. Ann Y, Reddy R, Delfi JJ. Influence of chemical amendments on phosphorus immobilization in soils from a constructed wetland. Ecological Engineering 2000;14 157–167.
132. Gutcho SJ. Waste Treatment with Polyelectrolytes and Other Flocculants No. 31. Noyes Data Corp 1977; Technology & Engineering. University of Minnesota, p24.
133. Ozbay G. Effects of Coagulant Treatments on Aquaculture Effluent Quality. Journal of Applied Aquaculture 2005;17(4) 1-23.
134. Wright RT, Coffin RB, Ersing CP, Pearson D. Field and laboratory measurements of bivalve filtration of natural marine bacterioplankton. Limnology and Oceanography 1982;27(1) 91-98.
135. Lin CK., Shrestha MK., Thakur DP., Diana JS. Management to minimize the environmental impacts of pond draining. In: Burke D., Baker J., Goetze B., Clair D. and Egna H. (eds.) Fifteenth Annual Technical Report; 1998. p132-139.
136. Shpigel M, Fridman R. Propigation of the Manila clam, Tapes semidecussatus, in the effluent of fish aquaculture ponds in Eliat, Israel. Aquaculture 1990;90 113-122.
137. Newell RIE, Cornwell JC, Owens M, Tuttle J. Role of oysters in maintaining estuarine water quality. National Shellfisheries Association. Journal of Shellfish Research 1999;18(1) 300-301.
138. Rensel J, Bright K, Siegrist Z. Integrated Fish-Shellfish Mariculture in Puget Sound. Final Report; 2011. p82.
139. Chrzanowski TH, Spurrier JD, Dame RF, Zingmark RG. Processing of microbial biomass by an intertidal reef community. Marine Ecology-Progress Series 1986;30 181-189.
140. Toro JE, Paredes PI, Villiagra JD, Senn CM. Seasonal variation in the planktonic community, seston and environmental variables during a 2-year period and oyster growth at two mariculture sites, southern Chile. Marine Ecology 1999;20(1) 63-89.
141. Ulanowicz RE, Tuttle JH. The trophic consequences of oyster stock rehabilitation in Chesapeake Bay. Estuaries 1992;15(3) 298-306.

142. Leffler M. Improving Bay Water Quality: The role of oyster reefs. Maryland Sea Grant. Maryland Aquafarmer 2001. http://www.mdsg.umd.edu/ Extension/Aquafarmer/ Winter01.html (accessed 10 October 2013).
143. Miura T, Yamashiro T. Size selective feeding of Anodonta calipygos, a phytoplanktivorous freshwater bivalve, and visibility of egested algae. Japanese Journal of Limnology 1990;51(2) 73-78.
144. Lowe R, Reich LA, Sierra J. Algal herbivory by the zebra mussel: Fate of algae in feces and pseudofeces. Journal of Shellfish Research 1991;10(1) 250-251.
145. Senichieva MI. Dynamics of phytoplankton community in the region with mussel mari-farming. Hydroores 1990;7(8) 86-89.
146. Santelices B, Martinez. The effect of filter feeders on macroalgae spores. 2. Indirect effects and landscape determination. Biota 1986;1:102.
147. Marenghi FP, Ozbay G. Floating Oyster, *Crassostrea virginica* Gmelin 1791, Aquaculture as habitat for fishes and macroinvertebrates in Delaware Inland Bays: The comparative value of oyster clusters and loose shell. Journal of Shellfish Research 2010a; 29(4) 889-904.
148. Marenghi FP, Ozbay G. Preliminary habitat assessment of floating oyster (Crassostrea virginica) gardens (Delaware). Ecological Restoration 2010b;28(3) 254-257.
149. Harding JM, Mann R. Estimates of naked goby (*Gobiosoma bosc*), striped blenny (*Chasmodes bosquianus*) and eastern oyster (*Crassostrea virginica*) larval production around a restored Chesapeake Bay oyster reef. Bulletin of Marine Science 2000;66(1) 29-45.
150. Tolley SG, Volety AK. The role of oysters in habitat use of oyster reefs by resident fishes and decapod crustaceans. Journal of Shellfish Research 2005;24(4) 1007-1012.
151. Breitburg DL. Are three-dimensional structure and healthy oyster populations the keys to an ecologically interesting and important fish community. Oyster reef habitat restoration: a synopsis and synthesis of approaches; 1999. Virginia Institute of Marine Science Press, Gloucester Point, Virginia. p239-250.
152. Posey DA. Developing sui generis options for the protection of living aquatic resources of indigenous and local communities. In: ICLARM Conference Proceedings, 1999.http://www.fao.org/ag/againfo/programmes/en/genetics/documents/ITWG -AnGR4/MontpellierAnGR-Papers.pdf (accessed 8 November 2013).
153. Peterson CH, Grabowski JH, Powers SP. Estimated enhancement of fish production resulting from restoring oyster reef habitat: quantitative valuation. Marine Ecology-Progress Series 2003;264 249–264.
154. The Nature Conservancy. 2013. Oyster Restoration Reaches New Depth. New Hampshire. Do you know how many gallons of water an oyster can filter in a day? http://www.nature.org/ourinitiatives/regions/northamerica/unitedstates/newha mpshire/explore/oyster-restoration-reaches-new-depths.xml.
155. Ozbay G, Reckenbeil B, Marenghi F, Erbland P. Can't Go Wrong With Eastern Oysters (*Crassostrea virginica*): Enhancing Biodiversity of the Delaware Inland Bays. Journal of Biodiversity and Endangered Species 2013;1 2.
156. Reckenbeil BA. Assessment of oyster restoration along human altered shorelines in the Delaware Inland Bays: An examination of riprap stocked with the eastern oyster (Crassostrea virginica). Master's thesis Delaware State University; 2013. p.174.
157. Rice MA. Environmental effects of shellfish aquaculture in the Northeast. NRAC Publication Fact Sheet 105; 2008.
158. Lin CK, Ruamthaveesub P, Wanuchsoontorn P. Integrated culture of the green mussel in wastewater from an intensive shrimp pond: concept and practice. World Aquaculture 1993;24 68-72.
159. Tseng KF, Huang JS, Liao I. Species control of microalgae in an aquaculture pond. Water Research 1991;25(11) 1431-1437.
160. Burke JS, Bayne DR. Impact of paddlefish on plankton and water quality of catfish ponds. The Progressive Fish Culturist 1986;48 177-183.

161. Smith NP. The suitability of routine weather data for estimating local estuarine heat energy fluxes. Estuaries 1985;8 270-278.
162. Fott J., L. Pechar and M. Prazakova. Fish as a factor controlling water quality in ponds. In: Barica J., Mur LR. (eds.) SIL Workshop on Hypertrophic Ecosystems, Vaxjo, Sweden; 1979. Sept. 10-14.
163. Canadian Aquaculture. Integrated Multi-trophic Aquaculture. What is it? Canadian Aquaculture Industry Alliance; 2012. http://www.aquaculture.ca/files/species-multi-trophic.php. (accessed 11 October 2013).
164. Gowen RJ. Aquaculture and the environment. In: De Pauw N., Joyce J. (eds.) Aquaculture and the Environment. European Aquaculture Society Special Publication 1991;16 23 – 48.
165. Filbrun JE, Culver DA. Can Reduced Provision of Manufactured Feed Improve Fish Production Efficiency in Ponds? North American Journal of Aquaculture 2013; 1 64-76.
166. Gross A, Boyd CE, Wood CW. Ammonia volatilization from freshwater fish ponds. Journal of Environmental Quality 1999;28(3) 793-797.
167. Abou Y, Saidou A, Mama D, Fiogbé ED, Micha JC. Evaluation of Nitrogen and Phosphorus Wastes Produced by Nile Tilapia (Oreochromis niloticus L.) Fed Azolla-Diets in Earthen Ponds. Journal of Environmental Protection 2012;3(6) 502-507.
168. Gross A, Boyd CE, Lovell RT, Eya JC. Phosphorus budgets for channel catfish ponds receiving diets with different phosphorus concentrations. Journal of the World Aquaculture Society 1998;29(1) 31-39.
169. Rangen. Salmon Feeds/ Trout Feeds/ Tilapia Feeds/ Quality Feed for Catfish. Rangen Inc; 2013 Buhl, ID, USA. http://www.rangen.com.
170. Boyd CE, Schmittou HR. Achievement of sustainable aquaculture through environmental management. Aquaculture Economics and Management 1999;3 59-69.
171. Goldburg R. Triplett T. Murky water: environmental effects of aquaculture in the United States. Environmental Defense Fund, New York, USA; 1999.
172. Frankic A, Hershner C. Sustainable aquaculture: developing the promise of aquaculture. Aquaculture International 2003;11 517-530.
173. Costa-Pierce, B. A. Ecology as the paradigm for future of aquaculture. In: Barry AC., (ed.) Ecological Aquaculture: The evolution of the blue revolution. Blackwell; 2002 Science, MA USA. p. 339-372.
174. Martins CIM, Eding EH, Verdegem MCJ, Heinsbroek LTN, Schneider O, Blancheton JP, Roque D'Orbcastel E, Verreth JAJ. New developments in recirculating aquaculture systems in Europe: a perspective on environmental sustainability. Aquaculture Engineering 2010;43 83-93.
175. Chen S., Summerfelt S., Losordo T., Malone R. Recirculating systems, effluents, and treatments. In: Aquaculture and the environment in the United States; 2002. p29.
176. Hollingsworth CS, Baldwin R, Wilda K, Ellis R, Soares S. Best management practices for finfish aquaculture in Massachusetts. Publication AG-BPFA; 2006. Western Massachusetts Center for Sustainable Aquaculture. University of Massachusetts Extension.
177. Badiola M, Mendiola D, Bostock J. 2012. Recirculating aquaculture systems (RAS) analysis: main issues on management and future challenges. Aquaculture Engineering 2012;51 26-35.
178. Timmons MB, Ebeling JM, Wheaton FW, Summerfelt ST, Vinci BJ. Recirculating Aquaculture Systems. Northeastern Regional Aquaculture Center. In Cayuga Aqua Ventures; 2001. Ithaca, NY 14850.
179. Timmons MB, Ebeling JM. Recirculating Aquaculture Systems. In Northeastern Regional Aquaculture Center; 2007. 2nd Ed. Cayuga, AquaVentures LLC. Ithaca, NY 14850.
180. Ebeling JM, Timmons MB. Recirculating Aquaculture Systems. In: James HT. (ed.) Aquaculture Production Systems. John Wiley and Sons; 2012. p245-277.

181. Summerfelt ST., Vinci. Better Management Practices for Recirculating Aquaculture Systems. In: Craig ST., John AH. (eds.) In. Environmental Best Management Practices for Aquaculture. Blackwell Publishing; 2008. p389-426.
182. Losordo T, Haider D, Tran H. Recirculation meets aquaponics: Engineering flexibility into fish/plant production systems. The Global Aquaculture Advocate 2013;16 88-90.
183. Miller D, Semmens K. Waste management in aquaculture. Publication # AQ02-1. Aquaculture information series: 2002. Agriculture and Resource Economics Program, Division of Resource Management. College of Agriculture, Forestry, and Consumer Sciences. West Virginia University, Morgantown, WV.
184. Valencia E, Adjei M, Martin J. Aquaculture effluent as a water and nutrient source for hay production in the seasonally dry tropics. Community Soil Science Plant Annals 2001;32 1293-1301.
185. Global Aquaculture Alliance. GAA: BAP Standards. www.galliance.org. (accessed 5 October 2013).
186. Hanna K, Wheaton F, Lazur A, Vankeuren S. Water quality in identical recirculating systems managed by different aquaculturists. International Journal of Recirculating Aquaculture 2012;11 55-74.
187. United States Department of Agriculture, National Institute of Food and Agriculture. USDA-NIFA: Legal definition of sustainable agriculture (US code Title 7, Section 3103). www.csrees.usda.gov/nea/ag_systems/in_focus/sustain_ag_if_legal.html (accessed 8 November 2013).
188. McMurtry MR, Nelson PV, Sanders DC, Hodges L. Sand culture of vegetables using recirculated aquaculture effluents. Applied Agriculture Research 1990;5 280-284.
189. Rakocy JE., Aquaponics-Integrating Fish and Plant Culture. In: James HT. (ed.) Aquaculture Production Systems. John Wiley and Sons; 2012. p343-386.
190. Rakocy JE, Losordo TM and Masser MP. Recirculating aquaculture tank production systems: integrating fish and plant culture. SRAC Publication 1992: Sourthern Region Aquaculture Center, Mississippi State University, Stoneville, MS. p1-11.
191. Rhodes MA, Hanson T, Alavi S, Davis DA. High-soy, fishmeal-free diets support Florida pompano growth. The Global Aquaculture Advocate 2013;16 100-103.
192. Watson AM, Barrows FT, Price AR. Taurine supplementation of plant derived protein and n-3 fatty acids are critical for optimal growth and development of cobia, *Rachycentron canadum*. Lipids 2013;48 819-913.
193. Ozbay G, Jackson AR. Aquaculture Effluents: Perspectives on Best Management Practices. Global Aquaculture Advocate 2006;9(5) 68-70.
194. Boyd CE. Guidelines for aquaculture effluent management at the farm-level. Aquaculture 2003;226(1) 101-112.
195. Villalón JR. Practical Manual for Semi-intensive Commercial Production of Marine Shrimp. Texas A&M University Sea Grant College Program; 1991. TAMU-SG-91-501. College Station, Texas, USA. p1-104.
196. Hopkins JS, Stokes AD, Browdy CL, Sandifer PA. The relationship between feeding rate, paddlewheel aeration rate and expected dawn dissolved oxygen in intensive shrimp ponds. Aquacultural Engineering 1991;10(4) 281-290.
197. United States Environmental Protection Agency. USEPA: Effluent Guidelines: Aquatic Animal Production Industry. http://water.epa.gov/scitech/wastetech /guide/ aquaculture/index.cfm (accessed 8 November 2013).
198. Northwest Earth Institute. 2012. Discussion Course on Choices for Sustainable Living. Northwest Earth Institute, Portland, OR USA. p138.
199. Newell CR, Hawkins AJS, Morris K, Richardson J, Davis C, Getchis T. ShellGIS: A Dynamic Tool for Shellfish Farm Site Selection. World Aquaculture 2013;44(3) 50-53.
200. ESRI News. 2013. Esri News for Agriculture. Clearwater Seafoods Achieves Sustainable Operations through GIS. Courtesy of Esri Canada Limited. Spring 2013.

201. Thailand DOF Information System Center. Fisheries Statistics of Thailand 2010; 2011. Department of Fisheries. Bangkok, Thailand. http://www. fisheries. go.th/dof/en/ (accessed 10 October 2013).
202. Bureau of Agricultural Economic Research.Situation of Agricultural products and Trend in the Year 2011. Office of Agricultural Economics, Bangkok, Thailand.
203. Center for Agricultural Information. Thailand Foreign Agricultural Trade Statistics 2011; 2011. Office of Agricultural Economics, Bangkok, Thailand.
204. Kongkeo H. Current Status and Development Trends of Aquaculture in the Asian Region. In: Subasinghe RP, Bueno P, Phillips MJ, Hough C, McGladdery SE, Arthur JR. (eds.) 2001: Aquaculture in the Third Millennium. Technical Proceedings of the Conference on Aquaculture in the Millennium, 20–25 February 2000, NACA, Bangkok and FAO, Rome, Italy.

# CHAPTER 9

# Environmental and Physiological Factors Shape The Gut Microbiota of Atlantic Salmon Parr (*Salmo Salar* L.)

**Carola E. Dehler, Christopher J. Secombes, Samuel A.M. Martin**

*Institute of Biological and Environmental Sciences, University of Aberdeen, Tillydrone Avenue, Aberdeen AB24 2TZ, UK*

## ABSTRACT

Gut microbes are key players in host immune system priming, protection and development, as well as providing nutrients to the host that would be otherwise unavailable. Due to this importance, studies investigating the link between host and microbe are being initiated in farmed fish. The establishment, maintenance and subsequent changes of the intestinal microbiota are central to define fish physiology and nutrition in the future. In fish, unlike mammals, acquiring intestinal microbes is believed to occur around the time of first feeding mainly from the water surrounding them and their microbial composition over time is shaped therefore by their habitat.

Here we compare the distal intestine microbiota of Atlantic salmon parr reared in a recirculating laboratory aquarium with that of age matched parr maintained in cage culture in an open freshwater loch environment of a commercial fish farm to establish the microbial profiles in the gut at the freshwater stage and investigate if there is a stable subset of bacteria present regardless of habitat type. We used deep sequencing across two variable regions of the 16S rRNA gene, with a mean read depth of 180,144 ± 12,096 raw sequences per sample. All individual fish used in this study had a minimum of 30,000 quality controlled reads, corresponding to an average of 342 ± 19 Operational Taxonomic Units (OTUs) per sample, which predominantly mapped to the phyla *Firmicutes*, *Proteobacteria*, and *Tenericutes*.

The results indicate that species richness is comparable between both treatment groups, however, significant differences were found in the compositions of the gut microbiota between the rearing groups. Furthermore, a core microbiota of 19 OTUs was identified, shared by all samples regardless of treatment group, mainly consisting of members of the phyla *Proteobacteria*, *Bacteroidetes* and *Firmicutes*. Core microbiotas of the individual rearing groups were determined (aquarium fish: 19 + 4 (total 23) OTUs, loch fish: 19 + 13 (total 32) OTUs), indicating that microbe acquisition or loss is occurring differently in the two habitats, but also that selective forces are acting within the host, offering niches to specific bacterial taxa.

The new information gathered in this study by the Illumina MiSeq approach will be useful to understand and define the gut microbiota of healthy Atlantic salmon in freshwater and expand on previous studies using DGGE, TGGE and T-RFPL. Monitoring deviations from these profiles, especially the core microbes which are present regardless of habitat type, might be used in the future as early indicator for intestinal health issues caused by sub optimal feed or infectious diseases in the farm setting.

## Statement of relevance

The Microbiome is central to gut health, local immune function and nutrient up take. We have used deep sequencing approach to show differences in rearing conditions of Atlantic salmon. This work is of interest to aquaculture nutritionists.

**Keywords:** Atlantic salmon Gut microbiota Next-generation sequencing

# 1. INTRODUCTION

The gut microbiota can be viewed as an "extra organ" due to its key role in intestinal development, homeostasis and protection and it is now becoming clear that the microbiota has major influences on growth, health and development in all vertebrates (O'Hara and Shanahan, 2006; Dhanasiri et al., 2011). The host may also exert direct control on community assembly by physiological and lifestyle forces, offering a niche for specialised bacteria that are beneficial for their dietary life style (Roeselers et al., 2011). In return, intestinal microbes supply the host with exogenous nutrients and extracellular enzymes, fatty acids and vitamins which otherwise would not be available to the host (Dhanasiri et al., 2011). Hence a balanced microbial community is central to the wellbeing of animals and humans. An established microbiome also can reduce the proliferation of pathogenic bacterial species and as such are central to ensuring disease control (Austin, 2006; Fjellheim et al., 2007). As intestinal microbiota of fish play a key role in health, understanding the interactions between fish and gut microbiota is an important research field for fish health and aquaculture (van Kessel et al., 2011). Initial studies have shown that fish harbour a diverse microbiota, dependent on host species (Sullam et al., 2012; Li et al., 2014; Givens et al., 2015), life cycle stage (Giatsis et al., 2014; Ingerslev et al., 2014a; Zarkasi et al., 2014, 2016) and diet (Navarrete et al., 2013; Wong et al., 2013; Ingerslev et al., 2014b). Fish present a unique case among vertebrates with microorganisms from environmental sources potentially having a much higher influence on bacterial composition and health status due to constant contact with water (Austin, 2006; Ringø et al., 2010).

The fish intestine can harbour $10^7$ to $10^{11}$ bacteria/g intestinal content, with aerobes and facultative anaerobes being more abundant than obligate anaerobes, and can be split between the autochthonous species that are attached to the intestinal mucosa and the allochthonous that do not attach either due to lack of ability or outcompetition by mucous attached bacteria (Nayak, 2010; Navarrete et al., 2012; Llewellyn et al., 2014; Givens et al., 2015).

Gut microbiota are detectable 1 day post hatching in rainbow trout, before first feeding commences, potentially due to contact with surrounding water and yok sac digestion (Ingerslev et al., 2014a). First feeding induces significant changes in the gut microbiota composition and higher transcription levels for immune system associated genes (Ingerslev et al., 2014a). Non-attached bacterial levels may be maintained artificially high in fish due to the constant contact with the surrounding water (Ringø et al., 2010). Genetic background and ontogenic factors are also factors that shape an individuals' gut microbiota (Benson et al., 2010; Ingerslev et al., 2014a). Previous studies showed that fish harbour more specialised gut microbiota than previously assumed (Abid et al., 2013). Furthermore, fish are able to retain specific bacteria, making gut microbial assemblages not a passive collection of seeding communities (Sullam et al., 2012).

Although intestinal microbiota profiles were found to cluster with their respective environmental water bacterial profiles, many bacterial species found in the intestine where not detected at all in the environmental water samples, indicating maintenance of key species derived from elsewhere

(Roeselers et al., 2011; Sullam et al., 2012). This suggests that a large percentage of gut microbiota may actually be specialised members of symbiotic communities rather than free-living microbes (Sullam et al., 2012).

Recent technological advances in sequencing have led to a much more comprehensive assessment of the intestinal microbiota. Studies until a few years ago relied on either growing bacteria *ex vivo*, for which only very small numbers of species could be assessed or using techniques such as DGGE and TGGE. Only about 1% of all gut bacteria in fish can be cultured, including *Enterobacteriaceae, Pseudomonas* spp., *Staphylococcus* spp., *Acinet obacter* spp., *Aeromonas* spp., *Vibrio* spp. and *Bacillus* spp., which severely limited early assessments of fish intestinal microbiota (Merrifield et al., 2009), with a limited number of species cultivated from zebrafish (Cantas et al., 2012) and trout (Skrodenyté-Arbačiauskiené et al., 2008). In Atlantic salmon limited species richness was found by TTGE (Navarrete et al., 2009a) and DGGE (Reveco et al., 2014).

Next-generation sequencing platforms including Roches' 454 and Illumin MiSeq, NextSeq 500, and HiSeq 2000 have overcome the limitations of the previous technologies (Van Kessel et al., 2011; Geraylou et al., 2013; Star et al., 2013; Ingerslev et al., 2014a,b; Zarkasi et al., 2014). The 16S rRNA gene is characterised by highly conserved and highly variable regions across the whole sequence and has emerged as popular sequencing target, facilitating taxonomic assessments *via* the use of universal primers (Baker et al., 2003; Rajendhran and Gunasekaran, 2011; Klindworth et al., 2013). Read lengths of between 250 and 500 bp have been shown sufficient for community comparisons and current studies often sequence to a depth of > 100,000 sequences per biological sample (Liu et al., 2007; Hamady and Knight, 2009). Sequencing at such depth allows for rare species/genera to be identified from a number of key sequence resources designed for such studies, in particular RDP (Wang et al., 2007), SILVA (Pruesse et al., 2007) and Greengenes (DeSantis et al., 2006), all containing a massive resource of bacterial species 16S rRNA sequences.

In this study we investigated the role of environmental rearing conditions in the composition of intestinal microbiota in Atlantic salmon parr. Our aim was to define the microbiota of healthy Atlantic salmon at high resolution and investigate how important the rearing water conditions are to the maintenance of microbes in the fish intestine. To achieve this, fish were sampled from two different holding conditions: an indoor recirculating aquarium facility and a cage culture in an open loch environment. Both groups of fish came from the same hatchery before being transferred to the respective rearing conditions. We hypothesised that the fish in the open water environment would have a greater diversity of bacterial species in the intestine due to the natural variation of microbial composition in the water column caused by currents, rain water influx, over-land water run-off and greater biological species diversity, thereby offering a higher variety of potential colonisers. Alpha diversity analysis based on the Chao 1 index confirms the hypothesis, showing a significantly higher number of OTUs in the intestines of fish kept in the open loch system. We found 328 Operational Taxonomic Units (OTUs) only present in fish of the open loch system and 306 OTUs that were unique to the recirculating aquarium. Furthermore we identified 71 OTUs that were significantly different in abundance between the rearing groups, with 61 OTUs being more abundant in open loch system. Additionally we found a core microbiota within the rearing groups and overall, regardless of source of the fish. To our knowledge, this is the first in depth study of bacterial OTUs and core microbiota harboured in the intestine of Atlantic salmon parr at high sequencing depth.

# 2. MATERIAL AND METHODS

## 2.1. Fish maintenance and sampling

Two groups of juvenile mixed sex Atlantic salmon were collected, one from open freshwater commercial cages on the West coast of Scotland and a second group from a recirculating aquarium system at the University of Aberdeen (n = 20 for each group). Both groups were derived from the same hatchery before being moved to the different rearing environments and were at the pre-smolt stage at the time of sampling. In the aquarium the temperature was kept at 12 °C, pH 7.6 under natural photoperiod and oxygen saturation of 90%. Fish were fed *ad libitum* by hand twice daily at 09:00 and 15:00 for eight months and feed rate was adjusted each month to ensure around 2% of their body weight was consumed per day. The fish in the open cage system were under natural ambient temperature and natural photoperiod. Both groups were fed a fish meal rich commercial diet (BioMar). Fish were killed by anaesthesia overdose (phenoxyethanol) followed by destruction of the brain. The final weights of the fish were 59.3 ± 9.2 g for the open loch group and 124.7 ± 33 g for the aquarium group. The digesta samples (100–200 mg) were collected in 1.5 ml Eppendorf tubes immediately following death and frozen on dry ice before longer term storage at − 80 °C.

## 2.2. DNA extraction

For DNA extraction, the QIAamp Fast DNA Stool Mini Kit (Qiagen) was used according to the manufacturers protocol with the following modifications.

InhibitEx buffer was directly applied to the frozen digesta samples and two tungsten beads were added. The samples were then pre-treated with mechanical lysis by TissueLyser for 5 min to avoid biases against tough-walled Gram-positive bacteria. Lysis temperature was 90 °C to allow for cell-wall break-down of difficult to lyse bacteria. DNA was quantified by NanoDrop spectrometry.

## 2.3. PCR amplification and sequencing

For primary PCR reactions, variable regions 3 and 4 of the 16S rRNA gene were targeted with the recently published primer pair by Klindworth et al. (2013) including a 5′ overhang as suggested by Illumina. The forward primer had the sequence 5′ TCGTCGGCAGCGTCAGATGT GTATAAGAGA CAG**CCTACGGGNGGCWGCAG**, and the reverse primer 5′GTCTCGTGGGCT CGGAGATGTGTATAAGAGACAG**GACTACHVGGGTATCTAATCC** with the bold underlined sequence being the locus-specific V3-V4 primers. PCR reactions were performed in triplicate for each sample and pooled after amplification to avoid biases in PCR efficiencies. PCR reactions were performed in 25 µl reactions including 5 µl of each forward and reverse primer (1 µM each, Sigma), 12.5 µl of 2 × Kapa HiFi HotStart ReadyMix including high-fidelity polymerase (Kapa Biosystems) and 2.5 µl of DNA (5 ng/µl). PCR conditions included an initial denaturation at 95 °C for 3 min, followed by 25 cycles of 30 s at 95 °C, 30 s at 55 °C and 30 s at 72 °C after which a final extension at 72 °C for 5 min was applied. The product (527 bp, 460 bp + 67 bases of overhang primer) was verified for each PCR reaction on a 1.5% agarose gel.

These PCR products were cleaned with AMPure XP (Agencourt) and verification of clean-up product with TapeStation (Agilent) was performed. This was followed by indexing PCR which attached Nextera XT barcodes and Illumina sequencing adapters to the 5′ overhangs. After another round of AMPure XP (Agencourt) clean-up, multiplexed amplicons were sequenced with Illumina MiSeq from both ends.

## 2.4. Bioinformatic analysis

Raw forward and reverse sequences were assembled for each sample into contigs using mothur (Schloss et al., 2009). The assembled reads were then quality-filtered one sample at a time using the mother script trim.seqs, excluding reads that were < 400 bp long, had a lower Phred quality score average than 25, had a homopolymer run longer than 10 bases and nucleotide differences of > 7 bases in the primer region. The resulting individual fasta files were assembled into one fasta file and all further analysis was performed in QIIME (Caporaso et al., 2010). Operational

Taxonomic Units (OTUs) were picked *de novo* using Uclust at the 97% similarity level. A reference set was picked randomly from the resulting OTU bins. Taxonomy was assigned using the RDP classifier (Wang et al., 2007) at an 80% confidence level. The resulting fasta file was summarised in an OTU table in biom format. Taxonomy-assigned sequences were aligned with PyNAST with a pairwise Uclust alignment algorithm against the Greengenes data base (Release GG_13_5, DeSantis et al., 2006). ChimeraSlayer was used to identify chimeric sequences, which were afterwards excluded from both OTU table and alignment. The known contaminant sequences of *Cyanobacteria* and singleton sequences were removed from the OTU table and cropped from the alignment. The final quality-filtered alignment was used to make a phylogenetic tree of the reference OTUs as a basis for alpha-diversity (diversity within treatment group) metrices. Shared OTUs between all samples and a core microbiome was computed and visualised in the Phyloseq package in R. Metastats (White et al., 2009) was used to identify which OTUs were significantly different between the environmental groups.

Alpha diversity metrics were calculated for Chao1 (assesses differences in species richness) and PD-whole-tree (assesses differences based on phylogenetic information). Rarefaction curves were computed for both metrics at sample and treatment level. The highest common rarefaction level of all samples was identified and used as the basis to test for statistical differences between aquarium and loch using Bonferroni-corrected non-parametric testing with 999 Monte-Carlo permutations. To assess beta-diversity (phylogenetic differences between treatment groups), distance matrixes were produced with an even sequence level at the highest common rarefaction level common for each sample. The distance matrixes were used to test for statistical differences between distances within and between treatments, both for weighted and unweighted UniFrac metrics and to produce Multiple dimension scale (MDS) plots (Lozupone and Knight, 2005; Lozupone et al., 2007). Visualisation of rarefaction plots and cluster analysis was done with the Phyloseq package in R (McMurdie and Holmes, 2013).

# 3. RESULTS

## 3.1. Sequencing overview and microbiota characterisation

DNA was extracted from 30 fish, 15 from the aquarium and 15 from the loch. However, several samples showed poor PCR amplification, most likely due to co-extraction of PCR inhibitors, and were not used for sequencing, leaving 8 fish from the aquarium and 12 fish from the loch for analysis. In total, 3,602,871 raw reads were obtained for both forward and reverse directions after sequencing. The mean read depth per sample was 180,144 ± 12,069 sequences per read direction per sample. No reads were lost after assembly of forward and reverse reads. After the initial quality filtering, 3,226,813 sequences passed with a mean of 161,341 ± 12,472 sequences per sample. Across all samples a total of 51,591 OTUs were detected as defined by initial identification by the RDP classifier against the unaligned Greengenes database. Within these OTUs all singletons and *Cyanobacteria*, a commonly observed single cell algae, were removed, leaving 2153 OTUs. A further 303 OTUs were eliminated as these were identified as chimeric sequences, leaving a total of 1850 OTUs for subsequent analysis.

Overall, nine phyla were identified at an abundance ≥ 0.1% (Fig. 1A). The only difference in terms of presence/absence of phyla at a ≥ 0.1% abundance was that loch samples showed only 8 phyla, lacking *Acidobacteria* at this abundance level. When samples are investigated individually, 16 phyla were identified at an abundance ≥ 0.1%. The intestinal microbiota of the fish examined was dominated by *Firmicutes* (50%), *Proteobacteria* (23.1%) and *Tenericutes* (6.9%), accounting for 80% of all sequences. OTUs mapped to "other Bacteria" (representing 15.5%) reflect a mix of different phylogenetic groups that could not be identified to a deeper taxonomic level and are not considered a homogenous group.

When the OTUs are considered at the genus level, a high diversity of microbes is identified. Overall, 43 genera were identified at ≥ 0.1% abundance (Fig. 1B). Samples from the aquarium group harboured 39 genera and loch samples harboured 44 genera at ≥ 0.1% abundance. When samples are considered individually, the number of genera increases to 145 genera at ≥ 0.1% abundance, reflecting the high inter-individual variation in microbiota composition. Across all samples, the most abundant genera were classified as "Other *Ruminococcaceae*" (46%), followed by "Other *Mycoplasmataceae*" and *Pseudomonas* sp. (6.9% and 6.1%).

## 3.2. Microbiota differences between recirculated aquarium and open loch

To compare differences in gut microbial community composition between fish reared in the recirculating aquarium and the open loch, Metastats analysis was performed (Fig. 2). OTUs were identified that were present in either group at a statistically significant level with OTUs only present in the aquarium reared fish (Supplementary Table S1), OTUs only present in the loch reared fish (Supplementary Table S2) and OTUs significantly more abundant in either group (Supplementary Table S3).

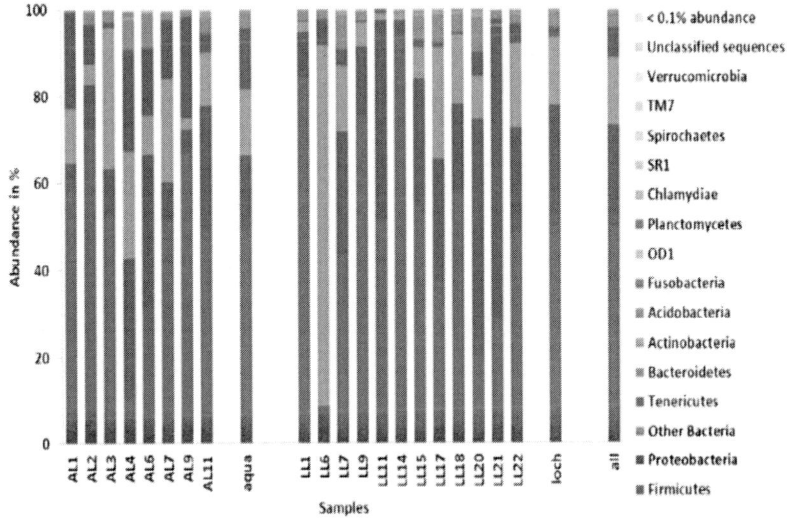
Abundance in %
100
80
60
40
20
0
AL1
AL2
AL3
AL4
AL6
AL7
AL9
AL11
aqua
LL1
LL6
LL7
LL9
LL11
LL14
LL15
LL17
LL18
LL20
LL21
LL22
loch
all
Samples
< 0.1% abundance
Unclassified sequences
Verrucomicrobia
TM7
Spirochaetes
SR1
Chlamydiae
Planctomycetes
OD1
Fusobacteria
Acidobacteria
Actinobacteria
Bacteroidetes
Tenericutes
Other Bacteria
Proteobacteria
Firmicutes

A

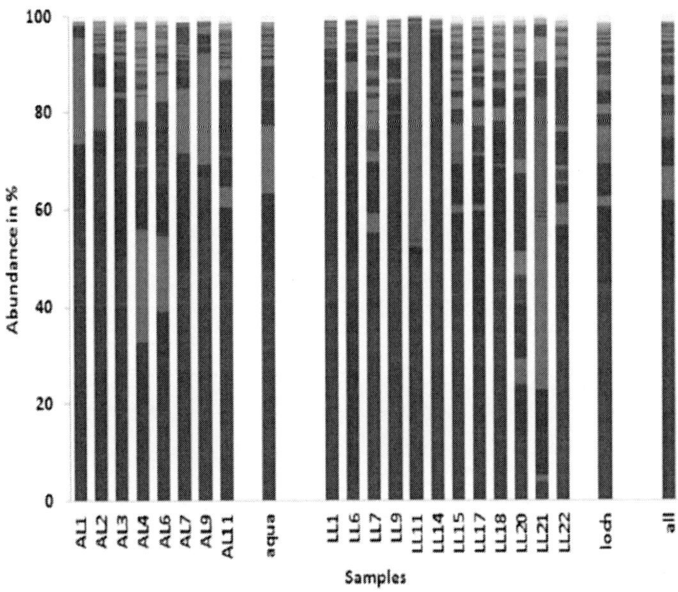
Abundance in %
100
80
60
40
20
0
AL1
AL2
AL3
AL4
AL6
AL7
AL9
AL11
aqua
LL1
LL6
LL7
LL9
LL11
LL14
LL15
LL17
LL18
LL20
LL21
LL22
loch
all
Samples

**Figure. 1.** Relative abundance (%) of main bacterial taxa found in digesta collected from the distal intestine of Atlantic salmon (*Salmo salar*) parr at phylum level (A) and genus level (B). Phyla and genera below an abundance of 0.1% are not shown but summarised in a mixed group "< 0.1% abundance". Bacterial profiles are shown on an individual level, a rearing group level (aqua = recirculating aquarium system, loch = open loch environment) and on an overall level (= all).

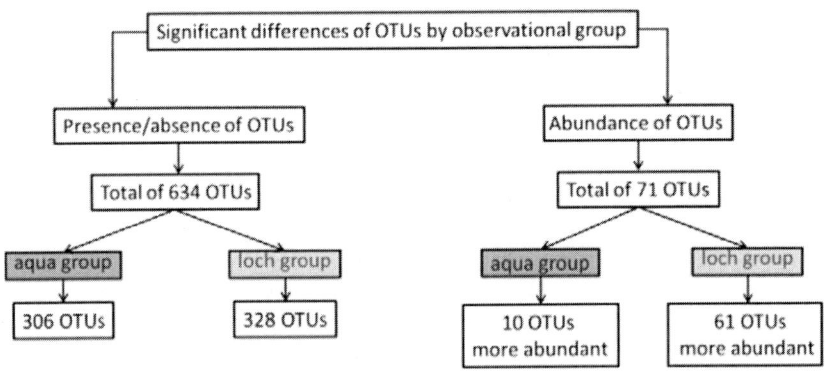


**Figure. 2.** Operational Taxonomic Units (OTUs) of microbes identified in the distal intestine digesta of Atlantic salmon (*Salmo salar*) parr found to be significantly different between rearing groups (recirculating aquarium group and open loch environment group) by Metastats analysis (see Supplementary Tables 1 and 2). 634 OTUs were found to be entirely significantly present, of which 306 were significantly present in the aquarium group and 328 significantly present in the loch group. Furthermore 71 OTUs were found to show significantly different

abundance between the observational groups, with 10 OTUs that were more abundant in the aquarium group and 61 OTUs that were more abundant in the loch group (see Supplementary Table 3). For a full list of OTUs found to differ significantly between observational groups refer to Supplementary Tables 1, 2 and 3.

Metastats analysis found that 306 OTUs were unique to the fish from the recirculating aquarium system, mapping to 16 phyla, and loch samples harboured 328 unique OTUs, mapping to 13 phyla. In samples from the aquarium, 68 OTUs could only be identified to "unclassified bacteria" (UB) (11 OTUs could not be assigned a Greengenes ID) and three OTUs to "unclassified sequence" (US), whereas in samples from the open loch, 99 OTUs could only be identified to UB (31 OTUs could not be assigned a Greengenes ID) and six OTUs to US. OTUs included in the phyla *Chloroflexi*, *Chloroplast*, *Tenericutes* and *Verrucomicrobia* were only found in samples from the recirculating aquarium, whereas OTUs included in the phylum *OP10* were only found in samples from the open loch.

A further 71 OTUs were significantly different in abundance between the environmental groups with higher abundance of 10 OTUs in fish from the aquarium and 61 OTUs more abundant in samples from the loch. Forty one OTUs could only be identified to an UB level, of which only one OTU had no match in the Greengenes database. Most OTUs with significant differences in abundance were associated with either the phylum *Firmicutes*(11 OTUs) or *Proteobacteria* (12 OTUs) (Supplementary Table S3).

## 3.3. Core microbiota

The core microbiome across all individuals examined was 19 shared OTUs, including one *Actinobacteria*, for *Bacteroidetes*, three *Firmicutes*, eight *Proteobacteria*, one *Tenericutes*and two "Other" bacteria (Fig. 3). Within this core microbiota, dominant members like *Rumino coccaceae* spp., *Mycoplasmataceae* spp. and *Pseudomonas* spp. were found, but also genera of much smaller relative abundance, for example, *Chryseobacterium* spp. and *Staphylococcus* spp. were identified. When separated by treatment group, the aquarium fish shared a core microbiota of 23 OTUs, with four additional OTUs to the above overall core microbiota (*Escherichia/Shigella*, *Brucella*, *Corynebacterium* and one "Other Bacteria"), whereas loch fish harboured additional 13 OTUs (two *Photobacterium*, *Aliivibrio*, *Moritella*, *Psychrobacter*, *Jeotgalicoccus*, two *Lactobacillus*, *Psychrilyobacter* and *Rubrobacter*, plus an additional OTU for *Pseudomonas*, *Stenotrophomonas* and *Streptococcus*) representing a core microbiota of 32 OTUs.

**Figure. 3.** Core microbiota of distal intestine digesta in Atlantic salmon (*Salmo salar*) parr. Core microbiota was identified at an overall level (regardless of rearing group), and at a rearing group level (aqua = recirculating aquarium facility, loch = open loch environment).

## 3.4. Alpha diversity

Samples had an average of 342 ± 19 OTUs, with the lowest being AL4 harbouring 176 OTUs and the highest LL7 with 525 OTUs. The Chao1 index found that the number of OTUs was significantly different ($P = 0.037$) with a higher number of OTUs in the loch samples. The PD_whole_tree metric, however, showed no significant differences in phylogenetic diversity between the two rearing groups, indicating that the differences in microbial species richness are due to enrichments of the same phylogenetic branches in the loch rearing group, rather than addition of novel phylogenetic groups. To test for robustness on even sequence levels per sample, rarefaction curves were produced based on the Chao1 metric with an iteration of 10 per rarefaction level on 100 levels from 10 sequences/sample to 134,410 sequences/sample. The rarefaction curves reach a plateau between 20,000 and 30,000 reads and confirm the higher diversity of OTUs found in the loch group, showing that evening the sequencing depth to this level for all samples is enough to identify patterns unique to the rearing groups (data not shown).

Rarefaction analysis identified the highest common rarefaction level of all samples is 33,610 sequences/sample. This limitation arose from sample AL6. Hence, all samples were limited to 33,610 sequences/sample randomly picked to facilitate evenness for beta diversity comparisons.

## 3.5. Beta diversity

Sample clustering was visualised with multiple dimension scaling (MDS) based both on unweighted and weighted UniFrac (Fig. 4A and B). Distance-based Redundancy Analysis (dbRDA) associated with MDS plots showed that based on the unweighted UniFrac distance matrix, there is a significant separation of the clusters by rearing group ($P = 0.006$). However, dbRDA performed on the weighted UniFrac distance matrix showed that there is no significant difference between the two rearing groups ($P = 0.44$). Additional statistical testing with PERMANOVA confirmed significant differences between the environments based on absence/presence of OTUs (Pseudo F-statistic 1.4599, $P = 0.006$), but not in terms of abundances of the OTUs (Pseudo F-statistic 0.9589, $P = 0.419$). The significance of the unweighted absence/presence approach reflects the results from the Metastats analysis, showing a high proportion of OTUs present in either one group or the other. The lack of significance seen in the weighted approach, which takes not only the absence/presence but also the abundance of each OTU into account, maybe due to the high individual variability of microbial composition masking the different patterns and introducing noise.

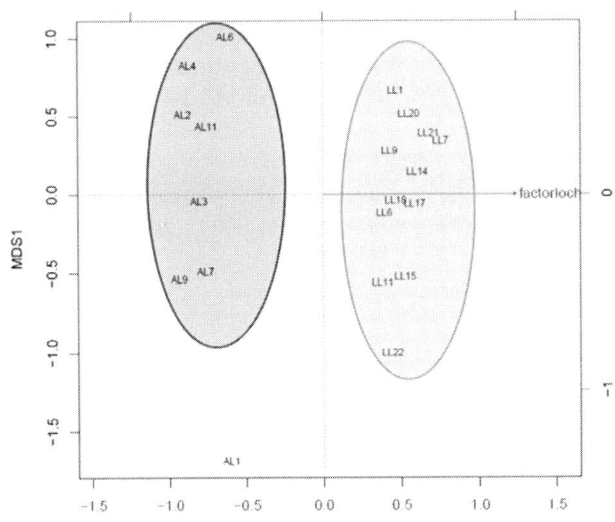


A

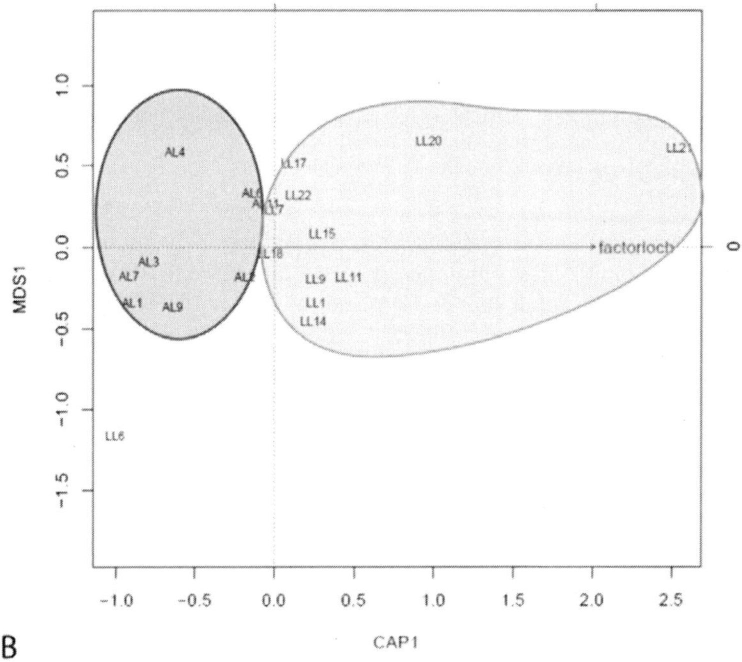


B

**Figure. 4.** Multiple dimension scale (MDS) plots of unweighted (A) and weighted (B) UniFrac distances of distal intestine digesta microbiota found in Atlantic salmon (*Salmo salar*) parr by rearing group. Beta diversity analysis of unweighted UniFrac (A, presence/absence of Operational Taxonomic Units (OTUs)) was found significantly different between observational groups (PERMANOVA, Pseudo F-statistic 1.4599, $P = 0.006$, based on 999 permutations). However, analysis of weighted UniFrac (B, presence/absence/abundance of OTUs) was not found significantly different (PERMANOVA, Pseudo F-statistic 0.9589, $P = 0.419$, based on 999 permutations). Red = recirculating aquarium facility, blue = open loch environment.

# 4. DISCUSSION

The intestinal microbiota of animals and humans has attracted much recent attention as it is believed to be a key factor in numerous animal functions including health, growth and disease status (Ley et al., 2008; Hamady and Knight, 2009; Lee and Mazmanian, 2010; Kau et al., 2011; Gevers et al., 2012; Lozupone et al., 2012). Recent advances in deep sequencing technologies have made the detailed study of fish microbiota a relatively straight forward procedure and no longer relies on the ability to grow bacteria *ex vivo*, which meant studies were highly limited (Austin, 2006; Navarrete et al., 2009b;Geraylou et al., 2013). In this study the intestinal microbiota of Atlantic salmon has been examined under two different environments, a recirculated aquarium facility and an open freshwater loch cage system in the West of Scotland using the 16S rRNA marker gene. Both groups of fish had been reared in a common flow through hatchery condition before being maintained in either the aquarium or loch. Our hypothesis was that a core microbiota would be maintained in both groups, but that the diversity might be altered by the environmental conditions.

The concept of a core set of microbial species fulfilling the minimal symbiotic functionality, has been suggested and promoted previously (Roeselers et al., 2011; Star et al., 2013). However, in humans, limited evidence has been given for universally abundant species at a fine taxonomic level (Lozupone et al., 2012). Recent studies in fish proposed a core microbiota of for example 10 OTUs in Atlantic cod (Star et al., 2013), 21 OTUs in zebrafish (Roeselers et al., 2011) and 52 OTUs in rainbow trout (Wong et al., 2013). In the present study 19 OTUs were shared among all samples regardless of holding condition, albeit at different proportional abundance for each sample. Several of those OTUs reflected some of the most dominant community members like *Ruminococcaceae*, *Pseudomonas* spp. and *Mycoplasmataceae*, but rarer sequences like *Propionibacterium* spp. and *Achromobacter*spp. were also found in the core group.

In the present study we also found accessory cores, with unique core OTUs found within treatment groups – aquarium fish harboured an additional 4 OTUs, whereas loch samples had an additional 13 unique core microbiota. These results may indicate that the environment shapes aspects of the gut microbiota of Atlantic salmon parr, with a higher shared number of OTUs within the semi-natural loch environment, but also confirms maintenance of bacteria regardless of the environment and may be dictated by the gut physiology of the host. Unfortunately, at this point, too little is known about intestinal bacteria in fish to speculate on the functional relevance of the group-specific microbiota. Future metagenomics studies will be crucial to link the knowledge of what bacterial species are present to what the functional relevance of these species is.

The present study is the first in-depth analysis of gut microbiota in Atlantic salmon at the freshwater stage. By using an Illumina NGS approach we were able to detect 1850 quality-controlled OTUs. This is a marked improvement compared to the detection ability of previously used techniques, such as direct bacterial culture, DGGE, TGGE and T-RFPL, which has limited the number of bacterial species identified in the fish intestine in the past (Romero and Navarrete, 2006; Hovda et al., 2007; Navarrete et al., 2009a; Green et al., 2013). The individual fish harboured a mean of 342 OTUs. For humans it has been suggested that between 1000 and 1150 species exist overall (Lee and Mazmanian, 2010). The higher number of OTUs we identified might be due to a different OTU picking method or to a lack of information for fish specific bacteria within the microbial databases. Here we show that the gut microbiota of Atlantic salmon parr is very diverse with a handful of phyla but deep-branching richness in potential species (OTUs), which is comparable to what has been found in NGS studies in humans (Benson et al., 2010) and other fish, like rainbow trout (Ingerslev et al., 2014b).

The most dominant phylum overall was *Firmicutes*, followed by *Proteobacteria* and *Tenericutes*. However, there was a greater presence of *Tenericutes* in samples from aquarium fish whereas *Proteobacteria* were more abundant in the loch sampled fish. The dominance of these phyla is in agreement with earlier studies on Atlantic salmon and rainbow trout using both DGGE and NGS (Abid et al., 2013; Green et al., 2013; Ingerslev et al., 2014b). This differs from mammals were *Firmicutes* and *Bacteroidetes* tend to dominate (Ley et al., 2008).

*Mycoplasmataceae* (phylum *Tenericutes*) was the second most abundant family in the aquarium fish samples, and were less abundant in loch sampled fish. Metastats analysis found two OTUs associated with the family *Mycoplasmataceae* significantly more abundant in aquarium samples than loch samples (Supplementary Table S3) and one OTU of the family *Mycoplasmataceae* unique in fish from the aquarium (Supplementary Table S1). Previous studies found *Mycoplasmataceae* to be dominant in Atlantic salmon reared in seawater by DGGE and TRF and hence they appear to be a common microbiota member across environments (Abid et al., 2013; Green et al., 2013). *Mycoplasmataceae* have therefore been suggested as a dominant community member in Atlantic salmon, although the species this family comprises may be host, tissue and habitat specific (Green et al., 2013; Givens et al., 2015). In this study, OTUs mapped to genera that are known to include pathogenic species, showed a significantly higher relative abundance in loch samples compared to aquarium samples, including *Photobacterium*, *Aliivibrio*, *Aeromonas*, *Yersinia*and *Chryseobacteri um* (Supplementary Table S3). Furthermore, OTUs belonging to the genera *Pseudomonas* and *Flavobacterium* were found more often at significant levels in loch samples (Supplementary Table S2). The presence of these genera shows that they can be present at low levels with no pathological effect on the fish, most likely as they are maintained at low levels by the healthy microbial community (Austin, 2006). It should be noted that sequencing short amplicons (such as Illumina) is only able to confidently assign taxonomy up to the genus level. Genera can include multiple species and strains with varying pathogenicity, for example, some non-pathogenic *Aeromonas* spp. may perform important roles in digestion by secreting several proteases (Nayak, 2010).

Lactic acid bacteria (LAB) can stimulate anti-inflammatory signals and are commonly used as probiotics to treat intestinal inflammatory disorders (Balcázar et al., 2007; Navarrete et al., 2013). We found several LAB associated OTUs that were at higher abundance in the loch fish, including 4 OTUs for *Lactobacillus*, 1 OTU for *Weissella*, and 1 OTU for *Vagococcus* (Supplementary Table S3). Furthermore we identified further OTUs mapping to the genera *Carnobacterium*, *Lactobacillus*, *Lactococcus* and *Bifidobacteria* that were only present in loch samples (Supplementary Table S2). The only LAB OTUs that were uniquely present in aquarium samples were 4 OTUs identified to the genera of *Lactobacillus* (Supplementary Table S1).

Although the NGS approach greatly improved the level of taxonomic information on the microbiota of Atlantic salmon parr, information stored in the microbial databases is not exhaustive and currently limiting microbial profiling. Not all OTUs found in this study could be confidently assigned to genus level, which is either due to lack of confidence as to where the OTU fits or it matches to a branch for which little taxonomic information is published (Goodrich et al., 2014). Members of the same genus are not necessarily closely related (Goodrich et al., 2014). Therefore, OTUs that are combined in "Other Bacteria" are an arbitrary mix (Goodrich et al., 2014). In the near future it is expected that these databases will greatly improve through the ongoing sequencing effort of bacterial genomes. Here these sequences were retained as different OTUs mapped to "Other Bacteria" were seen to be differentially present between the test groups.

To assess the diversity between the loch and aquarium groups, α-diversity metrics were calculated to identify species richness and diversity within each sample. Chao1 is a taxa-based metric used to estimate species richness in a sample and showed a significant difference between the two environmental groups, and is used extensively for intestinal microbiota studies (Chao, 1984; Hamady and Knight, 2009). Chao1 species richness was significantly greater in loch samples than aquarium samples, likely reflecting the greater diversity of bacterial species in the wild environment than in a recirculating system. A similar finding was observed in a comparative study of wild and captive killifish (*Fundulus heteroclitus*), with wild fish having higher species richness and phylogenetic diversity in gut than captive fish (Givens et al., 2015). In our data, the phylogenetic method PD_whole_tree did not reveal significant differences between loch and aquarium groups, indicating whilst there may be differences in species richness based on OTUs, when taxonomy is taken into account those differences even out as the different OTUs may be mapped to the same or similar taxonomic assignment (Hamady and Knight, 2009).

To assess β-diversity we used unweighted and weighted UniFrac (Lozupone and Knight, 2005; Lozupone et al., 2007). Whilst unweighted UniFrac detects differences in presence or absence of bacterial OTUs in different communities, the weighted UniFrac also includes the difference of abundances even when the overall groups of organisms that are present in each sample remain the same (Lozupone et al., 2007). We found that unweighted UniFrac showed clear statistically significant clustering by environmental group in both MDS plot (Fig. 4) and UPGMA tree (data not shown). This reflects the findings of the Metastats analysis showing a number of OTUs only present in either group, albeit most of these OTUs were of low relative abundance (Supplementary Tables S1, S2 and S3), and the indication of higher species richness of loch samples suggested by Chao1. However, the weighted UniFrac approach did not reveal any significant clustering. Although differences in the abundance of several taxa were observed, it is very likely that the high inter-individual variation regardless of environmental group hampered any detection of a pattern.

The detected differences in community composition between groups and individuals are most likely a reflection of the environment rather than host physiological pressures, as previously suggested (Star et al., 2013). The loch fish are likely to be continually exposed to a much greater variety of bacteria as there is a full natural community with fish also being able to eat additional food items above and beyond the commercial salmon feed presented. It is also presumed that the water itself has a greater bacterial content and diversity than in a recirculated system. All fish in the study are a farmed strain and have a similar genetic background, and since both groups were reared in a similar hatchery environment it can be expected that similar initial communities were established. Both groups of fish were fed similar diets which represent commercial salmon feeds for Atlantic salmon parr in freshwater.

However, the identification of a shared core microbiota as well as differences in core microbiota and presence/absence of certain OTUs within and between the two groups does not only indicate that habitat shapes the gut microbiota but also that there appears to be a need for colonisation by certain microbes indicating host physiological selection. Environmental rearing conditions on gut microbiota have been examined for zebrafish that indicate host selective processes play a role in the bacterial community (Roeselers et al., 2011). Zebrafish from different lab facilities and the wild shared a core of 21 OTUs, a similar observation to our data for Atlantic salmon (Roeselers et al., 2011). Furthermore several taxa were enriched in the fish gut when compared to water samples, further indicating selection mechanisms by the host (Roeselers et al., 2011).

The mechanisms and selective pressures that are most likely involved in establishment of the gut microbiota are complex (Roeselers et al., 2011).

Stochastic processes may also markedly influence the inter-individual variation in a founder takes all scenario, where a random first coloniser determines the composition of later arriving species at first colonisation at hatching (Star et al., 2013). The higher bacterial species richness in the intestine of fish reared in the open loch supports this random stochasticity, whereas the identified core microbiota indicates active maintenance of crucial OTUs by the host. Further in-depth studies are needed to resolve the interplay of environmental pressures and host selective mechanisms.

In conclusion, this is the first in-depth NGS study of the intestinal microbiota of Atlantic salmon parr. The findings support previous studies, and expand on their limitations caused by technical inferior methods. The taxonomic information gathered here will be of great use to define the healthy gut microbiota and the identification of indicator species that could be used in the future to monitor the gut health status of farmed fish at this developmental stage. NGS studies investigating the microbiome of Atlantic salmon at the seawater stage have been undertaken previously (Zarkasi et al., 2014, 2016) and are of great importance to understand and define the shifts of healthy gut microbiota throughout all life stages and thereby aquaculture production stages.

Furthermore we showed that there are significant differences between microbiota profiles of fish kept in different rearing environments, with higher species richness in fish exposed to a more natural open loch environment. Interestingly, some of those significant differences were found in genera containing potentially pathogenic species and LAB, both with higher abundances and even exclusive presence in fish from the semi-natural environment. This may lead to further investigations into a hypothetical "clean world" hypothesis applicable to fish, potentially indicating that fish in the open loch environment could be more resilient to certain pathogens by maintaining a bigger repertoire of LAB to keep them in check. Although difference in absence/presence of bacterial taxa was found significant between the groups, all fish shared a core microbiota of 19 OTUs that reflects some highly abundant taxa. These results strongly indicate that the acquisition and maintenance of gut microbiota is a complex process, which is dictated by both environmental availability of potential microbial colonisers and host physiological pressures leading to maintenance of specific bacterial taxa.

## ACKNOWLEDGEMENTS

This study was funded by the BBSRC Eastbio PhD studentship of CED and BBSRC grant BB/M026604/1. The authors wish to thank the team at the Centre for Genome Enabled Biology and Medicine (CGEBM, Aberdeen) for their help and support, especially Brennan Martin for technical support and Matthew Gemmel for bioinformatics support.

## APPENDIX A. SUPPLEMENTARY DATA

Supplementary tables: S1 OTUs unique to the distal intestine of Atlantic salmon (*Salmo salar*) parr kept in a recirculating aquarium facility. S2 OTUs unique to the distal intestine of Atlantic salmon (*Salmo salar*) parr kept in an open loch environment. S3 OTUs found at significantly different abundance in the distal intestine of Atlantic salmon (*Salmo salar*) parr kept in a recirculating aquarium facility and the distal intestine of Atlantic salmon (*Salmo salar*) parr kept in an open loch system.

## REFERENCES

1. Abid, A., Davies, S.J., Waines, P., Emery, M., Castex, M., Gioaccini, G., Carnevali, O., Bickerdike, R., Romero, J., Merrifield, D.L., 2013. Dietary symbiotic application modulates Atlantic salmon (Salmo salar) intestinal microbial communities and intestinal immunity. Fish Shellfish Immunol. 35, 1948–1956
2. Austin, B., 2006. The bacterial microflora of fish, revised. Sci. World J. 6, 931–945
3. Balcázar, J.L., de Blas, I., Ruiz-Zarzuela, I., Vendrell, D., Gironés, O., Muzquiz, J.L., 2007. Enhancement of the immune response and protection induced by probiotic lactic acid bacteria against furuncolosis in rainbow trout (Oncorhynchus mykiss). FEMS Immunol Med Microbiol 51, 185–193
4. Baker, G.C., Smith, J.J., Cowan, D.A., 2003. Review and re-analysis of domain-specific 16S primers. J. Microbiol. Methods 55, 541–555
5. Benson, A.K., Kelly, S.A., Legge, R., Ma, F., Low, S.J., Kim, J., Zhang, M., Oh, P.L., Nehrenberg,D., Hua, K., Kachman, S.D., Moriyama, E.N., Walter, J., Peterson, D.A., Pomp, D., 2010. Individuality in gut microbiota composition is a complex polygenic trait shaped by multiple environmental and host genetic factors. PNAS 107, 18933–18938
6. Cantas, L., Sørby, J.R.T., Aleström, P., Sørum, H., 2012. Culturable gut microbiota diversity in zebrafish. Zebrafish 9, 26–37
7. Caporaso, J.G., Kuczynski, J., Stombaugh, J., Bittinger, K., Bushman, F.D., Costello, E.K., Fierer, N., Peña, A.G., Goodrich, J.K., Gordon, J.I., Huttley, G.A., Kelley, S.T., Knights, D., Koenig, J.E., Ley, R.E., Lozupone, C.A., McDonald, D., Muegge, B.D., Pirrung, M., Reeder, J., Sevinsky, J.R., Turnbaugh, P.J., Walters, W.A., Widmann, J., Yatsunenko, T., Zaneveld, J., Knight, R., 2010. QIIME allows analysis of high-throughput community sequencing data. Nat. Methods 7, 335–336
8. Chao, A., 1984. Non-parametric estimation of the number of classes in a population. Scand. J. Stat. 11, 265–270
9. DeSantis, T.Z., Hugenholtz, P., Larsen, N., Rojas, M., Brodie, E.L., Keller, K., Huber, T., Dalevi, D., Hu, P., Andersen, G.L., 2006. Greengenes, a chimera-checked 16S rRNA gene database and workbench compatible with ARB. Appl. Environ. Microbiol. 72, 5069–5072

10. Dhanasiri, A.K., Brunvold, L., Brinchmann, M.F., Korsnes, K., Bergh, Ø., Kiron, V., 2011. Changes in the intestinal microbiota of wild Atlantic cod Gadus morhua L. upon captive rearing. Microb. Ecol. 61, 20–30
11. Fjellheim, A.J., Playfoot, K.J., Skjermo, J., Vadstein, O., 2007. Vibrionaceae dominates themicroflora antagonistic towards Listonella anguillarum in the intestine of cultured Atlantic cod (Gadus morhua) larvae. Aquaculture 269, 98–106
12. Geraylou, Z., Souffreau, C., Rurangwa, E., Maes, G.E., Spanier, K.I., Courtin, C.M., Delcour, J.A., Buyse, J., Ollevier, F., 2013. Prebiotic effects of arabinoxylan oligosaccharides on juvenile Siberian sturgeon (Acipenser baerii) with emphasis on the modulation of the gut microbiota using 454 pyrosequencing. FEMS Microbiol. Ecol. 86, 357–371
13. Gevers, D., Pop, M., Schloss, P.D., Huttenhower, C., 2012. Bioinformatics for the human microbiome project. PLoS Comput. Biol. 8. http://dx.doi.org/ 10.1371/journal.pcbi. 1002779.
14. Giatsis, C., Sipkema, D., Smidt, H., Verreth, J., Verdegem, M., 2014. The colonisation dynamics of the gut microbiota in tilapia larvae. PLoS One 7. http://dx.doi.org/10. 1371/journal.pone.0103641
15. Givens, C.E., Ransom, B., Bano, N., Hollibaugh, J.T., 2015. Comparison of the gut microbiomes of 12 bony fish and 3 shark species. Mar. Ecol. Prog. Ser. 518, 209–223
16. Goodrich, J.K., Di Rienzi, S.C., Poole, A.C., Koren, O.,Walters,W.A., Caporaso, J.G., Knight, R., Ley, R.E., 2014. Conducting a microbiome study. Cell 158, 250–262
17. Green, T.J., Smullen, R., Barnes, A.C., 2013. Dietary soybean protein concentrate-induced intestinal disorder in marine farmed Atlantic salmon, Salmo salar is associated with alterations in gut microbiota. Vet. Microbiol. 166, 286–292.
18. Hamady,M., Knight, R., 2009.Microbial community profiling for humanmicrobiome projects: tools, techniques and challenges. Genome Res. 19, 1141–1152
19. Hovda, M.B., Lunestad, B.T., Fontanillas, R., Rosnes, J.T., 2007. Molecular characterisation of intestinal microbiota of farmed Atlantic salmon (Salmo salar L.). Aquaculture 272, 581–588
20. Ingerslev, H.-C., Jorgensen, L.V., Strube, M.L., Larsen, N., Dalsgaard, I., Boye, M., 2014a. The development of the gut microbiota in rainbow trout is affected by first feeding and diet type. Aquaculture 424, 24–34
21. Ingerslev, H.-C., Strube, M.L., von Gersdorff Jørgensen, L., Dalsgaard, I., Boye, M., Madsen, L., 2014b. Diet type dictates the gut microbiota and the immune response against Yersinia ruckeri in rainbow trout (Oncorhynchus mykiss). Fish Shellfish Immunol. 40, 624–633
22. Kau, A.L., Ahern, P.P., Griffin, N.W., Goodman, A.L., Gordon, J.I., 2011. Human nutrition, the gut microbiome and the immune system. Nature 474, 327–336
23. Klindworth, A., Pruesse, E., Schweer, E., Peplies, J., Quast, C., Horn,M., Glöckner, F.O., 2013. Evaluation of general 16S ribosomal RNA gene PCR primers for classical and nextgeneration sequencing-based diversity studies. Nucleic Acids Res. 7. http://dx.doi. org/10.1093/nar/gks808
24. Lee, Y.K., Mazmanian, S.K., 2010. Has the microbiota played a critical role in the evolution of the adaptive immune system? Science 330, 1768–1773
25. Ley, R.E., Hamady, M., Lozupone, C., Turnbaugh, P.J., Ramey, R.R., Bircher, J.S., Schlegel, M.L., Tucker, T.A., Schrenzel,M.D., Knight, R., Gordon, J.I., 2008. Evolution ofmammals and their gut microbes. Science 20, 1647–1651.
26. Li, J., Ni, J., Li, J.,Wang, C., Li, X., Wu, S., Zhang, T., Yu, Y., Yan, Q., 2014. Comparative study on gastrointestinal microbiota of eight fish species with different feeding habits. J. Appl. Microbiol. 227, 1750–1760
27. Liu, Z., Lozupone, C., Hamady, M., Bushman, F.D., Knight, R., 2007. Short pyrosequencing reads suffice for accuratemicrobial community analysis. Nucleic Acids Res. 35, 120–129.
28. Llewellyn, M.S., Boutin, S., Hoseinifar, S.H., Derome, W., 2014. Teleost microbiomes: the state of the art in their characterisation,manipulation and importance in aquaculture and fisheries. Front. Microbiol. 5, 207–223

29. Lozupone, C., Knight, R., 2005. UniFrac: a new phylogenetic method for comparing microbial communities. Appl. Environ. Microbiol. 71, 8228–8235
30. Lozupone, C.A., Hamady, M., Kelley, S.T., Knight, R., 2007. Quantitative and qualitative β diversity measures lead to different insights into factors that structure microbial communities. Appl. Environ. Microbiol. 73, 1576–1585
31. Lozupone, C.A., Stombaugh, J.I., Gordon, J.I., Jansson, J.K., Knight, R., 2012. Diversity, stability and resilience of the human gut microbiota. Nature 489, 220–230
32. McMurdie, P.J., Holmes, S., 2013. phyloseq: an R package for reproducible interactive analysis and graphics of microbiome census data. PLoS ONE 8. http://dx.doi.org/10.1371/ journal.pone.0061217.
33. Merrifield, D.L., Dimitroglou, A., Bradley, G., Baker, T.M., Davies, S.J., 2009. Soybean meal alters autochthonousmicrobial populations,microvilli morphology and compromises intestinal enterocyte integrity of rainbow trout, Oncorhynchus mykiss (Walbaum). J. Fish Dis. 32, 755–766.
34. Navarrete, P., Espejo, R.T., Romero, J., 2009a. Molecular analysis ofmicrobiota along the digestive tract of juvenile Atlantic salmon (Salmo salar L.). Microb. Ecol. 57, 550–561.
35. Navarrete, P.,Magne, F., Mardones, P., Riveros,M., Opazo, R., Suau, A., Pochart, P., Romero, J., 2009b.Molecular analysis of intestinalmicrobiota of rainbow trout (Oncorhynchus mykiss). FEMS Microbiol Ecol 71, 148–156
36. Navarrete, P.,Magne, F., Araneda, C., Fuentes, P., Barros, L., Opazo, R., Espejo, R., Romero, J., 2012. PCR-TGGE analysis 16S rRNA from rainbow trout (Oncorhynchus mykiss) gut microbiota reveals host-specific communities of active bacteria. PLoS One 7. http:// dx.doi.org/10.1371/journal.pone.0031335.
37. Navarrete, P., Fuentes, P.,De La Fuente, L., Barros, L., Magne, F.,Opazo, R., Ibacache, C., Espejo
38. R., Romero, J., 2013. Short-term effects of dietary SBM and lactic acid bacteria on intestinal morphology and microbiota of Atlantic salmon. Aquac. Nutr. 19, 827–836
39. Nayak, S.K., 2010. Role of gastrointestinal microbiota in fish. Aquac. Res. 41, 1553–1573. O'Hara, A.M., Shanahan, F., 2006. The gut flora as forgotten organ. EMBO Rep. 7, 688–693
40. Pruesse, E., Quast, C., Knittel, K., Fuchs, B.M., Ludwig, W., Peplies, J., Glöckner, F.O., 2007. SILVA: a comprehensive online resource for quality checked and aligned ribosomal RNA sequence data compatible with ARB. Nucleic Acids Res 35, 7188–7196
41. Rajendhran, J., Gunasekaran, P., 2011.Microbial phylogeny and diversity: small subunit ribosomal RNA sequence analysis and beyond. Microbiol. Res. 166, 99–110
42. Reveco, F.E., Øverland, M., Romarheim, O.H.,Mydland, L.T., 2014. Intestinal bacterial community structure differs between healthy and inflamed intestines in Atlantic salmon (Salmo salar L.). Aquaculture 420–421, 262–269
43. Ringø, E., Løvma, L., Kristiansen, M., Bakken, Y., Salinas, I., Myklebust, R., Olsen, R.E., Mayhew, T.M., 2010. Lactic acid bacteria vs. pathogens in the gastrointestinal tract of fish: a review. Aquac. Res. 41, 451–467
44. Roeselers,G., Mittge, E.K., Stephens,W.Z., Parichy, D.M., Cavanaugh, C.M., Guillemin, K., Rawls, J.F., 2011. Evidence for a core gut microbiota in the zebrafish. ISME J. 5, 1595–1608.
45. Romero, J., Navarrete, P., 2006. 16S rDNA-based analysis of dominant bacterial populations associated with early life stages of coho salmon (Oncorhynchus kisutch). Microbial Ecology 51, 422–430
46. Schloss, P.D., Westcott, S.L., Ryabin, T., Hall, J.R., Hartmann, M., Hollister, E.B., Lesniewski, R.A., Oakley, B.B., Parks, D.H., Robinson, C.J., Sahl, J.W., Stres, B., Thallinger, G.G., Van
47. Horn, D.J., Weber, C.F., 2009. Introducing mothur: open-source, platform-independent, community-supported software for describing and comparing microbial communities. Appl. Environ. Microbiol. 75, 7537–7541

48. Skrodenytė-Arbačiauskienė, V., Sruoga, A., Butkauskas, D., Skrupskelis, K., 2008. Phylogenetic analysis of intestinal bacteria of freshwater salmon Salmo salar and sea trout Salmo trutta trutta and diet. Fish. Sci. 74, 1307–1314
49. Star, B., Haverkamp, T.H., Jentoft, S., Jakobsen, K.S., 2013. Next generation sequencing shows high variation of the intestinal microbial species composition in Atlantic cod caught at a single location. BMC Microbiol. 13, 248. http://dx.doi.org/10.1186/1471- 2180-13-248
50. Sullam, K.E., Essinger, S.D., Lozupone, C.A., O'Connor, M.P., Rosen, G.L., Knight, R., Kilham, S.S., Russell, J.A., 2012. Environmental and ecological factors that shape the gut bacterial communities of fish: a meta-analysis. Mol. Ecol. 21, 3363–3378
51. Van Kessel, M.A.H.J., Dutilh, B.E., Neveling, K., Kwint, M.P., Veltman, J.A., Flik, G., Jetten, M.S.M., Klaren, P.H.M., Op den Camp, H.J.M., 2011. Pyrosequencing of 16S rRNA gene amplicons to study themicrobiota in the gastrointestinal tract of carp (Cyprinus carpio L.). AMB Express 1, 41–49
52. Wang, Q., Garrity, G.M., Tiedje, J.M., Cole, J.R., 2007. Naïve Bayesian classifier for rapid assignment of rRNA sequences into the new bacterial taxonomy. Appl. Environ. Microbiol. 73, 5261–5267
53. White, J.R., Nagarajan, N., Pop, M., 2009. Statistical methods for detecting differentially abundant features in clinical metagenomics samples. PLoS Comput. Biol. 5. http:// dx.doi.org/10.1371/journal.pcbi.1000352
54. Wong, S., Waldrop, T., Summerfelt, S., Davidson, J., Barrows, F., Kenney, P.B., Welch, T., Wiens, G.D., Snekvik, K., Rawls, J.F., Good, C., 2013. Aquacultured rainbow trout (Oncorhynchus mykiss) possess a large core intestinal microbiota that is resistant to variation in diet and rearing density. Appl. Environ. Microbiol. 79, 4974–4984
55. Zarkasi, K.Z., Abell, G.C.J., Taylor, R.S., Neumann, C., Hatje, E., Tamplin, M.L., Katouli, M., Bowman, J.P., 2014. Pyrosequencing-based characterisation of gastrointestinal bacteria of Atlantic salmon (Salmo salar L.) within a commercial mariculture system. J. Appl. Microbiol. 117, 18–27
56. Zarkasi, K.Z., Taylor, R.S., Abell, G.C.J., Tamplin, M.L., Glencross, B.D., Bowman, J.P., 2016. Atlantic salmon (Salmo salar L.) gastrointestinal microbial community dynamics in relation to digesta properties and diet. Microb. Ecol. 71, 589–603

# CHAPTER 10

# Impacts of Rearing Techniques on Growth, Survival and Bacterial Microbiota of Carpet Shell Clam (*Ruditapes Decussatus*) Larvae

**Amel Medhioub[1]• Badreddine Mechri1,[2]• Sondes Bchir[1]• Younes Limeyem[1]• Wissem Slimani[1]• Mahjoub Aouni[2]•Mohamed Nejib Medhioub[1]**

[1]*Laboratoire d'Aquaculture - Institut National des Sciences et Technologies de la Mer, Route de Khniss, B.P. 59, 5000 Monastir, Tunisia*

[2]*Laboratoire des Maladies Transmissibles et Substances Biologiquement Actives, Faculte´ de Pharmacie, Universite´ de Monastir, rue Avicenne, 5000 Monastir, Tunisia*

## ABSTRACT

In this study, we investigate the impacts of two water treatment regimes, the closed aquaculture system (CAS) and flow-through system (FTS), on the carpet shell clam (*Ruditapes decussatus*) larval microbiota, growth and survival. A bacteriological barrier (B) was created among breeder stage treated by florfenicol (FF, 8 mg $L^{-1}day^{-1}$ per 5 days). After spawning, larvae were maintained in five different rearing techniques: CAS; FTS; B + CAS; B + FTS and B + CAS + FF (8 mg $L^{-1}$ $day^{-1}$ per 3 days). Significant decrease was observed in both total bacterial count and presumptive *Vibrio* count (PVC) in FF-treated breeders. Besides, larvae reared in B + FTS showed significantly lower PVC than larvae maintained in other rearing techniques.

It was also determined that growth and survival of larvae reared in B + CAS + FF and B + FTS were significantly better than those reared in other systems. One hundred and twelve *Vibrio* strains, isolated from breeders, eggs and larvae, were subjected to some bacteriological tests (API 20E strips, vibriostatic agent O/129, growth at different temperatures and salinity). *Vibrio alginolyticus* was the predominant isolated species (36.61 %), followed by *V. fluvialis* (25 %), *V. vunificus* (17.86 %), *V. cholera* (13.39 %) and *V. paraheamolyticus* (7.14 %). These results may contribute to the understanding of the effect of rearing techniques on the bacterial load, survival and growth of early life stages of *R. decussatus* larvae.

**Keywords:** *Ruditapes decussatus* Larval rearing Flow-through system Closed aquaculture system *Vibrionaceae*

## INTRODUCTION

The carpet shell clam, *Ruditapes decussatus* (Linnaeus 1758), is the most abundant and commercially important species of *Veneridae* in different North African estuaries and lagoons. In Tunisia, it is found predominantly in the Gulf of Gabes (Medhioub 1983) and its collecting represents a traditional activity that has a significant socioeconomic impact in terms of both income and employment. The development of this economic sector involves not only the rational management of beds but also the breeding of this species, which requires mass production of juveniles. To this aim, several basic and applied research programs have been conducted to develop new techniques for improvement in *R. decussatus* larval and post-larval reared in the experimental bivalve hatchery of the National Institute of Marine Sciences and Technology.

In general, *R. decussatus* larvae are produced in closed aquaculture systems (CASs), which represent the most common rearing method used in bivalve hatcheries (Castagna et al. 1996; Robert and Gérard 1999; Helm and Bourne 2004; Rico-Villa et al. 2006). However, there are some drawbacks, such as microbial contamination risk and disease transmission. In fact, larval culture tanks water is completely exchanged at least three times per week, allowing microbes to increase their biomass in the rearing water (Andersen et al. 2000; Torkildsen and Magnesen 2004). Despite the use of antibacterial agents as prophylactic therapy, we have been confronted with recurrent epizootic disease outbreaks in larval and early post-larval stages, causing high mortality (Mechri et al. 2015).

Nowadays, large varieties of water circulation systems are used in bivalve hatcheries, such as the flow-through systems (FTSs) and the recirculation aquaculture systems (RASs). The FTS is based on regular water exchange in order to reduce bacteria as well as toxic substances, which accumulate in the rearing water (New and Valenti 2000). This method has been successfully used in the nursery culture of oysters, scallops, clams, mussels and in numerous fish hatcheries (Andersen et al. 2000; Ritar 2001; Tieman and Goodwin 2001; Otoshi et al. 2003; Sarkis et al. 2006; Rico-Villa et al. 2008). Recirculating systems operate by water undergoing treatment, allowing to the hatcheries production system's waters to be partially reused (Rosenthal et al. 1986). This technology has been widely used in fish culture for more than three decades (Rosenthal 1980; Verreth and Eding 1993; Martins et al. 2005), but only recently has it been experimentally tested in nursery culture of some mollusk species such as *Mercenaria mercenaria* (Pfeiffer and Rusch 2000), *Argopecten purpuratus* (Merino et al. 2009) and *Pecten maximus* (Magnesen and Jacobsen 2012).

The aim of this paper is to enhance our understanding about the eventual effect of larvae rearing techniques and antimicrobial treatments at the breeder stage on the bacterial load in the rearing system to fulfill the goal of eliminating the antibacterial use at the larval stage. In this study, we investigate the impacts of the CAS and FTS on the growth, survival and microbiota associated with *R. decussatus* larvae.

# MATERIALS AND METHODS

## Location of sampling area

The study was carried out at the experimental bivalve hatchery of the National Institute of Marine Sciences and Technology (INSTM), Monastir, Tunisia, which is located on the edge of Monastir lagoon called Lake of Khniss. This lagoon communicates with the sea by a canal, of which a large part is artificial.

## Hatchery design and operation

### Water treatment

The water that feeds the experimental hall of bivalve mollusks and fish is a mixture between water lagoon and seawater at the proportion of 30 and 70 %, respectively. This water undergoes a sequential treatment: first sand filtration followed by a series of filtration pockets 25, 5 and 1 µm. UV treatment was used for algae culture up to 5 L and for larvae culture up to the metamorphosis.

**Broodstock conditioning**

*R. decussatus* broodstock from the south of Tunisia (Gabes gulf) was conditioned according to Medhioub et al. (2006) protocol. The reproducers were treated with florfenicol (FF) at the dose of 8 mg $L^{-1}$ $day^{-1}$ during 5 days before the spawning in order to create a barrier which limits the contamination of eggs and larvae. This treatment was incorporated with the food.

**Spawning, fertilization and incubation**

The spawning was realized according to Medhioub et al. (2006) protocol. Egg suspensions from the different adults are kept separate and should be filtered through 100-μm nylon mesh prior to the addition of the recent dense sperm at the ratio 1–2 mL $L^{-1}$ of egg suspension. Fertilization was achieved at 22–24 °C. All fertilized eggs were pooled after 2 h.

The embryos were transferred to a rectangular flat bottom plastic tank (50 L) at the density of $20 \times 10^3$–$40 \times 10^3$ $L^{-1}$ and at 22–24 °C. After 24–30 h, most larvae reached D-shaped stage; the culture water was well exchanged and sifted through a 30-μm nylon mesh. The D larvae were then counted before being transferred to a culture tank.

**Larvae culture method**

Two methods with different system of water circulation as described by Helm and Bourne (2004) were used for larvae up to the early pediveliger stage: traditional method with closed aquaculture system (CAS) and flow-through system (FTS). Both technics were carried in 210-L cylindroconic fiberglass tank filled with seawater at 20–21 °C and a mean salinity of 38.

The experiment was conducted in a series of five tanks with three replicates, respectively: six tanks with FTS, three of which contained larvae produced by untreated breeders and the other three contained larvae obtained from florfenicol (8 mg $L^{-1}$ $day^{-1}$ per 5 days)-treated breeders. Nine tanks with CAS, three of which received florfenicol treatment at 8 mg $L^{-1}$ $day^{-1}$ during 3 days and the other six did not received treatment. In the closed aquaculture system, larvae were reared at a density of 5 larvae $mL^{-1}$, and culture water was 100 % renewed every 2 days where larvae received full food ration. Food ration will be at the half when the water is not renewed the following day. In flow-through system, 25 larvae $mL^{-1}$ was used with water renewal equivalent at 6–7 time $day^{-1}$ with continuous and constant distribution of feeding. Survival was scored at the end of the cylindroconical experiments.

When larvae reach the early pediveliger stage, they must be transferred from the both system to settling in tube cylinder of (sieves tube) 60 l capacity with a circulating seawater system through 90-µm nylon mesh where water renewal was equivalent to 0.14 $m^3 h^{-1}$.

## Feeding

In this study, four species of microalgae were used: *Chaetoceros* sp, *Isochrysis* sp, *Tetraselmis*sp and *Chaetoceros calcitrans*. The first three strains were isolated from Tunisian coast, respectively, in 1998, 2002 and 2008 at INSTM Monastir Center.

In CAS and FTS, larvae were fed a mixture of *Chaetoceros* sp and *Isochysis* sp at the ration of 40 cells $\mu L^{-1}$ as an initial ration during 3–4 days and 80 cells $\mu L^{-1}$ as maximum ration up to the end of metamorphosis. This ration was equivalent to 40–80 cells of *Isochrysis* and calculated on a cell volume equivalency basis where 1.0 cell of *Isochrysis* sp = 2.08 cells *Chaetoceros* sp.

From early pediveliger stage to metamorphosis, three species of microalgae were used *Isochrysis* sp, *Cheatoceros calcitrans* and *Tetraselmis* sp. The ration 80 cells of *Isochrysis* $\mu L^{-1}$was calculated in a cell volume equivalency basis where 1.0 cell of *Isochrysis* sp = 0.87 *Chaetoceros calcitrans*, 1.0 cell of *Isochrysis* sp = 0.23 *Tetraselmis* sp.

## Measurement of larval length

Sixty *R. decussatus* larvae were picked every 2 days from each culture, fixed in formol-saline for 30 min and conserved at 4 °C. Larval specimens were photographed using a microscope (Zeiss, Germany) equipped with a digital camera. The anteroposterior length (µm) was measured using Image J software 1.45.

## Collection of samples

Breeders, fertilized eggs and larvae were collected during all three ponds. Samples were immediately packed in UV sterilized oxygen-filled polythene bags separately, placed in an insulated container, and transported to the laboratory for analysis within 2 h.

## Bacteriology

### Bacterial count

In this study, we used Marine agar (Ma) medium, for the enumeration of the moderately halophiles heterotrophic bacterial count, and thiosulfate citrate bile sucrose agar (TCBS modified agar; Scharlau Microbiogy, Spain) supplemented with 2 % NaCl, specific for the *Vibrionaceae* biota. The enumeration of bacteria was carried out by using serial dilutions and the plating method to establish viable bacterial cell counts.

The water samples were tenfold serially diluted with artificial sterile seawater (ASSW), 1 mL aliquot of each dilution sample was spread out over the two media in duplicate. Colonies were counted after 24–48 h for the TCBS medium and 24–72 h for the marine agar medium, with incubation at 23 °C. The results were expressed in colony forming units CFU $mL^{-1}$ of water.

The breeder samples were washed thoroughly under running water, sprinkled with 70 % ethanol and allowed to dry. Subsequently, the clams were opened aseptically using a sterile scalpel. Five grams of the intervalvar liquid and soft meat were homogenized with 5 mL of ASSW, and then, 1 g of the mixture was diluted with 4 mL of ASSW to obtain $10^{-1}$ diluted homogenate. Samples were tenfold serially diluted with artificial sterile seawater (ASSW), 1 mL aliquot of each dilution sample was spread out over the two media in duplicate. Colonies were counted after 24–48 h for the TCBS medium and 24–72 h for the marine agar medium, with incubation at 23 °C. The results were expressed in CFU $g^{-1}$ of clam-wet weight.

The fertilized eggs and larvae were enumerated by cell count under an optical microscope and were crushed in 1 mL of ASSW using an Eppendorf tube sterile pestle. The homogenate samples were serially diluted in the same solution, plated and counted as for the water samples. The count was multiplied by the dilution factor, divided by the number of the larvae and multiplied by $10^5$. The results were expressed in CFU × $10^{-5}$ eggs or larvae.

### Bacterial characterization

Three suspected colonies from typical *Vibrio* colonies on TCBS medium were picked up, purified on TSA agar (+2 % NaCl) and then identified through different physiological and biochemical tests such as Gram nonstaining (KOH) method, cytochrome oxidase activity, motility (Mannitol-Motility agar; Pronadisa, Madrid, Spain), resistance to vibriostatic O129 (10 and 150 µg), salt requirement (growth on 0, 2, 4, 8 and 10 % NaCl medium) and growth at 23 and 37 °C. The species identification of *Vibrio* was performed using API 20 E strips (bioMerieux, Marcy l'Etoile, France). The strains were conserved as frozen stocks at −80 °C in tryptic soy broth (TSB; Bio-Rad, France) with 2 % NaCl plus 15 % (v/v) glycerol.

### Statistical analysis

All data were analyzed with SPSS version 21.0 software. Significance differences between bacterial counts were determined by the nonparametric Kruskal–Wallis test and Mann–Whitney test. Tukey test was used to obtain comparison between the larval growth means. Significance levels for all analysis were set at $P < 0.05$.

# RESULTS

## Control of the bacterial biota

The bacteriological control of *R. decussatus* reproducer showed in all three ponds treated with FF, a significant decrease ($P < 0.05$) in total viable counts (TVC) in Ma medium (Pond1, $T_0$: $1.80 \times 10^9$ – $T_{96}$: $3.00 \times 10^5$ CFU g$^{-1}$) as well as in presumptive *Vibrio* counts (PVC) in TCBS medium (Pond1, $T_0$: $6.10 \times 10^6$ – $T_{96}$: $5.00 \times 10^2$ CFU g$^{-1}$). However, these bacterial counts increased progressively in untreated ponds, respectively, by a factor of 2.00 and 3.50 in Pond 1. In addition, the fertilized eggs obtained from treated reproducers were free from any *Vibrio*contamination compared with those produced by untreated breeders (Table 1).

**Table 1** Bacteriological control of *Ruditapes decussatus* reproducers biota by florfenicol treatment (8 mg L$^{-1}$)

| | Bacterial count (CFU g$^{-1}$) | | | | | | | | | | Spawning | |
|---|---|---|---|---|---|---|---|---|---|---|---|---|
| | $T_{0h}$ | | $T_{24h}$ | | $T_{48h}$ | | $T_{72h}$ | | $T_{96h}$ | | FE | |
| | TCV | PVC | TCV | PVC | TCV | PVC | TCV | PVC | TCV | PVC | TCV | PVC |
| Pond 1 | | | | | | | | | | | | |
| T | $1.80 \times 10^9$ | $6.10 \times 10^6$ | $6.00 \times 10^8$ | $1.79 \times 10^5$ | $2.10 \times 10^7$ | $6.90 \times 10^4$ | $1.60 \times 10^6$ | $3.30 \times 10^3$ | $3.00 \times 10^5$ | $5.00 \times 10^2$ | $3.10 \times 10^2$ | 0.00 |
| NT | $4.50 \times 10^{9(a)}$ | $2.70 \times 10^6$ | $4.50 \times 10^{9(a)}$ | $6.50 \times 10^{6a}$ | $5.20 \times 10^{9(a)}$ | $7.20 \times 10^{6a}$ | $6.00 \times 10^{9(a)}$ | $8.70 \times 10^{6a}$ | $9.10 \times 10^9$ | $9.40 \times 10^{6a}$ | $1.20 \times 10^5$ | $4.50 \times 10^2$ |
| Pond 2 | | | | | | | | | | | | |
| T | $5.20 \times 10^9$ | $8.00 \times 10^6$ | $5.00 \times 10^8$ | $8.20 \times 10^3$ | $8.00 \times 10^7$ | $4.40 \times 10^3$ | $1.50 \times 10^7$ | $7.00 \times 10^2$ | $5.60 \times 10^6$ | $3.20 \times 10^2$ | $5.50 \times 10^2$ | 0.00 |
| NT | $1.20 \times 10^{9(a)}$ | $6.40 \times 10^{6a}$ | $2.00 \times 10^{9(a)}$ | $6.60 \times 10^{6a}$ | $3.50 \times 10^{9(a)}$ | $7.80 \times 10^{6a}$ | $3.00 \times 10^{9(a)}$ | $8.00 \times 10^{6a}$ | $4.90 \times 10^{9(a)}$ | $8.80 \times 10^{6a}$ | $7.00 \times 10^5$ | $1.00 \times 10^3$ |
| Pond 3 | | | | | | | | | | | | |
| T | $2.00 \times 10^9$ | $2.70 \times 10^6$ | $7.00 \times 10^8$ | $2.90 \times 10^4$ | $1.20 \times 10^8$ | $1.70 \times 10^4$ | $3.20 \times 10^7$ | $5.00 \times 10^3$ | $7.90 \times 10^6$ | $9.60 \times 10^2$ | $8.00 \times 10^2$ | 0.00 |
| NT | $6.30 \times 10^{9(a)}$ | $5.50 \times 10^{6a}$ | $7.10 \times 10^{9(a)}$ | $6.00 \times 10^{6a}$ | $7.80 \times 10^{9(a)}$ | $6.60 \times 10^{6a}$ | $8.50 \times 10^{9(a)}$ | $6.60 \times 10^{6a}$ | $9.20 \times 10^{9(a)}$ | $7.90 \times 10^{6a}$ | $2.00 \times 10^6$ | $4.70 \times 10^3$ |

The count for breeder samples (CFU g$^{-1}$ wet weight soft tissues) is the average of 5 clam samples per test. The count for fertilized eggs (CFU $\times 10^{-5}$ eggs) is the average of 3 samples per pond

TVC Values in the same row with the same superscript in parentheses are not significantly different ($P > 0.05$)

PVC Values in the same row with the same superscript are not significantly different ($P > 0.05$)

TVC total viable counts g$^{-1}$ wet weight soft tissues ($10^{-5}$ eggs), PVC presumptive *Vibrio* counts g$^{-1}$ wet weight soft tissues ($10^{-5}$ eggs) TVC Values in the same row with

*T* treated, *NT* non-treated, *FE* fertilized eggs, *h* hours

## Influence of rearing techniques on larval microbiota

The assessments of the effects of rearing techniques on the bacterial load of *R. decussatus*larvae are presented in Tables 2 and 3, respectively. The total bacterial count for larvae reared in CAS without antimicrobial treatment ranged from $10^4$ to $10^8$ CFU × $10^{-5}$ larvae was significantly higher ($P < 0.05$) on the fourth day ($CAS_{D4}$: $10^7$–$10^8$ CFU × $10^{-5}$ larvae) than in larvae maintained in other rearing systems (D4: $10^3$–$10^6$ CFU × $10^{-5}$ larvae). Moreover, a total larval mortality has occurred by the sixth day (Pond 1 and 3) and the eighth day (Pond 2) of rearing in closed system. The bacterial biota of larvae obtained after microbiological breeder barrier and reared in closed-circuit seawater system ($B + CAS_{D6}$: $10^6$–$10^8$ CFU × $10^{-5}$ larvae) was significantly higher ($P < 0.05$) than in larvae reared in CAS and treated three times with FF at 8 mg $L^{-1}$ ($B + CAS + FF_{D6}$: $10^3$ CFU × $10^{-5}$ larvae).

**Table 2** Effect of *Ruditapes decussatus* larval rearing techniques on total bacterial count

| | Total viable count (CFU × $10^{-5}$ larvae) | | | | | | | | | |
|---|---|---|---|---|---|---|---|---|---|---|
| | POND 1 | | | | | POND 2 | | | | |
| | CAS | FTS | B + CAS | B + FTS | B + CAS + FF | CAS | FTS | B + CAS | B + FTS | B + CAS + FF |
| D0 | $2.00 \times 10^6$ | $5.20 \times 10^6$ | $5.10 \times 10^{4(a)}$ | $2.20 \times 10^{4a}$ | $6.10 \times 10^{4(a)}$ | $6.70 \times 10^5$ | $7.10 \times 10^{5a}$ | $7.30 \times 10^{4(b)}$ | $3.50 \times 10^4$ | $6.00 \times 10^{4(b)}$ |
| D2* | $4.90 \times 10^7$ | $7.00 \times 10^6$ | $6.90 \times 10^5$ | $3.90 \times 10^{4a}$ | $2.00 \times 10^5$ | $8.20 \times 10^6$ | $7.50 \times 10^{5a}$ | $4.00 \times 10^5$ | $6.00 \times 10^4$ | $2.70 \times 10^3$ |
| D4* | $5.90 \times 10^8$ | $9.80 \times 10^6$ | $8.20 \times 10^6$ | $8.20 \times 10^{4b}$ | $3.90 \times 10^4$ | $7.00 \times 10^7$ | $9.00 \times 10^5$ | $9.60 \times 10^5$ | $8.80 \times 10^4$ | $7.00 \times 10^4$ |
| D6* | X | $3.60 \times 10^7$ | $1.70 \times 10^8$ | $9.30 \times 10^{4b}$ | $9.00 \times 10^3$ | $4.90\,10^8$ | $3.20 \times 10^6$ | $3.20 \times 10^6$ | $1.60 \times 10^5$ | $8.10 \times 10^3$ |
| D8 | | $6.20 \times 10^5$ | X | $3.20 \times 10^5$ | $5.40 \times 10^3$ | X | $5.70 \times 10^6$ | $5.10 \times 10^5$ | $4.50 \times 10^5$ | $3.30 \times 10^3$ |
| D10 | | $8.10 \times 10^7$ | | $6.90 \times 10^{5c}$ | $8.90 \times 10^3$ | | $8.80 \times 10^6$ | $3.50 \times 10^8$ | $7.70 \times 10^5$ | $5.00 \times 10^3$ |
| D12 | | $1.00 \times 10^8$ | | $8.00 \times 10^{5c}$ | $5.90 \times 10^3$ | | $1.70 \times 10^7$ | X | $9.10 \times 10^5$ | $7.10 \times 10^3$ |

| | Total viable count (CFU × $10^{-5}$ larvae) | | | | |
|---|---|---|---|---|---|
| | POND 3 | | | | |
| | CAS | FTS | B + CAS | B + FTS | B + CAS + FF |
| D0 | $5.00 \times 10^5$ | $1.60 \times 10^5$ | $1.70 \times 10^{3(b)}$ | $4.20 \times 10^{3a}$ | $1.50 \times 10^{4(b)}$ |
| D2* | $7.10 \times 10^5$ | $3.70 \times 10^5$ | $9.50 \times 10^4$ | $5.50 \times 10^{3a}$ | $7.00 \times 10^5$ |
| D4* | $1.20 \times 10^8$ | $6.00 \times 10^5$ | $5.00 \times 10^5$ | $8.00 \times 10^3$ | $8.20 \times 10^4$ |
| D6* | X | $8.80 \times 10^5$ | $3.40 \times 10^6$ | $2.00 \times 10^4$ | $7.10 \times 10^3$ |
| D8 | | $2.10 \times 10^6$ | $7.20 \times 10^8$ | $5.10 \times 10^4$ | $2.30 \times 10^3$ |
| D10 | | $4.60 \times 10^6$ | X | $7.30 \times 10^{4b}$ | $3.70 \times 10^3$ |
| D12 | | $7.20 \times 10^6$ | | $9.80 \times 10^{4b}$ | $2.10 \times 10^7$ |

The count for larvae samples (CFU × $10^{-5}$ larvae) is the average of duplicate samples for each rearing technique values in the same column with the same superscript are not significantly different ($P > 0.05$)

Values in the same row with the same superscript in parentheses are not significantly different ($P > 0.05$)

*FF* florfenicol, *BB* breeders barrier, *CAS* closed aquaculture system, *FTS* flow-through system, *D* day

*Florfenicol treatment (8 mg $L^{-1}$)

**Table 3** Effect of *Ruditapes decussatus* larval rearing techniques on presumptive *Vibrio* count

| | Presumptive *Vibrio* count (CFU × $10^{-5}$ larvae) | | | | |
|---|---|---|---|---|---|
| | POND 3 | | | | |
| | CAS | FTS | B + CAS | B + FTS | B + CAS + FF |
| D0 | $8.80 \times 10^{3}$ | $5.00 \times 10^{3a}$ | $6.20 \times 10^{1(d)}$ | $2.20 \times 10^{1a}$ | $6.00 \times 10^{1(d)}$ |
| D2* | $6.70 \times 10^{5a}$ | $6.20 \times 10^{3a}$ | $5.10 \times 10^{2}$ | $2.80 \times 10^{1a}$ | $4.20 \times 10^{1}$ |
| D4* | $4.20 \times 10^{6}$ | $8.00 \times 10^{3}$ | $2.60 \times 10^{3}$ | $4.40 \times 10^{1}$ | $8.90 \times 10^{2}$ |
| D6* | X | $9.50 \times 10^{3}$ | $5.40 \times 10^{4}$ | $7.50 \times 10^{1}$ | $8.00 \times 10^{1}$ |
| D8 | | $2.80 \times 10^{4}$ | $3.80 \times 10^{6}$ | $1.20 \times 10^{2b}$ | $5.10 \times 10^{1}$ |
| D10 | | $4.70 \times 10^{4}$ | X | $2.70 \times 10^{2b}$ | $3.20 \times 10^{3}$ |
| D12 | | $8.20 \times 10^{4}$ | | $5.00 \times 10^{2}$ | $5.30 \times 10^{3}$ |

| | Presumptive *Vibrio* count (CFU × $10^{-5}$ larvae) | | | | | | | | | |
|---|---|---|---|---|---|---|---|---|---|---|
| | POND 1 | | | | | POND 2 | | | | |
| | CAS | FTS | B + CAS | B + FTS | B + CAS + FF | CAS | FTS | B + CAS | B + FTS | B + CAS + FF |
| D0 | $6.20 \times 10^{4}$ | $3.00 \times 10^{4}$ | $3.50 \times 10^{2(ab)}$ | $2.50 \times 10^{2a(a)}$ | $4.00 \times 10^{2(b)}$ | $4.40 \times 10^{3}$ | $7.20 \times 10^{3a}$ | $2.10 \times 10^{2(c)}$ | $1.50 \times 10^{2a(c)}$ | $2.20 \times 10^{2(c)}$ |
| D2* | $3.70 \times 10^{5}$ | $5.90 \times 10^{4a}$ | $5.20 \times 10^{3}$ | $3.50 \times 10^{2ab}$ | $7.40 \times 10^{3}$ | $6.90 \times 10^{3}$ | $8.00 \times 10^{3ab}$ | $3.50 \times 10^{3}$ | $2.20 \times 10^{2a}$ | $1.20 \times 10^{1}$ |
| D4* | $6.60 \times 10^{6}$ | $6.30 \times 10^{4a}$ | $7.00 \times 10^{5}$ | $4.30 \times 10^{2b}$ | $1.40 \times 10^{2}$ | $2.70 \times 10^{4}$ | $9.60 \times 10^{3b}$ | $9.90 \times 10^{3}$ | $5.50 \times 10^{2b}$ | $8.90 \times 10^{2}$ |
| D6* | X | $8.80 \times 10^{4}$ | $9.90 \times 10^{6}$ | $6.90 \times 10^{2c}$ | $7.20 \times 10^{1}$ | $2.00 \times 10^{6}$ | $2.20 \times 10^{4}$ | $3.20 \times 10^{4}$ | $6.70 \times 10^{2b}$ | $6.70 \times 10^{1}$ |
| D8 | | $3.30 \times 10^{5}$ | X | $8.20 \times 10^{2c}$ | $3.60 \times 10^{3}$ | X | $5.00 \times 10^{4b}$ | $7.50 \times 10^{5}$ | $9.30 \times 10^{2}$ | $4.70 \times 10^{1}$ |
| D10 | | $6.00 \times 10^{5}$ | | $2.10 \times 10^{3d}$ | $5.10 \times 10^{3}$ | | $6.60 \times 10^{4b}$ | $4.10 \times 10^{6}$ | $3.00 \times 10^{3c}$ | $6.20 \times 10^{3}$ |
| D12 | | $9.10 \times 10^{5}$ | | $3.90 \times 10^{3d}$ | $6.30 \times 10^{5}$ | | $8.20 \times 10^{4}$ | X | $4.60 \times 10^{3c}$ | $8.00 \times 10^{3}$ |

The count for larvae samples (CFU × $10^{-5}$ larvae) is the average of duplicate samples for each rearing technique Values in the same column with the same superscript are not significantly different ($P > 0.05$)

Values in the same row with the same superscript in parentheses are not significantly different ($P > 0.05$)

*FF* florfenicol, *B* breeders barrier, *CAS* closed aquaculture system, *FTS* flow-through system, *D* day

*Florfenicol treatment (8 mg $L^{-1}$)

The presumptive *Vibrio* spp. count for larvae obtained from FF-treated breeders and reared in FTS (B + $FTS_{D12}$: $10^2$–$10^3$ CFU × $10^{-5}$ larvae) was significantly lower ($P > 0.05$) than in FF-treated *R.decussatus* larvae and maintained in CAS (B + CAS + $FF_{D12}$: $10^5$ CFU × $10^{-5}$ larvae). On the eighth day of rearing, a total mortality was recorded in larvae produced in closed-circuit water system. In addition, larvae obtained from treated reproducers and reared in FTS (BB + $FTS_{D12}$: $10^2$–$10^3$ CFU × $10^{-5}$ larvae) showed significantly lower ($P > 0.05$) PVC than in larvae maintained in FTS ($FTS_{D12}$: $10^4$–$10^5$ CFU × $10^{-5}$ larvae).

## Influence of rearing techniques on larval growth and survival

The results showed that larval length was significantly affected by the rearing techniques (Table 4). In fact, larvae grew successfully and completed their metamorphosis with 100 % survival (data not presented) only when they were reared in B + FTS and B + CAS + FF. *R. decussatus* larvae maintained in FTS showed significantly better length (Pond 2, B + $FTS_{D10}$: 181.10 µm) than those produced in closed-circuit seawater system (Pond 2, B + $CAS_{D10}$:145.10 µm). Moreover, Larvae produced by untreated breeders and reared in FTS showed significantly lower length (Pond 3, $FTS_{D12}$: 168.40 µm) than those obtained from FF-treated reproducers (Pond 3, B + $FTS_{D12}$: 201.90 µm). Growth of larvae maintained in B + FTS did not differ from growth of larvae reared in CAS and treated with FF (Table 4). Besides, survival was better in larval group reared in B + CAS + FF (up to 69 %) compared to those maintained in other rearing systems (0–63 %).

**Table 4** Effect of rearing techniques on *Ruditapes decussatus* larval growth and survival

| | Mean ± SD (μm) | | | | | | | | | |
|---|---|---|---|---|---|---|---|---|---|---|
| | POND 1 | | | | | POND 2 | | | | |
| | CAS | FTS | B + CAS | B + FTS | B + CAS + FF | CAS | FTS | B + CAS | B + FTS | B + CAS + FF |
| D0 | 94.10 ± 2.2$^{a}$ | 94.90 ± 3.5$^{a}$ | 94.10 ± 3.8$^{a}$ | 94.10 ± 4.4$^{a}$ | 94.10 ± 3.8$^{a}$ | 93.10 ± 2.5$^{a}$ | 93.10 ± 5.2$^{a}$ | 93.10 ± 3.4$^{a}$ | 93.10 ± 2.1$^{a}$ | 93.10 ± 4.0$^{a}$ |
| D2* | 96.20 ± 3.5$^{a}$ | 97.21 ± 5.0$^{a}$ | 97.00 ± 4.8$^{a}$ | 96.90 ± 3.5$^{a}$ | 97.20 ± 4.8$^{a}$ | 95.50 ± 3.8$^{a}$ | 96.00 ± 4.8$^{a}$ | 95.20 ± 5.0$^{a}$ | 100.43 ± 4.2$^{b}$ | 105.10 ± 4.9$^{b}$ |
| D4* | 97.00 ± 3.1$^{a}$ | 100.88 ± 2.1$^{a}$ | 98.12 ± 3.5$^{a}$ | 106.00 ± 5.0$^{b}$ | 108.56 ± 3.5$^{b}$ | 96.90 ± 6.0$^{a}$ | 99.90 ± 4.0$^{a}$ | 99.50 ± 6.3$^{a}$ | 117.53 ± 5.1$^{b}$ | 120.31 ± 3.5$^{b}$ |
| D6* | X | 106.44 ± 4.0 | 99.45 ± 2.7 | 126.50 ± 11.3$^{b}$ | 129.23 ± 2.7$^{b}$ | 104.00 ± 5.5$^{a}$ | 108.20 ± 7.4$^{a}$ | 106.10 ± 6.9$^{a}$ | 131.46 ± 9.4$^{b}$ | 132.97 ± 2.7$^{b}$ |
| D8 | | 128.40 ± 5.5 | X | 156.70 ± 12.0$^{b}$ | 159.21 ± 5.0$^{b}$ | X | 125.40 ± 8.2$^{a}$ | 123.00 ± 8.0$^{a}$ | 161.00 ± 12.6$^{b}$ | 162.18 ± 5.8$^{b}$ |
| D10 | | 150.00 ± 8.0 | | 183.40 ± 10.0$^{b}$ | 185.90 ± 5.9$^{b}$ | | 149.10 ± 7.5$^{a}$ | 145.10 ± 10.2$^{a}$ | 181.10 ± 11.6$^{b}$ | 185.00 ± 5.9$^{b}$ |
| D12 | | 172.20 ± 10.2 | | 201.15 ± 12.6$^{b}$ | 205.00 ± 11.1$^{b}$ | | 169.00 ± 9.0 | X | 200.10 ± 14.2 | 206.50 ± 11.1 |
| S (%) | 0 | 15 | 0 | 50 | 69 | 0 | 10 | 0 | 56 | 70 |

| | Mean ± SD (μm) | | | | |
|---|---|---|---|---|---|
| | POND 3 | | | | |
| | CAS | FTS | B + CAS | B + FTS | B + CAS + FF |
| D0 | 93.90 ± 4.5$^{a}$ | 93.90 ± 5.8$^{a}$ | 93.90 ± 3.7$^{a}$ | 93.90 ± 2.6$^{a}$ | 93.90 ± 5.5$^{a}$ |
| D2* | 96.00 ± 5.0$^{a}$ | 97.90 ± 6.4$^{a}$ | 97.10 ± 4.5$^{a}$ | 98.10 ± 4.3$^{a}$ | 100.87 ± 3.0$^{a}$ |
| D4* | 98.20 ± 7.2$^{a}$ | 100.00 ± 4.0$^{a}$ | 98.90 ± 4.9$^{a}$ | 108.00 ± 3.5$^{b}$ | 110.64 ± 4.2$^{b}$ |
| D6* | X | 104.00 ± 5.3$^{a}$ | 103.50 ± 6.2$^{a}$ | 130.10 ± 7.7$^{b}$ | 133.63 ± 6.0$^{b}$ |
| D8 | | 129.30 ± 7.8 | 121.00 ± 7.0 | 160.30 ± 9.8$^{a}$ | 164.10 ± 8.0$^{a}$ |
| D10 | | 141.90 ± 9.0 | X | 180.00 ± 10.8 | 186.87 ± 9.7 |
| D12 | | 168.40 ± 10.5 | | 201.90 ± 11.5 | 208.31 ± 10.3 |
| S (%) | 0 | 18 | 0 | 63 | 75 |

The measurement of larval length (μm) is the average triplicate for each rearing technique and for each spawning. Values in the same row with the same superscript are not significantly different ($P < 0.05$)

*S* survival, *FF* florfenicol treatment, *BB* breeders barrier, *CAS* closed aquaculture system, *FTS* flow-through system, *SD* standard deviation, *D* day

* Florfenicol treatment (8 mg $L^{-1}$)

## Bacterial characterization

A total of 112 strains were isolated from breeders ($n = 42$) and larvae ($n = 70$) collected from the INSTM clam hatchery and biochemically identified as members of the family *Vibrionaceae* by API 20 E miniaturized system (Table 5). *Vibrio alginolyticus* was the most predominant species (41/112) followed by *V. fluvialis* (28/112), *V. vunificus* (20/112), *V. cholera* (15/112) and *V. paraheamolyticus* (8/112). High phenotypic heterogeneity was observed among the tested isolates. In fact, the majority of *Vibrio* strains were positive for lysine decarboxylase, indole production and glucose fermentation. Only, *V. fluvialis* isolates gave positive results with adenine dehydrolase and tryptophan deaminase tests. About 75 % (6/8) of the *V. paraheamolyticus* strains gave positive results with amygdalin and arabinose. *Vibrio* strains were able to tolerate low and moderate concentrations of NaCl (2–8 %), while only *V. alginolyticus* isolates were able to grow over a wide range of salinities (0–10 % NaCl).

# DISCUSSION

Members of the genus *Vibrio* are commonly found as part of the normal flora of molluscan intestinal tract (Oxley et al. 2002; Davis et al. 2013), thus making broodstock one of the principal sources of contamination in marine bivalve hatcheries. Several studies have been carried out on the gonads of reared mollusks revealed that they may be strongly infested with *Vibrio* (Lodeiros et al. 1987; Llanos et al. 2002; Sainz-Hernández et al. 2005) which leads to the hypothesis of their vertical transmission between reproducers and gametes. Widman (2001) suggested that *Vibrio* contamination into bivalve molluscan hatcheries can be controlled by relaying pre-spawning breeders for 24 h at 17–20 °C in filtered seawater. Conversely, Sainz-Hernández et al. (2005) revealed that this technique would not be sufficient for avoiding the circulation of vibrios into the clam system, since they found *Vibrio* in the gonadal tissue far from the digestive tract. In this study, we managed to create a bacteriological barrier among the breeder stage by FF treatments (8 mg $L^{-1}$) for five successive days before spawning in order to reduce *Vibrio* dissemination and to eliminate the use of the antimicrobial agents for *R. decussatus* larval rearing. TCBS data showed a significant decreasing of PVC in broodstock after antibiotic treatments. Moreover, fertilized eggs originated from treated reproducers presented relatively better microbiological quality than those obtained from control breeders.

This study clearly demonstrated that *R. decussatus* larvae obtained from FF-treated breeders and reared in FTS showed significantly lower bacterial load, especially for vibrio count, than other techniques. These findings are in agreement with Andersen et al. (2000) who showed that great scallop larvae rearing water in flow-through tanks had fewer bacterial counts than both rearing water in antibiotic-treated and untreated static tanks. In addition, only larvae maintained in both CAS + FF and B + FTS succeeded to develop beyond the stage of metamorphosis. In the literature, it is reported that antibiotic treatments during larval rearing prevent bacterial contamination, which increases successful larval rearing (Robert et al. 1996; Uriarte et al. 2001; Torkildsen et al. 2005). However, the overuse of antibiotics in treatment of mollusk aquaculture may contribute to an increase in the level of antimicrobial resistance (Mechri et al. 2013, 2015). The introduction of a flow-through system instead of closed-circuit water system in larval rearing tanks eliminated the use of antibiotics by improving water quality (Southgate and Ito 1998; Magnesen et al. 2006).

It was also noted that growth of larvae reared in B + CAS + FF and B + FTS was significantly better than those maintained in the other rearing systems. Phenicol are common broad-spectrum antibiotics used for aquacultural practice, as chemotherapeutic agents to manage and prevent diseases (Uriarte et al. 2001; Campa-Cordova et al. 2006; Mechri et al. 2012). Some countries have restricted the use of chloramphenicol and thiamphenicol in aquaculture because of their highly toxic potential on consumers (Van de Riet et al. 2003; Sarmah et al. 2006). FF is a fluorinated derivative of thiamphenicol with an expanded spectrum of activity and enhanced potency but with less harmful effects (Samuelsen and Bergh 2004; Xu et al. 2005). Thus, FF is often used to control bacterial disease outbreaks in mollusk larval hatcheries (Torkildsen et al. 2002; Torkildsen and Magnesen 2004; Miranda et al. 2013). However, several reports have demonstrated that the extensive use of antibiotics in aquaculture could facilitate the emergence and the selection of resistant bacteria (Laganà et al. 2011; Mechri et al. 2015). Previous studies of antimicrobial resistance in fish pathogenic bacteria have reported the occurrence of the *floR*gene as the major facilitator superfamily and code for efflux proteins that export florfenicol out of the cell (Miranda et al. 2013; Schwarz et al. 2004). Recently, Miranda et al. (2015) found that most of gram-negative bacteria associated with scallop hatcheries carried the *floR* gene.

Sarkis et al. (2006) showed that calico scallop growth for FT-reared larvae was comparable or significantly higher than in the static system, which corroborate with our findings. Similarly, Rico-Villa et al. (2008) reported significant growth and competence improvements in *Crassostria gigas* larvae reared in the flow-through system compared to static system. These authors also revealed that the use of FTS with constant flows of high-quality water and a daily cleaning of the sieve clogs facilitated the circulation in the rearing tank and removal of metabolic waste products providing favorable conditions for larval development. Several studies suggested an association between the stocking density and the water exchange rate (Magnesen et al. 2006; Rico-Villa et al. 2008). Furthermore, Andersen et al. (2000) and Sarkis et al. (2006) reported a reduction in growth performance at high stocking density when they reared scallop (*Pecten maximus*) and oyster (*C. gigas*) larvae in FTS.

Survival of *R. decussatus* larvae reared in B + CAS + FF was better than for those maintained in other rearing techniques. These findings are in agreement with Andersen et al. (2000) who showed that in spite of low bacterial counts in great scallop larvae, increased larval survival was not demonstrated in the FTS compared to the antibiotic-treated CAS. Torkildsen and Magnesen (2004) reported that *P. maximus* larval survival in FTS was approximately similar to survival in CAS with prophylactic addition of chloramphenicol. Besides, Temperature and algal concentration were the most important factors that affect growth and survival (Le Pennec et al. 2003; Pernet and Tremblay 2004).

The predominant species isolated from almost all of broodstock and larvae samples was *V. alginolyticus* (36.6 %). This species has become increasingly recognized as an important opportunistic pathogen causing epizootic outbreaks in cultured fish (Ben Kahla-Nakbi et al. 2006), shellfish (Beaz Hidalgo et al. 2008) and crustaceans (Gomez-Gil et al. 1998). In our previous study, we found that *V. alginolyticus* species was the most commonly isolated (55 %) from the different compartments of *R. decussatus* hatchery (Mechri et al. 2012). Castro et al. (2002) characterized 123 *Vibrio* isolates from cultured *Ruditapes philippinarum* using phenotypic and genotypic methods and showed that *V. harveyi, V. splendidus and V. tubiashii*were the most abundant species. Moreover, Beaz Hidalgo et al. (2008) reported that 67.7 % of the *Vibrio* species associated with cultured carpet shell clams (*R. decussatus*) and Manilla clams (*R. philippinarum*) belonged to *V. splendidus, V. alginolyticus* and *V. diazotrophicus.*

Phenotypic characteristics of *Vibrio* isolates recovered from INSTM hatchery revealed high levels of identity. In fact, 112 *Vibrio* strains isolated on TCBS agar were further biochemically identified using the commercial miniaturized Api 20E kit and showed biochemical profiles similar to those reported earlier by Mechri et al. (2012). Conversely, these finding disagree those of Al-Sunaiher et al. (2010) who showed that the majority of *V. fluvialis* strains, isolated during fish outbreak, were positive to indole test and were able to ferment mannitol. These authors also reported that all *V. vulnificus* strains gave negative result with lysine decarboxylase and were unable to ferment glucose and amygdalin.

## CONCLUSION

In this paper, we demonstrate that the prophylactic use of FF at breeder stage improve the bacteriological quality of *R. decussatus* eggs. Despite considerable variability of the *Vibrio* load, no significant differences in growth performance were noted among larvae reared in B + FTS and B + CAS + FF. Thus, FTS appears to be an alternative that can eliminate the need for antibiotic treatments during larval rearing. However, several important parameters, such as larval densities, temperature and alimentation, need optimization to improve *R. decussatus*rearing performance. We also reported that *V. alginolyticus* strains were predominantly isolated from breeders, eggs and larvae. Further research is needed to understand the boundaries between commensalism and pathogenicity in this species.

# REFERENCES


1. Al-Sunaiher AE, Abdelnasser S, Ibrahim S, Al-Salamah A (2010) Association of *Vibrio* species with disease incidence in some cultured fishes in the Kingdom of Saudi Arabia. World Appl Sci J 8(5):653–660
2. Andersen S, Burnell GG, Bergh OO (2000) Flow-through systems for culturing great scallop larvae. Aquacult Int 8:249–257
3. Beaz Hidalgo R, Cleenwerck I, Balboa S, De Wachter M, Thompson FL, Swings J, De Vos P, Romalde JL (2008) Diversity of Vibrios associated with reared clams in Galicia (NW Spain). Syst Appl Microbiol 31:215–222
4. Ben Kahla-Nakbi A, Chaieb K, Besbes A, Zmanter T, Bakhrouf A (2006) Virulence and enterobacterial repetitive intergenic consensus PCR of *Vibrio alginolyticus* strains isolated from Tunisian cultured gilthead sea bream and sea bass outbreaks. Vet Microbiol 117:321–327
5. Campa-Cordova AI, Luna-Gonzalez A, Ascencio F, Cortes-Jacinto E, Caceres- Martınez CJ (2006) Effects of chloramphenicol, erythromycin, and furazolidone on growth of *Isochrysis galbana* and *Chaetoceros gracilis*. Aquaculture 260:145–150
6. Castagna M, Gibbons MMC, Kurkowski KK (1996) Culture: application. In: Kennedy VS, Newell RIE, Eble AF (eds) The eastern oyster, *Crassostreaa virginica*. Maryland Sea Grant, College Park, pp 675–690
7. Castro D, Pujalte MJ, Lopez-Cortes L, Garay E, Borrego JJ (2002) Vibrios isolated from the cultured Manila clam (*Ruditapes philippinarum*): numerical taxonomy and antibacterial activities. J Appl Microbiol 93:438–447
8. Davis J, Fricke WF, Hamann MT, Esquenazi E, Dorrestein PC, Hill RT (2013) Characterization of the bacterial community of the chemically defended Hawaiian sacoglossan *Elysia rufescens*. Appl Environ Microbiol 79:7073–7081Central
9. Gomez-Gil B, Tron-Mayén L, Roque A, Turnbull JF, Inglis V, Guerra-Flores AL (1998) Species of *Vibrio* isolated from hepatopancreas, haemolymph and digestive tract of a population of healthy juvenile *Penaeus vannamei*. Aquaculture 163:1–9
10. Helm MM, Bourne N (2004) Hatchery culture of bivalves. A practical manual. In: Lovatelli A (ed) FAO Fisheries Technical Paper 471. Food and Agriculture Organization of the United Nations Publishing, Rome, p 177
11. Laganà P, Caruso G, Minutoli E, Zaccone R, Santi D (2011) Susceptibility to antibiotics of *Vibrio* spp. and *Photobacterium damsela* ssp. piscicida strains isolated from Italian aquaculture farms. New Microbiol 34(1):53–63
12. Le Pennec M, Paugam A, Le Pennec G (2003) The pelagic life of pectinid *Pecten maximus*—a review. ICES J Mar Sci 60:211–223
13. Llanos J, Cid M, Navarro S, Dinamarca A, García-Tello P (2002) Atypical bacteria accompanying the scallop *Argopecten purpuratus*. Invest Mar Valparaíso 30(2):57–59
14. Lodeiros C, Bolinches J, Dopazo CP, Toranzo AE (1987) Bacillary necrosis in hatcheries of *Ostrea edulis* in Spain. Aquaculture 65:15–29
15. Magnesen T, Jacobsen A (2012) Effect of water recirculation on seawater quality and production of scallop (*Pecten maximus*) larvaee. Aquacult Eng 47:1–6
16. Magnesen T, Bergh O, Christophersen G (2006) Yields of great scallop, *Pecten maximus*, larvaee in a commercial flow through rearing system in Norway. Aquacult Int 14:377–394
17. Martins CIM, Eding EH, Schneider O, Rasmussen R, Olesen B, Plesner L, Verreth JAJ (2005) Recirculation aquaculture systems in Europe. CONSENSUS. Oostende, Belgium, Consensus working Group, European Aquaculture Society: 31
18. Mechri B, Medhioub A, Medhioub MN, Aouni M (2012) Diversity of *Vibrionaceae* associated with *Ruditapes decussatus* hatchery in Tunisia. Ann Microbiol 61:597–606

19. Mechri B, Medhioub A, Medhioub MN, Aouni M (2013) Genotypic diversity, antimicrobial resistance and screening of *Vibrio cholerae* molecular virulence markers in *Vibrio alginolyticus*strains recovered from a Tunisian *Ruditapes decussatus* hatchery. Pol J Microbiol 62(3):263–272
20. Mechri B, Ben Salem I, Medhioub A, Medhioub MN, Aouni M (2015) Isolation and genotyping of potentially pathogenic *Vibrio alginolyticus* associated with *Ruditapes decussatus* larvae and juvenile mass mortalities. Aquacult Int 23(4):1033–1047
21. Medhioub MN (1983) La conchyliculture en Tunisie. Projet Tunis/92/002. Direction Générale de la Pêche et de l'Aquaculture. PNUD/FAO. pp 83
22. Medhioub MN, Lymayem Y, Fathallah S, Abed MM, Medhioub A (2006) Cycle d'élevage de la palourde *Ruditapes decussatus* (mollusque lamellibranche) produite en écloserie. Actes des 13ème Journées Scientifiques sur les Résultats de la Recherche Agricoles, pp 107–117
23. Merino G, Uribe E, Soria G, von Brand E (2009) A comparison of larval production of the northern scallop, *Argopecten purpuratus*, in closed and recirculating culture systems. Aquacult Eng 40:95–103
24. Miranda CD, Tello A, Keen PL (2013) Mechanisms of antimicrobial resistance in finfish aquaculture environments. Front Microbiol 4:233Central
25. Miranda CD, Rojas R, Geisse J, Romero J, González-Rochac G (2015) Scallop larvae hatcheries as source of bacteria carrying genes encoding for non-enzymatic phenicol resistance. Marine Poll Bull 95:173–182
26. New MB, Valenti WC (2000) Freshwater prawn culture: the farming of *Macrobrachium rosenbergii*. Blackwell Science, Oxford, p 52
27. Otoshi CA, Arce SSM, Moss SSM (2003) Growth and reproductive performance offbroodstock shrimp reared in a biosecure recirculating aquaculture system versus a flow-through pond. Aquacult Eng 29:93–107
28. Oxley APA, Shipton W, Owens L, McKay D (2002) Bacterial flora from the gut of the wild and cultured banana prawn, *Penaeus merguiensis*. J Appl Microbiol 93:214–223
29. Pernet F, Tremblay R (2004) Effect of varying levels of dietary essential fatty acid during early ontogeny of the sea scallop *Placopecten magellanicus*. J Exp Mar Biol Ecol 310:73–86
30. Pfeiffer TJ, Rusch KA (2000) An integrated system for microalgal and nursery seed clam culture. Aquacult Eng 24:15–31
31. Rico-Villa B, Woerther PP, Mingant CC, Lepiver DD, Pouvreau SS, Hamon MM, Robert RR (2008) A flow-through rearing system for ecophysiological studies of Pacific oyster *Crassostrea gigas* larvaee. Aquaculture 282:54–60
32. Rico-Villa B, Le Coz JR, Mingant C, Robert R (2006) Influence of phytoplankton diet mixtures on microalgae consumption, larval development and settlement of the Pacific oyster *Crassostrea gigas* (Thunberg). Aquaculture 256:377–388
33. Ritar AJ (2001) The experimental culture of phy llosoma larvaee of southern rock lobster (*Jasus edwardsii*) in a flow-through system. Aquacult Eng 24:149–156
34. Robert R, Gérard AA (1999) Bivalve hatchery technology: the current situation for the Pacific oyster *Crassostrea gigasa* and the scallop *Pecten maximus* in France. Aquat Living Resour 12:121–130
35. Robert R, Miner P, Nicolas JL (1996) Mortality control of scallop larvaee in the hatchery. Aquacult Inter 4(4):305–313
36. Rosenthal H (1980) Recirculation systems in Western Europe. World symposium on aquaculture in heated effluents and recirculation system, Stavanger, Institut für Kuesten-und Binnefischerei, Bundesforschungsanstalt Hamburg, BRD
37. Rosenthal H, Castell JD, Chiba K, Forster JRM, Hilge V, Hogendoorn H,Mayo RD, Muir JF, Murray KR, Petit J, Wedemeyer GA, Wheaton F, Wickins J (1986) Flow-through and recirculation systems, EIFAC, p 100
38. Sainz-Hernández JC, Martínez M, Alfonso N (2005) Sources of *Vibrio* bacteria in mollusc hatcheries and control methods: a case study. Aquacult Res 36:1611–1618

39. Samuelsen OB, Bergh O (2004) Efficacy of orally administered florfenicol and oxolinic acid for the treatment of vibriosis in cod (*Gadus morhua*). Aquaculture 235:27–35
40. Sarkis S, Hohn CC, Helm MM (2006) Larval rearing of calico scallops, *Argopecten gibbus*, in a flow—through system. Aquacult Int 14:527–538
41. Sarmah AK, Meyer MT, Boxall ABA (2006) A global perspective on the use, sales, exposure pathways, occurrence, fate and effects of veterinary antibiotics (VAs) in the environment. Chemosphere 65:725–759
42. Schwarz S, Kehrenberg C, Doublet B, Cloeckaert A (2004) Molecular basis of bacterial resistance to chloramphenicol and florfenicol. FEMS Microbiol Rev 28:519–542
43. Southgate PC, Ito MM (1998) Evaluation of a partial flow-through culture technique for pearl oyster (*Pinctada margaritifera L.*) larvaee. Aquacult Eng 18:1–7
44. Tieman DM, Goodwin AAE (2001) Treatments for ich infestations in channel catfish evaluated under static and flow-through water conditions. N Am J Aquacult 63:293–299
45. Torkildsen L, Magnesen T (2004) Hatchery production of scallop larvaee (*Pecten maximus*) survival in different rearing systems. Aquacult Int 12:489–507
46. Torkildsen L, Coyne R, Samuelsen OB, Magnesen T, Bergh Ø (2002) Treatment of the early life stages of scallop (*Pecten maximus)* with antimicrobial agents; searching for an alternative to chloramphenicol. Aquacult Int 10(5):399–409
47. Torkildsen L, Lambert C, Nylund A, Magnesen T, Bergh Ø (2005) Bacteria associated with early life stages of the great scallop, *Pecten maximus*: impact on larval survival. Aquacult Int 13(6):575–592
48. Uriarte I, Farías A, Castilla JC (2001) Effect of antibiotic treatment during larval development of the Chilean scallop *Argopecten purpuratus*. Aquacult Eng 25(3):139–147
49. Van de Riet JM, Potter RA, Christie-Fougere M, Garth-Burns B (2003) Simultaneous determination of residues of chloramphenicol, thiamphenicol, florfenicol and florfenicol amine in farmed aquatic species by liquid chromatography/mass spectrometry. J AOAC Int 86:510–514
50. Verreth JAJ, Eding EH (1993) European farming industry of African catfish (*Clarias gariepinus*) facts and figures. J World Aquacult Soc 24:6–13
51. Widman JC (2001) Manual for hatchery culture of the bay scallop, *Argopecten irradians irradians*. National Marine Fisheries Service, Connecticut Sea Grant College Program, Connecticut, p 50
52. Xu LW, Liao CR, Liu GF (2005) Security about florfenicol used in aquaculture. J Fish Sci China 12:512–518

# CHAPTER 11

# Application of Biological Agents in Abalone Aquaculture: A South African Perspective

**Ghaneshree Moodley[1, 2], Lethabo Mashigo[1], Rajesh Lalloo[1] and Suren Singh[2]**

[1] *CSIR Biosciences, Pretoria, South Africa*

[2] *Department of Biotechnology & Food, Faculty of Applied Sciences, Durban University of Technology, Durban, South Africa*

## 1. INTRODUCTION

### 1.1. The Impact of Abalone Mariculture on Developing Economies

Aquatic animals are nutritionally important for human consumption, as they are an excellent source of proteins, trace elements, and polyunsaturated fatty acids [1]. There has been a significant increase in the demand for an array of both fish and shellfish products as a result of growth in the global population [2]. Fisheries itself, cannot provide sufficient amounts of aquatic products to fulfil the demands of the consumer; therefore, aquaculture provides a crucial alternative resource [1,4]. Aquaculture has become more significant and intensive over the last few decades and is presently the fastest growing food production industry. An average industrial growth of more than 6% in the period between 1985 and

2005 has been reported, with an annual increase of approximately 3.2% per annum during the period up to 2009 [1,3].

Modern aquaculture involves the intensive production of finfish, crustaceans, molluscs, and algal plants under controlled conditions [5]. Aquaculture yields far exceeds that of natural fishing, and provides an effective means for a constant, year round supply for good quality seafood and seafood products [6-8]. The practice of aquaculture not only provides local food security, but also improves the livelihoods of people in many poorly developed coastal regions [2].

Commercial abalone mariculture has become a thriving, global industry. It has a promising future due to the high prices being paid for abalone, coupled to a worldwide decline in fisheries production because of overfishing and poaching [8,9]. Abalone is one of the most valuable seafood species in the world, whereby demand far exceeds supply, especially in Asian countries such as; Hong Kong, China, Japan, Taiwan and Singapore which are major destination markets (6-8,10). Abalone is used primarily as a celebration dish, especially during weddings and other special occasions such as the Chinese New Year [11]. On account of the ever growing demand of live, dried and canned abalone, already high prices of this seafood delicacy continue to escalate.

Abalone (family *Haliotidae*) belongs to a class of marine vetigastropod molluscs, which are distributed along rocky shores and reefs of coastal temperate and tropical waters [11, 12]. The abalone family consists of about 56 species all belonging to the genus *Haliotis* [13,14]. Many members of this family have achieved commercial status as fishery and/or aquaculture species, and are of major economic importance (Table 1). In 2007, it was reported that abalone was supplied to export markets in the following product forms; dried (7%), frozen (24%), live (18%) and canned (51%). Live abalone achieves higher revenues however, it does deem problematic in terms of transportation and related logistics [15]. Due to the demand of this prestigious seafood, supply of abalone is under severe pressure; and has led to the increase in the occurrence of abalone farming facilities around the world.

**TABLE 1.** Globally farmed abalone species and their location adapted from [16,17].

| SPECIES NAME | COMMON NAME | LOCATION | TYPE OF FISHERY |
|---|---|---|---|
| *Haliotis rufescens* | Red | N. America | Farmed/ |

| SPECIES NAME | COMMON NAME | LOCATION | TYPE OF FISHERY |
|---|---|---|---|
| | abalone | | Recreational |
| *Haliotis rufescens* | Red abalone | Chile | Farmed |
| *Haliotis cracherodii* | Black abalone | N. America | Farmed |
| *Haliotis fulgens* | Green abalone | N. America | Wild/Farmed (Mexico) |
| *Haliotis corrugata* | Pink abalone | N. America | Wild/Farmed (Mexico) |
| *Haliotis kamtschatkana* | Pinto abalone | N. America | Farmed |
| *Haliotis midae* | Perlemoen | South Africa | Wild/Farmed |
| *Haliotis laevigata* | Green-lip abalone | S. Australia | Wild |
| *Haliotis rubra* | Black-lip abalone | S. Australia | Wild/Farmed |
| *Haliotis roei* | Roe's abalone | Australia | Wild |
| *Haliotis iris* | Black footed paua | New Zealand | Wild |

| SPECIES NAME | COMMON NAME | LOCATION | TYPE OF FISHERY |
|---|---|---|---|
| *Haliotis diversicolor supertexta* | Small abalone | Taiwan | Wild/Farmed |
| *Haliotis discus hannai* | Disk abalone | Japan, China | Wild/Farmed |

Cultivation of abalone is widespread in many countries, including USA, Mexico, South Africa, Australia, New Zealand, Japan, Taiwan, China, Ireland, Chile and Iceland [17-20] China is the largest producer in the world with over 300 farms and a total production of approximately 4500 metric tonnes [9]. China, Taiwan and South Africa are considered as the key production powerhouses in the abalone industry (Table 2). China is the highest contributor of live product annually, and is still the major market for abalone produced world-wide [8]. This occurrence is closely related to the economic growth and the increase in personal wealth exhibited by the Chinese population as well as the growth of the Chinese middle class population [4,15]. It has been reported that the total abalone produce reaching markets through harvesting, illegal poaching and natural supply, does not meet demand for this seafood delicacy [9,18].

**TABLE 2.** Global production of abalone indicating world leaders in abalone production, quantity of produce legally harvested and sold, and quantity of product illegally harvested; 2004-2005 data [20].

| COUNTRY | CULTURE | LEGAL HARVEST | ILLEGAL HARVEST |
|---|---|---|---|
| **China** | 4500 | - | - |
| **Taiwan** | 3000 | - | - |
| **South Africa** | 600 | 237 | 1850 |

| COUNTRY | CULTURE | LEGAL HARVEST | ILLEGAL HARVEST |
|---|---|---|---|
| **Japan** | 200 | 2200 | 536 |
| **USA** | 170 | 0 | 250 |
| **Australia** | 290 | 5128 | 1000 |
| **Chile** | 200 | - | - |
| **Mexico** | 50 | 1066 | 550 |
| **New Zealand** | 3 | 1078 | 400 |
| **Other** | 30 | 442 | 110 |
| **Total** | 9 043 | 10 151 | 4696 |
| **Grand total in 2005:** 24 040 tonnes live weight | | | |

The South African abalone industry continues to establish itself as a premium brand in Asia, and is a good example of mariculture in a developing country. Abalone farming in SA is a relatively new but dynamic industry and has demonstrated a high production capacity [15]. One of the main challenges faced by the SA industry is the loss in revenue experienced due to poaching. Reports suggest that approximately 2000 tonnes are lost to the economy [4]. The abalone mariculture industry started developing in South Africa during the 1990's and has been gaining popularity. As a result, an economic environment whereby abalone aquaculture has become increasingly attractive as a financial investment has been established [17]. Abalone rearing facilities employs an intensive system in which abalone is reared at high densities in shore-based aquaculture systems [21].

The South African abalone, *Haliotis midae*, locally known as "perlemoen" is the only one of six indigenous species that is of commercial importance

[8,22]. The abalone *H. midae*, takes over 30 years to reach a maximum size of 200mm (shell length) in natural habitats [21]. Even under farmed conditions, abalone growth is slow and often varies with size and age [23]. *H. midae* takes approximately 4 to 5 years to reach a marketable size of 100 mm (shell length) before it can be sold between US$ 34 to 36 per kg on international markets [23,24]. Mariculture of abalone is thus important to ensure market supply and it is for these reasons that alternate approaches involved in the promotion of abalone growth and an increased immunity to disease of farmed abalone are required.

**FIGURE 1.** Holding tanks containing farm-produced abalone on the West coast of South Africa.

Land based aquaculture of abalone has increased over the last decade in South Africa (Figure 1), and commercially produced abalone has almost completely replaced the wild harvested product [4]. In 2010 the output of all facets of abalone harvest totalled 1015.44 metric tons.

The former status of abalone aquaculture in South Africa is outlined in Table 3. These farms produced 890 tonnes of abalone, and created direct employment to about 840 people. There was an increase in skilled individuals of approximately 7.6% over the 2 year period. Due to the high demand for this seafood delicacy, a gross turnover of approximately R200 million per annum was achieved [26]. The industry has demonstrated continued growth. In 2003/4; 19 enterprises secured permits to culture this species and by 2007, this number had increased to 24, further highlighting the growth potential of this particular sector [27]. It was estimated that by 2020 the production of abalone would amount 2895 tons with a value of R551 million, making abalone mariculture the leading subsector contributor in the aquaculture industry [4]. This growth has had a direct impact on the socio-economic growth of the country, whereby more than 1200 people with necessary skills are currently employed in the industry. Global aquaculture initiatives have shown that the success of the technology is largely dependent on government sectors for support to enable the creation of a robust and sustainable industry [15]. The mariculture of abalone and on-going growth of this industry is extremely important, as it addresses a number of challenges faced by the South African nation, which are also common to many developing countries. This practice will contribute to a number of strategic imperatives including economic and enterprise development, job creation, food security as well as the adoption of sustainable mariculture practices [15].

**TABLE 3.** The status of abalone aquaculture and total investment in the South African abalone industry between 2004 and 2006 [26].

| Year | No. of producing farms | Investment (R-million) | Tonnes produced | Annual increase in industry (%) | No. of employees |
|---|---|---|---|---|---|
| 2004 | 13 | - | 576 | - | 556 |
| 2005 | 13 | 197 | 745 | 27 | 776 |
| 2006 | 13 | 182 | 890 | 21 | 840 |

# 2. CHALLENGES FACED IN ABALONE MARICULTURE AND CONVENTIONALLY USED MITIGATION STRATEGIES

Many aquaculture farmers, including those in the abalone mariculture sub-sector are faced with a myriad of challenges [28]. The challenges are further exacerbated as abalone mariculture activities become more intensified to optimise efficiencies in land usage and productivity. Adversities faced include slow growth rate of abalone, the outbreak of diseases, waste accumulation and deterioration of environmental conditions within the culture system [29,30]. Disease occurrence is usually associated with primary invasion by pathogenic strains as well as mechanical injury coupled to stressful environmental conditions such as physiochemical changes and poor water quality [31]. These factors, in an interactive way, challenge the health and immune response of the abalone and can lead to poor growth, ill health and increased mortality. This predicament has become one of the main barriers towards the successful development in the aquaculture industry, given that it limits the production of aquaculture products in terms of quality, quantity, and regularity [23].

Disease control is an inherent part of any animal production system, however, in the aquatic environment, the intimate relationship between bacteria and their host, and the use of open production systems adds to this challenge [5]. Unpredictable mass mortalities still occur in the early life stages as a result of the proliferation of pathogens and opportunistic microorganisms, which are responsible for major economic losses [1]. Abalone like other aquatic species is susceptible to common marine pathogenic organisms such as *Vibrio parahaemolyticus, Vibrio anguillarum* and *Vibrio carchariae*, as well as prokaryotes and viruses [23,32,33]. When pathogenic bacteria or viruses are detected, farmers usually apply antimicrobial compounds to the feed and the rearing water [34]. Broad-spectrum anti-microbials have been extensively used as a means of disease control on many aquaculture facilities and unfortunately remains the method of choice for many farmers [23]. Some farmers also use antibiotics as prophylactics in large quantities, even when pathogens are not evident. This ill-advised practice has led to an increase in *Vibrios*, and other opportunistic pathogens, which possess multiple antibiotic resistance and as a result leads to the emergence of more virulent pathogens [28,35]. Plasmid-carrying resistance determinants have been transferred *in-vitro* from aquatic pathogens to human pathogens, such as from *V. cholerae* and *V. parahaemolyticus* to *Escherichia coli* by the horizontal spread of plasmids [36]. Furthermore, the presence of antibiotic residues in the tissues of animals, an imbalance of microorganisms in the gastrointestinal tract of aquatic species and the release of antibiotics into natural waters, and thereby poses further challenges. Consequently, the indiscriminate use of antibiotics confers a negative effect on the health of aquatic host species, the environment and consumers of food products [37]. Due to these concerns, more stringent regulation of antibiotic use in aquaculture has been imposed by the European Union [38]. Since the application of antibiotics is problematic, a strong demand for alternative methods of disease control is required in abalone mariculture.

Abalones are generally regarded as opportunistic herbivores that readily accept a wide range of diets. In natural ecosystems, abalone feed primarily on seaweed or kelp. This food contains a high degree of alginolytic material that is not readily digestible; as a result, enteric microflora is relied upon to effectively digest this material. If the host intestinal flora lacks the ability to produce beneficial enzymes, a very slow digestion process would result, and consequently hinder the growth of the abalone itself. The proper nutrition and resultant growth of cultured abalone are critical factors that require insight in order to successfully culture this mollusc. Appropriate mechanisms for feeding of abalone are therefore very important and it has been shown that different diets results in different growth rates [39]. Growth rates, especially at the early life stages of abalone are affected considerably by the diet and the ability of the individuals to utilize available food with a high resultant feed conversion ratio [40]. In abalone production systems abalones are fed either formulated diets or seaweed/kelp, and in some instances, a combination of both [25]. An optimum formulated diet should enable more efficient digestion consequently resulting in higher feed conversion ratios, and ultimately boost the growth of the abalone, but the reality is that diets are based on raw material availability and minimum cost formulation models. This presents a challenge in digestibility, feed conversion efficiency, animal health and waste generation into the culture environment. The development of artificial feeds and specialized feeding regimes to improve the growth of abalone has assisted in developing this practice into a more cost-effective and manageable industry [21]. It has been reported that abalone fed an artificial diet, have better canning characteristics than that of wild abalone, and canning yields have shown an increase of up to 15% [15].

Incorrectly formulated diets, may also lead to the accumulation of waste in the culture system which could cause the deterioration of water quality in the culture environment. The propensity of algal blooms and the proliferation of disease-causing parasites and pathogens increases in the event of waste accumulation due to poor husbandry and poor feed digestibility. The abalone itself then becomes highly susceptible to disease due to these negative conditions in the mariculture water and succumbs to such challenging conditions. Additionally, the digestive systems of these aquatic hosts are in constant contact with the rearing water, making the host more prone to infection.

In conventional mariculture operations, due to the high stocking densities, the generation of elevated stressful conditions in the culture environment is a frequent occurrence [41]. During the sorting process, abalones are presented with further stresses due to excessive handling and may sustain mechanical damage. Both disease and the deterioration of the environmental conditions are the most significant contributors to mass mortalities in mariculture operations [42]. Most operations employ land-based cultivation systems and use pump ashore technology which is energy intensive and costly [15]. The dilution of culture water, to reduce waste concentrations, by increasing flow rates is therefore not a feasible option. Regulatory authorities are also becoming more stringent on the poor quality of farm effluent that is returned to the sea, as a result, preservation of the surrounding environment also becomes a serious challenge to abalone farmers. Bearing in mind that these factors are interactive and ultimately; either as singular occurrences or in combination, may result in decreased production and potential negative impact on the entire aquatic system. Improving digestion, reducing the concentration of waste and disease causing agents in the surrounding water and a heightened immune response are logical mitigation considerations to address the challenges of abalone mariculture. However, classical interventions are costly and mass mortalities continue to occur, resulting in severe setbacks on both economic and social fronts. In more serious instances, some farms have had no other option but to cease operations. The abalone mariculture industry is therefore in dire need of suitable interventions that can address these challenges in an affordable and sustainable manner.

# 3. BIOLOGICAL AGENTS AS AN OPTION TO ADDRESS THE CHALLENGES IN ABALONE AQUACULTURE

During the past two decades, the use of biological agents, particularly in feed and as water additives, as an alternative to the use of antibiotics and chemicals has shown to be promising in aquaculture, particularly in fish and shellfish larviculture [43]. The concept of biological agents has been traditionally associated with the use of beneficial microorganisms to restore the microbial balance in the gastro-intestinal tract of the host and to treat or prevent diseases and/or disorders [44]. Biological agents are emerging as a significant microbial supplements in the field of prophylaxis [36]. Many studies to date have revealed the potential of these beneficial organisms to combat disease in an aquaculture environment [5, 45-51].

In aquatic ecosystems there is an intimate relationship between microorganisms and other biota in the environment [47]. Apart from the aquatic animal being surrounded by water, there is also a constant flow of water through the digestive tract of the aquatic animal. This consequently affects the synergistic balance of indigenous microflora associated with the cultured animal. The classical definition of a probiotic being that of microbes added to food, has become modified with respect to aquaculture, whereby a biological agent is used as a wider term and is defined as "a live microbial adjunct which has a beneficial effect on the host by modifying the host-association or ambient microbial community, by ensuring improved use of the feed or enhancing its nutritional value, by enhancing the host response towards disease, or by improving the quality of its ambient environment" [47]. Some studies have shown that as a result of intensification of aquaculture farms a negative impact has been conferred on the composition of the different protective microbial flora interacting with the host [5]. This occurrence results in an increase in susceptibility of the host to diseases. It has become evident that augmentation of aquaculture systems with biological agents can lead to growth of beneficial bacteria thus improving overall health of the culture system and the host [5].

The use of biological agents in disease control and improvement of aquaculture is important as demand for environmentally friendly aquaculture practices is on the rise. Biological agents that may be applied in aquaculture comprise of isolates belonging to a wide range of yeast, bacteria and even phytoplankton species [52]. In abalone aquaculture, potential probionts listed to date include, *Vibrio* spp., [23,53-55] *Debaryomyces sp*., *Cryptococcus sp*., and *Pseudoalteromonas sp*., [23,24,30], *Lactobacillus* and *Enterococcus sp*., [56]; *Pediococcus* sp. Strain Ab1 [57], *Agarivorans albus* F1-UMA [58] and *Shewanella sp* [51].

Biological agents have been found to confer beneficial effects on the host by various modes of action. These may occur as a singular or combined effect, and thus far the following have been reported; (1) the production of antimicrobial products; (2) competitive exclusion; (3) colonisation of the gut and improving microbial balance; (4) enhancement of the host immune response; (5) detoxification of harmful compounds; (6) improved growth rate of the host;(7) antiviral effects, (8) provision of nutrients and enzymatic functions; and (9) improved water quality. Further reports by [24] stated that the addition of probiotics to the diet of farmed abalone, could possibly lead to a boost in abalone growth by a number of potential strategies. Some of which include (1) increasing the nutrients accessible to the abalone for absorption in the gut, (2) increasing the pool of secreted digestive enzymes in the gut of abalone, and (3) use of bacterial supplements as an additional nutrient source.

In many instances, pathogen inhibition and/or disease control has been observed as a consequence of the release of chemical substances with bactericidal effects by probiotic bacteria [47]. The production of antibiotics, bacteriocins, enzymes, hydrogen peroxide, siderophores and the altering of the pH levels due to the generation of organic acids are all traits displayed by biological agents [47,59-61]. In addition, these biological agents compete with pathogens based on intrinsic growth rate and spacial attachment. Microbial colonisation is characterised by the attachment of the biological agent to the mucosal surface and epithelial cells of the host. This prevents the proliferation of opportunistic pathogens thereby preventing infection [62]. It is common knowledge that for a pathogen to be active and replicate in a host system, it requires attachment to these surfaces [62]. When probiotics are administered over a long period, they successfully colonize the gastrointestinal tract, even after cessation of feed supplemented with probiotics. This occurs since the multiplication rate of these probiotics is higher than the rate at which they are removed, thus a build-up in the intestinal mucosa of the host is observed [62].

Host nutrition is improved, as the applied probionts secrete high levels of hydrolytic enzymes such as amylases, proteases and lipases; as well as the provision of growth factors such as fatty acids, amino acids and vitamins [63,64]. Some isolates also have the ability to break down potentially indigestible components of the feed thus reducing toxicity and improving feed conversion efficiency [23,63]. Abalones are in most instances, fed a diet consisting mainly of kelp, which is a complex macroalgal polysaccharide deficient in many essential nutrients [64]. It is therefore imperative that enteric bacteria in the abalone gut are present in sufficient amounts which will adequately facilitate digestion by supplying highly effective polysaccharolytic enzymes [23]. Many bacteria displaying these properties have been found to exist throughout the digestive tract of *H. midae* [23,40]. Some findings indicated that enteric bacteria isolated from the gastrointestinal tract of abalone were capable of degrading agar, carrageenan, laminarin, and alginate. It was also shown that 70 - 90% of the enzyme activity was extracellular suggesting that bacterial enzymes were secreted into the lumen of the gut where they were able to hydrolyse complex algal polysaccharides [40].

Related studies have indicated that *Debaryomyces hansenii* HF1; isolated from larvae of European bass (*Dicentrarchus labrax*) demonstrated high levels of amylase and trypsin; which aided in the digestion of feed [65]. Similar studies on a combination of 3 potential probiotic strains

(*Agarivorans albus* F1-UMA, *Vibrio sp.* C21-UMA and *Vibrio sp.* F15-UMA) showed significant increases in growth of abalone over a 210 day period [58]. An average monthly improvement in growth of 9.58% of length and 15.94% in weight was observed in relevant test systems. Probiotic organisms persisted in the gut up to a concentration of $10^6$ CFU.$g^{-1}$ and also remained present for 16 to 19 days in juvenile and adult abalone after cessation of feeding with a probiotic supplemented diet. Authors, [40] and [66] also reported that when probiotics were applied to a host, a higher growth rate was observed, as isolated gut bacteria produced enzymes that were able to aid in digestion thus improving the health of abalone.

An inaugural application of probiotics in abalone aquaculture was demonstrated by [23]. They reported that microbes isolated from the gastrointestinal tract of *H. midae* demonstrated an ability to improve digestion, growth and immunity of abalone. From their study it was discovered that *D. hansenii, Cryptococcus sp., V. midae, and Pseudoalteromonas sp.* reside in the intestinal tract of *H. midae* and have the ability to improve the nutritional status of the abalone feed. Further research demonstrated that these probionts were able to breakdown complex proteins and starches, hence making the subsequent assimilation by abalone easier. Studies conducted by [23] indicated that abalones that had been supplemented with probiotics had a survival rate of 62% compared to 25% of untreated abalones; in challenge trials against bacterium *V. anguillarum*. They later formulated a mixture of probiotics using two yeasts and one bacterial strain (SS1, AY1 and SY9) respectively for abalone. The probiont cocktail was added to dry feed to a final concentration of $1 \times 10^7$ cells.$g^{-1}$. The growth rate of small abalone (20 mm) improved by 8% and large abalone (60 mm) increased by 34%. In addition, increases in intestinal proteolytic and amylolytic activity were observed, in probiotic fed abalone when compared to abalone fed the standard feed devoid of probiotics [30].

Lactic acid bacteria (LAB) from different sources and evaluated potential probiotic effects in abalones *in-vitro*, *Lactobacillus sp.* strain a3 and *Enterococcus sp.* strain s6, was isolated by [56], and were shown to inhibit the growth of three abalone pathogens *viz.*, *Listonella anguillarum*, *V. carchariae* and *V. harveyi*. Furthermore these organisms were able to colonize the gut of *Haliotis gigantea* thus enhancing the production of volatile short chain fatty acids (VSCFA) such as acetic acid. They later showed that by supplementing commercially available abalone feed with a potential probiotic organism, *Pediococcus sp.* Ab 1, a change in host intestinal flora was observed. In addition, higher levels of alginate lyase activity and VSCFAs were recorded. All of these factors led to a combined impact by enhancing the growth of the abalone, *H. gigantea* [57].

Studies conducted by [51] revealed that within a week of supplementing the feed of *Haliotis discus hannai Ino* with two probiotic organisms, *Shewanella colwelliana* WA64 and *Shewanella olleyana*WA65, increases in cellular and humoral immune response, higher haemocytes, respiratory burst activity, serum lysozyme activity and total levels of protein were observed. It was therefore concluded that both strains may be used as a dietary probiotic supplement to improve innate immunity and disease resistance in abalone.

Studies on feed probiotics for abalone aquaculture show much promise, however the use of water bioremediation bacteria has been neglected. With intensification of abalone culture activities, increased energy costs of pumping sea water and stricter regulation on environmental pollution, the need for such biological agents will become obvious in the near future. A study based on carp was done by [67] where a consortium of three *Bacillus* isolates demonstrated the ability to reduce diseases and improve water quality. Additionally, studies revealed that when a three organism consortium was added to a culture environment, a decrease in the prevalence of pathogenic bacteria was observed. Moreover it was found that nitrate, nitrite and ammonium concentrations were significantly lower as compared to the control treatments and that the applied treatment did not alter the health, growth and oxygen sufficiency of the test systems negatively. The attractive nature of *Bacillus* spp. as biological agents should be considered for application in abalone mariculture.

# 4. RATIONALE USED FOR THE PRODUCTION OF PROBIOTICS AND BIOLOGICAL AGENTS

The use of biological agents in aquaculture has over the years gained momentum. It is thus, imperative that these micro-organisms be commercially produced in order to meet market needs. A comprehensive production process needs to be developed and optimised for each biological agent. This will facilitate the commercial roll-out of probiotic products of this nature, but will be largely dependent on (1) the efficiency of the production process and (2) the ability to produce large quantities of the probiotic in a suitable form with practical shelf stability [69]. Important criteria influencing the commercial use of biological products are cost, efficacy, shelf life and convenience to the end user [70,71]. The cultivation of microorganisms at a large scale is influenced by various factors such as the composition of the media, physical and chemical variables, substrate feed, oxygen availability and many others [72]; each of which have to be optimized to ensure a cost effective production process.

The growth medium that is used to support high productivities in commercial bioprocesses is predominantly formulated with inexpensive nutrient sources [73]. The choice of medium to be used in production is an essential aspect of process development as it influences the economic competitiveness of the bioprocess technology [74]. Nutrient sources generally play a dominant role in the productivity of the production process since they supply nutrient and growth factors that are directly linked with the formation of biomass and metabolites [75]. It has been suggested that economical and commercially available medium options be investigated in order to reduce production costs [76,77]. The growth medium used can be either a defined or undefined medium. A defined medium has known quantities of all the ingredients that constitute the formulation. An undefined medium contains complex ingredients such corn steep liquor (CSL), which consist of a mixture of chemicals in unknown quantities that vary according to supplier and production batches. The undefined medium option is usually applied in industrial processes based on its low cost [74].

Yeast extract is a commonly used growth medium component, and has been used extensively in many production processes. It is an important nitrogen and nutrient source as it contains an array of amino acids, vitamins and other growth factors required for microbial growth [78-80]. Several studies have indicated that high cell yields and productivities have been obtained with the use of yeast extract in various production processes [81-83]. However, the disadvantage of using this nutrient source is that it results in high-priced production processes due to its associated cost [78]. The use of yeast extract in a production process is therefore regarded as a major technical hurdle that should be overcome in order to successfully minimize production costs [79]. Other nutrient sources that have been used include casamino acids and peptone, which are produced via the enzymatic digestion of meat. The use of these products results in expensive production processes; even though these have been shown to be highly effective nutrient sources. Furthermore, these nutrients sources have negative market acceptance as they are animal by-products [80]. Regulations have also exerted significant pressure on the use of these animal by-products, which have limited their availability. It is therefore imperative that cheaper, safer and readily accessible nutrient sources, capable of supporting production of biological agents, be used in order to ensure that a production process is economically attractive.

CSL has been identified as a lower cost, more effective growth medium that can be used in production, in comparison to conventional nutrient substrates such as yeast extract, peptone and casamino acids [50]. It is produced by immersing corn into dilute sulphur oxide during the starch-manufacturing processes and is a major by-product of the corn-starch processing industry [84]. It has also been shown to be a supplementary source of vitamins and nitrogen to the culture medium [85,86]. The use of CSL has had numerous successes in diverse industrial fermentation processes [76] with high cell yields and productivities being major benefits [87]. Other than the assessment of a suitable nitrogen source, alternative carbohydrate sources also need to be reviewed as they play a dominant role in the productivity of a production process. These nutrient sources are directly linked with energy provision for the formation of biomass and metabolites [75]. Different microbes utilise carbohydrate sources in varying ways. Glucose is a relatively expensive carbohydrate source, and its use in large scale process is limited as a result of high subsequent production costs [88]. When developing efficient bioprocesses, attempts are made to obtain economical and commercially available carbohydrate sources such that the production costs are minimised [74,76,77,89,90]. High test molasses (HTM) is a valuable carbohydrate used commercially due to its local availability and cost competitiveness. It has been applied extensively as an alternative carbohydrate source in various production processes [74,76,77,80,90,91]. HTM, unlike conventional molasses, is a purer product form that enhances mass transfer and reduces impurities in a production process. HTM has been used as a carbohydrate source because it consists of glucose, fructose and sucrose. Inverted HTM is also readily accessible, which contains mainly glucose and fructose in equal proportions with a small amount of residual sucrose. Other than being a carbohydrate source, HTM also provides abundant vitamins and other growth factors required for microbial growth [80,92].

In some instances, microorganisms may require vitamins to be present in the cultivation medium which can be found in the supplemented complex nutrient sources, whereas others can be cultivated in a medium devoid of vitamins [93]. Vitamins are growth factors required by most microorganisms for the production of enzyme cofactors [74,94]. Other than vitamins, microorganisms also require trace elements for their growth. Trace elements form part of enzymes and co-factors and they aid in the catalysis of reactions and maintenance of protein structures [74,95]. Supplementation of exotic trace elements and vitamins can be costly and are therefore avoided if cheaper nutrient sources can satisfy the essential requirements for growth of biological agents in production processes.

Other than an influence on growth, the type of growth medium used in a production process also influences physical parameters such as mass transfer and the formation of foam. Growth media rich in nitrogen sources usually result in increased foam formation [96]. In addition the sparging of gas through the growth medium and agitation at high speeds results in excess foam formation, in oxygen intensive processes. This reduces the efficiency of gaseous exchange at the surface of the culture, as a barrier is formed between the culture and the gases present in the headspace of the vessel [97,98]. Additionally, cells and the culture medium can be lost in the foam phase in the event of vessel overflow. The sensitivity of microorganisms to antifoam toxicity is an important factor that must be considered during the development of production processes; as it can result in a significant decrease of the process performance [98].

Once a suitable fermentation medium has been developed, optimization of physiological growth conditions such as temperature, pH and oxygen sufficiency are imperative, in order to successfully produce biological agents on a large scale. Temperature and pH have been reported to be amongst the most important environmental parameters which control the activities and growth rates of many microorganisms as it governs all the physiological processes. The impact of temperature has been observed at the cellular level, and can either increase or decrease the catalytic activity of pertinent metabolic and digestive enzymes [99-102]. It has been reported that the alteration of growth conditions to an unsuitable range can significantly increase the lag phase of a wide range of micro-organisms, which is highly undesirable when designing an efficient bioprocess strategy [104]. Since temperature affects microbial growth rate, it also affects the growth yield of a culture because the relative energy requirements for cell maintenance increases; when growth rates are reduced [105]. pH homeostasis is another important factor that needs to be considered during the growth of microorganisms [106]. For most microorganisms, there is an increase in growth rate between the minimum and the optimum pH levels and a corresponding decrease in growth rate between the optimum and the maximum pH value [107]. It is well known that pH is important in controlling initiation of growth by microorganisms [108]. The effect of pH on growth include: (1) affecting the production and activity of enzyme systems controlling growth and division, (2) altering the solubility of essential nutrients, (3) modifying the permeability of cells to substances essential for growth, (4) changing the nature of cell surfaces of envelope materials and cell morphology, and (5) modifying the composition of the cultivation medium [108-111]. Oxygen sufficiency is an additional factor to be considered in the design of an optimum bioprocess strategy. In high-cell density cultivations, oxygen limitation can be very challenging, and prevents attainment of high cell titres [112]. The method of oxygenation must be given a high degree of consideration, as excessive rates of agitation and sparging will encourage foam formation and initiate cell sheer. On the contrary, inadequate aeration causes oxygen limitation, and has been reported to be highly detrimental to process productivity, in terms of growth rates and product formation as well as cell viability [96,112]

These factors have an impact on the improved yield and productivity of a process and as a result the overall cost of the production process. In addition they also confer information of the functionality of the probiotic once it enters the host environment [50,90]. Therefore, bioprocesses are designed such that the overall process has increased cell yields, productivities and a lowered cost, which ultimately results in a feasible and economically attractive production process. It is essential that these requirements are met to ensure that biological agents can be affordably adopted for use in abalone aquaculture.

# 5. PROCESSING OF PROBIOTICS INTO MARKET ACCEPTABLE PRODUCTS

Once the relevant biological agents have been successfully cultivated at a large scale, the resultant fermentation broth needs to be recovered efficiently to be utilized in subsequent processing and formulation steps [113,114]. The downstream process has a major influence on product commercialization as major constraints in most processes are embedded in harvesting and formulation costs [115,116]. This includes key aspects such as maximising recovery and preservation of viability, which are essential, in terms of applying an effective biological agent, especially in aquatic systems [117]. In addition, it is also vital to ensure that the final product to be administered to the host aligns with end user requirements such as stability, consistency, easy application, efficacy and affordability [115,118,119]. As a consequence; robust cost-effective choices of process steps and ingredients, dictated by the end product characteristics, are necessary to improve the commercial success of newly developed biological agents [115].

The main objective for downstream processing is to minimise the number of unit operations involved in the process, thus reducing overall process and validation costs, while also simplifying ease and economy of process automation [115]. An additional consideration is the final anticipated form of the end product which has implications on the choice of process options while meeting customer expectations (Figure 2) [120]. The downstream process unit operations, completes the process chain; from the upstream fermentation to the end product. It is therefore considered to be an extremely important prerequisite for commercialization of probiotic technologies. Regrettably, published literature regarding downstream processing and formulation for commercially available products is very limited [115,119].

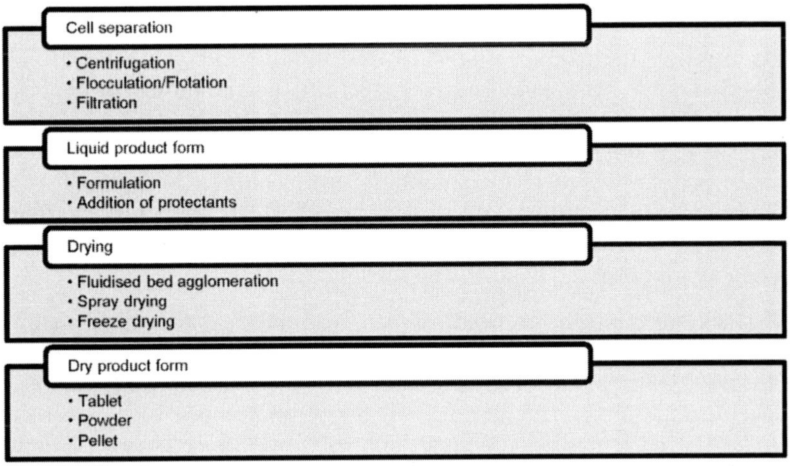


**FIGURE 2.** Schematic illustration of potential downstream process unit options [121].

The harvesting efficacy of a unit operation governs the marketability of a product, as it affects the potency and aids in any potential further processing during formulation and product development. The goal of the recovery process is to produce a product of acceptable quality, in compliance with any regulatory and safety requirements, at an acceptable cost [120]. Process options for cell harvesting from fermentation broth include microfiltration, sedimentation, flocculation and ultrafiltration (Figure 2)[116,122]. Flocculation and flotation using surface action or electrical charge have been reported to be inefficient in the separation of bacterial cells [120]. Although there have been some positive reports for harvesting using ultra-filtration [116], the most widely used process remains centrifugation [123-125]. Among all options, centrifugation appears to be the most viable alternative for cell harvesting resulting in recoveries of ~ 99% [115,126]. Usually, product intermediates are anticipated to be a high cell concentration paste containing the biological agent. Tube centrifugation has been considered to be a useful process that can yield a lower moisture paste thus minimizing the energy required in later stage drying steps if necessary [127,128].

Subsequent to cell separation, formulation generates a crucial link between production and application of probiotics (Figure 2). This crucial step dictates process-ability, economy, shelf life, and efficacy as well as ease of application and provision of a product form that commands customer appeal. Intelligent formulations will allow innovation in application techniques using unique combinations of active ingredients, adjuvants or inerts and the end product should seamlessly integrate into standard food production and farming activities [115,129]. Formulations of biological agents can be broadly classified into either dry solids in various forms, or liquid suspensions or emulsions [130]. The inclusion of additives that enhance process-ability eco-friendliness and customer acceptance of the final product are also important considerations [115]. Any potential impact to the host, environment and even end product consumer, must be thoroughly investigated prior to application [131].

In the case of a probiotic product, the formulation needs to encompass ingredients that aid viability and growth of the cells in its intended application. Sugars and proteins are normally the key nutrients that support the stability of cell preparation. It also further provides a protective layer for the cells, preventing death and assists in the recovery of injured cells during processing [115,132,133]. The addition of nutrients was also shown to improve storage of a *Pseudomonas fluorescens* F113 strain [134] and a *Bacillus megaterium* [135] for use in biocontrol applications. Appropriate formulations can facilitate easier processing and influence the stability and appeal of an end product in large scale production [70].

Processing options for abalone biological agents, will include, both dry and liquid product forms (Figure 2). Due to the intended use of the selected isolates as a living cell preparation, product options with a high stability are considered to be most appealing. The application of fresh cells that need to be routinely produced is not attractive as there is significant risk in ensuring consistent inclusion into the abalone feed [135]. This complicates the processing segment of the technology to a large extent, as innovative ways of ensuring and maintaining cell viabilities are required. Potential options for commercial processes to stabilize these probiotic products include refrigeration, freezing, freeze drying, spray drying and low temperature fluidised bed agglomeration. Refrigerated and frozen cultures occupy large storage volumes and demand higher storage and shipment costs in contrast to dry cultures which are an economic and practical alternative; however, some microorganisms are highly vulnerable to death when any form of drying is carried out [117,136,137]. An alternate approach is to concentrate the product into a convenient dosage quantity and form that reduces the bulk logistics burden for products that are not amenable to drying. Low temperature drying processes such as freeze drying are suitable for higher value, heat labile bio-products, but is costly, time consuming and discontinuous for bulk production compared with moderate temperature drying processes [131,138]. Spray drying processes are widely used for large scale drying of products; however, higher drying temperatures decrease the viability of microbes faster than lower drying temperatures [131,139]. Spray drying requires high temperatures to facilitate water evaporation, which can cause irreversible changes to structural and functional integrity of the intended biological product and reduce viability and activity of the organisms itself [140,141]. Spray drying also has a high energy demand requiring 2500 to 10 000 $J.g^{-1}$ of evaporated water and is therefore not the most attractive process option for drying of abalone probiotics [122].

There are several reports on the use of agglomeration as a commercially viable process option for moderate temperature drying of biological material, mainly due to excellent mass and heat transfer characteristics [133,139,142]. During agglomeration, a mixture is atomised to form droplets at lower temperature (typically 30-40°C) which results in coating of the probiotic cells on the surface of suitable carrier particles. Probiotic cell cultures are subjected to evaporative cooling during the warming up and constant-rate drying periods and therefore have a substantially lower temperature than the drying air, resulting in increased viability [139.]. Advantages of fluidised bed drying over freeze and spray drying include lower investment and maintenance costs, ease of large scale continuous production, rapid exchange of heat, minimising heat damage, rapid mixing providing near isothermal conditions and uniform end product [122,139,143]. Due to these reasons fluidized bed drying has become an accepted method for large scale production of heat labile biological materials, however, viability losses have still been reported [142].

In addition to production and formulation of user-friendly product through a downstream process, the stability and consistency of product intermediates and the end product itself are crucial requirements for successful commercialization [144]. A loss of bioactivity in a product, that is intended to be applied in a viable state, will definitely incur a great deal of process complications and as a result impart a direct increase in production cost [141]. In a typical production process, the lag time between process operations can vary due to process integration and scheduling during manufacture. Thus storage conditions and the addition of specific stabilizers may be required to prevent vegetative growth or the appearance of contamination in the probiotic product or its relevant intermediates [145].

The problems of stability during processing, storage and after application have stalled development of biologically based products [146]. Accelerated aging studies based on the methodology of death rate plots at different temperatures to generate thermal resistance curves have been shown to be a useful technique for predicting stability [147]. Temperature dependant half-life plots can be generated to predict stability of the probiotic product intermediates as well as the end product. This approach has however only been used to a limited extent [121,123,135].

After addressing the considerations of the actual production process, success of the technology is still not a certainty. It is imperative, that in order to realise the success of using this new technology, the probiotic product must be supplied as a live cell preparation, and must be able to survive not only the feed production process, but also maintain viability in the digestive tract of the host [44,47]. Many probiotics have been successfully applied to land-based animal production practices, however, the aquaculture industry are faced with further limitations as a result of continuous water exposure [148]. The method of incorporation selected must overcome challenges faced in feed production and the mariculture system itself; such as major losses of viability, in order to achieve the desired effect of the probiotic technology.

There have been various methods applied to successfully administer viable probiotics to a host in aquaculture environments. These include mixing, soaking, spraying, vacuum infusion, extrusion and bathing [148]. Incorporation of the probiotic into the feed is almost always the method of choice, except when bioremediation agents are added directly to the water. Mixing is the most commonly used method and involves the incorporation of the probiotic into the dry ingredients of the feed during the feed production process. Many researchers [56,150,151] have successfully used this method; however, probiotics that are susceptible to excessive heating and drying during the feed production process do not show high rates of survival [117]. The soaking method uses preformed feed pellets, which are soaked in a saline broth containing the probiotic organism at a desired concentration [57,149]. Soaked pellets are then dried and stored appropriately for further use. A modified method of soaking, whereby actual fronds of macroalgal species, *Macrocystis integrifolia* were soaked in preconditioned tanks containing bacterial cells were used by [58]. Upon aeration of these tanks, the bacteria were allowed to colonize the surface of the fronds, and were thereafter fed to test abalone. In other studies, the spraying of feeds with probiotic cells was also carried out [153]. In addition, [154] described methods of spraying dried feeds with cells that were placed onto plastic trays. Lastly, the bathing option of probiotics involving the application of living cells directly to the rearing water has been used [155]. All potential mechanisms of probiotic inclusion into the feed must be suitably ratified in order to maximize the potential of the technology. The method selected, should have the ability to integrate easily into existing feed production process, and should in no way negatively impact on the host or the rearing environment. The journey taken to produce a commercially viable probiotic product is by no means forthright. It encompasses innovative process design, effective cell production and formulation technologies, as well as successful maintenance of cell viability and stability. Once all the identified challenges have been effectively overcome, the uptake of this technology and the associated boom in abalone export by means of aquaculture will be inevitable.

# 6. CONSIDERATIONS FOR APPLICATION OF BIOLOGICAL AGENTS IN ABALONE AQUACULTURE

Over the past two decades, the applicability of probiotics as solutions to various aquaculture related challenges have been widely reported. However, it is still imperative to consider the safety issues associated with the use of these probiotic products [29,52,156]. Safety is the state of being certain that a biological agent used will not have undesirable effects under defined conditions. The production system in which the cell cultivations are conducted must also maintain high levels of sterility to easily facilitate a monoseptic culture, as well as reduce any potential contamination by common food pathogenic bacteria [68].

Once the selected culture has been accurately identified and deposited into a culture repository, extensive literature searches and relevant scientific experimentation must be carried out in order to obtain information on the biological agent of interest. As the number of isolated probiotic species increase, it is important not to assume biosafety levels and characteristics of each probiotic strain. Furthermore, it is recommended that the exact mode of action of the probiotic organisms be elucidated, in order to achieve the desirable effect when applied to the host system. It has been suggested that prior to incorporation of these organisms in abalone aquaculture, it is important to carefully assess the probiotics for pathogenicity, infectivity, toxicity and their resultant metabolite production for quality assurance [157]. These critical factors have sometimes been overlooked, and have consequently led to the failure of probiotic technologies in some instances [41]. In many case studies, the use of LAB as probiotics have been rendered safe, however, in recent times there have been reports of disease-causing members belonging to *Lactococcus*, *Vagococcus* or *Carnobacterium* families [158]. Additionally, strain testing of potential probiotics should encompass the robustness of the product against process fluctuations under farm conditions and confirmation of non-transmission of drug resistance genes or virulence plasmids [159]. Another barrier preventing the worldwide adoption of this technology, relates to the absence of efficacy data, which as a result casts a shadow of doubt over the technology, thereby hindering its uptake by the aqua-culturists.

# 7. THE IMPACT AND BENEFITS OF THE APPLICATION OF BIOLOGICAL AGENTS IN ABALONE AQUACULTURE

Most aquaculture industries are leaning towards the use of probiotic technology as a solution to many of the challenges faced by the industry. The basis for the inclusion of probiotics into the farming environments include higher survival rate of juvenile and adult abalone, improved feed uptake and conversion ratios resulting in faster growth rates, improved resistance to disease and reduced contribution to water pollution [47]. Using probiotics is more environmentally friendly because the effluent water is cleaner and there are significant improvements in the gut flora thus enhancing the overall immune response of the host and an increase in food assimilation [160]. However other factors such as temperature, enzyme levels, water quality and genetic resistance may have an effect on the success of the technology in the farming facility [29].

Thus far, the uptake of the technology is slow-moving. This is due to the fact that farmers expect the probiotic technology to operate using the same basis as antibiotic treatment technology, in that they require and anticipate fast rapid results [160]. However changes in the microbial ecosystems present in the environment is a gradual one; and requires the continuous addition of beneficial microorganisms to compound the desired effect [158]. In addition, ineffective and costly probiotic products previously offered in aquaculture has negatively tainted the impact of this technology. Some products include *Clostridium spp.*, *Pseudomonas putida* and other potential human pathogens, and others consist of cell densities that are too low to deliver any sort of benefit to the host [160].

The commercial aquaculture sector will make a notable difference in terms creating jobs and economic development in most developing countries by embracing this activity. To date, South Africa has validated itself to be a key player in the abalone mariculture arena. With support from government, this industry could experience a further boom, and as a result, assist in reducing the high levels of unemployment that exists [27], particularly in coastal areas that can effectively participate in aquaculture practices. Abalone industries not only include direct employment at the farm level, it also indirectly supports interlinked businesses such as the seaweed and abalone processing industries [9]. The challenge is to ensure long term sustainable growth of the abalone mariculture industry. The use of appropriate and safe biological agents in abalone mariculture has excellent potential to meet the new challenges of this important industry.

# REFERENCES


1. Marques A., Dinh T., Ioakeimidis C., Huys G., Swings J., Verstraete W., Dhont P., Sorgeloos, P. and Bossier, P. Effects Of Bacteria On *Artemia*Cultured In Different Gnotobiotic Environments. Journal of Applied and Environmental Microbiology 2005; 71(8) 4307-4317.
2. Subasinghe RP. Epidemiological Approach to Aquatic Animal Health Management: Opportunities and Challenges for Developing Countries to Increase Aquatic Production through Aquaculture. Preventative Veterinary Medicine 2005; 67 117-124.
3. FAO. The State Of World Fisheries and Aquaculture. Rome. 2012; 209pp.
4. Britz PJ. A study on the Status of Aquaculture Production and Trade in South Africa. Volume 2: Growth Potential of the South African Aquaculture Industry and Recommendations for Sector Development. A report for the Department of trade and Industry produced by Enviro-fish Africa (Pty.) Ltd. 2007; 24p.
5. Olafsen JA. Interactions between Fish Larvae and Bacteria in Marine Aquaculture. Aquaculture 2001; 200(1-2) 223-247.
6. Gordon HR and Cook PA. World Abalone Fisheries and Aquaculture Update: Supply and Market Dynamics. Journal of Shellfish Research 2004; 23 935-939.
7. Stanford J. Aqua Culture Ambition: South African Aquaculture Industry Is Heading For Export Boom. Engineering News 16-17 August. 2004; 6-12.
8. Reddy-Lopata K., Auerswald L. and Cook P. Ammonia Toxicity and Its Effect on the Growth of the South African Abalone *Haliotis Midae Linnaeus*. Aquaculture 2006; 261(2) 678-687.
9. Troell M., Robertson-Andersson D., Anderson R.J., Bolton J.J., Maneveldt G., Halling C and Probyn T. Abalone Faming In South Africa: An Overview With Perspectives On Kelp Resources, Abalone Feed, Potential For On-Farm Seaweed Production And Socio- Economic Importance. Aquaculture 2006; 257 266-281.
10. Aquaculture Association of South Africa. AASA. www.aasa-aqua.co.za (accessed on 10 May 2013).
11. Raemaekers S and Britz PJ. Profile of the Illegal Abalone Fishery (*Haliotis Midae*) In the Eastern Cape Province, South Africa: Organised Pillage and Management Failure. Fisheries Research 2009; 97(3) 183-195.
12. Degnan SM, Imron Geiger DL and Degnan BM. Evolution in Temperate and Tropical Seas: Disparate Patterns in Southern Hemisphere Abalone (Mollusca: Vetigastropoda: Haliotidae). Molecular Phylogenetics and Evolution 2006; 41(1) 249-256.
13. Geiger DL. A Total Evidence Cladistic Analysis Of The Family Haliotidae (Gastropoda: Vetigastropoda). Ph.D. Thesis, University Of Southern California, Los Angeles. 1999; 423.
14. Geiger DL and Poppe GT and Groh K. Haliotidae. In: (Eds), *Conchological Iconography,* Conchbooks, Hackenheim 2000; 135 83.
15. Western Cape aquaculture development initiative; WCADI (2012) Draft Western Cape aquaculture market analysis and development programme/strategy. Draft Document. 1-167.
16. Fishtech Inc: www.fishtech.com (accessed on 10 May 2013).
17. Oakes FR and Ponte RD. The Abalone Market: Opportunities for Cultured Abalone. Aquaculture 1996; 140(1-2) 187-195.
18. Gordon HR and Cook PA. World Abalone Supply, Markets and Pricing: Historical, Current and Future. Journal of Shellfish Research 2001; 29(3) 569-571.
19. Flores-Aguilar RA., Gutierrez A., Ellwanger AS and Searcy-Bernal R. Development and Current Status of Abalone Aquaculture in Chile. Journal of Shellfish Research 2007 26(3) 705-711.
20. PAUA Industry Council Limited. www.paua.org.nz (Accessed 23 November 2009.).

21. Sales J and Britz PJ. Research on Abalone Cultivation in South Africa. Aquaculture Research 2001; 32(11) 863-874.
22. Evans BS., Sweijd NA., Bowie RCK., Cook PA and Elliott NG. Population Genetic Structure of the Perlemoen *Haliotis Midae* in South Africa: Evidence of Range Expansion and Founder Events. Marine Ecology Progress Series 2004; 270 163-172.
23. Macey BM and Coyne VE. Improved Growth Rate and Disease Resistance in Farmed *Haliotis midae* through Probiotics Treatment. Marine Biotechnology 2005; 245(1-4) 249-261.
24. ten Doeschate K and Coyne VE. Improved Growth Rate in Farmed *Haliotis midae* through Probiotic Treatment. Aquaculture 2008; 284(1-4) 174-179.
25. Aquaculture Annual Report: South Africa (2011) ISBN: 973-1-868-71-355-4. 1-40
26. Andersson D, Troell M, Halling C., Anderson R., Maneveldt G. and Bolton JJ. Abalone (*Haliotis midae*) Farming and Seaweed Harvesting in South Africa: Industry Interdependencies and Socio-Economic Importance. Slideshow Presented at the Phycological Society of Southern Africa Meeting in Maputo, Mozambique, 2006.
27. Shipton T and Britz PJ. A Study on the Status of Aquaculture Production and Trade in South Africa. Volume 1: Industry Status and Diagnostic Report. A Report for the Department of Trade and Industry produced by Enviro-Fish Africa (Pty.) Ltd. 2007 90p.
28. Moriarty DJW. 1999. Disease Control In Shrimp Aquaculture with Probiotic Bacteria, Microbial Interactions in Aquaculture, In: Bell CR, Brylinsky M (Eds) Proceedings of the 8th International Symposium on Microbial Ecology, Canada.
29. Balcázar JL., Blas ID., Ruiz ZI., Cunningham D., Vendrell D., & Múzquiz JL. The Role of Probiotics in Aquaculture. Veterinary Microbiology 2006; 114(3-4) 173-186.
30. Macey MB and Coyne VE. Colonization of the Gastrointestinal Tract of the Farmed South African Abalone *Haliotis midae* by the Probionts *Vibrio midae* SY9, *Cryptococcus Sp.* SS1 and *Debaryomyces hansenii* AY1. Marine Biotechnology 2006; 8 246-259.
31. Jeney Z and Jeney G. Recent Achievements in Studies of Common Carp (*Cyprinus carpio* L.). Aquaculture 1995; 129(1-4) 397-420.
32. Nicolas JL., Basuyaux O and Mazurie J. The bault, a *Vibrio carchariae*, a Pathogen of the Abalone, *Haliotis tuberculata*. Diseases of Aquatic Organisms. 2002; 50 35-43.
33. Bower SM. Update on Emerging Abalone Diseases and Techniques for Health Assessment. Journal of Shellfish Research. 2003; 22 805-810.
34. Gram L, Løvald T., Nielsen J., Melchiorsen J and Spanggaard B. In-vitro Antagonism of Probiont *Pseudomonas Fluorescens* against *Aeromonas Salmonicida* Does Not Confer Protection of Salmon Against Furunculosis. Aquaculture 2001; 199(1-2) 1-11.
35. Nomoto K. Prevention of Infections by Probiotics. Journal of Bioscience and Bioengineering 2005; 100(6) 583-592.
36. Gomez GRD, Balcazar JL and Shen MA. Probiotics as Control Agents in Aquaculture. Journal of Ocean University of China 2007; 6(1) 76-79.
37. Moriarty, DJW. Control of Luminous *Vibrio* Species in Aquaculture Ponds. Aquaculture 1998; 164(1-4) 351-358.
38. Ronson P.J and Medina R. *Probioticos en al Acuicultura, Ciencia y Mar Notas.* 2002.
39. Naidoo K, Maneveldt G, Ruck K and Bolton JJ. A Comparison of Various Seaweed-Based Diets and Formulated Feed on Growth Rate of Abalone in a Land-Based Aquaculture System. Journal of Applied Phycology 2006; 18(3-5) 437-443.
40. Erasmus JH, Cook PA and Coyne VE. The Role of Bacteria in the Digestion of Seaweed by the Abalone *Haliotis midae*. Aquaculture 1997; 155 377-386.
41. Wang Y, Li J, and Lin J. Probiotics in Aquaculture: Challenges and Outlook. Aquaculture. 2008; 281(1-2) 1-4.
42. Bondad-Reantuso MG, Subasinghe RP and Arthur JR. Disease and Health Management in Asian Aquaculture. Veterinary Parasitology 200; 132(3-4) 249-272.
43. Tinh NTN, Dierckens K, Sorgeloos P. and Bossier P. A Review of the Functionality of Probiotics in the Larviculture Food Chain. Marine Biotechnology. 2007; 10 1-12.

44. Gatesoupe FJ. The Use of Probiotics in Aquaculture. Aquaculture 1999; 180(1-2) 147-165.
45. Rengpipat S, Rukpratanporn S, Piyatiratitivorakul S and Menasaveta P. Immunity Enhancement in Black Tiger Shrimp (*Penaeus monodon*) by a Probiont Bacterium (*Bacillus* S11). Aquaculture 2000; 191(4) 271-288.
46. Robertson PAW, O'Dowd C, Burrells C, Williams P and Austin B. Use of *Carnobacterium* sp. As a Probiotic for Atlantic Salmon *Salmo salar* L. and rainbow Trout *Oncorhynchus mykiss,* Walbaum. Aquaculture 2000; 185(3-4) 235-243.
47. Vershuere L, Rombant G, Sorgeloos P. and Verstraete W. Probiotic Bacteria as Biological Agents in Aquaculture. Microbiology and Molecular Biology Reviews 2000; 64(4) 655-671.
48. Chythanya R, Karunasagar I and Karunasagar I. Inhibition Of Shrimp Pathogenic *Vibrios* By a Marine *Pseudomonas* I-2 Strain. Aquaculture 2002; 208(1-2) 1-10.
49. Bautista-Teruel MN, Fermin AC and Koshio SS. Diet Development and Evaluation for Juvenile Abalone, *Haliotis asinina*: Animal and Plant Protein Sources. Aquaculture 2003; 219(1-4) 645-653.
50. Lalloo R, Maharajh D, Gorgens J and Gardiner N. Functionality of a *Bacillus Cereus* Biological Agent in Response to Physiological Variables Encountered in Aquaculture. Applied Microbiology and Biotechnology 2008; 79 112- 113.
51. Jiang HF, Lui XL. Chang YQ, Lui MT and Wang GX. Effects of Dietary Supplementation of Probiotic *Shewanella colwelliana* WA64, *Shewenella olleyana* WA65 on the Innate Immunity and Disease Resistance of Abalone, *Haliotis discus hannai Ino.* Fish and Shellfish Immunology 2013; 1-4.
52. Kesarcodi-Watson A, Kaspar H, Lategan MJ and Gibson L. Probiotics in Aquaculture: The Need, Principles and Mechanisms of Action and Screening Processes. Aquaculture 2008; 274(1) 1-14.
53. Sawabe T, Sugimura I, Ohtsuka M, Nakano K, Tajima K, Ezura Y and Christen R. *Vibrio halioticoli* sp. nov., A Non-Motile Alginolytic Marine Bacterium Isolated From The Gut of the Abalone *Haliotis discus hannai*. International Journal of Systematic Bacteriology 1998; 48(2) 573-580.
54. Tanaka R, Sugimura I, Sawabe T, Yoshimizu M and Ezura Y. Gut Microflora of Abalone *Haliotis discus hannai* in Culture Changes Coincident with a Change in Diet. Fish Science 2003; 69(3) 951-958.
55. Sawabe T, Inoue S, Fukui Y, Yoshie K, Nishihara Y and Miura H. Mass Mortality of Japanese Abalone *Haliotis discus hannai* Caused by *Vibrio harveyi* Infection. Microbes Environment 2007; 22(3) 300-308.
56. Iehata S, Inagaki T, Okunishi S, Nakano M, Tanaka R and Maeda H. Colonization and Probiotic Effects of Lactic Acid Bacteria in the Gut of the Abalone *Halotis gigantea.* Fish Science 2009; 75 1285-1293.
57. Iehata S, Ingaka T, Okunishi S, Nakano M, Tanaka R and Maeda H. Improved Gut Environment ff Abalone *Haliotis gigantea* Through *Pediococcus*sp. Ab1 treatment. Aquaculture 2010; 245 249-261.
58. Silva-Aicares FR, Carvajal PO, Mejías CA and Riquelme CE. Use of Macroalgae Supplemented with Probiotics in the *Haliotis rufescens* Culture in Northern Chile. Aquaculture Research 2011; 42 953-961.
59. Hong HA, Duc LH and Cutting SM. The use of bacterial spore formers as probiotics. Federation of European Microbiological Societies Microbiology Review 2005; 29(4) 813-835.
60. Lalloo R, Moonsamy G, Ramchuran S, Görgens and Gardiner N. Competitive Exclusion as a Mode of Action of a Novel *Bacillus cereus* Aquaculture Biological Agent. Letters in Applied Microbiology 2010; 50(6) 563-567.
61. Pandiyan, P., Deivasigamani, B., Thirunavukkarasu, R., George, E.G.J., Subaramaniyan, K., Manikkam, S and Sadayappan, B. Probiotics in aquaculture. Drug Invention Today 2003; 5(1) 55-59.

62. Ouwehand AC, Kirjavainen PV, Grönlund MM, Isolauri E and Salminen S. Adhesion of Probiotic Microorganisms to Intestinal Mucus. International Dairy Journal 1999; 9(9) 623-630.
63. Irianto A and Austin B. 2002. Probionts in Aquaculture. Review Journal of Fish Diseases 2002; 25(11) 633-642.
64. Ghosh S, Sinha A and Sahu C. Effect of Probiotic on Reproductive Performance in Female Livebearing Ornamental fish. Aquaculture Research 2007; 38 518-526.
65. Simpson BJA. An Investigation of the Diet Management Strategies for the Culture of the South African Abalone, *Haliotis midae*. MSc Thesis. University of Cape Town; 1994.
66. Tovar D, Zambonino J, Cahu, Gatesoupe FJ, Vázquez-Juárez R and Lèsel R. Effect of Live Yeast Incorporation in Compound Diet on Digestive Enzyme Activity in Sea Bass (*Dicentrachus labrax*) Larvae. Aquaculture 2002; 204(1-2) 113-123.
67. El-Shanshoury AR, Mona MH, Shoukr FA and El-Bossery, AM. The Enumeration and Characterization of Bacteria and Fungi Associated with Marine Wood-Boring Isopods, and the Ability of these Microorganisms to Digest Cellulose and Wood. Marine Biology 1994; 119 321-326.
68. Lalloo R, Ramchuran S, Ramduth D, Görgens J and Gardiner N. Isolation and Selection of *Bacillus* Spp. as Potential Biological Agents for Enhancement of Water Quality in Culture of Ornamental Fish. Journal of Applied Microbiology 2007; 103(5) 1471-1479.
69. Patel VJ, Tendulkar SR and Chattoo BB. Bioprocess Development for the Production of an Antifungal Molecule by *Bacillus licheniformis* BC98. Journal of Biosciences and Bioengineering 2008; 98(4) 231-235.
70. Amer GA and Utkhede RS. Development of Formulations of Biological Agent for Management of Root Rot of Lettuce and Cucumber. Canadian Journal of Microbiology 2000; 46(9) 809-816.
71. Keller K, Friedmann T and Boxman A. The Bioseparation Needs for Tomorrow. Trends in Biotechnology 2001; 19(11) 438-441.
72. Maharajh D, Roth R, Lalloo R, Simpson C, Mitra R, Gorgens J and Ramchuran S. Multi-Copy Expression And Fed Batch Production Of *Rhodotorula araucariae* Epoxide Hydrolase in *Yarrowia lipolytica*. Applied Microbiology and Biotechnology 2008; 79 235-244.
73. Zhang J and Greasham R. Chemically Defined Media for Commercial Fermentations. Applied Microbial Biotechnology 1999; 51 407-421.
74. Prescott LM, Harley JP and Klein DA. Microbiology. New York: Mc Graw Hill 2005.
75. Lopez JLC, Perez JAS, Sevilla JMF, Fernandez FGA, Grima EM and Chisti Y. Production of Lovastatin by *Aspergillus terreus*: Effects of the C:N Ratio and the Principal Nutrients on Growth and Metabolite Production. Enzyme and Microbial Technology 2003; 33(2-3) 270-277.
76. Lawford HG and Rousseau JD. Corn Steep Liquor as a Cost Effective Nutrition Adjunct in High-Performance *Zymomonas* Ethanol Fermentations. Applied Biochemistry and Biotechnology 1997; 63-65 287-304.
77. Kona RP, Qureshi N and Pai JS. Production of glucose oxidase using Aspergillus niger and corn steep liquor. Bioresource technology 2001; 78(2) 123-126.
78. Payot T, Chemaly Z and Fick M. Lactic Acid Production by *Bacillus coagulans* – Kinetic Studies and Optimization of Culture Medium for Batch and Continuous Fermentations. Enzyme and Microbial Technology 1998; 24(3-4) 191-199.
79. Salgado JM, Rodriguez N, Cortes S and Dominguez JM. Development of Cost-Effective Media to Increase the Economic Potential for Larger-Scale Bioproduction of Natural Food Additives by *Lactobacillus rhamnosus*, *Debaryomyces hansenii* and *Aspergillus niger*. Journal of Agriculture and Food Chemistry 2009; 57(21) 10414-10428.
80. Lalloo R, Maharajh D, Görgens J and Gardiner N. High-Density Spore Production of a *Bacillus cereus* Aquaculture Biological Agent by Nutrient Supplementation. Applied Microbiology and Biotechnology 2009; 83 59-66.

81. Nohata Y and Kurane R. Complete Defined Medium for Large-Scale Production of Polysaccharide Bioabsorbent from *Alcaligenes latus* B-16. Journal of Fermentation and Bioengineering 1997; 83(1) 116-117.
82. Vuolanto A, von Weymarn N, Kerovuo J, Ojamo H and Leisola M. Phytase Production by High Cell Density Culture or Recombinant *Bacillus subtillus*. Biotechnology Letters 2001; 23 761-766.
83. Prabakaran G, Balaraman K, Hoti SL and Manonmani AM. A Cost-Effective Medium for the Large-Scale Production of *Bacillus sphaericus* H5a5b (VCRC B42) for Mosquito Control. Biological Control 2007; 41(3) 379-383.
84. Niwa T, Doi U, Kato Y and Osawa T. Antioxidative Properties of Phenolic Antioxidants Isolated From Corn Steep Liquor. Journal of Agricultural and Food Chemistry 2001; 49(1) 177-182.
85. Amartey S. and Jeffries TW. Comparison of Corn Steep Liquor with Other Nutrients in the Fermentation of D-Xylose by *Pichia stipitis* CBS 6054. Biotechnology Letters 1994; 16(2) 211-214.
86. Silveira MM, Wisbeck E, Hoch I and Jonas R. Production Of Glucose-Fructose Oxidoreductase and Ethanol by *Zymomonas mobilis* ATCC 29191 in Medium Containing Corn Steep Liquor as a Source of Vitamins. Applied Microbial Biotechnology 2001; 55 442-445.
87. Srivastava RAK, Baruah JN. Culture conditions for production of thermostable amylase by *Bacillus stearothermophilus*. Applied Environmental Microbiology 1986; 52(1) 179-184.
88. Yahiro K, Shibata S, Jia SR, Park YS and Okabe M. Efficient Itaconic Acid Production from Raw Corn Starch. Journal of Fermentation Bioengineering 1997; 84(4) 375-377.
89. Lee SY and Chang HN. Production of Poly(hydroxyalkanoic acid). Advances in Biochemical Engineering 1995; 52 27-58.
90. Burkert JFM, Maugeri F and Rodrigues MI. Optimization of Extracellular Lipase Production by *Geotrichum sp.* using Factorial Design. Bioresource Technology 2004; 91(1) 77-84.
91. Gouda MK, Swellam AE, Omar SH. Production of PHB by a *Bacillus megaterium* Strain Using Sugarcane Molasses and Corn Steep Liquor as Sole Carbon and Nitrogen Sources. Microbiology Research 2001; 156(3) 201–207.
92. Malathi S. and Chakraborty R. Production of Alkaline Protease by a New *Aspergillus flavus* Isolate Under Solid-Substrate Fermentation Conditions for Use as a Depilation Agent. Applied Environmental Microbiology 1991; 57(3) 712-716.
93. Kask S, Laht TM, Pall T. and Paalme T. A Study on Growth Characteristics and Nutrient Consumption of *Lactobacillus plantarum* In A-Stat Culture. Antonie van Leeuwenhoek 1999; 75 309-320.
94. Chan ECS. Microbial Nutrition and Basic Metabolism: in the Handbook of Water and Wastewater Microbiology. London UK: Academic Press. 2003.
95. Burrows W. The Nutritional Requirements of Bacteria, Chicago: The University of Chicago Press. 1936.
96. Morao A, Maia CI, Fonseca MMR, Vasconcelos JMT and Alves SS. Effect of Antifoam Addition on Gas-Liquid Mass Transfer in Stirred Fermenters. Bioprocess Engineering 1999; 20(2) 165-172.
97. Huang TK, Wang PM. and Wu WT. Cultivation of *Bacillus thuringiensis* in an Airlift Reactor with Wire Mesh Draft Tubes. Biochemical Engineering Journal 2001; 7(1) 35-39.
98. Holmes W, Smith R. and Bill R. Microbial Cell Factories, Aston Academy of Life Sciences, Birmingham, UK, Aston University, Aston Triangle. 2006.
99. Ratkowsky DA, Olley J, McMeekin TA and Ball A. Relationship between Temperature and Growth Rate of Bacterial Cultures. Biotechnology 1982; 149(1) 1-5.
100. Mayo AW and Noike T. Effects of Temperature and pH on the Growth of Heterotrophic Bacteria in Waste Stabilisation Ponds. Water Research 1995; 30(2) 447-455.

101. Betts GD, Linton P. and Betteridge RJ. Synergistic Effect of Sodium Chloride, Temperature and pH on Growth of a Cocktail of Spoilage Yeast: A Research Note. Food Microbiology 2000; 17(1) 47-52.
102. Kurita O and Yamazaki E. Growth under Alkaline Conditions of Salt Tolerant Yeast *Debaryomyces hansenii* IFO 10939. Current Microbiology 2002; 45 277-280.
103. Bonaiti C, Leclercq-Perlat MN, Latrille E and Corrieu G. Deacidification by *Debaryomyces hansenii* of Smear Soft Cheeses Ripened under Uncontrolled Conditions: Relative Humidity and Temperature Influences. Dairy Science 2004; 87(11) 3976-3988.
104. Xiang G, Li J, Duan J, Shao F, Xu J, Fu, S, and Gong H. Acceleration Effect of Amino Acid Supplementation on Glycerol Assimilation by *Escherichia Coli* in Minimal Medium. Biotechnology Letters 2013; DOI 10.1007/s10529-013-1232-4.
105. Knoblauch C. and Jorgensen BB. Effect of Temperature on Sulphate Reduction, Growth Rate and Growth Yield in Five Psychrophilic Sulphate Reducing Bacteria from Arctic Sediments. *Environmental Microbiology* 1999; 1 457-567.
106. Hutkins RW. and Nannen NL. pH Homeostasis in Lactic Acid Bacteria. Journal Series Nebraska Agricultural Experiment Station 1992; 76 2354-2365.
107. Todar K. Online Textbook of Bacteriology. University of Wisconsin-Madison Department of Bacteriology. 2007.
108. Lichstein HC. Symposium on Initiation of Bacterial Growth III. Physiological Aspects of Growth Initiation. Bacteriology Reviews 1959; 23 261-266.
109. Cochrane VW. Physiology of Fungi. John Wiley and Sons, Inc. New York. 1958.
110. Mitchell P. Physical Factors Affecting the Growth and Death in Bacterial Physiology Werkman CH and Wilson PW. New York: Academic Press, Inc. 1951.
111. Pirt SJ. Principles of Microbe and Cell Cultivation. Oxford, London: Blackwell Scientific Publications, Inc. 1975.
112. Fike R. Nutrient Supplementation Strategies for Biopharmaceutical Production, Part 3. Scaling Strategies for Rapid Nutrient Supplement Prototyping. 2010. Bioprocess Technical, Bioprocess International, January 2010; 24-31.
113. Tsun HY, Liu CM and Tzeng YM. Recovery and Purification of Thuringiensin from the Fermentation Broth of *Bacillus thuringiensis*. Bioseparation 1999; 7 309-316.
114. Rowe GE. and Margaritis A. Bioprocess Design and Economic Analysis for the Commercial Production of Environmentally Friendly Bio-Insecticides from *Bacillus thuringiensis* HD-1 Biotechnology and Bioengineering 2004; 86(4) 377-388.
115. Brar SK, Verma M, Tyagi RD and Valéro JR. Recent Advances in Downstream Processing and Formulations of *Bacillus thuringiensis* based on Biopesticides. Process Biochemistry 2006; 41(2) 323-342.
116. Prabakaran G, Balaraman K, Hoti SL. and Manonmani AM. A Cost-Effective Medium for the Large-Scale Production of *Bacillus sphaericus* H5a5b (VCRC B42) For Mosquito Control. Biological Control 2007; 41(3) 379-383.
117. Lacroix C. and Yildirim S. Fermentation Technologies for the Production of Probiotics with High Viability and Functionality. Current Opinion in Biotechnology 2007; 18 176-183.
118. Burges HD. and Jones KA. Formulation of Bacteria, Viruses and Protozoa to Control Insects H.D. Burges (Ed.), Formulation of Microbial Biopesticides: Beneficial Microorganisms, Nematodes and Seed Treatments, Kluwer, Dortrecht 1998. pp. 33-127.
119. Schisler DA, Slininger PJ, Behle RW, Jackson MA. Formulation of *Bacillus spp.* for Biological Control of Plant Diseases. Phytopathology 2004; 94 1267-1271.
120. Lydersen BK, D'Elia NA, Nelson KL. Bioprocess Engineering: Systems, Eequipment and Facilities. Chemical Engineering Science. 1994; 50(6) 1069 1070
121. Lalloo R, Maharajh D, Gorgens J. and Gardiner N. A Downstream Process for Production of A Viable and Stable *Bacillus Cereus* Aquaculture Biological Agent. Applied Microbiology and Biotechnology 2010; DOI 10.1007/s00253-009-2294-2.
122. Luna-Solano G, Salgado-Cervantes MA, Rodrígu

123. ez-Jimenes GC and García-Alvarado MA. Optimization of Brewer's Yeast Spray Drying Process. Journal of Food Engineering 2005; 68(1) 9-18.
124. Puziss M, Manning LC, Lynch JW, Barclay E, Abelow I, Wright GG. Large-Scale Production of Protective Antigen of *Bacillus anthracis* in Anaerobic Cultures. Applied Microbiology 1963; 11(4) 330-334.
125. Zamola B, Valles P, Meli G, Miccoli P, Kajfez F. Use Of The Centrifugal Separation Technique in Manufacturing a Bioinsecticide Based on *Bacillus thuringiensis.* Biotechnology Bioengineering 1981; 23(5) 1079-1086.
126. Rojas JV, Gutierrez E, de la Torre M. Primary Separation of the Entomopathogenic Products Of *Bacillus thuringiensis*. Biotechnology Progress 1996; 12(4) 564-566.
127. Torres-Anjel MJ and Hedrick TI. Spore Removal by Centrifugation and Its Effect on Ultra-High Temperature Commercial Sterilization of Milk. Journal of Dairy Research 1970; 54(3) 326-330.
128. Rivière J.Industrial applications of microbiology. Wiley, London 1977
129. Berovic M. New Products and New Areas of Bioprocess Engineering. Chemical Engineering Science 1998; 50(6)1069-1070
130. de Medeiros FPM, de Melo Santos MAV, Regis L, Rios EMM, Neto PJM. Development of a *Bacillus sphaericus* Tablet Formulation and its Evaluation as a Larvicide in the Biological Control of *Culex Quinquefasciatus.* Memórias do Instituto Oswaldo Cruz 2005; 100(4) 431-434.
131. Rhodes DJ. Exploitation of Microorganisms, London: Chapman & Hall; 1993.
132. Werner L, Latzko F, Hampel W. Spray Drying of Yeast-Lytic Enzymes From *Arthrobacter* sp. Biotechnology Techniques 1993; 7(9) 663-666.
133. Costa E, Teixidó N, Usall J, Fons E, Gimeno V, Delgado J, Viñas I. Survival of *Pantoea agglomerans* Strain CPA-2 in Spray-drying Process. Journal of Food Protection 2001; 65(1) 185-191.
134. Larena I, de Cal A, Liñán M, Melgarejo P. Drying of *Epicoccum nigrum conidia* for Obtaining a Shelf-Stable Biological Product against Brown Rot Disease. Journal of Applied Microbiology 2003; 94(3) 508-514.
135. Moene-Loccoz Y, Tichy HV, O'Donnell A, Simon R and O'Gara F. Impact of 2,4-Diacetylphloroglucinol-Producing Biocontrol Strain *Pseudomonas fluorescens* F113 on Intraspecific Diversity of Resident Culturable Fluorescent Pseudomonads Associated with the Roots of Field-Grown Sugar Beet Seedlings. Applied Environmental Microbiology. 1999; 67(8) 3418-3425.
136. Wiwattanapatapee R, Pengnoo A, Kanjanamaneesathian M, Matchavanich W, Nilratana L, Jantharangsri A. Floating Pellets Containing Bacterial Antagonists for Control Sheath Blight of Rice: Formulations, Viability and Bacterial Release Studies. Journal Controlled Release 2004; 95(3) 455-462.
137. Klein N. and Lortal S. Attenuated Starters: An Efficient Means to Influence Cheese Ripening-A Review. International Dairy Journal 1999; 9(11) 751-762.
138. Meng XC, Stanton C, Fitzgerald GF, Daly C, Ross RP. Anhydrobiotics: The Challenges of Drying Probiotic Cultures. Food Chemistry 2008; 106 1406-1416.
139. Knorr D. Technology Aspects Related to Microorganisms in Functional Foods. Trends in Food Science and Technology 1998; 9(8-9) 295-306.
140. Bayrock D. and Ingledew WM. Mechanism of Viability Loss during Fluidized Bed Drying of Baker's Yeast. Food Research International 1997; 30(6) 417-425.
141. Tamez-Guerra P, McGuire MR, Medrano-Roldan H. and Galan-Wong LJ. Sprayable Granule Formulations of *Bacillus thuringiensis.* Journal of Economic Entymology 2000; 93(2) 219-225.
142. Chen XD, and Patel KC. Microorganism Inactivation During Drying of Small Droplets or Thin-Layer Slabs-A Critical Review of Existing Kinetics Models and an Appraisal of the Drying Rate Dependent Model. Journal of Food Engineering 2007; 82(1) 1–10.
143. Grabowski JA, Truong VD. and Daubert CR. Spray-Drying of Amylase Hydrolyzed Sweet Potato Puree and Physicochemical Properties of Powder. Journal of Food Science 2009; 71 (5) E209–217.

144. Mille Y, Obert J, Beney L. and Gervais P. New Drying Process for Lactic Bacteria Based on Their Dehydration Behaviour in Liquid Medium. Biotechnology and Bioengineering 2004; 88(1) 71-76.
145. Lewis JA. and Papavizas GC. Biocontrol of Plant Diseases: The Approach for Tomorrow. Crop Protection 1991; 10(2) 95-105.
146. Soper RS. and Ward MG. Beltsville Symposia In Agricultural Research. Biological Control in Crop Production 1981; 5 161-180.
147. Tsuji K, Watanuki T, Kondo F, Watanabe MF, Nakazawa H, Suzuki M, Uchida H and Harada K. Stability of Microcystins from Cyanobacteria-iv. Effect of Chlorination on Decomposition. Toxicon 1997; 35(7) 1033-1041.
148. Ramaswamy H. and Marcotte M (Eds.), Food Processing: Principles and Applications, CRC Press, Taylor and Francis Group, Boca Raton, FL, USA, 2006 (xvi C420 ISBN 1-58716-008-0)
149. Venkat HK, Sahu NP. and Jain KK. Effect of Feeding *Lactobacillus*-Based Probiotics on the Gut Microflora, Growth And Survival Of Postlarva of *Macrobrachium rosenbergii* (de Man). Aquaculture Research 2004; 35(5) 501-507.
150. Keysami MA, Mohammadpor M. and Saad CR. Probiotic Activity of *Bacillus subtilis* In Juvenile Freshwater Prawn, *Macrobrachium rosenbergii*(de Man) at different Methods of Administration to the Feed. Aquaculture International. 2012; 20 499-511.
151. Nikoskelainen S, Ouwenhand A, Salminen S. and Bylund G. Protection of Rainbow Trout *Oncorhynchus Mykiss* from Furunculosis by *Lactobacillus rhamnosus.* Aquaculture 2001; 198(3-4) 229-236.
152. Meunpol O, Lopinyosiri K. and Menasveta P. The Effects of Ozone and Probiotics on The Survival of Black Tiger Shrimp (*Penaeus monodon*). Aquaculture 2003; 220(1-4) 437-448.
153. Robertson, P.A.W., O'Dowd, C., Burrells, C., Williams, P and Austin, B. Use of Carnobacterium sp. as a probiotic for Atlantic salmon *Salmo salar* L. and rainbow trout *Oncorhynchus mykiss*, Walbaum. Aquaculture 2000; 185(3-4) 235–243.
154. García de La Banda, I., Loboa, C., León-Rubiob, J.M., Tapia-Paniagua, S., Balebona, M.C., Moriñigo, M.A., Moreno-Ventas, X., Lucas L.M., Linares,F., Arce, F., and Arijo S. Influence Of Two Closely Related Probiotics On Juvenile Senegalese Sole (*Solea Senegalensis*, Kaup 1858) Performance And Protection Against *Photobacterium Damselae Subsp. Piscicida.* Aquaculture 2012; 306 281-288.
155. Gildberg A. and Mikkelsen H. Effects of Supplementing the Feed of Atlantic Cod (*Gadus morhua*) Fry with Lactic Acid Bacteria and Immune-Stimulating Peptides During A Challenge Trial With *Vibrio anguillarum*. Aquaculture 1998; 167(1-2) 103-113.
156. Gullian M, Thompson F. and Rodriguez J. Selection of Probiotic and Study of the Immunostimulatory Effect on *Penaeus vannamei.* Aquaculture 2004; 233(1-4) 1-14.
157. Courvalin P. Antibiotic Resistance: The Pros and Cons of Probiotics. Digestive and Liver Disease 2006; 38(2) S261-S265.
158. Ishibashi N. and Yamazaki S. Probiotics and Safety. American Journal of Clinical Nutrition 2001; 73(2) S465-S470.
159. Ringo E. and Gatesoupe FJ. Lactic Acid Bacteria in Fish: A Review. Aquaculture 1998; 160(3-4) 177-203.
160. Moubareck C, Gavini F, Vaugien L, Butel MJ. and Doucer-Popularie F. Antimicrobial susceptibility of *Bifidobacteria.* Journal of Antimicrobial Chemotherapy 2005; 55(1) 38-44.
161. Moriarty DJW, Decamp O. and Lavens P. Probiotics In Aquaculture. Shrimp Culture management in AQUA Culture Asia Pacific Magazine. September/October 2005. 14-16

# CHAPTER 12

# Fisheries and Aquaculture in Bangladesh: Challenges and Opportunities

**Bishwajit Ghose**

*Institute of Nutrition and Food Science, University of Dhaka, Ramna 1000, Dhaka, Bangladesh*

## ABSTRACT

The purpose of this paper is to review the performance of fisheries sector in Bangladesh and the challenges it is facing. Data and information were sourced from the publication of the Department of Fisheries (DoF) and related non-published grey literature. Bangladesh is predominantly an agrarian economy and is naturally endowed with a huge sweet water resources and the world's longest continuous sea beach. With the world's largest flooded wetland, the third largest aquatic biodiversity in Asia behind only to China and India, Bangladesh is considered as one the most suitable region for aquaculture and fisheries in the world. The country has an inland water area of about 45,000 km2 and about 710 km long coastal belt. Given this extensive water resource, it is evident that fisheries play an important role in the economy and the diet of the population. Fish and fish products supply 60 percent of animal protein and around three percents of total export earnings. In recent years, however, the fisheries sector is confronted with challenges posed by numerous natural and anthropogenic causes such as climate change, natural disasters, unbalanced urbanization and industrialization, overfishing and environmental pollution. The combined effect of these factors is posing significant threat to the income and food security of the population and urges for immediate actions by government and policymakers.

**Keywords:** Fisheries; Bangladesh; Food security; Nutrition security; water pollution; Community based fisheries management

## ABBREVIATIONS

CBFM: Community Based Fisheries Management; CBOs: Community Based Organizations; DfID: Department For International Development; IUCN: International Union for Conservation of Nature; DOF: Department of Fisheries; NGOs: Non-Governmental Organizations

## INTRODUCTION

Fish is the second most valuable agricultural crop in Bangladesh and its production contributes to the livelihoods and employment of millions of people. The culture and consumption of fish therefore has important implications for national income and food security. Bangladeshi people are popularly referred to as "Mache Bhate Bangali" or "fish and rice makes a Bengali".

The fisheries sector in Bangladesh is broadly divided into four sub-sectors- inland capture, inland culture, mariculture (artisanal fisheries) and marine industrial fisheries. Bangladesh is one of the world's leading inland fisheries producers and has a huge water resource all over the country in the form of small ponds, ditches, lakes, canals, small and large rivers, and estuaries covering about 4.34 million hectares. Freshwater aquaculture involves pond aquaculture especially the polyculture of native and exotic species. The country also has a coastal area of 2.30 million ha and a coastline of 714 km along the Bay of Bengal, which supports a large artisanal and coastal fisheries. Bangladesh is considered one of the most suitable countries in the world for farming of freshwater prawn (Macrobrachium rosenbergii) [11].

Bangladesh is one of the resourceful countries with its wide range of marine aquatic bio-diversities. There are about 1093 marine aquatic organisms where 44.35% are finfish, 32.23% shellfish, 15.10% seaweeds and only 8.32% are other organisms including shrimps. In 2011, an area of around 276,492 ha was used for prawn and shrimp cultivation [3]. Aquaculture production in Bangladesh has shown an average growth of 28 percent from 0.12 million tonnes to 0.66 million tonnes during the period 1985 to 2000. Despite that, Bangladesh has so far realized only a fraction of its production potential.

Bangladesh has some 130 deep-sea fishing trawlers, 22000 mechanized fishing boats, and 25000 non mechanized fishing boats. Currently there are 133 fish processing plants in Bangladesh which are mostly located in port cities(Khulna and Chittagong) of which 74 processing plants are EU approved. Though the country is endowed with enormous fishery resources which are vital to the livelihood of millions of people and national food and nutrition security, the sector is facing major constraints including climate change, poor fisheries infrastructure, resource mismanagement, water and environmental population, natural disasters such as recurrent flood and cyclones, and lack of knowledge among farmers. Bangladesh is working with close collaboration with Department for International Development(DfID), World Fish Center and other international organization to develop the sector by building research partnerships and increasing investment. Community based management of fisheries is proving its potential to avert the longstanding political challenges farmers have been facing. The country, however, faces urgent imperatives to strengthen environmental laws to curb pollution which is significantly compromising the performance of the fisheries sector.

## Importance of fish export in the economy

Fisheries and aquaculture sector have emerged as the second most important contributors in export earnings of Bangladesh. It is the second largest export industry in Bangladesh and produces 2.5 percent of the global production of shrimp. Though rice is the most widely produced agricultural crop in Bangladesh, fisheries has a unique feature for its role in providing an important source of animal protein and essential elements for the population. In 2010-2011, some 5.5 million people were directly involved in fisheries as the main source of earning. Table 1 shows the trend in export of fish products.

**Table 1** Export of Fish and Fish-Product from Bangladesh. Source: DoF, 2009.

| Year | Frozen Shrimp/Prawn | | Frozen Fish (Others) | | Dry fish | |
|---|---|---|---|---|---|---|
| | Quantity (MT) | Value(Crore taka) | Quantity (MT) | Value(Crore taka) | Quantity (MT) | Value(Crore taka) |
| 1992 | 19224 | 604.03 | 2704 | 38.31 | 1042 | 12.26 |
| 1995 | 25225 | 1106.39 | 8827 | 176.62 | 521 | 8.39 |
| 2000 | 29713 | 1885.15 | 9484 | 137.19 | 215 | 3.65 |
| 2005 | 46533 | 2281.6 | 15763 | 256.20 | 272 | 3.71 |
| 2009 | 50368 | 2744.12 | 19294 | 450.89 | 341 | 11.99 |

Source: DoF, 2009.

Bangladesh is considered one of the most suitable countries in the world for prawn and shrimp farming, because of its vast resources of shallow water bodies which provide a unique opportunity for prawn and shrimp production [1,2]. Prawn farming has brought about dramatic improvements in the livelihoods of the coastal poor, including women. During the 1990s, the rapid development of prawn farming in southwest Bangladesh has been likened to a "blue revolution [7]. The migration of poor communities form the coastal region to the urban areas has reduced thanks to improved livelihood opportunities brought about by prawn and shrimp culture. The industry accounted for 22.21% of total agricultural GDP and 2.73% of total export income during the year 2011. USA and EU together account for around 80% of the total shrimp export. In 2011-2012, the European Union made up most of Bangladesh's yearly US $500,000 shrimp export.

## Trend of fish production in Bangladesh

Bangladesh ranks third among the world's largest inland fish producing countries after China and India. Table shows the trend in aquaculture production since. Around three quarters of rural households practice some form of freshwater aquaculture covering some 10 million ponds and most of which measure less than 400 m2.

Inland pond culture represents the most important part of aquaculture in Bangladesh contributing to around 86% of total production. Aquaculture accounted for about 43.5 percent of the total fish production during 2003–04, with inland open water fisheries contributed 34.8 percent. Fisheries in Bangladesh are diverse and there are about 795 native species of fish and shrimp in the fresh and marine waters. Besides that, there are 10 species of pearl bearing bivalves, 12 species of edible tortoise and turtle, 15 species of crab and 3 species of lobster. Most of the prawn and shrimp farms(~75% ) are located in southwest part of the country, mainly Bagerhat, Khulna and Satkhira districts, with the remainders in the southeast region including Cox's Bazar and Noakhali district. In 2003, more than 0.6 people are engaged in shrimp farming activities [21].

Although inland capture dominate total production, aquaculture production has increased significantly with a threefold rise during the period between 1989 to 1999 [15]. The contribution of cultured ponds also increased substantially from 27% in 1984 to 52% in1996 [16]. Production of shrimp has increased from 11,000 to 94,000 tonnes, recording an average annual growth of 45 percent during the same period [25]. In 2012, fish production reached a new a record 3.22 million tonnes which was 5.2% higher than the previous year .

Hilsha(Tenualosa ilisha) is one of the most favorite fish species in South Asian fish consumers and is recognized as the national fish of Bangladesh. Despite being a marine fish, the Padma-Meghna-Jamuna delta is the main site for hilsha capture where they migrate for laying eggs. Hilsha contributes to nearly 16.4% of the country's total fish production. Though Bangladesh produces about 350,000 tonnes of hilsha fish annually, the amount is found to be decreasing in recent years due to conservation purposes.

## Role of fish in diet

Fish is the primary source of animal protein for Bangladeshi population, especially poor rural households. Fishes are the major source of animal protein providing 80% of the animal protein intake and 7% of total protein supplies. Rice and fish constitute such an important part of Bangladeshi food culture that it has become a popular proverb- "mache bhate bangali," which means "fish and rice make a Bengali. In terms of weight, fish is the third most widely consumed food nationwide. But consumption pattern tends to vary among urban and rural areas and is shown to be lower among female members [13]. Per capita annual fish consumption in Bangladesh is about 14 kg against a recommended minimum requirement of 18 kg/year. From the last national survey in rural Bangladesh, the mean total protein intake was 48 g/person/d, of which fish contributed 3 g [17]. Table 2 shows the contribution of fisheries sector in national food supply.

**Table 2** National Fish food supply (tonnes).

Source: DoF, 2010.

| Items | 1980 | 1990 | 2000 | 2005 | 2007 | 2009 |
|---|---|---|---|---|---|---|
| Freshwater Fish | 521,450 | 675,596 | 1,376,081 | 1,810,458 | 2,015,331 | 2,235,742 |
| Marine Fish | 40,106 | 51,018 | 72,506 | 130,618 | 120,840 | 203,513 |

Source: DoF, 2010.

## Prospects of integrated fish farming

Though rice monoculture is the main characteristic of Bangladeshi agriculture, rice-fish farming began to receive attention in the 1980s [23]. Integrated rice-fish farming offers better resource utilization, diversity and food supply as well. In Mymensingh district, which is considered as one of the rice bowls of the country, has been identified as the most important region for integrated rice-fish culture due to its favorable climatic conditions and availability of low-lying agricultural land.Though currently a small number of farmers are practicing this method, it has been reported that the cultivation of fish in rice fields increases rice yields by 8 to 15% and thus the scope of rice cropping with integrated fish farm is remains considerably wide [25]. Integrated fish farming can fully utilize the water body, the water surface, the land, and the pond silt to increase the food available for human consumption. Integrated farming reduces the need for pelleted grains, which is both economical and create less pressure on total grain supply for human consumption. Rice-fish farming is also being regarded as an important approach to integrated pest management (IPM). Integrated rice-fish farming is most technically and cost efficient, using the least inputs, in particular fertilizer and provides a sustainable alternative to rice monoculture. Researchers also suggest that integrated rice-fish farming system is better than rice monoculture in terms of a range of social, economic and environmental measures [24].

Green revolution(GR) worldwide has increased staple crop production and greatly contributed to combat global hunger. However, now when the long term repercussions of GR are understood such monoculture, soil degradation, and extensive use of agrochemicals, all of which have had serious negative impacts on fisheries production, there is an imperative felt by agriculturists to find more sustainable strategies to reduce these negative impacts. Bangladesh has seen a dramatic rise in national rice production. Many watery areas have been brought under staple crop production to meet the demand of the huge population. Though fish production has also increased, but the its still far below the real potential of the country. Rice-fish farming can boost the production of both items which will help the country to improve food and nutrition security. The demand for rice and fish is constantly rising in Bangladesh with nearly three million people being added each year to its population [21]. Integrated rice-fish farming can help Bangladesh keep pace with the current demand for food through rice and fish production.

Increasing population translates to increasing demand for food and more pressure on land and water. If sustainable policies are not taken, rice demand may continue to compromise fisheries output in Bangladesh. Although official figures show that the number of floodplain fisheries increased in the 1990s, after decreasing in the 1980s, it is widely held that floodplain catches have been falling. Agricultural intensification has been proposed by researchers to meet the rising food demand for the huge population in Bangladesh. While rice production is still likely to increase, its also possible that it'll come at the expense of diminishing resources for fish production.

**Figure 1** Inland aquaculture has improved the economic condition of many rural households.

## Challenges for fisheries sector in Bangladesh

The fisheries industry in Bangladesh is confronted with a range of economic, institutional and environmental concerns. According to a IUCN(International Union for Conservation of Nature) study, 54 floodplain fish species are in danger of extinction and the pressure of fishing is so heavy in the floodplains that less than 2% of produced fish survives the end of each year. Table 3 shows that growth in fish production has slowed since 2000. Recurrent floods and natural disasters are believed to be main underlying causes behind this slump. Bangladesh is a low-lying land which makes it extremely vulnerable to sea-level rise, and is ranked first among countries to be affected by the adverse effects of climate change [4]. Nearly 80% of total area in Bangladesh is regarded as floodplains, and its precarious geographical position makes it highly prone to natural disasters as well. Climate change have devastating impacts on fishery-based livelihoods and on domestic food supply. Vulnerability of fishery-based livelihoods may substantially increase in the coming decades due to climate change, and in the absence of adaptation, increased frequency and intensity of cyclones and floods would result in greater damage to fishing materials and loss of fish [27].

**Table 3** Annual growth rate of fisheries sector, 1994-2003. Source: DoF 2005.

| Year | 1994-95 | 1996-97 | 1998-1999 | 2001-02 | 2002-03 |
|---|---|---|---|---|---|
| Growth(%) | 6.79 | 7.6 | 9.96 | 2.22 | 2.33 |

Source: DoF 2005.

Mariculture is also at risk of increasing salinity and over-fishing. According to FAO, globally, around a quarter of all fish stocks are overexploited and half of them are fully exploited. In Bangladesh, marine capture represents about 20% of total fish production. The floodplain and marine fisheries are under serious threat from overfishing. Overexploitation in the coastal region poses significant challenges on marine living resources and increases the dependency on distant water fishing in the long run.

Water pollution is another growing threat for the future of fisheries sector in Bangladesh and is fast becoming a serious public health issue and a constraint for food production. Industrial(especially textile and tannery) effluent, fertilizer and pesticide run-off, poor sewerage infrastructure and improper disposal of household waste are the major causes of water pollution in Bangladesh. Rivers and canals near the urban areas are threatened by sedimentation and siltation due mainly to soil erosion, and compounded by industrial expansion, most of these water bodies have already become to polluted to support biological system. Poor urban and industrial management and lack of enforcement of environmental laws are contributing to this pollution spree. The Buriganga river that flows through the capital city is the most polluted river in the country, many parts of which have already turned coal black.

Besides these natural and chemical events, inadequate financial capacities, poor resources management and lack of research facilities are also responsible for underperformance of the fisheries sector and environmental degradation [20]. Researchers have shown that poor management of prawn and shrimp culture is having devastating effects on the Sundarbans (the largest mangrove forest in the world) [5] where an estimated 9700 ha of the forest-mass has been lost as a result of intense shrimp farming [6]. Conversion of many natural wetlands to prawn farms has resulted in impediment of water flows and also decreased the scope of migration for many fish species [8].

## Community-based fisheries management, window of Opportunities

The Community Based Fisheries Management (CBFM) project was founded with an ambition to promote sustainable use of inland capture fisheries by empowering fisher's communities to manage their own aquatic resources [19]. Most fishers in rural Bangladesh operate on a small-scale basis(Table 4). Poor fishers suffer disadvantaged situations due policies that favor powerful players in the sector [18]. The CBFM initiative has developed a series of fisheries management approaches for ensuring equitable access to fisheries resources for community-based organisations (CBOs). Founded in 1994, the project supported by Department for International Development(DfID) and implemented by Bangladesh's Department of Fisheries(DoF) in partnership with the WorldFish Center and 11 NGOs adopted a research-based approach to promote equitable access and sustainable management of inland fisheries resources and to be run by the CBOs [14]. So far, a total of 164 fish sanctuaries have been established in over 80 water bodies under the CBFM approach. The project involved 14,000 CBO group members and a further 9,000 direct beneficiaries. In a survey conducted in 2007, the majority of CBOs reported increases in total production despite a 30 per cent increase in the number of fishers by then. Thus it appears that community-based fisheries management can offer significantly better utilization of the public water bodies covering over 4 million ha of floodplain wetlands in the country.

**Table 4** Shows the contribution of homestead pond aquaculture to household income and mean size of operation.

| Source | Aquaculture as % HH income | Mean pond size (ha) |
|---|---|---|
| Thompson et al. (2006) | 3 | 0.09 |
| Jahan et al. (2010) | 10 | 0.1 |
| Hossain et al. (2010) | 10 | 0.04 |
| Winrock International (2004) | 13.2 | 0.08 |
| Karim (2006) | 15.5 | 0.1-0.2 |

## CONCLUSIONS

This article gives an overview of the fisheries sector and its challenges and opportunities in Bangladesh. It is evident that the performance of fisheries sector is crucial from a national macroeconomic and food and nutrition security perspective. Therefore it proposes that a more efficient and sustainable management of the aquatic resources will contribute greatly to health and economy of the country. Policy makers must spare no effort to ensure the functioning of this sector in full swing by enhancing investment and research infrastructure, more strict environmental policies, and introducing better storage and marketing facilities. More importantly, the situation of fishers must be taken into account and special task force should be build to assess their vulnerability and strategies to tackle them. In order to meet the soaring demand for food for the burgeoning population, there is a need for increased rice and fish production in Bangladesh. But attention should also be given to the negative environmental externalities such as land and water biodiversity and water and air pollution which is inextricably linked with the success of agricultural sector.

## ACKNOWLEDGEMENTS

We extend sincere thanks to World Fish Center for giving permission to reuse the photographs.

## REFERENCES

1. Ahmed N, Demaine H, Muir JF. Freshwater prawn farming in Bangladesh: history, present status and future prospects. Aquaculture Research. 2008; 39: 806–819.
2. Islam MS. From pond to plate: towards a twin-driven commodity chain in Bangladesh shrimp aquaculture. Food Policy. 2008; 33: 209–223.
3. Fisheries Statistical Yearbook of Bangladesh. Fisheries Resources Survey System (FRSS), Department of Fisheries, Bangladesh. 2012.
4. Harmeling S. Global Climate Risk Index 2010: Who Is the Most Vulnerable? Weather-related Loss Events Since 1990 and How Copenhagen Needs to Responds. Germanwatch Bonn. 2010.
5. Hoq ME. An analysis of fisheries exploitation and management practices in Sundarbans mangrove ecosystem, Bangladesh. Ocean and Coastal Management. 2007; 50: 411–427.
6. Paul BG, Vogl CR. Impacts of shrimp farming in Bangladesh: challenges and alternatives. Ocean and Coastal Management. 2011; 54: 201–211.
7. Ahmed N, Allison EH, Muir JF. Rice-fields to prawn farms: a blue revolution in southwest Bangladesh? Aquaculture International. 2010; 18: 555–574.
8. Ahmed N, Demaine H, Muir JF. Freshwater prawn farming in Bangladesh: history, present status and future prospects Aquaculture Research. 2008; 39: 806–819.
9. Mohanty RK, Verma HN, Brahmanand PS. Performance evaluation of rice–fish integration system in rainfed medium and ecosystem. Aquaculture. 2004; 23: 125–135.
10. Mishra A, Mohanty RK. Productivity enhancement through rice–fish farming using a two-stage rainwater conservation technique. Agricultural Water Management. 2004; 67:119–131.
11. World Fish Center. Fish and Human Nutrition. The World Fish Center. Penang. Malaysia 2011.

12. Nesar A, Harvey D, James FM. Freshwater prawn farming in Bangladesh: history, present status and future prospects. Aquaculture Research. 2008; 39: 806–819.
13. Overseas Development Administration of United Kingdom. Impacts of flood control and drainage with or without irrigation (FCD/I). Project on fish resources and fishing community in Bangladesh: executive summary of Flood Action Plan (FAP)-17. Tsai C, Ali MY, editors. In: Openwater Fisheries of Bangladesh. The University Press Limited, Dhaka, Bangladesh. 1997; 199–212.
14. Liquid assets: community based Fisheries management in Bangladesh. The WorldFish Center. 2010.
15. Fish catch statistics of Bangladesh 1998–1999. Department of Fisheries (DOF), 2000; Dhaka, Bangladesh.
16. Fish catch statistics of Bangladesh 1995–1996. Department of Fisheries (DOF), 1996; Dhaka, Bangladesh.
17. Ahmad K, Hassan N. Nutritional Survey of Rural Bangladesh 1981–82. Institute of Nutrition and Food Sciences. 1983; Dhaka University, Dhaka, Bangladesh.
18. Parvin S, Paul MT. Human Ecology. Community Based Fisheries Management and Fisher Livelihoods: Bangladesh Case Studies. 2007; 35: 527-546.
19. Mustafa MG, Halls AS. Impact of the Community-Based Fisheries Management on sustainable use of inland fisheries in Bangladesh. Conference Paper 03. WorldFish Center.
20. Ferdous A, Kenneth J T. Current constraints and future possibilities for Bangladesh fisheries. Food Policy. 2001; 26: 297–313.
21. Karim MR. Present status and strategies for future development of shrimp farming in Bangladesh. Wahab MA, editor. In: Environmental and socioeconomic impact of shrimp farming in Bangladesh. Technical Proceeding BAU- NORAD Workshop. BRAC Center, Dhaka, Bangladesh. 2003; 1–8.
22. Chowdhury MR. Population Challenge Facing Bangladesh. Long Island University. CW Post Campus. USA. 2009.
23. Nabi R. Constraints to the adoption of rice-fish farming by smallholders in Bangladesh: a farming systems analysis. Aquaculture Economics & Management. 2008; 12: 145–153.
24. Nesar A, Kerstin K, Zander S, Garnett T. Socioeconomic aspects of rice-fish farming in Bangladesh: opportunities, challenges and production efficiency. Australian Journal of Agricultural and Resource Economics. 2011; 55: 199–219.
25. Mishra A, Mohanty RK. Productivity enhancement through rice-fish farming using a two-stage rainwater conservation technique. Agricultural Water Management . 2004; 67, 119–131.

# CHAPTER 13

# Use of Probiotics in Aquaculture

**Rafael Vieira de Azevedo[1] and Luís Gustavo Tavares Braga[2]**

*[1] State University of Norte Fluminense Darcy Ribeiro, Center for Agricultural Science and Technology, Campos dos Goytacazes, Rio de Janeiro,, Brazil*

*[2] State University of Santa Cruz, Department of Agricultural and Environmental Sciences, Ilhéus, Bahia,, Brazil*

## 1. INTRODUCTION

The term probiotics was first used by Lilly & Stillwell in 1965. Probiotic was defined as the microbiological origin factor that stimulates the growth of other organisms. In 1989 Roy Fuller introduced the idea that probiotics generate a beneficial effect to the host. He defined probiotics as live microorganisms which, when administered in adequate amounts, confer benefit to the host's health, improving the balance of the microbiota in the intestine.

Probiotics are defined by Food and Agriculture Organization/World Health Organization as "live microorganisms which when administered in adequate amounts confer a health benefit on the host" [1].

The purpose of its use is to install, improve or compensate for the functions of the indigenous microbiota that inhabit the digestive tract or the surface of the body.

The idea of using fermented foods for some health benefits is not new, being mentioned in the Persian version of the Old Testament (Genesis 18:8) that "Abraham attributed his longevity to the consumption of sour milk". Later, in 76 BC, a Roman historian, Pline, recommended the use of fermented milk products for the treatment of gastroenteritis cases [2].

However, a scientific approach, recognizing the beneficial role of certain microorganisms was applied only in the first decades of the 20th century, with the suggestion of using *Lactobacillus* (in 1907 Elie Metchnikoff attributed the longevity of Bulgarian populations to yoghurt consumption); *Bifidobacterium* (in 1906 Henri Tissier observed a greater presence of *Bifidobacteria* in the feces of breastfed healthy children); and *Saccharomyces boulardii* (Henri Boulard emphasized the use of a tropical fruit colonized by this yeast to treat diarrhea of local populations in the East during an episode of cholera in 1920) [3].

Several clinical studies have shown the benefits of probiotics to human health. For example, diarrhea treatment [4]; lactose intolerance [5]; irritable bowel syndrome [6]; allergies [7]; cancer [8]; among others.

The use of growth promoters allows improving the zootechnical performance of animals. Initially a large variety of substances with antibiotic function was used to improve performance of poultry, pigs and cattle, especially penicillin and tetracycline.

The use of antibiotics as additives to feeds showed great benefits to animal husbandry, expressed primarily in improved weight gain and feed conversion.

Antibiotics were used for decades, but are being banished from the zootechnical activity, mainly due to the risks posed by antibiotic-resistant bacteria, which can result in problems for animal and human health.

Accordingly, probiotics have deserved attention from researchers seeking alternatives to the use of traditional growth promoters in the field of animal nutrition.

Probiotics have also received special attention from researchers seeking animal nutrition alternatives to the use of traditional growth promoters (antibiotics). Therefore, the use of probiotics is being increasingly seen as an alternative to the use of antibiotics in animal production.

Many scientific papers show the beneficial effects of supplementation with probiotic strains in diets for poultry, pigs, cattle, fish, crustaceans, mollusks and amphibians [9-13].

Probiotics have been incorporated through diet in order to maintain the balance of the intestinal flora of animals, preventing digestive tract diseases, improving the digestibility of feed, leading to increased use of nutrients and causing better zootechnical performance of animals [14, 15].

## 2. PROBIOTIC ORGANISMS

The requirements that a probiotic organism must meet are [16]:

i. Resistance to the acid stomach environment, bile and pancreatic enzymes;

ii. Accession to the cells of the intestinal mucosa;

iii. Capacity for colonization;

iv. Staying alive for a long period of time, during the transport, storage, so that they can colonize the host efficiently;

v. Production of antimicrobial substances against the pathogenic bacteria; and

vi. Absence of translocation.

The species normally used as probiotics in animal nutrition are usually non-pathogenic normal microflora, such as lactic-acid bacteria (*Bifidobacterium, Lactobacillus, Lactococcus, Streptococcus* and *Enterococcus*) and yeasts as *Saccharomyces* spp. (Table 1).

## 3. MECHANISMS OF ACTION

The mechanisms of action of bacteria used as probiotics, although not yet fully elucidated, are described as [14, 15, 18]:

a. Competition for binding sites: also known as "competitive exclusion", where probiotics bacteria bind with the binding sites in the intestinal mucosa, forming a physical barrier, preventing the connection by pathogenic bacteria;

b. Production of antibacterial substances: probiotic bacteria synthesize compounds like hydrogen peroxide and bacteriocins, which have antibacterial action, mainly in relation to pathogenic bacteria. They also produce organic acids that lower the environment's pH of the gastrointestinal tract, preventing the growth of various pathogens and development of certain species of *Lactobacillus*;

c. Competition for nutrients: the lack of nutrients available that may be used by pathogenic bacteria is a limiting factor for their maintenance;

d. Stimulation of immune system: some probiotics bacteria are directly linked to the stimulation of the immune response, by increasing the production of antibodies, activation of macrophages, T-cell proliferation and production of interferon.

**TABLE 1.** Microorganisms recognized as safe and used as probiotics in animals. Source: [**17**]

| | |
|---|---|
| Aspergillus | *A. niger, A. orizae* |
| *Bacillus* | *B. coagulans, B. lentus, B. licheniformis, B. subtilis* |
| *Bifidobacterium* | *B. animalis, B. bifidum, B. longun, B. thermophylum* |
| *Lactobacillus* | *L. acidophillus, L. brevis, L. bulgaricus, L. casei, L. cellobiosis, L. fermentarum, L. curvatus, L. lactis, L. plantarum, L. reuterii, L. delbruekii,* |
| *Pediococcus* | *P. acidilacticii, P. cerevisae, P. pentosaceus, P. damnosus* |
| *Saccharomyces* | *S. cerevisiae, S. boulardii* |
| *Streptococcus* | *S. cremoris, S. faecium, S. lactis, S. intermedius, S. thermophyllus, S. diacetylatis* |

The mechanism of action of yeasts still needs substantiation by means of research. A likely mechanism of action of yeasts is related to total inhibition (*in vitro*) or partial inhibition of pathogens. Inactive yeasts contain large quantities of protein and polysaccharides in its walls, which can act positively in the immune system and in the absorption of nutrients. In addition, yeasts produce nutritious metabolites in digestive tract that boost animal performance, besides possessing minerals (Mn, Co, Zn) and vitamins (A, $B_{12}$, $D_3$) that enhance the action of beneficial microorganisms [19].

Although some mechanisms had been suggested on the action of probiotics, they are not completely clarified, but it is known that they inhibit growth of pathogenic microorganism by producing antimicrobial compounds; they compete with pathogens for adhesion sites and nutrients; and they model immune system of the host [20].

## 4. SELECTION OF PROBIOTICS

Briefly, for the use of a given microorganism as probiotic, it is necessary its isolation, characterization and testing certifying its probiotic efficiency (Figure 1).

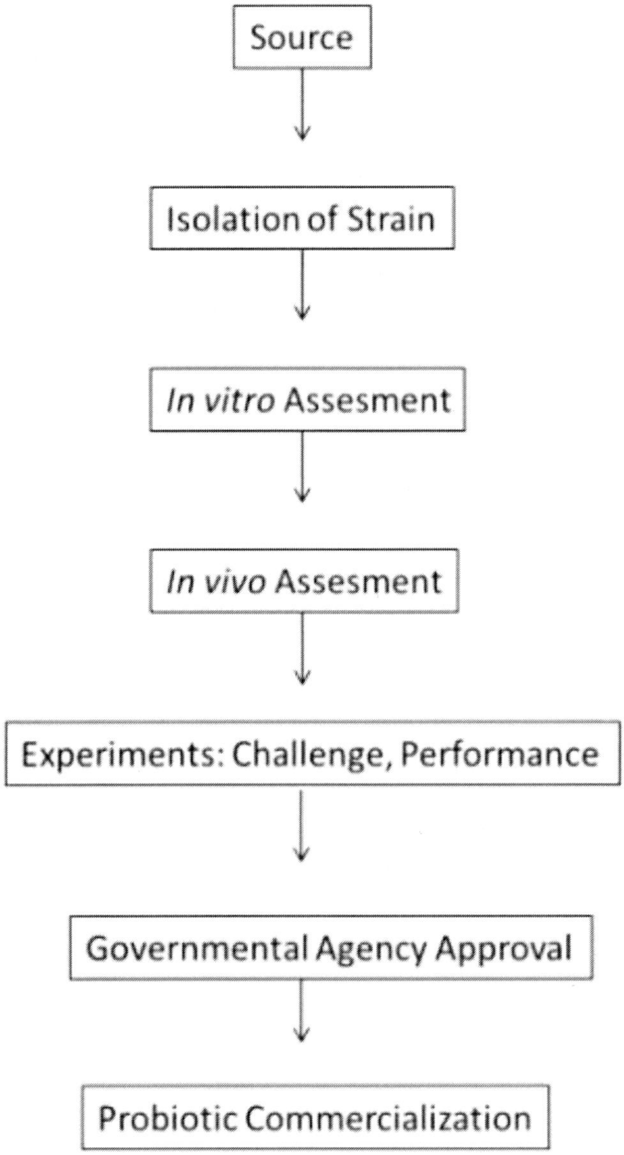


**FIGURE 1.**Diagram for selection of probiotics

First a source of microorganisms (e.g. digestive tract of healthy animals) must be selected.

After, the microorganisms with which the work is to be carried out are isolated and identified by means of selective culture.

Then a new culture with only the colonies of interest for conducting *in vitro* evaluations (inhibition of pathogens; pathogenicity to target species; resistance conditions of host; among others) is performed.

In case of the absence of restrictions on the use of the target species, experiments with *in vivo*supplementation, and small and large scale, are carried out to check if there are real benefits to the host.

Finally, the probiotic that presented significantly satisfactory result can be produced commercially and utilized.

## 5. USE OF PROBIOTIC IN AQUACULTURE

Probiotics in aquaculture may act in a manner similar to that observed for terrestrial animals.

However, the relationship of aquatic organisms with the farming environment is much more complex than the one involving terrestrial animals.

Because of this intimate relationship between animal and farming environment, the traditional definition of probiotics is insufficient for aquaculture.

In this sense, Verschuere and colleagues [21] suggest a broader definition:

"It is a microbial supplement with living microorganism with beneficial effects to the host, by modifying its microbial community associated with the host or its farming environment, ensuring better use of artificial food and its nutritional value by improving the host's response to diseases and improving the quality of the farming environment."

The microorganisms present in the aquatic environment are in direct contact with the animals, with the gills and with the food supplied, having easy access to the digestive tract of the animal.

Among the microorganisms present in the aquatic environment are potentially pathogenic microorganisms, which are opportunists, i.e., they take advantage of some animal's stress situation (high density, poor nutrition) to cause infections, worsening in zootechnical performance and even death.

For this reason, the use of probiotics for aquatic organisms aims not only the direct benefit to the animal, but also their effect on the farming environment.

Bergh and colleagues [22] observed that, when starting its first feeding, the intestinal flora of the Atlantic halibut (*Hippoglossus hippoglossus*) changed from a prevalence of *Flavobacterium* spp. to *Aeromonas* spp./*Vibrio* spp. showing the influence of the external environment and food on the microbial community of this fish.

*Vibrio* spp., *Plesiomonas shigelloides*, and *Aeromonas* spp. are the main causative agents of diseases in aquaculture, and may even cause food infections in humans.

The interaction between the environment and the host in an aquatic environment is complex. The microorganisms present in the water influence the microbiota of the host's intestine and vice versa.

Makridis and colleagues [23] demonstrated that the provision of two strains of bacteria via food directly into the farming water of the incubators of turbot larvae (*Scophthalmus maximus*) promoted the maintenance of the bacteria in the environment, as well as the colonization of the digestive tract of the larvae.

Changes in salinity, temperature and dissolved oxygen variations, change the conditions that are favorable to different organisms, with consequent changes in dominant species, which could lead to the loss of effectiveness of the product.

Accordingly, the addition of a given probiotic in the farming water of aquatic organisms must be constant, because the conditions of environment suffer periodic changes.

Thus, the variety of microorganisms present must therefore be considered in the choice of probiotic to be used in aquaculture.

Intensive farming systems utilize high stocking densities, among other stressors (e.g. management), which often end up resulting in low growth and feed efficiency rates, besides of weakness in the immune system, making these animals susceptible to the presence of opportunistic pathogens present in the environment.

In this sense, the effect of probiotics on the immune system has led to a large number of researches with beneficial results on the health of aquatic organisms, although it has not yet been clarified how they act.

In addition, probiotics can also be used to promote the growth of aquatic organisms, whether by direct aid in the absorption of nutrients, or by their supply.

Probiotics most used in aquaculture are those belonging to the genus *Bacillus* spp. (*B. subtilis, B. licheniformis* and *B. circulans*), *Bifidobacterium* spp. (*B. bifidum, B. lactis, and B. thermophilum*), lactic-acid bacteria (*Lactobacillus* spp. e *Carnobacterium* spp.) and yeast *Saccharomyces cerevisiae*[24,25].

The benefits observed in the supplementation of probiotics in aquaculture include [21, 26-28]:

1. Improvement of the nutritional value of food;
2. Enzymatic contribution to digestion;
3. Inhibition of pathogens;
4. Growth promoting factors;
5. Improvement in immune response; and
6. Farming water quality.

Among the most recent studies that point to the effect of the use of probiotics for various aquatic organisms stand those for fish [21], shrimps [26], mollusks [30] and frogs [29].

## 5.1. Results of Probiotics in Fish Farming

### 5.1.1. Immune System

Gatesoupe [31] observed that turbot larvae (*Scophthalmus maximus*) fed rotifera enriched with lactic-acid bacteria increased resistance against infection by *Vibrio* spp.

The joint administration of *Lactobacillus fructivorans* and *Lactobacillus plantarum* through dry or live feed promoted the colonization of the intestine of sea bream larvae (*Sparus aurata*) and the decrease in mortality of animals during larviculture and nursery [32].

Gram and colleagues [33] showed that the use of *Pseudomonas fluorescens* AH2 as probiotics decreased the mortality of juveniles of rainbow trout (*Oncorhynchus mykiss*) exposed to *Vibrio anguillarum*.

Kumar and colleagues [34] observed higher survival rate of carp *Labeo rohita* fed *Bacillus subtilis*, submitted to intraperitoneal injection with *Aeromonas hydrophila*.

Oral administration of *Clostridium butyricum* increased phagocytic activity of leucocytes of rainbow trout [35].

Nikoskelainen and colleagues [36] observed that the administration of *Lactobacillus rhamnosus* at 105 UFC $g^{-1}$, stimulated the respiratory burst in rainbow trout.

Other studies showed an increase in immune response with the use of probiotics for different species, such *Carnobacterium maltaromaticum* B26 and *Carnobacterium divergens* B33 for rainbow trout [38], *Lactobacillus belbrüeckii*, *Bacillus subtilis* and *Debaryomyces hansenii* for gilthead seabream [39-41], *B. subtilis* and *Pseudomonas aeruginosa* for *Labeo rohita* [42,43], *Lactococcus lactis* for Nile tilapia (*Oreochromis niloticus*) [44] and *B. simplex* DR-834 to carp (*Cyprinus carpio*) [45].

### 5.1.2. Performance

Tovar and colleagues [37] incorporated the yeast *Debaryomyces hansenii* to the feed of sea bass larvae and observed improvement in the maturation of the digestive tract of this species. According to the authors this satisfactory effect was due to the high secretion rate of spermine and spermidine by yeasts.

Increase of weight gain and survival was observed for turbot larvae fed rotifera enriched with acid-lactic bacteria [31].

Queiroz and Boyd [46] observed enhancement of the zootechnical performance and survival of channel catfish (*Ictalurus punctatus*) when a mixture of *Bacillus* spp. was added to the farming water.

Using yeast *Saccharomyces cerevisiae* as probiotic for Israeli carp, Noh and colleagues [47] observed an increase in the food efficiency of this species.

Lara-Flores and colleagues [48] concluded that the use of *Saccharomyces cerevisiae* as probiotic for fry of Nile tilapia resulted in better growth and food efficiency, suggesting that this yeast promotes adequate growth in tilapia farming. In this study it was observed that fish fed control diet showed reduced survival and digestibility of feed with increased storage density, considered a stressful factor for growing fish. This result highlighted the efficiency of the use of this probiotic in stressful situations.

Other positive results of the probiotic on the performance of fish are found for *Labeo rohita* fingerlings [49], Nile tilapia [50] and common carp [51].

## 5.2. Results of The Use of Probiotics in Shrimp Farming

### 5.2.1. Immune System

In relation to farmed shrimp, bacterial diseases are considered as the largest cause of mortality in larvae.

The administration of a mixture of bacteria (*Bacillus* spp. and *Vibrio* spp.) positively influenced on survival and had protective effect against *Vibrio harveyi* and the white spot syndrome virus (WSSV) [15]. This result was due to stimulation of the immune system, by increasing phagocytosis and antibacterial activity.

The administration of a commercial probiotic for the larvae of *Marsupenaeus japonicus* resulted in increased survival (97%) being significantly higher than the control treatment [52].

Thus, the use of *Bacillus coagulans* SC8168 as probiotic for postlarvae of *Litopenaeus vannamei*resulted in higher survival of animals [53].

In a study with tiger shrimp (*Penaeus monodon*), the inoculation of *Bacillus* S11, a saprophyte strain, resulted in higher survival of postlarvae challenged by a luminescent pathogenic bacterial culture [54].

*Bacillus subtilis* and *Lactobacillus plantarum* for *Litopenaeus vannamei* [55-58], *Pediococcus acidilactici* to *Litopenaeus stylirostris* [59] and *Bacillus* NL110 and *Vibrio* NE17 for *Macrobrachium rosenberguii* [60] also proved effective in improving the immune system of these animals.

### 5.2.2. Performance

Lin and colleagues [61] used *Bacillus* spp. in the diet of *Litopenaeus vannamei* enhancing digestibility rates of the feed.

Ziaei-Nejad and colleagues [26] added the probiotic *Bacillus* spp. in the farming of *Fenneropenaeus indicus* larvae and observed survival increase, and also an increase in the activities of lipase, protease and amylase enzymes in the digestive tract of shrimps.

Several studies have shown that the bacteria of the genus *Bacillus* spp. secrete exoenzymes (proteases, lipases and carbohydrases) that can help improve digestion and nutrient absorption increase, resulting in better use of food and animal growth [62].

## 5.3. Results From The Use Of Probiotics In The Farming Of Others Aquatic Organisms

### 5.3.1. Mollusks

The culture of oysters and scallops has been introduced in many countries, however, mass mortalities of larvae have frequently occurred and to prevent these mortalities, most farmers use antibiotics [63]. Thus, the use of probiotic bacteria has been fueled, especially during the hatchery [64].

Riquelme and colleagues [65] identified a bacteria (*Alteromonas haloplanktis*) capable of reducing the mortality of Chilean scallop larvae (*Argopecten purpuratus*) when exposed to $10^3$ colony forming units per milliliter (UFC $ml^{-1}$) of *Vibrio anguillarum*.

Cultures of *Alteromonas media* control *Vibrio tubiashii* infections in larvae of Pacific oysters (*Crassostrea gigas*) [66].

Other bacteria with probiotic potential for mollusks such as Pacific oysters (*Alteromonas* spp.) [67, 68], Scallop larvae (*Roseobacter* spp., *Vibrio* spp., *Pseudomonas* spp., *Arthrobacter* spp.) [69-71], promoted growth, survival and immune response of animals.

### 5.3.2. Frogs

For Bull Frog (*Lithobates catesbeianus*) with an average weight of 3.13 g, the addition of probiotic *Bacillus subtilis* in different doses (2.5, 5.0 and 10 g $kg^{-1}$ feed) resulted in improved weight gain, feed conversion and apparent survival, when compared to control treatment (without added probiotic); however, the immunostimulant effect was demonstrated through the increased phagocytic capacity of animals [72].

Likewise, Dias and colleagues [29] observed the beneficial effect of two commercial probiotics on the immune system of *L. catesbeianus*.

## 5.4. Probiotics and Quality of Water in Aquaculture

Another aspect of the use of probiotics in aquaculture is the improvement of the quality of the water in the farming nurseries. Increases in organic load, levels of phosphorous and nitrogen compounds are growing concerns in aquaculture.

Boyd [73] noted the beneficial effect of probiotics on organic matter decomposition and reduction of the levels of phosphate and nitrogen compounds.

Aerobic denitrifying bacteria are considered good candidates to reduce nitrate or nitrite to $N_2$ in aquaculture waters.

To this end some bacteria were isolated in shrimp farming tanks. *Acinetobacter, Arthrobacter, Bacillus, Cellulosimicrobium, Halomonas, Microb acterium, Paracoccus, Pseudomonas, Sphingobacterium* and *Stenotrophomas* are some of the denitrifying bacteria already identified [28].

Reduction in levels of phosphorous and nitrogen compounds in the farming water of shrimp *Litopenaeus vannamei* was also observed when commercial probiotics were added to the water [27].

Similarly, for the shrimp *Penaeus monodon*, an improvement in the quality of farming water was observed with the addition of *Bacillus* spp. as probiotic [74].

Gram-positive bacteria are better converting organic matter into $CO_2$ than gram-negative bacteria. Thus, during a production cycle, higher levels of these bacteria can reduce the accumulation of particulate organic carbon. Thus, maintaining higher levels of these gram-positive bacteria in production pond, farmers can minimize the buildup of dissolved and particulate organic carbon during the culture cycle while promoting more stable phytoplankton blooms through the increased production of $CO_2$ [21].

## 6. CONCLUSION

The results reported so far with the use of probiotics for aquatic organisms are promising. However, many works have not achieved satisfactory results.

Sometimes in experiments in which aquatic organisms are challenged by some pathogenic agent, the probiotic organism does not exhibit inhibiting action against the pathogen, resulting in mortality.

Similarly, the conditions to which the animals are subjected during farming may directly influence the effectiveness of probiotics. Thus, when not subjected to stressful situations, the results often do not show a significant effect of probiotics on the performance of animals.

In general, the effects of adding probiotics tend to be most striking in unsuitable operating conditions or in conditions of stress, when the microflora is unbalanced, primarily in young animals.

Among these factors, the most commonly featured are: temperature above or below the thermal comfort zone; presence of pathogens; poor sanitary conditions; stressful management; change in nutrition; transport; high storage density; after treatment with antibiotics; sudden change of environment.

Also, the results obtained in experiments with probiotics may be affected by factors such as: type of probiotic microorganism; method and quantity administered; condition of the host; condition of intestinal microbiota; age of the animal.

# REFERENCES


1. WHO/FAO.Joint World Health Organization/Food and Agricultural Organization Working Group. Guidelines for the Evaluation of Probiotics in Food 2002. Ontario, Canada
2. J. Schrezenmeir, M. de Vrese, prebiotics. Probiotics, a. symbiotic-approaching, definition, The American Journal of Clinical Nutrition 2001S-364S.
3. C. Shortt, probiotic. The, historical. century, perspectives. current, Trends in Food Science & Technology19991012411417
4. M. De Vrese, P. R. Marteau, Probiotics, Effects. Prebiotics, diarrhea. on, The Journal of Nutrition 2007S-811S.
5. T. He, M. G. Priebe, Y. Zhong, C. Huang, H. J. Harmsen, G. C. Raangs, J. M. Antoine, G. W. Wellingand, R. J. Vonk, Effects of yogurt and bifidobacteria supplementation on the colonic microbiota in lactose-intolerant subjects. Journal of Applied Microbiology20081042595604
6. R. Nagala, C. Routray, Clinical case study multispecies probiotic supplement minimizes symptoms of irritable Bowel Syndrome. US Gastroenterology & Hepatology Review 2011713637
7. S. Jain, H. Yadav, P. R. Sinha, S. Kapila, Y. Naito, F. Marotta, Anti-allergic effects of probiotic Dahi through modulation of the gut immune systemTurkish Journal of Gastroenterology2010213244250
8. X. Chen, J. Fruehauf, J. Goldsmith, H. Xu, K. Katchar, H. Koon, D. Zhao, . E. Kokkotou, C. Pothoulakis, C. Kelly, Saccharomyces boulardii inhibits EGF receptor signaling and intestinal tumor growth in Apc(min) mice. Gastroenterology 20091373914923
9. S. M. Aly, A. M. Abd-Rahman El, G. John, M. F. Mohamed, Characterization of some bacteria isolated from Oreochromis niloticus and their potential use as probioticsAquaculture20081 EOF
10. M. Ignatova, V. Sredkova, V. Marasheva, Effect of dietary inclusion of probiotic on chickens performance and some blood. Biotechnology in Animal Husbandry 2009
11. N. A. Soleimani, R. K. Kermanshahi, B. Yakhchali, T. N. Sattari, Antagonistic activity of probiotic lactobacilli against Staphylococcus aureus isolated from bovine mastitisAfrican Journal of Microbiology Research201042021692173
12. E. Veizaj-Delia, T. Piub, P. Lekajc, M. Tafaj, Using combined probiotic to improve growth performance of weaned piglets on extensive farm conditionsLivestock Science20101341249251
13. Gatesoupe FJ.The use of probiotics in aquaculture.Aquaculture19991801147165
14. R. Fuller, Problems, In. prospects, R. Fuller, (ed, scientific. Probiotics-The, London. basis, Chapman, Hall, 1992377386
15. J. L. Balcázar, I. Blas, I. Zarzuela-Ruiz, D. Cunningham, D. Vendrell, J. L. Músquiz, The role of probiotics in aquacultureVeterinary Microbiology20061141173186
16. V. D. Capriles, K. E. A. Silva, M. Fisberg, probióticos. e. Prebióticos, Nova. simbióticos, no. tendência, alimentos. mercado de, funcionais, Revista Nutrição Brasil 200546327335
17. P. Lyons, out. Yeast, the. of, box. black, Feed Manangement 19863710814
18. L. Z. Jin, Y. W. Ho, N. Abdullah, S. Jalaludin, in. Probiotics, modes. poultry, action. of, World's Poultry Science Journal 1997534351368
19. J. Hill, S. V. Tracey, M. Willis, L. Jones, A. D. Ellis, Yeast culture in equine nutrition and physiology. In: Proceedings of Alltech's Annual Symposium 2006. Available from: http://eNo.engormix.com/MA-equines/nutrition/articles/yeast-culture-equine-nutrition-t279/0 htmacessed 31 April 2012
20. T. Oelschlaeget, Mechanisms of probiotic actions- A review. International Journal of Medical Microbiology 201030015762

21. L. Verschuere, G. Rombaut, P. Sorgeloos, W. Verstraete, Probiotic bacteria as biological control agents in aquacultureMicrobiology and Molecular Biology Review 2000644655671
22. Ø. Bergh, G. H. Hansen, R. E. Taxt, Experimental infection of eggs and yolk sac larvae of halibut, Hippoglossus hippoglossus L.Journal of Fish Diseases1992155379391
23. P. Makridis, A. J. Fjellheim, J. Skjermo, O. Vadstein, Colonization of the gut in first feeding turbot by bacterial strains added to the water or biencapusated in rotifers. Aquaculture International 200085367380
24. Y. K. Lee, K. Nomoto, S. Salminen, S. L. Gorbash, of. Handbook, New. probiotics, York, Wiley; 1999
25. Sanders ME, Klaenhammer TR. Invited review: the scientific basis of Lactobacillus acidophilus NCFM functionality as a probiotic.Journal of Dairy Science 2001842319331
26. S. Ziaei-Nejad, M. Rezaei, G. Takami, D. Lovett, A. Mirvaghefi, M. Shakouri, The effect of Bacillus spp. bacteria used as probiotics on digestive enzyme activity, survival and growth in the Indian white shrimp Fenneropenaeus indicus Aquaculture 2006516 EOF
27. Y. Wang, Z. Xu, M. Xia, The effectiveness of commercial probiotics in Northern White Shrimp Penaeus vannamei pondsFisheries Science200571510341039
28. Wang, G. Zheng, S. Liao, H. Huang, R. Sun, Diversity analysis of bacteria capable of removing nitrate/nitrite in a shrimp pond. Acta Ecologica Sinica 200727519371943
29. Dias DC, De Stéfani, MV, Ferreira CM, França FM, Ranzani-Paiva MJT, Santos AA.Hematologic and immunologic parameters of bullfrogs, Lithobates catesbeianus, fed probiotics. Aquaculture Research 201041710641071
30. Macey BM, Coyne VE.Improved growth rate and disease resistance in farmed Haliotis midae through probiotic treatmentAquaculture2005249 EOF261 EOF
31. Gatesoupe FJ.Lactic acid bacteria increase the resistance of turbot larvae, Scophthalmus maximus, against pathogenic VibrioAquatic Living Resource 199471277282
32. O. Carnevali, M. C. Zamponi, R. Sulpizio, A. Rollo, M. Nardi, C. Orpianesi, S. Silvi, M. Caggiano, A. M. Polzonetti, A. Cresci, Administration of probiotic strain to improve sea bream wellness during developmentAquaculture International2004377 EOF386 EOF
33. L. Gram, J. Melchiorsen, B. Spanggard, I. Huber, T. Nielsen, Inhibition of Vibrio anguillarum by Pseudomonas fluorescens AH2, a possible probiotic treatment of fish.Applied and Environmental Microbiology1999653969973
34. R. Kumar, S. C. Mukherjee, K. P. Prasad, A. K. Pal, Evaluation of Bacillus subtilis as a probiotic to indian major carp Labeo rohita (Ham.)Aquaculture Research2006371212151221
35. M. Sakai, T. Yoshida, S. Astuta, M. Kobayashi, Enhancement of resistance to vibriosis in rainbow trout, Oncorhynchus mykiss (Walbaum) by oral administration of Clostridium butyricum bacterin.Journal of Fish Disease 1995182187190
36. S. Nikoskelainen, A. C. Ouwehand, G. Bylund, S. Salminen, E. M. Lilius, Immune enhancement in rainbow trout (Oncorhynchus mykiss) by potential probiotic bacteria (Lactobacillus rhamnosus).Fish & Shellfish Immunology2003155443452
37. D. Tovar, J. Zambonino, C. Cahu, F. J. Gatesoupe, R. Vázquez-Juárez, R. Lésel, Effect of live yeast incorporation in compound diet on digestive enzyme activity in sea bass (Dicentrarchus labrax) larvaeAquaculture2002113 EOF
38. D. Kim, B. Austin, Innate immune responses in rainbow trout (Oncorhynchus mykiss, Walbaum) induced by probioticsFish & Shellfish Immunology 2006215513524
39. Salinas, A. Cuesta, Meseguer. J. Esteban, Dietary administration of Lactobacillus delbrüeckii and Bacillus subtilis, single or combined, on gilthead seabream cellular innate immune responses.Fish & Shellfish Immunology20051916777
40. M. Reyes-Becerril, I. Salinas, A. Cuesta, J. Meseguer, D. Tovar-Ramirez, Ascencio, F. Valle, Esteban, 2008256731739

41. Salinas, L. Abelli, F. Bertoni, S. Picchietti, A. Roque, D. Furones, A. Cuesta, J. Meseguer, Esteban, Monospecies and multispecies probiotic formulations produce different systemic and local immunostimulatory effects in the gilthead seabream (Sparus aurata L.).Fish & Shellfish Immunology2008114 EOF23 EOF
42. S. K. Nayak, P. Swain, S. C. Mukhrjee, Effect of dietary supplementation of probiotic and vitamin C on the immune response of Indian major carp, Labeo rohita (Ham.)Fish & Shellfish Immunology2007234892896
43. S. S. Giri, S. S. Sen, V. Sukumaran, Effects of dietary supplementation of potential probiotic Pseudomonas aeruginosa VSG-2 on the innate immunity and disease resistance of tropical freshwater fish, Labeo rohitaFish & Shellfish Immunology 201232611351140
44. X. Zhou, Y. Wang, J. Yao, W. Li, Inhibition ability of probiotic Lactococcus lactis, against A. hydrophila and study of its immunostimulatory effect in tilapia (Oreochromis niloticus).International Journal of EngineeringScience and Technology 2010277380
45. G. Wang, Y. Liu, F. Li, H. Gao, Y. Lei, X. Liu, Immunostimulatory activities of Bacillus simplex DR-834 to carp (Cyprinus carpio)Fish & Shellfish Immunology 2010293378387
46. F. Queiroz, C. Boyd, Effects of a bacterial inoculum in channel catfish ponds.Journal of the World Aquaculture Society19982916773
47. S. H. Noh, K. Han, T. H. Won, Y. J. Choi, of. Effect, enzyme. antibiotics, culture. yeast, on. probiotics, growth. the, of. performance, carp. Israeli, Korean Journal of Animal Science 1994361480486
48. M. Lara-Flores, Guzman. Olvea-Novoa-Mendez, B. E. López-Madrid, W. , Use of the bacteria Streptococcus faecium and Lactobacillus acidophilus, and the yeast Saccharomyces cerevisiae as growth promoters in Nile tilapia (Oreochromis niloticus)Aquaculture2003193 EOF
49. Ghosh, S. K. Sen, A. K. Ray, Supplementation of Lactobacillus acidophilus in compound diets for Labeo rohita fingerlings. Indian Journal of Fisheries 2004514521526
50. El-Haroun ER, Goda AMAS, Chowdhury MAK.Effect of dietary Biogen® supplementation as a growth promoter on growth performance and feed utilization of Nile tilapia Oreochromis niloticus (L.). Aquaculture Research 20063714 14731480
51. M. Faramarzi, S. Kiaalvandi, F. Iranshahi, The effect of probiotics on growth performance and body composition of common carp (Cyprinus carpio). Journal of Animal and Veterinary Advances 2011101824082413
52. El-Sersy NA, Abdel-Razek FA, Taha SM.Evaluation of various probiotic bacteria for the survival of Penaeus japonicus larvaeFresenius Environmental Bulletin2006A) 15061511
53. X. Zhou, Y. Wang, W. Li, Effect of probiotic on larvae shrimp (Penaeus vannamei) based on water quality, survival rate and digestive enzyme activities Aquaculture2009349 EOF353 EOF
54. S. T. Rengpipat, W. Phianphak, S. Piyatiratitivorakul, P. Menasaveta, Effects of a probiotic bacterium in black tiger shrimp Penaeus monodon survival and growthAquaculture1998301 EOF
55. D. Tseng, P. Ho, S. Huang, S. Cheng, Y. Shiu, C. Chiu, C. Liu, Enhancement of immunity and disease resistance in the white shrimp, Litopenaeus vannamei, by the probiotic, Bacillus subtilis E20Fish & Shellfish Immunology2009262339344
56. Liu, C. Chiu, Y. Shiu, W. Cheng, C. Liu, Effects of the probiotic, Bacillus subtilis E20, on the survival, development, stress tolerance, and immune status of white shrimp Litopenaeus vannamei larvae.Fish & Shellfish Immunology 2010837 EOF844 EOF
57. E. F. Silva, Calazans. N. F. Soares, J. L. Vogeley, B. C. Valle, R. Soares, S. Peixoto, Effect of probiotic (Bacillus spp.) addition during larvae and postlarvae culture of the white shrimp Litopenaeus vannameiAquaculture Research201119

58. Kongnum, T. Hongpattarakere, Effect of Lactobacillus plantarum isolated from digestive tract of wild shrimp on growth and survival of white shrimp (Litopenaeus vannamei) challenged with Vibrio harveyiFish & Shellfish Immunology2012321170177
59. Castex, P. Lemaire, N. Wabete, L. Chim, Effect of Pediococcus acidilactici on antioxidant defences and oxidative stress of Litopenaeus stylirostris under Vibrio nigripulchritudo challenge.Fish & Shellfish Immunology2010284622631
60. M. M. Rahiman, Y. Jesmi, A. P. Thomas, A. A. M. Hatha, Probiotic effect of Bacillus NL110 and Vibrio NE17 on the survival, growth performance and immune response of Macrobrachium rosenbergii (de Man)Aquaculture Research2010419120134
61. H. Z. Lin, Z. Guo, Y. Yang, W. L. I. Z. J. Zheng, Effect of dietary probiotics on apparent digestibility coefficients of nutrients of white shrimp Litopenaeus vannamei BooneAquaculture Research2004351514411447
62. Selvin. J. Ninawe, Probiotics in shrimp aquaculture: Avenues and challenges. Critical Reviews in Microbiology 2009; 35 1 4366 .
63. Farzanfar, The use of probiotics in shrimp aquacultureFEMS Immunology and Medical Microbiology2006482149158
64. Vaseeharan, P. Ramasamy, Control of pathogenic Vibrio spp. by Bacillus subtilis BT23, a possible probiotic treatment for black tiger shrimp Penaeus monodonLetters in Applied Microbiology20033628387
65. Riquelme, G. Hayashida, R. Araya, A. Uchida, M. Satomi, Y. Ishida, Isolation of a native bacterial strain from the scallop Argopecten purpuratus with inhibitory effects against pathogenic vibriosJournal of Shellfish Research1996152369374
66. F. Gibson, J. Woodworth, A. M. George, Probiotic activity of Aeromonas media on the Pacific oyster, Crassostrea gigas, when challenged with Vibrio tubiashiiAquaculture1998111 EOF120 EOF
67. P. Douillet, C. J. Langdon, Effects of marine bacteria on the culture of axenic oyster Crassostrea gigas (Thunberg) larvaeBiological Bulletin199318413651
68. Douillet AP, Langdon CJ.Use of a probiotic for the culture of larvae of the Pacific oyster (Crassostrea gigas, ThunbergAquaculture199411912540
69. C. Ruiz-Ponte, J. F. Samain, J. L. Sánchez, J. L. Nicolas, The benefit of a Roseobacter species on the survival of scallop larvaeMarine Biotechnology1999115259
70. C. Riquelme, R. Araya, R. Escribano, Selective incorporation of bacteria by Argopecten purpuratus larvae: implications for the use of probiotics in culturing systems of the Chilean scallopAquaculture200025 EOF
71. Jorquera. Riquelme, A. I. Rojas, R. E. Avendaño, N. Reyes, Addition of inhibitor-producing bacteria to mass cultures of Argopecten purpuratus larvae (Lamarck, 1819)Aquaculture2001111 EOF
72. F. M. França, D. C. Dias, P. C. Teixeira, Stéfani. M. V. Marcantônio De, A. Antonucci, G. Rocha, M. J. T. Ranzani-Paiva, C. M. Ferreira, Efeito do probiótico Bacillus subtilis no crescimento, sobrevivência e fisiologia de rãs-touro (Rana catesbeiana). Boletim do Instituto de Pesca 2008343403412
73. Boyd CE.Aquaculture sustainability and environmental issuesWorld Aquaculture19993021072
74. G. Dalmin, K. Kathiresan, A. Purushothaman, Effect of probiotics on bacterial population and health status of shrimp in culture pond ecosystem.Indian Journal of Experimental Biology2001399939942

# CHAPTER 14

# Time-To-Response Toxicity Analysis as a Method for Drug Susceptibility Assessment in Salmon Lice

**Greta Carmona-Antoñanzas[a], Joseph L. Humble[a], Stephen N. Carmichael[a], Jan Heumann[a], Hayden R.L. Christie[a], Darren M. Green[a], David I. Bassett[b], James E. Bron[a], Armin Sturm[a,*]**

[a]*Institute of Aquaculture, School of Natural Sciences, University of Stirling, Stirling, FK9 4LA, Scotland, UK*

[b]*Marine Environmental Research Laboratory, University of Stirling, Machrihanish, Argyll, PA28 6PZ, Scotland, UK*

## ABSTRACT

The salmon louse *Lepeophtheirus salmonis* (Krøyer, 1837) is an ectoparasite causing infections of wild and farmed Atlantic salmon (*Salmo salar* L.) in the Northern hemisphere. While *L. salmonis* control at commercial mariculture sites increasingly employs non-medicinal approaches, such as cage designs reducing infection rates and biological control through cleaner fish, anti-parasitic drugs are still a requirement for effective fish health care. With only a limited range of salmon delousing agents available, all of which have been in use for more than a decade, drug resistance formation has been reported for different products. Successful resistance management requires reliable susceptibility assessment, which is usually achieved through *L. salmonis* bioassays. These tests involve the exposure of parasites to different drug concentrations and require significant numbers of suitable *L. salmonis* stages. The present study reports an alternative bioassay that is based on time-to-response toxicity analyses and can be carried out with limited parasite numbers. The assay determines the median effective time ($ET_{50}$), i.e., the time required until impaired swimming and/or attachment behaviour becomes apparent in 50% of parasites, by conducting repeated examinations of test animals starting at the time point where exposure to a set drug concentration commences. This experimental approach further allows the estimation of the apparent drug susceptibility of individual *L. salmonis* by determining their time to response, which may prove useful in experiments designed to elucidate associations between genetic factors and the drug susceptibility phenotype of parasites. Three laboratory strains of *L. salmonis* differing in susceptibility to emamectin benzoate were characterised using standard 24 h bioassays and time-to-response toxicity assays. While both the median effective concentration ($EC_{50}$) and the $ET_{50}$ showed variability between experimental repeats, both types of bioassay consistently discriminated susceptible and drug-resistant *L. salmonis* laboratory strains.

**Statement of relevance:** Infections by sea lice cause significant costs to the global salmon farming industry, which have been estimated to exceed €300 million per year worldwide. Control of sea lice still relies to a significant extent on chemical delousing; however, chemical control is threatened by resistance formation. Resistance can be combated by rotation between different drugs and strategic implementation of non-medicinal strategies. However, resistance management requires reliable and feasible methods of susceptibility assessment.

The present study is a technical note introducing a novel approach to susceptibility assessments in sea lice. The method can be applied in susceptibility assessments on farms, where it offers the advantage of a reduced requirement of parasites for testing. In addition, the novel method allows deriving the times of parasite require to show a response after drug treatment has started, thus providing a variable characterizing the drug susceptibility phenotype of individual parasites. Accordingly, the bioassay approach presented here will be useful for studies aiming at unravelling the genetic determinants of drug resistance.

# 1.INTRODUCTION

Sea lice (Caligidae: Copepoda) are marine fish ectoparasites feeding on the host's mucus and skin tissues (Boxaspen, 2006). Sea louse infections of farmed Atlantic salmon (*Salmo salar*) mostly involve *Lepeophtheirus salmonis* (salmon louse) and *Caligus elongatus* in the Northern hemisphere, and *Caligus rogercresseyi* in Chile (Costello, 2006). Sea louse control on salmon farms relies mainly on the use of veterinary drugs, supplemented by farm management measures (Costello, 2009) and non-medicinal approaches such as the deployment of cleaner fish (Sayer et al., 1996) and modified cage designs reducing the likelihood of infection (Stien et al., 2016).

Only a restricted range of licensed anti-sea louse drugs are currently available (Burridge et al., 2010). The continued use of the same or a few types of control agents, however, can favour the development of drug resistance in parasites (Denholm et al., 2002). Drug resistance is well documented in *L. salmonis*, and compounds for which loss of efficacy has been reported include the organophosphates dichlorvos and azamethiphos (Jones et al., 1992; Kaur et al., 2015; Roth et al., 1996), the pyrethroid deltamethrin (Sevatdal and Horsberg, 2003), the non-specific oxidising agent hydrogen peroxide (Treasurer et al., 2000) and the macrocyclic lactone emamectin benzoate (EMB) (Lees et al., 2008). Resistance to different drugs has further been found in *C. rogercresseyi* (Agusti et al., 2016; Bravo et al., 2010). Accordingly, there is an urgent need for efficient resistance prevention and management strategies in sea lice (Aaen et al., 2015).

A key requirement for such effective sea louse control strategies is an accurate assessment of the drug susceptibility status of sea louse populations. Such assessments are usually achieved by conducting bioassays, which typically involve subjecting batches of preadult or adult parasites to a dilution series of the drug (Sevatdal and Horsberg, 2003; Westcott et al., 2008). The internal exposure of an aquatic organism taking up a toxicant from water will increase both with increasing toxicant concentration and increasing length of exposure. Traditional aquatic bioassays typically employ different toxicant concentrations to achieve gradually varied exposures, while keeping the length of the exposure period constant. Results are expressed as the median lethal or median effective concentration ($LC_{50}$, $EC_{50}$), i.e., the concentration theoretically causing a toxic effect in 50% of the tested population. In an alternative approach called time-to-response toxicity analysis, gradual exposure levels are achieved by combining a fixed toxicant concentration with variable exposure periods. In this approach, the susceptibility of the population to the toxicant is expressed as the median lethal or effective time ($LT_{50}$, $ET_{50}$) (Robertson and Preisler, 1991). Time-to-response bioassays have been used in susceptibility assessments of terrestrial arthropod pests (Robertson and Preisler, 1991) and ecotoxicity studies in aquatic invertebrates and fish (Pascoe and Edwards, 1989; Rand, 2008).

The availability of sea louse stages suitable for bioassays can be restricted at production sites. Alternative bioassays involving only one drug concentration and a fixed exposure period have been proposed to allow for drug susceptibility assessment under such circumstances (Helgesen and

Horsberg, 2013a; Whyte et al., 2013). While single-dose bioassays can be highly useful as a tool in fish health management, their ability to resolve susceptibility differences between parasite populations is by design limited. Time-to-response toxicity analyses could provide a complementary approach allowing characterisation of the susceptibility status at greater depth when sea louse availability is restricted. In addition to supporting veterinary decisions on fish farms, drug susceptibility assessments in sea lice are central to experimental plans aiming at identifying genetic determinants of drug resistance, which often require the determination of the susceptibility phenotypes of individuals (Besnier et al., 2014). Differentiation between susceptible and resistant parasites has been previously achieved by rating toxic responses following exposure to a diagnostic drug concentration for a set time period (Ljungfeldt et al., 2014). Using a similar approach, but additionally implementing repeated observations to determine the time to response for individual parasites, would permit a more graduated characterisation of the drug susceptibility phenotype than achievable with a test design employing a one concentration/one exposure time cut-off criterion to define resistance/susceptibility.

The aim of the present study was to investigate the potential of time-to-response toxicity analyses as an alternative approach to conducting sea louse drug sensitivity assessments. Time-to-response toxicity analyses were compared to standard bioassays with respect to their ability to differentiate between well-characterised laboratory strains of *L. salmonis*showing different degrees of resistance to the salmon delousing agent EMB.

# 2. MATERIALS AND METHODS

## 2.1. Salmon louse (L. salmonis) strains and husbandry

Three *L. salmonis* laboratory-maintained strains established from field isolates of egg strings without further selection in the laboratory were used in this study. The drug-susceptible strain IoA-00 (previously called "S") (Heumann et al., 2012) was established in 2003 from a Scottish farm site where no chemical control agents other than hydrogen peroxide had been used. The EMB-resistant strain IoA-01 (previously called "PT" or "R") (Heumann et al., 2012, 2014) and the multi-resistant strain IoA-02, which is hyposensitive to both EMB and deltamethrin, were created in December 2008 and September 2011, respectively, from other Scottish sites where there had been reports of variable treatment efficacies. These strains have since been cultured under identical laboratory conditions using Atlantic salmon as host fish, as described in detail elsewhere (Heumann et al., 2012). To propagate cultures, *L. salmonis* egg strings were collected from gravid females and allowed to hatch and develop to infective copepodids, which were used to inoculate tanks containing fresh host fish. To collect *L. salmonis*, host fish were either euthanised under a UK Home Office approved Schedule 1 method, or anaesthetised with 100 mg $L^{-1}$ 2-phenoxyethanol (99%; Sigma-Aldrich, Dorset, UK) in seawater for 3 min. Previous experiments assessing the effect of anaesthesia on bioassay results did not find significant differences between the two sea lice collection methods (data not shown). Parasites were removed from fish into clean aerated seawater using fine forceps, and fish were transferred into clean seawater with aeration for recovery. Infection rates were maintained at levels compatible with good fish welfare according to MERL Good Laboratory Practise (GLP) Standard Operating Procedure (SOP) protocols. All laboratory infections were carried out under UK Home Office licence and appropriate veterinary supervision.

## 2.2. Standard bioassays

Experiments used adult male or preadult II female *L. salmonis*. After collection from host fish, parasites were allowed to recover for 2–4 h in filtered aerated seawater at 12 °C before use in bioassays. Exposures were performed in a temperature-controlled incubator set to 12 °C, using 150 mL plastic Petri dishes holding 70 mL of exposure solutions and containing ten sea lice. EMB (technical grade, a gift from MSD Animal Health) was solubilised using polyethylene glycol of a number average molecular weight ($M_n$) of 300 (PEG 300, pH. Eur., Merck Chemicals, UK) before being diluted in seawater. Exposure solutions contained a final concentration of 0.05% (*v*/v) PEG 300. Each test comprised a geometrical dilution series of EMB of at least five concentrations in addition to seawater and solvent controls, the latter containing 0.05% (v/v) PEG 300. Sea lice were assigned to treatments randomly. At least two replicate Petri dishes were used per combination of strain and drug or control treatment. After 24 h of exposure, sea lice were visually examined and rated according to their attachment and mobility behaviour. Prior to rating, beakers were re-labelled with codes by laboratory staff not involved in the recording of experimental outcomes to allow for observer-blinded rating. In experiments conducted in 2011 and before, salmon lice were rated as normally motile (unaffected) or immotile (affected) upon visual examination and stimulation with a fine brush (Heumann et al., 2012). Later experiments used rating criteria initially proposed by Sevatdal and Horsberg (2003) and Westcott et al. (2008) and modified by Igboeli et al. (2012), where parasites are rated "live" when firmly attached to the surface of the Petri dish or swimming normally, "weak" when swimming irregularly and failing to attach to surfaces firmly (animals may attach briefly, but dislodge again instantly), "moribund" when incapable of swimming away or attaching (animals may twitch appendages or move uncoordinatedly in a close circle) and "dead" when showing no movements in extremities, gut or other organs as apparent from examination under a microscope. Using this set of criteria, parasites were considered unaffected when rated "live" or "weak" and affected when rated "moribund" or "dead". Bioassays were considered invalid if > 10% of parasites in control treatments were deemed affected.

## 2.3. Time-to-response bioassays

Sea lice were exposed for 24 h to 400 μg $L^{-1}$, 800 μg $L^{-1}$ or 1200 μg $L^{-1}$ of EMB, using similar general procedures and involving the same types of controls and levels of replication as described in Section 2.1.2. At set time intervals throughout the exposure period, parasites were visually examined and rated as described in Section 2.1.2. Typically, this involved hourly examinations for the beginning of the exposure period (1–15 h after addition of toxicants) followed by further examinations at 18, 21 and 24 h. Experimentation thus involved repeated examination of the same sea lice. In time-to-response bioassays, the main factor controlling toxic response is exposure time, which cannot be concealed to the experimenter by coded labelling. To ensure impartial rating in these assays, animals were rated by a second observer unaware of the nature of the experiment at selected time points. Ratings of immotile parasites per Petri dish never differed by more than one between observers. All other experimental conditions were identical to those described in Section 1.1.2 for the standard bioassay.

## 2.4. Statistics

In standard bioassays, median effective concentrations ($EC_{50}$) of EMB and 95% confidence intervals were generally calculated by probit analysis using the statistical programme Minitab 16.1.0 (Minitab Inc.). In cases where data were not suitable for probit analysis, the trimmed Spearman–Karber method was used to derive the $EC_{50}$ and 95% confidence limits (Hamilton et al., 1977). In time-to-response bioassays, median effective times ($ET_{50}$) and 95% confidence intervals were estimated using the survival time module of Minitab, assuming a log-normal distribution. In all tests associated to the above estimations, results were considered significant if the probability value ($P$) was $< 0.05$.

# 3. RESULTS

## 3.1. Standard bioassay

Three laboratory-maintained salmon louse strains differing in susceptibility to EMB were used in this study. When applied at concentrations > 100 µg $L^{-1}$, EMB induced behavioural toxic responses in strain IoA-00 in a concentration-dependent manner (Fig. 1). Higher EMB concentrations were required to provoke similar toxic responses in strains IoA-01 and IoA-02 (Fig. 1). Strain differences in EMB susceptibility are illustrated by median effective concentrations ($EC_{50}$) obtained in different experiments, with estimates for the $EC_{50}$ in adult males ranging from 74.3 to 159.3 µg $L^{-1}$ for IoA-00, from 553 to 780 µg $L^{-1}$ in IoA-01 and 445 to 675 µg $L^{-1}$ in IoA-02 (Table 1). With strain IoA02, preadult II females were significantly more susceptible to EMB than adult males. By contrast, gender differences in EMB susceptibility were not significant in strain IoA-00 (Table 1).

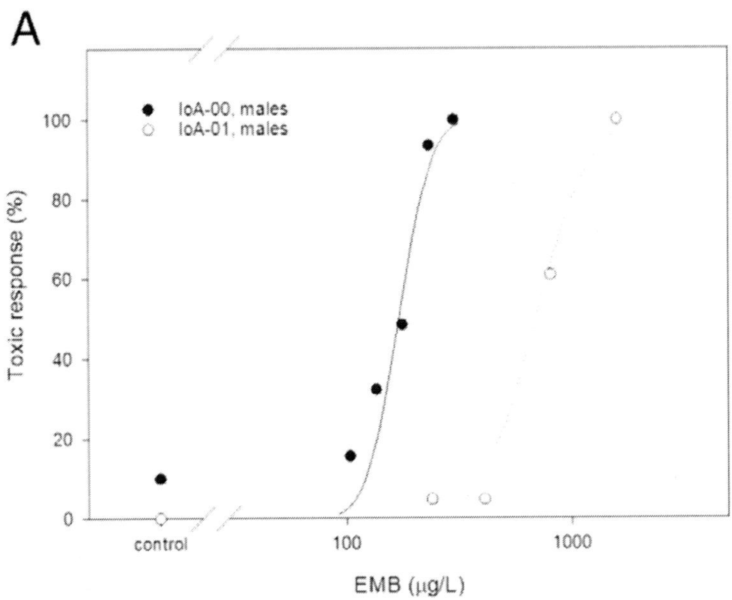

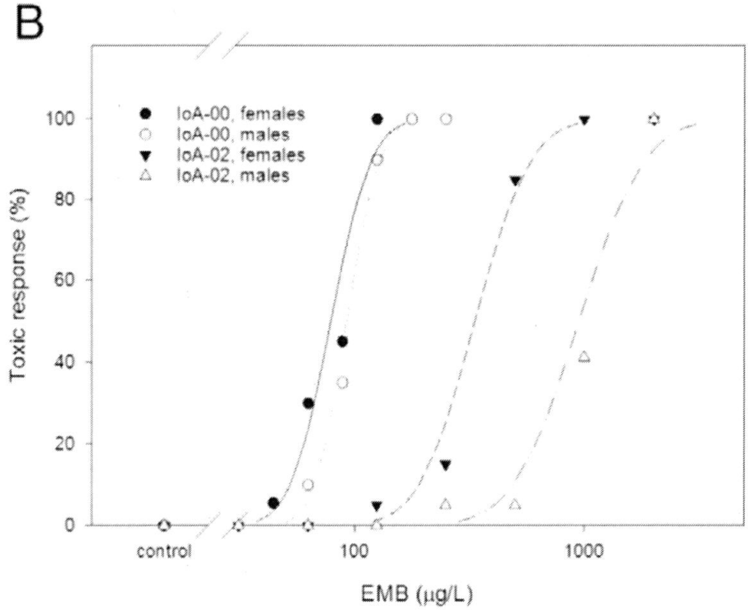


**Figure. 1**. Standard emamectin benzoate (EMB) bioassay with *Lepeophtheirus salmonis*. Symbols represent the average behavioural toxic response of at least two batches of ten animals, recorded after 24 h of exposure to EMB in plastic Petri dishes. A, EMB toxicity in salmon louse strains IoA-00 and IoA-01. B, EMB toxicity in strains IoA-00 and IoA-02.

**Table 1**. Susceptibility of *Lepeophtheirus salmonis* strains to emamectin benzoate. The toxicity of EMB to defined parasite stages of different laboratory-maintained strains was determined in 24 h bioassays. Results are expressed as EMB median effective concentrations ($EC_{50}$), determined by probit analysis except where noted otherwise.

| Strain | Sex/stage | Response rate in controls | $EC_{50}$ (µg $L^{-1}$) (95% confidence limits) | Date of experiment |
|---|---|---|---|---|
| **IoA-00** | Male, adult | 10% | 159.3 (145.4–175.6)[1] | 01/2011 |
| | Male, adult | 0% | 74.3 (59.0–94.1)[2] | 05/2011 |
| | Male, adult | 0% | 92.4 (83.2–106.7) | 03/2015 |
| | Female, preadult II | 0% | 78.6 (70.0–88.3) | 03/2015 |
| **IoA-01** | Male, adult | 0% | 695.5 (583.5–841.1)[1] | 11/2010 |
| | Male, adult | 0% | 780.3 (642.9– | 05/2011 |

| | | | | |
|---|---|---|---|---|
| | | | 967.4)[2] | |
| | Male, adult | 2.5% | 553.0 (486.1–624.7) | 07/2011 |
| IoA-02 | Male, adult | 0% | 949.8 (767.8–1204.8)[3] | 11/2011 |
| | Male, adult | 0% | 675.0 (618.0–738.0)[4] | 10/2014 |
| | Female, preadult II | 0% | 335.3 (277.0–409.4)[3] | 11/2011 |
| | Female, preadult II | 0% | 355.1 (289.9–521.1) | 03/2015 |

[1]Data shown in Fig. 1A.

[2]Data from Carmichael et al. (2013).

[3]Data shown in Fig. 1B.

[4]Trimmed Spearman–Karber statistics.

## 3.2. Time-to-response bioassay

To investigate the potential of time-to-response salmon louse bioassays to assess strain and gender differences in EMB susceptibility, experiments were first conducted using an exposure concentration of 800 μg $L^{-1}$ (Fig. 2). Apparent response levels in exposed parasites increased in a time-dependent fashion, allowing the median effective time ($ET_{50}$) to be derived for the populations tested (Table 2). While $ET_{50}$ values determined at an exposure level of 800 μg $L^{-1}$ EMB showed some variation between experiments, bioassays with male salmon lice consistently found greater $ET_{50}$ values in strains IoA-01 and IoA-02 than strain IoA-00 (Table 2). During exposure to 800 μg $L^{-1}$ EMB, response levels in the IoA-00 strain increased very rapidly after first signs of toxicity became apparent, with few or no partial responses being observed at set examination times (Fig. 2). This precluded the estimation of 95% confidence limits in some experiments (Table 2). Subsequent time-to-response experiments employed EMB levels of 400 or 1200 μg $L^{-1}$. Changing the EMB exposure concentration within the above limits had little effect on the time course of toxicity in the IoA-00 strain, as apparent from $ET_{50}$ values showing overlapping 95% confidence intervals (Table 2). Similarly, $ET_{50}$ values for IoA-01 males were similar between trials using 800 and 1200 μg $L^{-1}$ of EMB (Table 2). By contrast, $ET_{50}$ values for IoA-02 females at 400 μg $L^{-1}$ were significantly greater than at 800 μg $L^{-1}$ EMB. In exposures with 400 μg $L^{-1}$ EMB, no $ET_{50}$ could be determined for IoA-02 males, as these salmon lice failed to show apparent signs of toxicity within the 24 h experiment (Table 2).

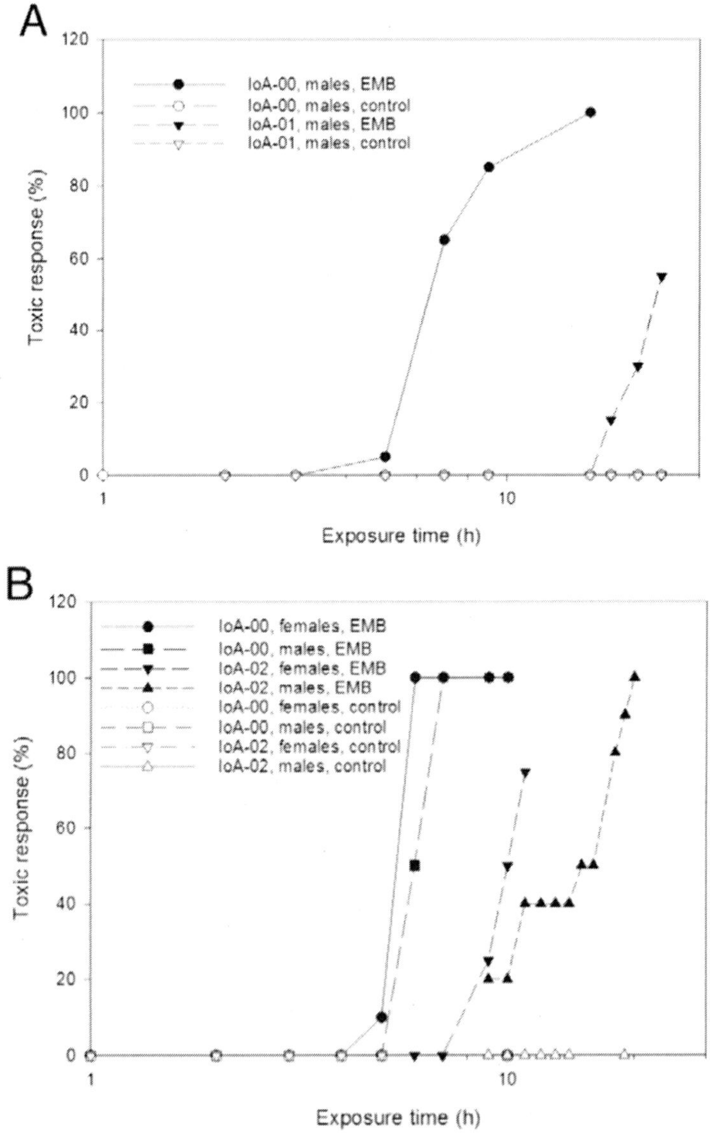


**Figure. 2.** Time-to-response bioassay with *Lepeophtheirus salmonis.* Symbols represent the average behavioural toxic response of at least ten animals. Salmon lice were exposed to 800 µg $L^{-1}$ of emamectin benzoate (EMB) for a total period of 24 h, during for which repeated observations of parasite motility and attachment behaviour were made. A, EMB toxicity in salmon louse strains IoA-00 and IoA-01; B, EMB toxicity in strains IoA-00 and IoA-02.

**Table 2.** Toxic response of *Lepeophtheirus salmonis* strains to emamectin benzoate in time-to-response bioassays. Toxicity is expressed as the median effective time ($ET_{50}$).

| Date | EMB (μg L[-1]) | $ET_{50}$ (h) (95% confidence limits) | | | | |
|---|---|---|---|---|---|---|
| | | IoA-00 | | IoA-01 | IoA-02 | |
| | | Males | Females | Males | Males | Females |
| **12/2011** | 800 | 5.86[1] | 4.03 (3.67–4.43) | 21.36 (16.82–27.13) | | |
| **07/2012[2]** | 800 | 6.77 (6.08–7.54) | | 28.46 (15.94–50.84) | | |
| **07/2012** | 1200 | 5.72 (5.07–6.44) | | 25.66 (17.49–37.65) | | |
| **03/2015[3]** | 800 | 6.00[1] | 5.08[1] | | 13.18 (10.51–16.51) | 9.98 (8.99–11.08) |
| **07/2015** | 400 | 6.82 (6.33–7.35) | 5.21 (4.68–5.81) | | > 24[4] | 16.50 (13.91–19.56) |

[1]Data do not allow calculation of 95% confidence limits.

[2]Data set shown in Fig. 2A.

[3]Data set shown in Fig. 2B.

[4]Data do not allow calculation of $ET_{50}$.

# 4. DISCUSSION

The assessment of drug susceptibility of salmon louse populations causing farm infections is essential to provide information to fish health professionals and aid the identification of effective therapeutic options. Susceptibility assessments are commonly based on the performance of salmon louse bioassays, which involve the exposure of batches of parasites to drug concentrations forming a geometrical series. While salmon louse bioassays are well standardised and field-tested (Sevatdal and Horsberg, 2003; Westcott et al., 2008), their requirement for significant numbers of test animals can be problematic in situations where the availability of suitable parasite stages is restricted. A typical bioassay with five test concentrations and one control and involving duplicate experimental and control treatments with ten animals each will require 120 parasites. By comparison, only 40 parasites are required in time-to-response bioassay at the same level of replication. While the time-to-response bioassay described here is more labour intensive than single-dose bioassays involving a single exposure time (Helgesen and Horsberg, 2013a; Whyte et al., 2013), the time-to-response bioassay characterises the drug susceptibility phenotype of the given parasite population at greater resolution. Thus, the time-to-response bioassay can be expected to complement existing methods of drug susceptibility assessment in salmon lice with deployment of the new bioassay depending on context of the study. In addition to the potential for implementation of this novel bioassay in the assessment of farm site infections, the time-to-response bioassay is capable of providing estimates of the drug susceptibility of individual parasites, which suggests its usefulness in the context of analysing molecular responses of sea lice to anti-parasitic drugs.

While salmon lice parasitising EMB-treated fish experience drug exposure both through contact with and ingestion of mucous and skin tissues, parasites are exposed solely through the aqueous route in 24 h bioassays, complicating the prediction of treatment efficiencies from bioassay findings. In an earlier study, EMB medication (50 μg $kg^{-1}$, seven days) of experimentally infected salmon resulted in 98% salmon lice clearance for strain IoA-00 parasites by day 21 after the start of treatment cycle but had no effect on IoA-01 salmon lice (SARF, 2012). Similarly, no clearance of strain IoA-02 was observed from a standard EMB treatment cycle (data not shown). Accordingly, the levels of resistance of parasite strains used in this study are clinically relevant. The fact that the time-to-response bioassay described in this work is able to differentiate between the above EMB-resistant strains and a susceptible reference strain suggests the usefulness of the time-to-response bioassay for susceptibility assessments.

Salmon louse bioassays have been initially defined using a parasite behaviour rating scale of three categories (Sevatdal and Horsberg, 2003; Westcott et al., 2008). According to this scale, animals are rated "live" at the end of the test when they attach firmly to the Petri dish or show normal locomotive behaviour, such as swimming in a straight line. By contrast, "moribund" salmon lice show impaired motility behaviours, ranging from irregular swimming to inability to move, as well as weak adhesion. Parasites are rated "dead" if they show no movement in the extremities or the gut (Sevatdal and Horsberg, 2003; Westcott et al., 2008). More recently, different groups have adopted a four category rating scheme, involving a more narrow definition of "moribund", which now includes only immotile parasites still showing twitching of appendages but not irregular swimmers, and the new category "weak", which comprises parasites displaying irregular swimming or weak attachment to the Petri dish wall

(Igboeli et al., 2012; Saksida et al., 2013). In the present study, this four criteria scale was adapted in experiments carried out in 2012 and thereafter, while earlier experiments used the three-category scale (Sevatdal and Horsberg, 2003; Westcott et al., 2008). The change in rating criteria had little effect on the outcome of most bioassays but markedly reduced the number tests that were invalid due to excessive response levels in controls, which were mostly due to parasites showing weak attachment behaviour but only slightly impaired swimming. The present investigation included standard and time-to-response bioassays conducted with both methodologies, so that changes in rating scale do not affect the study's findings.

EMB $EC_{50}$ values obtained in different experiments with male parasites ranged from 74.3 to 159.3 μg $L^{-1}$ for IoA-00 and 553.0 to 780.3 μg $L^{-1}$ for IoA-01. The lower EMB $EC_{50}$ of 289.0 μg $L^{-1}$ obtained for IoA-01 males in a previous study has been excluded from data considered in this study, as DMSO was used as solvent carrier in that study (Heumann et al., 2014), which may have influenced toxicity (Tanneberger et al., 2010). While results obtained in the current study illustrate variability among $EC_{50}$ estimates, standard bioassays consistently differentiated the EMB-resistant strains IoA-01 and IoA-02 from the susceptible strain IoA-00. Paralleling the results obtained in the present study, Espedal et al. (2013)observed consistent strain differences in EMB susceptibility despite between-test variability in absolute EMB bioassay estimates. Similar to standard bioassays, the time-to-response bioassay introduced in the present report displays a certain degree of variation in $ET_{50}$values between repeated experiments. Factors that might be responsible for the variability observed in this study include seasonal effects, as well as variability in the actual test concentrations of EMB, known for its hydrophobicity. An earlier study from this laboratory detected 49.7% of a nominal EMB concentration of 200 μg $L^{-1}$ 3 h after the start of exposures (Carmichael et al., 2013). Similarly, in another study (Helgesen and Horsberg, 2013b), 48.9% of an nominal concentration of 150 μg $L^{-1}$ EMB was detected in the seawater phase at the end of 24 h salmon louse bioassays performed in high-density polyethylene dishes. In the same study, recoveries were slightly less in tests using glass dishes (41.9%) and substantially inferior when using Teflon™ or silanised glass containers (27.4% and 24.0%, respectively) (Helgesen and Horsberg, 2013b).

## 4.1. Conclusion

In conclusion, the present study describes a novel salmon louse bioassay based on time-to-response analyses, which requires less test animals than current standard salmon louse bioassays. Our results indicate that time-to-response bioassays are powerful tools to assess the susceptibility of salmon louse populations to control agents such as EMB, particularly in situations where the availability of test animals is restricted. A further advantage of this novel bioassay is its ability to provide estimates of the drug susceptibility of individual parasites, which is expected to be useful in the context of analysing molecular responses of salmon lice to anti-parasitic drugs.

## ACKNOWLEDGEMENTS

This study was supported by United Kingdom Biotechnology and Biological Sciences Research Council grant BB/L022923/1 awarded to Armin Sturm. The authors further gratefully acknowledge funding received from the MASTS pooling initiative (The Marine Alliance for Science and Technology for Scotland). MASTS is funded by the Scottish Funding Council (grant reference HR09011) and contributing institutions. The authors are further thankful to the Scottish Salmon Producer's Organisation and the Moredun Foundation for financial support to Stephen N. Carmichael, and to Merck Animal Health for financial support to Jan. Heumann. The authors further gratefully acknowledge the Fisheries Society of the British Isles for granting support for a summer internship to Hayden Christie. Gregg Arthur of the North Atlantic Fisheries College, Shetland, UK, is thanked for critical comments and discussion of the time-to-response bioassay.

## REFERENCES

1. Aaen, S.M., Helgesen, K.O., Bakke,M.J., Kaur, K., Horsberg, T.E., 2015. Drug resistance in sea lice: a threat to salmonid aquaculture. Trends Parasitol. 31, 72–81. http://dx.doi.org/ 10.1016/j.pt.2014.12.006.
2. Agusti, C., Bravo, S., Contreras, G., Bakke, M.J., Helgesen, K.O., Winkler, C., Silva, M.T., Mendoza, J., Horsberg, T.E., 2016. Sensitivity assessment of Caligus rogercresseyi to anti-louse chemicals in relation to treatment efficacy in Chilean salmonid farms. Aquaculture 458, 195–205. http://dx.doi.org/10.1016/j.aquaculture.2016.03.006.
3. Besnier, F., Kent, M., Skern-Mauritzen, R., Lien, S., Malde, K., Edvardsen, R.B., Taylor, S., Ljungfeldt, L.E., Nilsen, F., Glover, K.A., 2014. Human-induced evolution caught in action: SNP-array reveals rapid amphi-Atlantic spread of pesticide resistance in the salmon ecotoparasite Lepeophtheirus salmonis. BMC Genomics 15, 937. http://dx.doi. org/10.1186/1471-2164-15-937.
4. Boxaspen, K., 2006. A review of the biology and genetics of sea lice. ICES J. Mar. Sci. 63, 1304–1316. http://dx.doi.org/10.1016/j.icesjms.2006.04.017.
5. Bravo, S., Sevatdal, S., Horsberg, T.E., 2010. Sensitivity assessment in the progeny of Caligus rogercresseyi to emamectin benzoate. Bull. Eur. Assoc. Fish Pathol. 30, 99–105.
6. Burridge, L., Weis, J.S., Cabello, F., Pizarro, J., Bostick, K., 2010. Chemical use in salmon aquaculture: a review of current practices and possible environmental

effects. Aquaculture 306, 7–23. http://dx.doi.org/10.1016 /j.aquaculture. 2010.05.020.

7. Carmichael, S.N., Bron, J.E., Taggart, J.B., Ireland, J.H., Bekaert, M., Burgess, S.T.G., Skuce, P.J., Nisbet, A.J., Gharbi, K., Sturm, A., 2013. Salmon lice (Lepeophtheirus salmonis) showing varying emamectin benzoate susceptibilities differ in neuronal acetylcholine receptor and GABA-gated chloride channel mRNA expression. BMC Genomics 14, 408.
8. Costello, M.J., 2006. Ecology of sea lice parasitic on farmed and wild fish. Trends Parasitol. 22, 475–483. http://dx.doi.org/10.1016/j.pt.2006.08.006.
9. Costello,M.J., 2009. The global economic cost of sea lice to the salmonid farming industry. J. Fish Dis. 32, 115–118. http://dx.doi.org/10.1111/j.1365-2761. 2008.01011.x.
10. Denholm, I., Devine, G.J., Horsberg, T.E., Sevatdal, S., Fallang, A., Nolan, D.V., Powell, R., 2002. Analysis and management of resistance to chemotherapeutants in salmon lice, Lepeophtheirus salmonis (Copepoda: Caligidae). Pest Manag. Sci. 58, 528–536. http://dx.doi.org/10.1002/ps.482.
11. Espedal, P.G., Glover, K.A., Horsberg, T.E., Nilsen, F., 2013. Emamectin benzoate resistance and fitness in laboratory reared salmon lice (Lepeophtheirus salmonis). Aquaculture 416-417, 111–118. http://dx.doi.org/10.1016/ j.aquaculture. 2013.09.001.
12. Hamilton, M.A., Russo, R.C., Thurston, R.V., 1977. Trimmed Spearman–Karber method for estimating median lethal concentrations in toxicity bioassays. Environ. Sci. Technol. 11, 714–719.
13. Helgesen, K.O., Horsberg, T.E., 2013a. Single-dose field bioassay for sensitivity testing in sea lice, Lepeophtheirus salmonis: development of a rapid diagnostic tool. J. Fish Dis. 36, 261–272. http://dx.doi.org/10.1111/jfd.12053.
14. Helgesen, K.O., Horsberg, T.E., 2013b. Influence of different materials on the concentration of delousing agents in sea water during bioassays. J. Fish Dis. 36, 529–532. http://dx. doi.org/10.1111/jfd.12046.
15. Heumann, J., Carmichael, S., Bron, J.E., Tildesley, A., Sturm, A., 2012. Molecular cloning and characterisation of a novel P-glycoprotein in the salmon louse Lepeophtheirus salmonis. Comp. Biochem. Physiol. C Toxicol. Pharmacol. 155, 198–205. http://dx. doi.org/10.1016/j.cbpc.2011.08.004.
16. Heumann, J., Carmichael, S.N., Bron, J.E., Sturm, A., 2014. Isolation and characterisation of four partial cDNA sequences encoding multidrug resistance-associated proteins (MRPs) in the salmon louse Lepeophtheirus salmonis (Krøyer, 1837). Aquaculture 424-425, 207–214.
17. Igboeli, O.O., Fast, M.D., Heumann, J., Burka, J.F., 2012. Role of P-glycoprotein in emamectin benzoate (SLICE®) resistance in sea lice, Lepeophtheirus salmonis. Aquaculture 344-349, 40–47. http://dx.doi.org/10.1016/j.aquaculture.2012.03.026.
18. Jones, M.W., Sommerville, C., Wootten, R., 1992. Reduced sensitivity of the salmon louse, Lepeophtheirus salmonis, to the organophosphate dichlorvos. J. Fish Dis. 15, 197–202. http://dx.doi.org/10.1111/j.1365-2761.1992.tb00654.x.
19. Kaur, K., Helgesen, K.O., Bakke, M.J., Horsberg, T.E., 2015. Mechanism behind resistance against the organophosphate azamethiphos in salmon lice (Lepeophtheirus salmonis). PLoS One 10, e0124220. http://dx.doi.org/10.1371/journal.pone.0124220.
20. Lees, F., Baillie, M., Gettinby, G., Revie, C.W., 2008. The efficacy of emamectin benzoate against infestations of Lepeophtheirus salmonis on farmed Atlantic salmon (Salmo salar L) in Scotland, 2002-2006. PLoS One 3, e1549. http://dx.doi. org/10.1371/ journal.pone.0001549.
21. Ljungfeldt, L.E.R., Espedal, P.G., Nilsen, F., Skern-Mauritzen, M., Glover, K.A., 2014. A common- garden experiment to quantify evolutionary processes in copepods: the case of emamectin benzoate resistance in the parasitic sea louse Lepeophtheirus salmonis. BMC Evol. Biol. 14, 108. http://dx.doi.org/10.1186/1471-2148-14-108.

22. Pascoe, D., Edwards, R.W., 1989. Single species toxicity tests. Aquatic Ecotoxicology: Fundamental Concepts and Methodologies vol. 2. CRC Press, Boca Raton, FL, USA, pp. 93–127.
23. Rand, G.M., 2008. Fish Toxicity Studies. In: Di Giulio, R.T., Hinton, D.E. (Eds.), The Toxicology of Fishes. CRC Press, Boca Raton, FL, USA, pp. 659–681.
24. Robertson, J.L., Preisler, H.K., 1991. Pesticide Bioassays with Arthropods. CRC Press, Boca Raton, FL, USA.
25. Roth,M., Richards, R.H., Dobson, D.P., Rae, G.H., 1996. Field trials on the efficacy of the organophosphorus bcompound azamethiphos for the control of sea lice (Copepoda: Caligidae) infestations of farmed Atlantic salmon (Salmo salar). Aquaculture 140, 217–239.
26. Saksida, S.M., Morrison, D., McKenzie, P., Milligan, B., Downey, E., Boyce, B., Eaves, A., 2013. Use of Atlantic salmon, Salmo salar L., farm treatment data and bioassays to assess nfor resistance of sea lice, Lepeophtheirus salmonis, to emamectin benzoate (SLICE(®) ) in British Columbia, Canada. J. Fish Dis. 36, 515–520. http://dx.doi.org/10.1111/jfd.12018.
27. SARF, 2012. Evaluation of sensitivity to chemotherapeutants in successive generations of Lepeoptheirus salmonis from a resistant population. Sarf069, Pitlochry, Scotland, UK.
28. Sayer,M.D.J., Treasurer, J.W., Costello, M.J. (Eds.), 1996.Wrasse: Biology and Use in Aquaculture. Wiley-Blackwell, Oxford, UK.
29. Sevatdal, S., Horsberg, T.E., 2003. Determination of reduced sensitivity in sea lice (Lepeophtheirus salmonis Krøyer) against the pyrethroid deltamethrin using bioassays and probit modelling. Aquaculture 218, 21–31. http://dx.doi.org /10.1016/S0044- 8486(02)00339-3.
30. Stien, L.H., Dempster, T., Bui, S., Glaropoulos, A., Fosseidengen, J.E.,Wright, D.W., Oppedal, F., 2016. "Snorkel" sea lice barrier technology reduces sea lice loads on harvest-sized Atlantic salmon with minimal welfare impacts. Aquaculture 458, 29–37. http://dx. ndoi.org/10.1016/j.aquaculture.2016.02.014.
31. Tanneberger, K., Rico-Rico, A., Kramer, N.I., Busser, F.J.M., Hermens, J.L.M., Schirmer, K., 2010. Effects of solvents and dosing procedure on chemical toxicity in cell-based in nvitro assays. Environ. Sci. Technol. 44, 4775–4781. http://dx.doi.org/10.1021/es100045y.
32. Treasurer, J.W., Wadsworth, S., Grant, A., 2000. Resistance of sea lice, Lepeophtheirus salmonis (Kroyer), to hydrogen peroxide on farmed Atlantic salmon, Salmo salar L. Aquac. Res. 31, 855–860.
33. Westcott, J.D., Stryhn, H., Burka, J.F., Hammell, K.L., 2008. Optimization and field use of a bioassay to monitor sea lice Lepeophtheirus salmonis sensitivity to emamectin benzoate. Dis. Aquat. Org. 79, 119–131. http://dx.doi.org/10.3354/dao01887.
34. Whyte, S.K., Westcott, J.D., Elmoslemany, A., Hammell, K.L., Revie, C.W., 2013. A fixeddose approach to conducting emamectin benzoate tolerance assessments on fieldcollected sea lice, Lepeophtheirus salmonis. J. Fish Dis. 36, 283–292. http://dx.doi.org/ 10.1111/jfd.12055.

# CHAPTER 15

# Nutrient Profiles of Rotifers (*Brachionus* Sp.) and Rotifer Diets from Four Different Marine Fish Hatcheries

**Kristin Hamre ***

*National Institute of Nutrition and Seafood Research (NIFES), PO Box 2029, 5817 Bergen, Norway*

## ABSTRACT

Cultured marine fish larvae are most often fed rotifers (*Brachionus* sp.) the first few weeks after first-feeding. However the nutritional value of rotifers is generally inferior to the natural diet of marine fish larvae which is comprised of zooplankton, mostly copepods. Commercial diets for rotifers vary considerably in nutrient composition and this variation is partly mirrored in rotifers. The present study was undertaken to investigate the variation in rotifers and rotifer diets used in commercial marine fish hatcheries and to identify possible deficiencies or excess of individual nutrients. Culture and enrichment diets and unenriched and enriched rotifers were sampled from four hatcheries for Ballan wrasse (*Labrus bergylta*) and analyzed for nutrients which can potentially fall outside the safe window of supplementation for marine fish larvae. It is concluded that rotifer diets generally contain appropriate levels of fatty acids and vitamins C and E, while vitamin A, iodine and selenium need more attention. For vitamins D and K and many of the micro-minerals, data on larval requirements are still lacking and these nutrients need further research. Protein and phospholipid levels are mainly determined by the rotifer's own metabolism and can be low compared to assumed requirements.

## Statement of relevance

- This study shows that there is a large variation in the nutrient composition of commercial rotifer diets and a resulting large variation in the nutrient composition of rotifers fed to marine fish larvae at an industrial scale.
- In some cases the levels of single nutrients in rotifers are below the requirements given for fish by the NRC (2011).
- This means that the relatively low predictability of fish larval culture can partly be explained by lack of knowledge in nutrition and inferior commercial live feed diets.
- This may inspire to more research into how the nutrient composition of live feeds can be improved and what are the nutrient requirements of marine fish larvae, in order to standardize diets.

**Keywords:** Rotifers, Rotifer diets, Nutrient composition

# 1. INTRODUCTION

Rotifers (*Brachionus* sp.) are used as first-feed for most cultured marine fish larvae. Partly dependent on the conditions for culture and enrichment, several nutrients may be present at potentially deficient levels in rotifers (Hamre et al., 2008b). The levels of some nutrients, such as protein, amino acids (Srivastava et al., 2006) and phospholipid (Helland et al., 2010), are largely determined by the metabolism of the feed organisms and are quite stable and independent of culture conditions. Other nutrients show a dose response relationship between diet concentration and rotifer concentration (Helland et al., 2010; Merchie et al., 1997; Nordgreen et al., 2013; Olsen, 2004; Penglase et al., 2011; Srivastava et al., 2011, 2012). There are many different rotifer diets on the market which vary in nutrient compositions (Hamre et al., 2008a; Maehre et al., 2013; Srivastava et al., 2006). Furthermore, the live feed culture and enrichment protocols vary from one hatchery to another. The purpose of the present study was to assess the nutritional quality of the rotifers offered to first-feeding Ballan Wrasse (*Labrus bergylta*) larvae and the composition of some of the rotifer diets. This was done by 1) investigating the variation in composition of diets used for rotifers, 2) investigating the variation in nutrient composition in unenriched and enriched rotifers and 3) identifying potentially limiting or excessive levels of individual nutrients in enriched rotifers.

# 2. MATERIALS AND METHODS

## 2.1. Culture and enrichment of rotifers

### 2.1.1. Hatchery A

The rotifers, strain *Brachionus plicatilis*, "Austria" (L strain, 130–340 μm, Florida Aqua Farms), were cultured with 2 mL day$^{-1}$ million rotifers$^{-1}$*Chlorella*, (*Chlorella* SV 12, PACIFIC TRADING CO., Fukuoka, Japan). The cultures were run in a flow through system in 2800 L tanks with exchange of 3 tank volumes of water per day, temperature 23 ± 0.5 °C, [$O_2$] > 80%, pH ~ 7.2, salinity ~ 36 g L$^{-1}$ and the rotifer density was held at 500 ind mL$^{-1}$. Every day 10–30% of each culture was harvested, depending on the culture growth rate. The cultures were pumped to a clean tank once a week. Approximately 16 h before enrichment, the harvested cultures were transferred to a waiting tank where they were fed ¼ of the dose of *Chlorella* used in the culture tanks. The rotifers were then washed, transferred to an 800 L enrichment tank (density 500–1500 ind mL$^{-1}$) and enriched for 2–3 h using a diet composed of the following ingredients: oil blend (50% of the diet): phospholipid (LC 60, Phosphotech); Croda oil (Incromega DHA 500TG, CRODA NORDICA, Yorkshire, G.B.); vitamin C (6-O-palmitoyl-L-ascorbic acid, Sigma-Aldrich); vitamin A (Retinol palmitate all trans, Sigma-Aldrich); vitamin E (α-tocopherol acetate, Sigma-Aldrich); Tween 80 (Polyoxyethylene-sorbitan monooleate, Sigma-Aldrich); vitamin B1 (thiamine HCL, Sigma-Aldrich); Astaxanthin (Aquasta, NATURXAN, Maryland CO). Protein blend (50% of the diet): microfeed (TROMSØ FISKEINDUSTRI, Tromsø, Norway,); ORI-GREEN (SKRETTING, Stavanger, Norway); Sel-plex 2000 (ALLTECH). The oil blend was added to the tank 30 min before the protein blend. The temperature, oxygen concentration, pH and salinity were the same as during culture. After enrichment, the rotifers were washed using water with the same temperature as used in enrichment. During washing, the rotifers were suspended between two slowly rotating screens of plankton net, taking care that the gut contents was not evacuated. The temperature was then slowly decreased to the level in the larval tanks to avoid settling of the rotifers, and samples were taken.

### 2.1.2. Hatchery B

The rotifers (probably "Cayman", lorica size 150–250 μm) were cultured for 3 days in batch culture (tank size 3800 L, 22–25 °C, [$O_2$] > 70%, density 400–1600 ind mL$^{-1}$). Aeration was set so that the water in the cone was stagnant for sedimentation, above the cone there was a slow movement of the water. *Chlorella* (PACIFIC TRADING CO., Fukuoka, Japan;

1.8 mL $day^{-1}$ mill $rotifers^{-1}$) was used as culture feed. The rotifers were transferred to 2000 L tanks, density 1000–2500 ind $mL^{-1}$ at the same conditions with respect to temperature and oxygen as during culture, and then enriched with Larviva Multigain (BIOMAR AS, Brande, Denmark, 0.4 g mill $rotifers^{-1}$) for 1.5 h, the last half hour with Pyceze (EUROPHARMA AS). Washing between culture and enrichment and after enrichment were performed with water temperated to 22 °C in a system where the rotifers were suspended. The samples were taken from the washer.

### 2.1.3. Hatchery C

The rotifers were *Brachionus*, mainly of the strain "Cayman" but with incidence of "Austria" (the lorica length should be 150–350 µm). They were cultured in 3500 L tanks, 500–2600 ind $mL^{-1}$ in continuous culture which lasted for 63 and 66 days for the two samples. The temperature varied from 24.1–27.0 °C, [$O_2$] from 72–125% and the culture feed was Rotifer diet (REED MARICULTURE, Campbel, CA; 1.1 mL mill $rotifers^{-1}$ $day^{-1}$) and baker's yeast (local supplier, 0.33 g mill $rotifers^{-1}$ $day^{-1}$). The rotifer diet was pumped continuously except for a 6 h pause every day, while the yeast was fed in 3 portions per day. The rotifers were enriched for 2 h with controlled supply of oxygen in 315 or 630 L tanks at 1000–5000 ind $L^{-1}$, dependent on the demand for rotifers. The enrichment diet was Larviva Multigain + ORI-GREEN (50/50) 0.35 g mill $rotifers^{-1}$. Washing was performed in 630 L tanks with suspended rotifers and continuous supply of temperated water for 30 min. The samples were taken from the washer.

### 2.1.4. Hatchery D

The rotifer stock was a blend of *Brachionus* "Nevada", lorica length ~ 250 µm and *Brachionus rotundiformis* which are slightly smaller. Culture of rotifers was performed in 2400 L tanks with temperature 22 °C, oxygen control and rotifer density 500–1000 ind $mL^{-1}$. The rotifers were cultured for 7 days, 50% of the volume was harvested on days 3, 5 and 7 and the remaining 50% of the culture on day 7 was washed before the procedure was restarted. Feeding of the rotifer cultures: Monday–Friday with dry yeast (0.225 g mill $rotifers^{-1}$ $day^{-1}$) and Larviva Multigain (0.04 g mill $rotifers^{-1}$ $day^{-1}$), Saturday and Sunday with dry yeast (0.25 g mill $rotifers^{-1}$ $day^{-1}$) and rotifer diet (0.2 g mill $rotifers^{-1}$ $day^{-1}$). Enrichment was performed over night (approximately 20 h) in 1200 L tanks at 18–20 °C and rotifer density 100–1500 ind $mL^{-1}$, alternating between three enrichment diets: S. Presso (INVE, Dendermonde, Belgium; 3 × 0.12 g mill $rotifers^{-1}$), Larviva Multigain (3 × 0.09 g mill $rotifers^{-1}$), Red Pepper (BERNAQUA, Olen, Belgium; 3 × 0.12 g mill $rotifers^{-1}$). The enriched rotifers were transferred to a washer where they were held in suspension, washed for 10 min in temperated water, and gradually cooled to 6 °C. Samples were taken from the washer after cooling.

## 2.2. Sampling

Samples of unenriched and enriched rotifers were taken at the hatcheries A–C. At hatchery D samples were taken only of enriched rotifers, but from rotifers enriched with each of the three different diets. Culture and enrichment diets were also sampled at the hatcheries. Water was silt from the rotifers using a plankton filter, the rotifers were rinsed in sea water and the filter was patted dry from underneath using a paper towel. The samples were filled in separate tubes for the different analyses and stored at – 80 °C, on dry ice or on liquid $N_2$, and sent to NIFES on a liquid $N_2$ transport tank or on dry ice. The analyses were done on wet material and recalculated to dry matter (DM) using the results from the dry matter analyses.

## 2.3. Analyses

The analyses were performed at NIFES using accredited methods (Table 1). There are two replicate treatments at hatcheries A and C, at the other hatcheries there is only one replicate for each treatment. The analyses were performed in duplicate and the results were reanalyzed if the deviation between replicates was above the method specific threshold. The measurement uncertainty is given in Table 1.

# 3. RESULTS

## 3.1. Protein, lipids and fatty acids

The culture diets, e.g., *Chlorella*, yeast and rotifer diet from Reed Mariculture, were characterized by relatively high levels of protein (37–51% DM) and relatively low levels of lipid (1.5–13% DM) accompanied by a high phospholipid to neutral lipid (PL:NL) ratio, PL amounting to 38–44% of total lipids (Table 2, data not given for yeast). Yeast was devoid of long chain fatty acids, *Chlorella*contained EPA and DHA at 4 and 9% of total fatty acids (TFA), while rotifer diet contained relatively high levels of ARA and EPA (3.4 and 24% of TFA, respectively) and insignificant levels of DHA. The enrichment diets, ORI-GREEN, Larviva Multigain and Red Pepper, contained 41–43% lipid, where 84–96% was NL. The fatty acid composition was characterized by relatively low levels of ARA and EPA, but high levels of DHA (24–34% of TFA). ORI-GREEN had a relatively high level of protein

(43% of DM), while Multigain and Red Pepper contained 12 and 17% protein of DM, respectively. The enrichment diet of hatchery A and the S. Presso diet used in one of the enrichments from hatchery D were not analyzed. PL and protein were added in the enrichment diet from hatchery A, but the precise levels are not known.

**Table 1.** Analytical methods.

| Analyte | Principle | 2RSD%[a] | Reference |
|---|---|---|---|
| Dry matter | Gravimetric after freeze drying | 3 | Hamre and Mangor-Jensen (2006) |
| Protein | N × 6.25, Leco N analyzer | 11 | Hamre and Mangor-Jensen (2006) |
| Lipid | Gravimetric after acid hydrolysis | 10 | (EU directive 98/64/EC 1998) |
| Fatty acids | Transmethylation, extraction and GC/FID | 10 | Lie and Lambertsen (1991) |
| Vitamin C | HPLC | 15 | Mæland and Waagbø (1998) |
| Vitamin A | HPLC | 20 | Moren et al. (2002) |
| Vitamin D | HPLC | 15 | CEN (1999) |
| Vitamin E | HPLC | 10 | Hamre et al. (2010) |
| Vitamin K | HPLC | [b] | CEN (2003) |
| Macrominerals | ICP-MS | 15 | Julshamn et al. (2007) |
| Microminerals | ICP-MS | <25 | Julshamn et al. (2004) |
| Iodine | IC-PMS | 15 | Julshamn et al. (2001) |

[a] 2RSD, Relative standard deviation of a long series of determination of a control sample, % of mean value.
[b] The method of vitamin K analysis is new and the internal reproducibility has not been determined.

**Table 2.** Macronutrients, lipid classes and essential fatty acids in rotifer diets.

| | DM | Protein | Lipid | Sum PL | Sum NL | ARA | EPA | DHA |
|---|---|---|---|---|---|---|---|---|
| | % WW | % DM | % DM | Wt.%TL | Wt.%TL | % TFA | % TFA | % TFA |
| Yeast | 33 | 37 | 1.5 | n.a. | n.a. | 0 | 0 | 0 |
| *Chlorella* 1 | 11 | 52 | 12 | 44 | 56 | 0.2 | 4 | 9.5 |
| *Chlorella* 2 | 11 | 51 | 11 | 42 | 58 | 0.1 | 4 | 9.2 |
| Rotifier diet | 22 | 40 | 13 | 38 | 62 | 3.4 | 23.6 | 0.6 |
| ORI-GREEN | 97 | 43 | 41 | 17 | 84 | 1.5 | 4.8 | 24.5 |
| Multigain | 98 | 12 | 44 | 9 | 91 | 0.5 | 1.4 | 33.7 |
| Red Pepper | 30 | 17 | 43 | 4 | 96 | 1.3 | 2.5 | 28.6 |

n.a., not analyzed; TL, total lipids; TFA, total fatty acids; *Chlorella* 1 and 2, batches from hatcheries A and B, respectively.

Unenriched rotifers contained 41, 43, and 45% protein at hatcheries A, B and C, respectively (Table 3). The protein content was stable or was reduced by 1–2% upon enrichment. The protein levels in enriched rotifers from hatchery D varied between 37 and 42% of DM. The lipid levels in unenriched rotifers varied between 7–8% of DM at hatcheries A and B and 11–13% of DM at hatchery C. Enrichment gave a 2% increase in lipid level at hatcheries A and B, while at hatchery C the lipid level was stable or decreased by 3% after enrichment. The lipid level in enriched rotifers at hatchery D was higher than at the other hatcheries, at 16–17% of DM. The fraction of PL in total lipid (TL) was 64–66% in unenriched rotifers from hatcheries A and B and 71–72% at hatchery C. Enrichment gave a decrease in PL fraction at all three hatcheries to between 52 and 58% of TL. The PL fraction in rotifers from hatchery D was 39–52%. Unenriched rotifers from hatcheries A and B had very similar fatty acid composition, 0.5–1% ARA, 8.9–9.3% EPA and 12.1–13.7% DHA of TFA. Enrichment gave very little change in ARA and EPA while DHA increased to between 19.4 and 23.7% of TFA. Unenriched rotifers from hatchery C were characterized by a high level of EPA (13–14% of TFA) and a low level of DHA (0.1–0.2% of TFA). The level of ARA was 1.5–2.2% of TFA. Enrichment gave a minimal change in ARA, EPA decreased to 8–9% and DHA increased to 11–14% of TFA. The enriched rotifers from hatchery D varied in ARA and DHA levels (0.8–1.4% and 22–35% of TFA, respectively), while the EPA level was stable at 4%.

**Table 3.** Macronutrients, lipid classes and essential fatty acids in unenriched and enriched rotifers sampled in hatcheries A–C and rotifers enriched using three different protocols from hatchery D, compared with fish requirements (NRC, 2011) and nutrient composition of copepods, mainly copepodites of *Temora longicornis* (Hamre et al., 2008a,b, 2013).

| | DM | Protein | Lipid | Sum PL | Sum NL | ARA | EPA | DHA |
|---|---|---|---|---|---|---|---|---|
| | % WW | % DM | % DM | Weight% | Weight% | % TFA | % TFA | % TFA |
| Unenriched | | | | | | | | |
| A 23/9 | 13.3 | 41.3 | 7.4 | 66 | 34 | 0.9 | 9.3 | 13.7 |
| A 26/9 | 13.0 | 41.3 | 8.1 | 64 | 36 | 0.7 | 9.3 | 13.4 |
| B | 15.8 | 42.6 | 7.1 | 65 | 35 | 0.5 | 8.9 | 12.1 |
| C 26/10 | 11.4 | 44.6 | 13.2 | 72 | 28 | 2.2 | 13.8 | 0.2 |
| C 31/10 | 18.3 | 45.1 | 10.9 | 71 | 29 | 1.5 | 13 | 0.1 |
| Enriched | | | | | | | | |
| A 23/9 | 15.1 | 40.3 | 9.8 | 52 | 48 | 1 | 10 | 22.1 |
| A 26/9 | 14.4 | 40.1 | 10.4 | 53 | 47 | 0.8 | 9.9 | 23.7 |
| B | 15.4 | 42.1 | 9.1 | 55 | 45 | 0.5 | 6.6 | 19.4 |
| C 26/10 | 24.3 | 44.0 | 9.9 | 58 | 42 | 2.1 | 9 | 11.1 |
| C 31/10 | 19.6 | 43.1 | 10.6 | 54 | 46 | 1.3 | 7.9 | 14.4 |
| D 1 | 14.2 | 41.5 | 17.1 | 52 | 48 | 1.4 | 4.3 | 22.6 |
| D 2 | 13.3 | 38.18 | 16.7 | 40 | 60 | 1 | 3.7 | 34.6 |
| D 3 | 12.5 | 37.20 | 16.1 | 39 | 61 | 0.8 | 4.2 | 22.4 |
| NRC (2011) | – | 30–60 | – | – | – | – | – | – |
| Copepod levels | – | 63 ± 9 | 16 ± 3 | 50 ± 12 | 50 ± 12 | 0.7 ± 0.2 | 19 ± 3 | 29 ± 5 |

–, not given.

## 3.2. Vitamins

Concerning the culture diets, yeast was low in vitamin C, while *Chlorella* and rotifer diet contained 221–234 mg $kg^{-1}$ DM vitamin C (Table 4). The vitamin C levels in the enrichment diets were measured to 15–38 mg $kg^{-1}$ which is probably a large underestimation caused by the fact that our analytical method does not function on all synthetic vitamin C formulas. Vitamin A was below the detection limit in the culture diets and between 33 and 80 mg $kg^{-1}$ in the enrichment diets, while vitamin D concentration varied both in the culture (0.09–1.15 mg $kg^{-1}$ DM) and enrichment (0.17–1.02 mg $kg^{-1}$ DM) diets. The biologically most active form of vitamin E, α-tocopherol (TOH), was almost absent in yeast. In the other culture diets, the concentrations varied between 177 and 244 mg $kg^{-1}$ DM, while the enrichment diets contained 2000–3000 mg $kg^{-1}$ DM of α-TOH. ORI-GREEN and Red Pepper also contained significant amounts of γ-TOH. The concentration of vitamin K was very variable between the diets, 0.12–3.6 mg $kg^{-1}$ DM in the culture diets and 0.07–1.62 mg $kg^{-1}$ DM in enrichment diets.

**Table 4.** Vitamins in rotifer diets.

| mg $kg^{-1}$ DM | Vitamin C | Vitamin A | Vitamin-D3 | α-TOH | γ-TOH | Vitamin K |
|---|---|---|---|---|---|---|
| Yeast[a] | 0.4 | <0.3 | n.a. | 0.77 | n.a. | n.a. |
| *Chlorella* 1 | 221 | <0.3 | 0.97 | 177 | <0.05 | 0.13 |
| *Chlorella* 2 | 229 | <0.3 | 1.15 | 168 | <0.05 | 0.12 |
| Rotifier diet | 234 | <0.3 | 0.09 | 244 | <0.05 | 3.60 |
| ORI-GREEN | 38 | 80 | 0.70 | 2002 | 105 | 0.99 |
| Multigain | 51 | 41 | 1.02 | 2991 | <0.05 | 0.07 |
| Red Pepper | 15 | 33 | 0.17 | 2004 | 825 | 1.62 |

n.a., not analyzed; *Chlorella* 1 and 2, batches from hatcheries A and B, respectively.
[a] Hamre et al. (2008a,b).

The vitamin C concentrations in unenriched rotifers was in the range of 750–850, 425 and 100–200 mg $kg^{-1}$ DM at hatcheries A, B and C, respectively, and increased to 1000–1200, 1750 and 450 after enrichment, supporting the hypothesis that the enrichment diets contained considerably more vitamin C than what was analyzed (Table 5). Vitamin C concentration in enriched rotifers from hatchery D varied between 500 and 1300 mg $kg^{-1}$ DM. The vitamin A concentrations at hatchery A were 1.5–1.6 mg $kg^{-1}$ DM in unenriched and 5.3–5.5 mg $kg^{-1}$ DM in enriched rotifers. In hatchery B the levels were 3.1 and 10.4 and in hatchery C, 5.5–6.4 and 17–24 mg $kg^{-1}$ DM, respectively. The vitamin A levels in the enriched rotifers from hatchery D varied between 0.6 and 2.0 mg $kg^{-1}$ DM. α-TOH in unenriched rotifers was 220–230, 89 and 55–70 mg $kg^{-1}$ DM at hatcheries A, B and C, respectively, while enriched rotifers contained 400, 280 and 310–510 mg $kg^{-1}$ DM. Low levels of γ-TOH were found in most of the rotifer samples (3–13 mg $kg^{-1}$ DM), while the Red Pepper enriched rotifers from hatchery D contained 200 mg $kg^{-1}$ DM γ-TOH. Vitamin D in unenriched rotifers from hatcheries A and C was not detected by the analytical method, while in those from hatchery B it was 0.06 mg $kg^{-1}$ DM. Vitamin D in enriched rotifers varied between 0.07 and 0.38 mg $kg^{-1}$ DM and rotifers from the same hatchery could vary within the whole range of vitamin D levels. Unenriched rotifers from hatchery A contained 0.2 mg $kg^{-1}$ DM vitamin K and the concentration increased to 0.3–0.4 mg $kg^{-1}$ DM by enrichment. Hatchery B and C had unenriched rotifers with 0.4 and 0.7–0.9 mg $kg^{-1}$ DM vitamin K and the concentrations decreased during enrichment to 0.1–0.2 mg $kg^{-1}$ DM. The rotifers from hatchery D contained 1.5–2.1 mg $kg^{-1}$ DM vitamin K. Vitamin K concentrations in enriched rotifers in this study seem to be low compared to requirements in fish (NRC, 2011).

**Table 5.** Vitamins in unenriched and enriched rotifers sampled in hatcheries A–C and rotifers enriched using three different protocols from hatchery D, compared with fish requirements (NRC, 2011) and nutrient composition of copepods, mainly copepodites of *Temora longicornis* (Hamre et al., 2008a,b, 2013).

| mg $kg^{-1}$ DM | Vitamin C | Vitamin A | Vitamin-D3 | α-TOH | γ-TOH | Vitamin K |
|---|---|---|---|---|---|---|
| Unenriched | | | | | | |
| A 23/9 | 745 | 1.58 | <0.08 | 226 | 4.5 | 0.17 |
| A 26/9 | 846 | 1.46 | <0.08 | 215 | 4.6 | 0.18 |
| B | 425 | 3.05 | 0.06 | 89 | 3.2 | 0.44 |
| C 26/10 | 167 | 6.15 | <0.08 | 70 | 3.4 | 0.87 |
| C 31/10 | 109 | 5.46 | <0.08 | 55 | 2.7 | 0.66 |
| Enriched | | | | | | |
| A 23/9 | 1194 | 5.31 | 0.13 | 398 | 11.9 | 0.41 |
| A 26/9 | 1040 | 5.54 | 0.07 | 395 | 13.2 | 0.34 |
| B | 1754 | 10.40 | 0.19 | 279 | 3.2 | 0.17 |
| C 26/10 | 453 | 24.3 | 0.12 | 313 | 6.6 | 0.10 |
| C 31/10 | 459 | 16.8 | 0.26 | 510 | 10.2 | 0.19 |
| D 1 | 1271 | 0.56 | 0.07 | 614 | 198 | 1.53 |
| D 2 | 980 | 1.96 | 0.38 | 588 | 3.8 | 1.56 |
| D 3 | 522 | 1.44 | 0.16 | 955 | 9.6 | 2.11 |
| NRC (2011) | 50 | 0.75 | 0.006–0.06 | 50 | – | 0.2–2 |
| Copepod levels | 500 | 0.00 | n.a. | 110 | n.a. | n.a. |

– not given; n.a. not analyzed.

## 3.3. Macrominerals

Ca varied between 0.42–0.88 g $kg^{-1}$ DM in the culture diets to 2–5.6 g $kg^{-1}$ in the enrichment diets. Sodium concentration in the diets varied between 12 and 18 g $kg^{-1}$, except in ORI-GREEN, where the sodium concentration was 97 g $kg^{-1}$. Dietary potassium varied between 2 and 16 g $kg^{-1}$, *Chlorella* was high at 15–16 g $kg^{-1}$ while rotifer diet was lowest at 2.7 g $kg^{-1}$. The variation in Mg was from 0.9 to 5.1 g $kg^{-1}$, while P varied from 2.6 to 16 g $kg^{-1}$ (Table 6).

**Table 6.** Macrominerals in rotifer diets.

| g $kg^{-1}$ DM | Ca | Na | K | Mg | P | Ca:P |
|---|---|---|---|---|---|---|
| *Chlorella* 1 | 0.88 | 17.7 | 15.0 | 3.71 | 14.1 | 1:16 |
| *Chlorella* 2 | 0.88 | 16.8 | 15.9 | 3.62 | 15.9 | 1:18 |
| Rotifer diet | 0.42 | 12.3 | 2.67 | 1.44 | 7.39 | 1:18 |
| ORI-GREEN | 1.98 | 96.6 | 9.66 | 5.06 | 9.66 | 1:5 |
| Multigain | 1.94 | 14.3 | 5.92 | 1.33 | 3.37 | 1:2 |
| Red Pepper | 5.61 | 17.5 | 3.63 | 0.92 | 2.64 | 2:1 |

Macrominerals in yeast were not analyzed; *Chlorella* 1 and 2, batches from hatcheries A and B, respectively.

The rotifers at hatcheries A and B seemed to increase slightly in Ca in response to enrichment and rotifers fed to fish larvae would contain 2–3 g $kg^{-1}$ Ca. Rotifers from hatchery C did not increase their Ca content upon enrichment and contained only 1.1–1.4 g $kg^{-1}$, perhaps reflecting the low level of Ca in the culture diets. In hatchery D, enriched rotifers contained 2.9–3.3 g $kg^{-1}$ Ca. Rotifers in hatcheries A, B and D contained 40–55 g $kg^{-1}$ Na, while three of the batches in hatchery C contained 20–27 g $kg^{-1}$ Na. K varied between 10 and 14 g $kg^{-1}$, except for the first batch of unenriched rotifers which contained 28 g $kg^{-1}$ K. Mg was present at 5.0–6.3 g $kg^{-1}$, except the three last batches from hatchery C which contained 2.5–3.5 g $kg^{-1}$ Mg and the first enriched batch from hatchery A which contained 4.3 g $kg^{-1}$, e.g., higher than or within the range of copepod levels. The level of P varied between 8.3 and 11.1, except for the first batch from hatchery C which contained 20 g $kg^{-1}$ P. Fish requirement for P is 4.5–6.0 g $kg^{-1}$ DM (NRC, 2011), while copepods contain 12–15 g $kg^{-1}$(Table 7).

**Table 7.** Macrominerals in unenriched and enriched rotifers sampled in hatcheries A–C and rotifers enriched using three different protocols from hatchery D, compared with fish requirements (NRC, 2011) and nutrient composition of copepods, mainly copepodites of *Temora longicornis* (Hamre et al., 2008a,b, 2013).

| $g\ kg^{-1}$ DM | Ca | Na | K | Mg | P | Ca:P |
|---|---|---|---|---|---|---|
| Unenriched | | | | | | |
| A 23/9 | 2.1 | 54 | 12.0 | 5.0 | 8.8 | 1:4 |
| A 26/9 | 2.1 | 55 | 11.3 | 5.1 | 9.2 | 1:4 |
| B | 1.8 | 46 | 12.5 | 5.3 | 8.3 | 1:5 |
| C 26/10 | 2.3 | 51 | 28.0 | 6.3 | 19.9 | 1:9 |
| C 31/10 | 1.3 | 29 | 13.3 | 3.5 | 11.1 | 1:8 |
| Enriched | | | | | | |
| A 23/9 | 2.4 | 47 | 11.1 | 4.3 | 9.0 | 1:4 |
| A 26/9 | 2.9 | 54 | 12.5 | 5.1 | 9.4 | 1:3 |
| B | 2.2 | 53 | 14.4 | 6.1 | 9.8 | 1:4 |
| C 26/10 | 1.1 | 22 | 10.0 | 2.5 | 8.3 | 1:8 |
| C 31/10 | 1.4 | 27 | 12.0 | 3.2 | 10.1 | 1:7 |
| D 1 | 3.0 | 40 | 11.2 | 5.0 | 10.1 | 1:3 |
| D 2 | 2.9 | 44 | 10.9 | 5.4 | 8.7 | 1:3 |
| D 3 | 3.3 | 49 | 10.3 | 6.0 | 9.3 | 1:3 |
| NRC (2011) | – | – | – | 0.4–0.6 | 4.5–6.0 | – |
| Copepod levels | 1.1–2.4 | n.a. | n.a. | 2.4–3.1 | 12.4–15.0 | 1:5–1:9 |

–, not given; n.a, not analyzed.

## 3.4. Microminerals

The culture diets and ORI-GREEN contained a maximum of 3.5 mg $kg^{-1}$ DM iodine, while Multigain and Red Pepper contained 87 and 31 mg $kg^{-1}$ DM iodine, respectively. Manganese was low in yeast (8 mg $kg^{-1}$ DM) compared to the algal based culture diets *Chlorella* and rotifer diet (48 and 97 mg $kg^{-1}$ DM). The enrichment diets ORI-GREEN, Multigain and Red Pepper contained 12, 42 and 530 mg $kg^{-1}$ DM manganese, respectively. The concentration of copper in the culture diets and in ORI-GREEN varied between 1.3 and 3.9 mg $kg^{-1}$ DM, while Multigain and Red Pepper contained 23 and 211 mg $kg^{-1}$ DM, respectively. The concentration of zinc in yeast was 127 mg $kg^{-1}$ DM, in the algal based culture diets, *Chlorella* and rotifer diet, 9 mg $kg^{-1}$ DM, while the enrichment diets ORI-GREEN, Multigain and Red Pepper contained zinc at 43, 143 and 990 mg $kg^{-1}$ DM, respectively. The concentration of selenium in yeast was 0.03, that in *Chlorella* and Multigain was 0.1, while the rotifer diet contained 0.5 mg $kg^{-1}$ DM of selenium. Selenium was high in Red Pepper at 4.0 mg $kg^{-1}$DM (Table 8).

**Table 8.** Microminerals in rotifer diets.

| mg kg$^{-1}$ DM | I | Mn | Cu | Zn | Se |
|---|---|---|---|---|---|
| Yeast[1] | 0.03 | 7.8 | 1.3 | 127 | 0.03 |
| *Chlorella* 1 | 0.0 | 49 | 3.9 | 8.8 | 0.1 |
| *Chlorella* 2 | 0.0 | 48 | 3.8 | 8.8 | 0.1 |
| Rotifer diet | 3.5 | 97 | 3.3 | 9.2 | 0.5 |
| ORI-GREEN | 0.7 | 12 | 2.8 | 43 | 0.2 |
| Multigain | 87 | 42 | 23 | 143 | 0.1 |
| Red Pepper | 31 | 528 | 211 | 990 | 4.0 |

[1] Hamre et al. (2008a,b).

Iodine varied between 0.36 and 0.73 mg kg$^{-1}$ DM in unenriched rotifers. Enriched rotifers from hatchery A did not have an increased iodine concentration, while at hatcheries B and C the iodine concentration increased to between 1.5 and 3.8 mg kg$^{-1}$ DM. Enriched rotifers from hatchery D contained 2.6–8.8 mg kg$^{-1}$ DM iodine. Manganese in unenriched rotifers at hatcheries A, B and C were 15–16, 11 and 12–20 mg kg$^{-1}$ DM, respectively. Enrichment did not give increase in rotifer manganese concentration, which was stable at hatcheries A and B and decreased to 7–9 mg kg$^{-1}$ DM at hatchery C. Manganese varied between 5.9 and 6.5 mg kg$^{-1}$ DM at hatchery D. Copper varied between 3.6 and 6.5 mg kg$^{-1}$ DM in enriched and unenriched rotifers from hatcheries A and B and was only marginally affected by enrichment. The copper concentrations in unenriched rotifers at hatchery C were 58 and 97 mg kg$^{-1}$ DM and were increased to 120 and 128 mg kg$^{-1}$ DM upon enrichment. At hatchery D, the rotifers in one of the batches contained 19 and the other two contained 6–9 mg kg$^{-1}$ DM copper. The concentrations of zinc in unenriched rotifers were 40–42 and 33 at hatcheries A and B and increased slightly upon enrichment to 45–47 and 41 mg kg$^{-1}$ DM, respectively. Hatchery C had varying zinc levels both in unenriched (91 and 60 mg kg$^{-1}$ DM) and enriched (47 and 66 mg kg$^{-1}$ DM) rotifers. The zinc concentration in rotifers from hatchery D was 59–64 mg kg$^{-1}$ DM. At hatchery A, selenium increased from 0.2–0.3 to 2.9–4.7 mg kg$^{-1}$ DM after enrichment. Rotifers from hatchery B had 0.08 and 0.07 mg kg$^{-1}$ DM selenium in unenriched and enriched rotifers, respectively. At hatchery C, selenium was not increased by enrichment and enriched rotifers contained 0.29–0.34 mg kg$^{-1}$ DM, while rotifers from hatchery D contained 0.12–0.57 mg kg$^{-1}$ DM selenium (Table 9).

**Table 9.** Microminerals in unenriched and enriched rotifers sampled in hatcheries A–C and rotifers enriched using three different protocols from hatchery D, compared with fish requirements (NRC, 2011) and nutrient composition of copepods, mainly copepodites of *Temora longicornis* (Hamre et al., 2008a,b, 2013).

| mg $kg^{-1}$ DM | I | Mn | Cu | Zn | Se |
|---|---|---|---|---|---|
| Unenriched | | | | | |
| A 23/9 | 0.70 | 15 | 6.5 | 43 | 0.23 |
| A 26/9 | 0.72 | 16 | 6.4 | 40 | 0.30 |
| B | 0.36 | 11 | 3.6 | 33 | 0.08 |
| C 26/10 | 0.47 | 20 | 97 | 91 | 0.75 |
| C 31/10 | 0.66 | 12 | 58 | 60 | 0.39 |
| Enriched | | | | | |
| A 23/9 | 0.66 | 13 | 4.9 | 45 | 2.9 |
| A 26/9 | 0.73 | 15 | 4.8 | 47 | 4.7 |
| B | 3.04 | 11 | 5.5 | 41 | 0.07 |
| C 26/10 | 1.52 | 7.3 | 120 | 47 | 0.29 |
| C 31/10 | 3.80 | 8.9 | 128 | 66 | 0.34 |
| D 1 | 2.62 | 6.5 | 19 | 64 | 0.57 |
| D 2 | 5.31 | 6.5 | 9.1 | 60 | 0.12 |
| D 3 | 8.83 | 5.9 | 6.1 | 59 | 0.36 |
| NRC (2011) | 0.6–1.1 | 13 | 3–5 | 20–30 | 0.25–0.3 |
| Copepod levels | 50–350 | 8–25 | 12–38 | 340–570 | 3–5 |

# 4. DISCUSSION

This study shows that culture and enrichment diets for rotifers vary considerably in nutrient composition and that the variation is partly mirrored in the nutrient composition of both unenriched and enriched rotifers. For some of the nutrients, such as Se, the variations are critical and the rotifers cultured on diets in the lower ranges will give larval mortalities. Other nutrients, such as vitamin A and copper, are sometimes present at excessive amounts. Protein and phospholipid levels are mainly determined by rotifer metabolism and are hard to manipulate through the diets (Hamre et al., 2013).

## 4.1. Macronutrients, fatty acids and lipid classes

Protein levels in rotifers are relatively stable at approximately 40%, as has been found before (Hamre et al., 2013). The protein level is largely determined by rotifers genetics, less by rotifer diets or culture conditions and protein in rotifers is lower than the 60% commonly found in copepods (Hamre et al., 2013; Karlsen et al., 2015). Changes in lipid level in rotifers will affect the protein content in DM, but not protein levels measured on a wet weight basis, since lipid replaces water in animal tissues (Srivastava et al., 2006). The small increase in lipid level in hatcheries A and B upon enrichment is therefore in line with the similar decrease in DM based protein level. The lipid level of enriched rotifers is determined both by the culture and enrichment diets, since a specific enrichment diet seems to raise the level of lipid with a fixed amount (hatcheries A and B, Table 3). The end concentration will therefore be dependent on the final concentration in unenriched rotifers and that in the enrichment diet.

Enrichment lead to lowered levels of PL to total lipids in all cases. The enrichment diets all contained high fractions of NL, which may have contributed to this result. However, enriching with PL will give a quick metabolic conversion of PL into NL so that only the PL still in the digestive tract will add to the rotifer PL fraction, which mainly consists of membrane lipids from the tissues (Rainuzzo et al., 1997). Rotifers with a low lipid level will have a smaller fraction of storage lipid and therefore a high PL fraction (Hamre et al., 2013).

The fatty acid composition of rotifers was dependent on the diets as has been shown before (see for example (Olsen, 2004)). In hatcheries A and B, the *Chlorella* culture diet contained EPA and DHA, which was mirrored in the unenriched rotifers (Table 3). Enrichment increased DHA from 12–14% of TFA to approximately 20%, in line with the high level of DHA of Multigain and Croda oil (50% of TFA). In hatchery C, the fatty acid composition of unenriched rotifers also mirrored the rotifer diet and yeast used for culture, with relatively high levels of EPA but less DHA than 1% of TFA. It seems again that the DHA increased upon enrichment at a fixed amount in response to enrichment with Multigain. Enriched rotifers from hatchery D contained more than 20% DHA. The fatty acid requirements of Ballan wrasse are not known, but if we hypothesize that copepod levels cover the requirements, all hatcheries except hatchery C should have sufficient DHA levels. The ARA levels also seem similar to copepod levels in rotifers from all hatcheries except hatchery C. EPA was lower than in copepods in all hatcheries, especially in hatchery D, but the significance of this finding is unclear since EPA can probably be synthesized by conversion of DHA.

## 4.2. Vitamins

The requirements of vitamins C and E in fish are both 50 mg $kg^{-1}$ according to (NRC, 2011). Yeast contained very little vitamins C and E while the concentrations in the other culture diets were above 200 and

150 mg kg$^{-1}$ for vitamins C and E, respectively. This is mirrored in vitamins C and E concentrations at or above copepod levels, already in the unenriched rotifers at hatcheries A and B. The lower concentrations in unenriched rotifers at hatchery C may be explained by the use of yeast as part of the culture diet. The vitamin C concentrations in the enrichment diets were probably greatly underestimated, since our analytical method does not detect some of the commercially available vitamin C preparations. This is underpinned by the fact that vitamin C showed a large increase after enrichment of rotifers and concentrations in enriched rotifers was very high in all samples, at or above copepod levels. Vitamin E concentrations were above 2 g kg$^{-1}$ in all enrichment diets and all hatcheries also had vitamin E levels far above both requirement in fish and copepod levels. Vitamins C and E at high doses are not thought to be toxic, rather, several studies indicate improved stress resistance and immune function of pharmacological levels of vitamins C and E (Verlac Trichet, 2010; Waagbø, 1994, 2006).

The minimum requirement and optimum level of vitamin A in fish are 0.75 and 2.4 mg kg$^{-1}$ DM, respectively (Moren et al., 2004; NRC, 2011), while high levels of vitamin A give skeletal deformities (Boglione et al., 2013; Dedi et al., 1997; Hamre et al., 2013). The exact upper limit for safe vitamin A supplementation is not known, but results in (Moren et al., 2004) indicate that 25 mg/kg may be too high. The vitamin A levels of one of the batches of rotifers from hatchery D and the enriched rotifers from hatchery C, therefore seem to be slightly too low and too high, respectively.

There was a large variation in the vitamins D and K concentrations of the rotifer diets in this study. However, if the larval requirement for vitamin D is similar to the requirements in fish (NRC, 2011), all enriched rotifers had vitamin D levels above the requirement. For vitamin K, rotifer diet and Red Pepper had 10–30 fold higher levels (1.6 and 3.6 mg kg$^{-1}$) than *Chlorella* and Larviva Multigain (0.12 and 0.07). In line with this, the unenriched rotifers at hatchery C cultured with rotifer diet, seemed to have higher levels of vitamin K than those from hatcheries A and B, using *Chlorella*. The levels of vitamin K in enriched rotifers from hatcheries A and D were within the range given by (NRC, 2011), but relatively low in hatchery A. Hatcheries B and C, were below this range. Hatcheries A, B and C all used Multigain for enrichment, and were low in vitamin K compared to hatchery D, where at least one of the enrichments was Red Pepper. Very few studies have been done on vitamins D and K requirements in marine fish larvae and it seems that these vitamins are given little attention when formulating rotifer diets.

## 4.3. Minerals

The Ca levels in rotifers seem to be higher than in copepods, except in rotifers from hatchery C, where the concentration was within the copepod range. Mineral uptake by fish from copepods has been demonstrated to occur largely from the soft body parts, but the majority of copepod minerals are bound in forms with low bioavailability in the chitin exoskeleton (Reinfelder and Fisher, 1994). The digestibility of minerals from rotifers therefore seems to be higher than from copepods (Karlsen et al., 2015), so the larval Ca requirement should be covered by these rotifers and sea water. Copepods have high concentrations of many minerals and a relevant question is whether larval requirements are similar to fish requirements or to copepod levels (Hamre et al., 2008b). We have shown for iodine that the requirement is nearer to fish requirement than to copepod levels (Penglase et al., 2013) and this may be the case for other minerals as well. The requirement for iodine in cod larvae was estimated to 3.5 mg $kg^{-1}$ DM (Penglase et al., 2013). Of the other microminerals, selenium was critically low in rotifers from hatchery B, which were cultured on *Chlorella*and enriched with Multigain, both diets very low in selenium. At hatchery A the rotifers were enriched with selenium according to the protocol of (Penglase et al., 2011) and reached the target level, while Se in the other hatcheries varied around the minimum requirement in fish (NRC, 2011). Copper in rotifers from hatchery C was very high. This hatchery turned out to have copper valves in their pipeline system. At hatchery D the first batch was probably enriched with Red Pepper, which had a very high copper level. Mn and Zn requirements in marine fish larvae are not known, but Mn levels seem to be low compared to the requirements in fish (NRC, 2011). However, adding extra Mn to rotifers fed to cod larvae did not enhance larval growth or survival (Penglase et al., 2015).

# 5. CONCLUSION

Manufacturers of diets for rotifers used in culture of marine fish larvae seem to have reasonably good control of fatty acid composition and concentrations of vitamins C and E in rotifers, while vitamin A, iodine and selenium need more attention. Culture and enrichment procedures and requirements for these nutrients have been published (Nordgreen et al., 2013; Penglase et al., 2011; Srivastava et al., 2011, 2012) and diet manufacturers and hatchery managers could improve the nutrient composition of the feed by analyzing their rotifers and use the recommended enrichments if necessary. For vitamins D and K and many of the micro-minerals, data on larval requirements are still lacking and these nutrients need further research. The B vitamins, except thiamine, are usually present in rotifers at sufficient levels (Hamre et al., 2008b) and were therefore not analyzed in the present study.

## ACKNOWLEDGMENTS

The present study was funded by the Norwegian Seafood Research Fund — FHF (Project 900554). I am grateful to the hatchery staff for supplying the samples used for analyses and for the information on culture and enrichment procedures: Erling Otterlei from Sagafjord Seafarms, Espen Grøtan and Karen Kvalheim from Marine Harvest Labrus, Helge Ressem and Else Marie Halse Ressem from Profunda and Elin Eidsvik and Mads Kristian Lenes from Nordland Leppefisk.

## REFERENCES

1. Boglione, C., Gisbert, E., Gavaia, P.,Witten, P.E., Moren, M., Fontagne, S., Koumoundouros, G., 2013. Skeletal anomalies in reared European fish larvae and juveniles. Part 2:main typologies, occurrences and causative factors. Rev. Aquac. 5, S121–S167.
2. CEN, 1999. Comitè Europèen de Normalisation: Foodstuffs — Determination of Vitamin D by High Performance Liquid Chromatography — Measurement of Cholecalciferol (D3) and Ergocalciferol (D2) (prEN12821).
3. CEN, 2003. Comite Europeen de Normalisation: Foodstuffs — Determination of Vitamin K1 by HPLC (Reference EN14148).
4. Dedi, J., Takeuchi, T., Seikai, T.,Watanabe, T., Hosoya, K., 1997. Hypervitaminosis A during vertebral morphogenesis in larval Japanese flounder. Fish. Sci. 63, 466–473.
5. EU Directive 98/64/EC of 3 September, 1998. Establishing communitymethods of analysis for the determination of amino acids, crude oils and fats. Off. J. Eur. Com. (L57/14).
6. Hamre, K.,Mangor-Jensen, A., 2006. Amultivariate approach to optimization ofmacronutrient composition in weaning diets for cod (Gadus morhua). Aquac. Nutr. 12, 15–24.
7. Hamre, K., Mollan, T.A., Saele, O., Erstad, B., 2008a. Rotifers enriched with iodine and selenium increase survival in Atlantic cod (Gadus morhua) larvae. Aquaculture 284, 190–195.
8. Hamre, K., Srivastava, A., Ronnestad, I., Mangor-Jensen, A., Stoss, J., 2008b. Several micronutrients in the rotifer Brachionus sp. may not fulfil the nutritional requirements of marine fish larvae. Aquac. Nutr. 14, 51–60.
9. Hamre, K., Kolas, K., Sandnes, K., 2010. Protection of fish feed, made directly from marine raw materials, with natural antioxidants. Food Chem. 119, 270–278.
10. Hamre, K., Yufera,M., Ronnestad, I., Boglione, C., Conceicao, L.E.C., Izquierdo,M., 2013. Fish larval nutrition and feed formulation: knowledge gaps and bottlenecks for advances in larval rearing. Rev. Aquac. 5, S26–S58.
11. Helland, S., Oehme, M., Ibieta, P., Hamre, K., Lein, I., Barr, Y., 2010. Effects of enriching rotifers Brachionus (Cayman) with protein, taurine, arginine, or

phospholipids — start feed for cod larvae Gadus morhua L. Aquaculture Europe 2010, October 6–10. European Aquaculture Society, Porto.

12. Julshamn, K., Dahl, L., Eckhoff, K., 2001. Determination of iodine in seafood by inductively coupled plasma/mass spectrometry. J. AOAC Int. 84, 1976–1983.
13. Julshamn, K., Lundebye, A.K., Heggstad, K., Berntssen, M.H.G., Boe, B., 2004. Norwegian monitoring programme on the inorganic and organic contaminants in fish caught in the Barents Sea, Norwegian Sea and North Sea, 1994–2001. Food Addit. Contam. 21, 365–376.
14. Julshamn, K., Maage, A., Norli, H.S., Grobecker, K.H., Jorhem, L., Fecher, P., 2007. Determination of arsenic, cadmium, mercury, and lead by inductively coupled plasma/mass spectrometry in foods after pressure digestion: NMKL1 interlaboratory study. J. AOAC Int. 90, 844–856.
15. Karlsen, Ø., van der Meeren, T., Rønnestad, I., Mangor-Jensen, A., Galloway, T., Kjørsvik, E., Hamre, K., 2015. Copepods enhance nutritional status, growth, and development in Atlantic cod (Gadus morhua L.) larvae — can we identify the underlying factors? PeerJ 3, e902. http://dx.doi.org/10.7717/peerj.902.
16. Lie, Ø., Lambertsen, G., 1991. Fatty acid composition of glycerophospholipids in seven tissues of cod (Gadusmorhua), determined by a combinedHPLC/GC method. J. Chromatogr. 565, 119–129.
17. Maehre, H.K., Hamre, K., Elvevoll, E.O., 2013. Nutrient evaluation of rotifers and zooplankton: feed for marine fish larvae. Aquac. Nutr. 19, 301–311.
18. Mæland, A., Waagbø, R., 1998. Examination of the qualitative ability of some cold water marine teleosts to synthesise ascorbic acid. Comp. Biochem. Physiol. A 121, 249–255.
19. Merchie, G., Lavens, P., Sorgeloos, P., 1997. Optimization of dietary vitamin C in fish and crustacean larvae: a review. Aquaculture 155, 165–181.
20. Moren, M., Naess, T., Hamre, K., 2002. Conversion of beta-carotene, canthaxanthin and astaxanthin to vitamin A in Atlantic halibut (Hippoglossus hippoglossus L.) juveniles. Fish Physiol. Biochem. 27, 71–80.
21. Moren, M., Opstad, I., Berntssen, M.H.G., Infante, J.L.Z., Hamre, K., 2004. An optimumlevel of vitamin A supplements for Atlantic halibut (Hippoglossus hippoglossus L.) juveniles. Aquaculture 235, 587–599.
22. Nordgreen, A., Penglase, S., Hamre, K., 2013. Increasing the levels of the essential trace elements Se, Zn, Cu and Mn in rotifers (Brachionus plicatilis) used as live feed. Aquaculture 380, 120–129.
23. NRC, 2011. Nutrient Requirements of Fish and Shrimp. The Natioanal Academic Press, Washington D.C.
24. Olsen, Y., 2004. Live food technology of cold-water marine fish larvae. In: Moksness, E., Kjørsvik, E., Olsen, Y. (Eds.), Culture of Cold-water Marine Fish. Blackwell Publishing Ltd., Oxford, UK, pp. 73–193.
25. Penglase, S., Hamre, K., Sweetman, J.W., Nordgreen, A., 2011. A new method to increase andmaintain the concentration of seleniumin rotifers (Brachionus spp.). Aquaculture 315, 144–153.
26. Penglase, S., Hamre, K., Olsvik, P.A., Grøtan, E., Nordgreen, A., 2015. Rotifers enriched with iodine, copper and manganese had no effect on larval cod (Gadus morhua) growth, mineral status or redox system gene mRNA levels. Aquac. Res. 46, 1793–1800.
27. Penglase, S., Harboe, T., Saele, O., Helland, S., Nordgreen, A., Hamre, K., 2013. Iodine nutrition and toxicity in Atlantic cod (Gadus morhua) larvae. PeerJ 1, e20.
28. Rainuzzo, J.R., Reitan, K.I., Olsen, Y., 1997. The significance of lipids at early stages of marine fish: a review. Aquaculture 155, 103–115.
29. Reinfelder, J.R., Fisher, N.S., 1994. Retention of elements absorbed by juvenile fish (Menidia menidia, Menidia beryllina) from zooplankton prey. Limnol. Oceanogr. 39, 1783–1789.

30. Srivastava, A., Hamre, K., Stoss, J., Chakrabarti, R., Tonheim, S.K., 2006. Protein content and amino acid composition of the live feed rotifer (Brachionus plicatilis): with emphasis on the water soluble fraction. Aquaculture 254, 534–543.
31. Srivastava, A., Stoss, J., Hamre, K., 2011. A study on enrichment of the rotifer Brachionus "Cayman" with iodine and selected vitamins. Aquaculture 319, 430–438.
32. Srivastava, A., Hamre, K., Stoss, J., Nordgreen, A., 2012. A study on enrichment of the rotifer Brachionus "Cayman" with iodine from different sources. Aquaculture 334, 82–88.
33. Verlac Trichet, V., 2010. Nutrition and immunity: an update. Aquac. Res. 41, 356–372.
34. Waagbø, R., 1994. The impact of nutritional factors on the immune system in Atlantic salmon, Salmo salar L.: a review. Aquac. Fish. Manag. 25, 175–197.
35. Waagbø, R., 2006. Feeding and disease resistance in fish. In: Mosenthin, R., Zentek, J., Zebrowska, T. (Eds.), Biology of Growing Animals. Elsevier Ltd., London, pp. 387–415.

# CHAPTER 16

# The Role of Shore Crabs and Mussel Density in Mussel Losses at a Commercial Intertidal Mussel Plot After Seeding

**Jacob J. Capelle[1,2,3] Gerard Scheiberlich[2] Jeroen W. M. Wijsman[2] Aad C. Smaal[1,2]**

[1]*Department of Aquaculture and Fisheries, Wageningen University, P.O. Box 338, 6700 AH Wageningen, The Netherlands*

[2]*IMARES Wageningen UR—Institute for Marine Resources and Ecosystem Studies, P.O. Box 77, 4400 AB Yerseke, The Netherlands*

[3]*HZ University of Applied Sciences, Edisonweg 4, 4382 NW Vlissingen, The Netherlands*

## ABSTRACT

Mussel losses peak after relaying seed on culture plots. The present paper is an attempt to examine the role of shore crab predation and initial mussel density on mussel losses in mussel bottom culture using an intertidal culture plot as a case study. Because of their small size and loose attachment, mussels are particularly vulnerable to predation at this stage, as well as to handling stress and intraspecific competition. In the experimental field plots (1 × 1 m) in the intertidal Oosterschelde, three different densities (1, 5 and 10 kg $m^{-2}$) of mussel seed are laid, with half of the experimental plots protected from predation by means of exclosures. Duration of the experiment was 5 weeks (August–September 2012) post-seeding. Protection was the major factor accounting for biomass production, followed by mussel density. Loss rates increased with mussel density, both in the exclosures and in the exposed plots. Losses in the exclosures with the lowest density were still 45 %. There are indications that handling stress prior to the start of the experiment played a major role in these losses. At the higher densities in the exclosures, losses increased to 72.1 % and were not significantly different between 5 and 10 kg $m^{-2}$. About one-third of the total loss (32.6 %) was attributed to shore crab predation. The number of shore crabs observed on the plots did not differ between treatments. Byssal thread development during the experimental period was followed and was found to be a slow process that was insufficient to protect mussels from crab predation at the sheltered experimental location.

**Keywords:** ***Mytilus edulis*** Mussel culture Byssus Attachment strength Mussel density ***Carcinus maenas***

# INTRODUCTION

The seeding of mussel seed dredged from natural beds or collected from submerged seed mussel collectors (SMCs) in the water column is the most critical step in the mussel bottom culture cycle (Calderwood et al. 2014; Capelle et al. 2014). Most of the mussel seed does not survive the cycle from seeding to harvest. Average survival of SMC-seed or seed relayed from wild mussel beds in autumn over the culture cycle is 8 and 15 %, respectively (Capelle et al. 2016). A substantial part of the losses is associated with seeding (Theisen 1968; Capelle et al. 2014, 2016). Such losses are a significant factor in biomass production (Capelle et al. 2016).

Losses on culture plots associated with or following seeding are clearly density dependent (Gascoigne et al. 2005; Capelle et al. 2014, 2016). Seeding practice in bottom culture leads to a highly heterogeneous distribution of mussels on culture plots, with high mussel densities within the space occupied by mussels (Capelle et al. 2014). Mussels are gregarious organisms and aggregate in patches, thereby competing for food and space (Fréchette and Bourget 1985; Liu et al. 2012).

Other factors that account for such losses are damaged mussels, handling stress (Calderwood et al. 2014) and predation. The main predators of young mussels are starfish in the subtidal area (Gallagher et al. 2008), and birds and tidal-moving shore crabs in the intertidal zone (Hilgerloh et al. 1997; Silva et al. 2014).

The shore crab (*Carcinus maenas*) is a well-studied predator of juvenile mussels. The feeding rate of shore crabs depends on the predator–prey size ratio, where the predation rate shows a rapid decrease with increasing mussel size (Crothers 1968; Mascaró and Seed 2001; Murray et al. 2007;

Kamermans et al. 2009). The impact of crab predation over the entire culture cycle has been estimated to range from 9.5 % (Murray et al. 2007) to 52 % mussel loss (Grosholz et al. 2011). These data are extrapolations of laboratory results on feeding rates to estimated field abundances. Exclusion of shore crabs from an intertidal culture plot (800 $m^2$) via installation of fences improved the yield by a factor of 4–5 for the entire culture cycle (Davies et al. 1980).

Mussels display several defence mechanisms against predation. In response to chemical cues from crabs, mussels increase their shell thickness, which significantly increases the handling time for foraging crabs (Freeman 2007). Mussels also increase the production of byssal thread in the presence of predators, thereby being more firmly attached and leading to reduced predation rates. This defence is widely used and has been known with regard to *Homarus* sp., *C. maenas* and *Cancer* sp. feeding on *Mytilus edulis* (Côté 1995; Reimer and Tedengren 1997; Garner and Litvaitis 2013), *Callinectes sapidus* feeding on *Ischadium recurvum* (Brown et al. 2011) and *Acanthocylus gayi* feeding on *Perumytilus purpuratus* and *Semimytilus algosus*(Caro et al. 2008), where *A. gayi* was selecting the mussels with the weakest attachment. Since shore crabs display competitive behaviour for food, dense mussel aggregations increase the intraspecific interactions between the crabs that might lead to a decrease in the predator–prey interactions in the field (Sneddon et al. 1997; Smallegange and Van Der Meer 2007). Interference between crabs is also indicated by experiments of Kamermans et al. (2009) where the rate at which 20 mm mussel seed was consumed, decreased from 6 seeds/day/crab to 3 seeds/day/crab when two crabs were placed together in one cage. However, when this hypothesis was tested on cockles, it was observed that the aggregation level had no effect on predation success of shore crabs (Whitton et al. 2012). When mussels were presented as a group rather than as a sequence of single individuals, predation rates increased, as the shore crabs were able to select smaller-sized individuals of a group more easily (Burch and Seed 2000).

Mussels attach to conspecific organisms using their byssus threads. The attachment strength of mussels increases with time and depends on the number of byssal threads that are produced (Kangeri et al. 2014). If shore crab predation rates are dependent on byssus attachment strength, mussels are most vulnerable during the short period after seeding, when the mussels redistribute (Capelle et al. 2014) and need to develop their attachment strength. Furthermore, damaged mussels or dead mussels, as a result of the seeding process (Calderwood et al. 2014) emit cues that enable shore crabs (which orientate themselves mainly via chemoreception) to locate a culture plot from a long distance shortly after seeding (Crothers 1968). Damaged or dead mussels are easy prey for shore crabs.

Losses peak on intertidal (Capelle et al. 2014) and subtidal plots (Capelle et al. 2016) shortly after seeding. The contribution of different loss factors to this loss is not known. It is hypothesised that density-dependent losses and shore crab predation are the major loss factors for mussel seed on intertidal culture plots.

We expect that shore crab predation will contribute significantly to losses shortly after seeding because of the small mussel size, weak byssal attachment and the high chemical attraction of damaged mussels. Therefore, the role of byssal attachment strength shortly after seeding as a protection mechanism against loss was also investigated.

## MATERIALS AND METHODS

A field experiment was conducted in order to estimate losses during the first 5 weeks post-seeding. In this experiment, the effects of mussel density and protection against predation on mussel condition and survival were measured.

### Site and experimental set-up

The field experiment was undertaken on an intertidal commercial mussel plot (51°N33.193′, 3°E53.327′) from the 15th of August 2012 to the 20th of September 2012. The site is located at a sheltered location within the Oosterschelde estuary in the Netherlands (Fig. 1). A section of this mussel plot was laid with mussel seed by a mussel farmer during high tide on the 14th of August 2012. Mussel seed used was harvested from seed mussel collectors (SMCs) from the western part of the Oosterschelde (Fig. 1) a day previously and was held overnight without water in the hold of a mussel vessel prior to seeding. A total of eighteen experimental units were set out in three rows, with six experimental units of 1 m × 1 m per row, with 1 m between each experimental unit. The experimental units were located approximately 10 m north-east of the area where the mussels were laid by the mussel farmer. Within each experimental unit of 1 m$^2$, mussels that were collected from the culture plot were evenly distributed in three different biomass densities over the experimental plots a day after seeding: 1, 5 and 10 kg m$^{-2}$, which corresponded to approximate mussel densities of 3000, 15,000 and 30,000 mussels m$^{-2}$, respectively. Half of the experimental units from each density were protected by cages Fig. 2). The cages were constructed using wooden frames (1 m × 1 m × 0.5 m) that were wrapped with chicken wire mesh (size: 0.8 mm). All the treatments (density and exclosure) were randomly assigned to the experimental units in triplicate. During the course of the experiment, due to an unknown reason, one of the exclosures of an initial biomass of 1 kg $m^{-2}$ of mussels was lost between the third and fourth week. At the start of the experiment, a random sub-sample ($n$ = 424) was taken from the culture plot in order to measure the length and the byssal attachment strength of the mussels. For a period of 4 weeks, three core samples ($d$ = 75 mm) were taken from each experimental unit haphazardly, once every week, and the collected mussels were pooled per experimental unit. In the fifth week of the experiment, all the remaining mussels were collected from each of the experimental units and were weighed for biomass, after which a random volumetric (100 ml) sub-sample was taken for analysis.

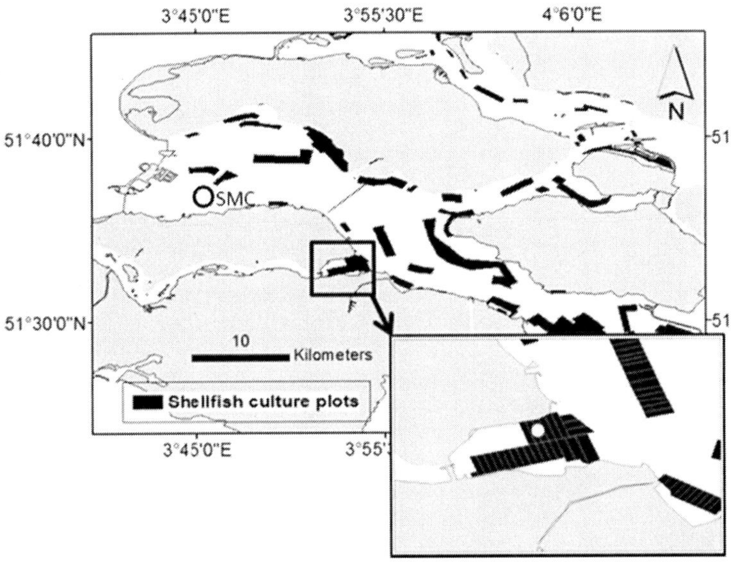


**Figure.** 1 Location of field experiment (white dot, 51°N33.193′, 3°E53.327′) and location of seed mussel collector (SMC) where mussel seed was collected prior to seeding on the experimental plot

**Figure.** 2 Picture of experimental design, showing cages as exclosures, open plots marked by wooden poles and L-shaped frames to which cameras were attached

Byssal attachment strengths of all the individual mussels, both within the sub-sample and within the intermediate core samples, were measured using a digital gauge (Sauter FK-10, 0–10 N). This device was attached to plastic tick tweezers, which were positioned around an individual mussel and pulled in a 180° direction from the point of attachment of the byssus threads until the mussel became loose. The force gauge displayed the maximum force (*N*) that was used during the experiment. The length of the individual mussel (*L*) was measured using a digital calliper (accurate to

0.1 mm), while the ash-free dry weight (AFDW) per sample was obtained by drying the sample at 80 °C and ashing it at 560 °C in a prepASH® 340 series until the difference in weight was observed to be less than 1 % per 30 min. Relative mussel loss rates ($r$, $d^{-1}$) were calculated for each of the experimental units as ($\ln(N_{end}) - \ln(N_{start})$) divided by the time of the experiment (days), where $N$ is the number of mussels $m^{-2}$. A condition index (CI, mg $cm^{-3}$) was calculated for each experimental unit via $AFDW/L^3$ (Beukema 1976). Biomass development over the experiment was expressed as $B_{end}/B_{start}$, where $B$ is biomass (kg $m^{-2}$).

Eight video cameras (wide-angle waterproof surveillance cameras) were placed above the exposed plots on the 22nd of August and nine cameras on the 29th of August, each covering an area of 0.6–0.7 $m^2$. Cameras recorded the activities on the plots continuously for 2.2 h, during high tide. The videos were used to estimate the average density of foraging shore crabs on the exposed plots in order to evaluate the relationship between the numbers of shore crabs and the mussel density, as well as to estimate the shore crab predation pressure. The crabs were counted and expressed as average number of individuals per $m^2$ per hour. Predation pressure was expressed as the number of foraging crabs per $m^2$ per hour. A crab was considered to be foraging when it stayed in one place for at least 30 s. Food processing movements were also observed in most such cases.

Average water temperature was 18 °C during the experimental period. Average wind speed was 4.0 m $s^{-1}$ (Beaufort 3); the maximum wind speed during the experimental period was recorded as 9.2 m $s^{-1}$ (Beaufort 5) on 1 day during the experiment (day 17 from the start). Wind direction was observed to range between 120° and 340°, with an average of 223° north. Experimental plots were completely dry only at water levels below about –1.50 m NAP (Amsterdam Ordnance Datum), which occurred for about 2 h out of every 24 h on average. No shore crabs were found in the exclosures at the end of the experiment.

In addition, we carried out a small predation experiment in the laboratory in which pre-aggregated mussels (four tanks) and loose mussels (four tanks) were exposed to an individual shore crab for a period of 8 h—the crabs had carapace width of 60–70 mm, and mussels ranged from 20 to 25 mm in shell length, from a different batch than used in the field experiment, and were fished from a subtidal commercial mussel plot. The experiment was repeated eight times over a period of 5 weeks using different crabs and mussels for each run. Average attachment strength of the pre-aggregated mussels was found to be 2.0 ± 1.0 N ($n = 185$), and mussels that were presented loose developed an attachment strength of 1.0 ± 0.7 N ($n = 83$) over the course of the 8 h for which the experiment lasted.

## Data analysis

The field experiment was set up as a random design. Differences in loss rates, biomass development and CI between treatments at the end of the experimental period were analysed using a two-way analysis of variance (ANOVA), with the factors of initial biomass and protection as response variables. The analyses were followed by a post hoc Tukey HSD (honest significant difference). The quantity of crabs that was measured from the video observations was tested against the initial mussel biomass (as factor) via the Kruskal–Wallis test. The level of significance was 0.05 for each analysis. Normality and homogeneity of the variances of the residuals were evaluated for all models by following Zuur et al. (2010).

The development of the attachment strength of the mussels over time was first analysed via a repeated-measures ANOVA; however, because of the heterogeneity of residual variances, a linear mixed-effects model (LME) was applied using 'lme' from the nlme library (Pinheiro et al. 2014). In this model, the mussel density, protection and time (number of weeks) were taken as fixed effects, and time nested in subject as a random effect. Model selection was carried out according to Zuur et al. (2009), and non-significant ($P > 0.05$) interactions were removed in a reverse stepwise manner. Restricted maximum likelihood (REML) estimators were used to estimate whether random structures improved model fit. When the model was improved via a random structure, the optimal fixed-effect structure was determined (with the same random structure) using maximum likelihood (ML) estimators. The final model was presented using an REML estimation. Normalised residuals of this fit were visually checked for normality using a histogram and were plotted against fitted values and against each explanatory variable to validate the model. The residuals closely followed a normal distribution, and no violation of homogeneity was indicated at this point. When significant overall effects were found, pairwise comparisons were made by means of a Tukey HSD test.

The number of mussels consumed by each shore crab per 8 h in the laboratory experiment was analysed using a generalised linear model (GLM), with the response factor ($Y$) containing the proportion dead and live mussels per observation, and a quasi-binomial variance model to account for over-dispersion. All analyses were carried out using R v.3.1.2 (R Core Team 2013).

# RESULTS

Video observations in the first week (22nd of August) showed significantly more ($F(1,11) = 8.33$, $P = 0.01$) (5.4 ± 2.8) shore crabs foraging for mussels (11.8 % of all passing crabs) than in the second week (29th of August, 1.7 ± 1.8 shore crabs, 7.4 % of all passing crabs). No statistical association between the number of foraging crabs and the initial mussel density was observed. Since crabs migrate with the tide, the measurements obtained during high tide will probably reflect the peak in numbers within each day.

Loss rates differed significantly between exclosures and exposed plots ($F(1,13) = 80.38$, $P < 0.001$) and between levels of initial density ($F(2, 13) = 8.71$, $P = 0.003$, Fig. 3); protection and levels of initial density showed no significant interaction. Loss increased with (ln) initial density. The differences in mussel loss between the exclosures and the exposed plots are expressed as percentages in Table 1.

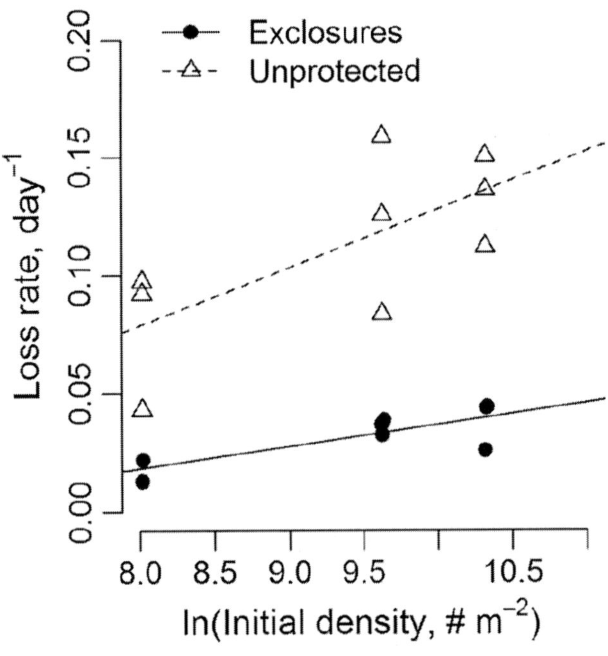


**Figure**. 3Mussel loss rates over a 5-week field experiment at three different initial densities on exclosures and unprotected plots. *Lines* show a significant relationship for exclosures ($r^2 = 0.57$, $F(1,6) = 10.37$, $P = 0.018$, coefficients: $a = 0.0092 \pm 0.0029$ (sem), $b = -0.055 \pm 0.027$ (sem)) and unprotected plots ($r^2 = 0.41$, $F(1,7) = 6.51$, $P = 0.038$, coefficients: $a = 0.025 \pm 0.0096$ (sem), $b = -0.12 \pm 0.090$ (sem))

**Table** 1 Percentage of mussel loss and differences for exclosures and exposed plots with different initial densities over a 5-week period after seeding

| Initial density Kg $m^{-2}$ | A. Exclosure % (SD) | B. Exposed plots % (SD) | B–A % % |
|---|---|---|---|
| 1 | 44.8 (12.2) | 90.2 (10.7) | 45.4 |
| 5 | 71.2 (3.3) | 97.7 (2.7) | 26.5 |
| 10 | 72.1 (11.1) | 98.9 (0.8) | 26.8 |

Biomass development, calculated as final mussel biomass/initial biomass over the 5-week experimental period, was affected by the factors of initial density ($F(2,11) = 16.22, P < 0.001$) and protection ($F(1,11) = 99.19$, $P < 0.001$). However, there was a significant interaction effect ($F(2,11) = 4.25, P = 0.04$) between the factors, indicating that the biomass development was higher on the protected plots (exclosures) than it was on the exposed plots (Fig. 4). In the exclosures, mussel biomass increased in 5 weeks with a factor of 1.5; this ratio was below 1 in all the plots with higher initial biomasses.

Condition indices of the mussels from all the treatments (Fig. 5) were subjected to a two-way analysis of variance test, with two levels of protection and three levels of initial density. However, the effects were not statistically different.

Attachment strength increased over time (Fig. 5) and was significantly affected by protection (Table 2). A regression tree model showed that protection did not matter before the fifth week. In other words, within the exclosures, byssal attachment strength of the mussels did not increase, whereas on the exposed plots, it increased in the fifth week. The initial biomass was found to have no effect on attachment strength. Results of the additional (controlled) laboratory experiment showed that the number of mussels consumed by each shore crab per 8 h (mean 10.15 ± 5.53) did not differ between the pre-aggregated or the loose mussels (GLM, $P > 0.05$). Average attachment strength of the pre-aggregated mussels was 2.0 ± 1.0 N ($n = 185$), and the mussels that were presented loose developed an attachment strength of 1.0 ± 0.7 N ($n = 83$) over the course of the 8 h that the experiment lasted. This shows that mussel attachment strength that developed in the first week in the field was not sufficient to protect them from crab predation (Fig. 6).

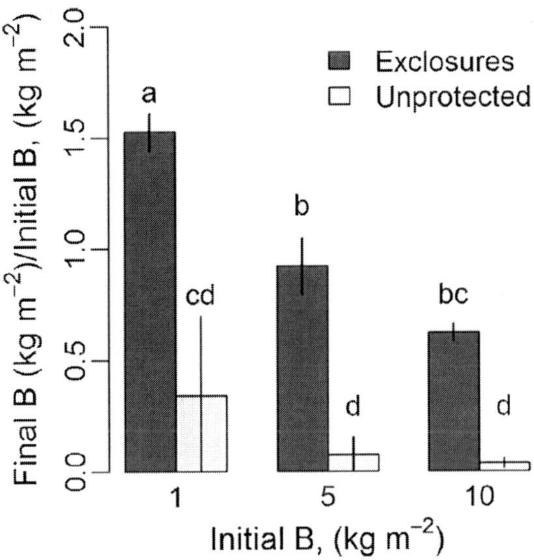


**Figure**. 4Average biomass development after 5 weeks as a ratio between the final biomass (Final B, kg $m^{-2}$) and the initial biomass (Initial B, kg $m^{-2}$), shown for the different treatments; *a*, *b*, *c* and *d* are statistically different from each other (Tukey HSD, $P < 0.05$)

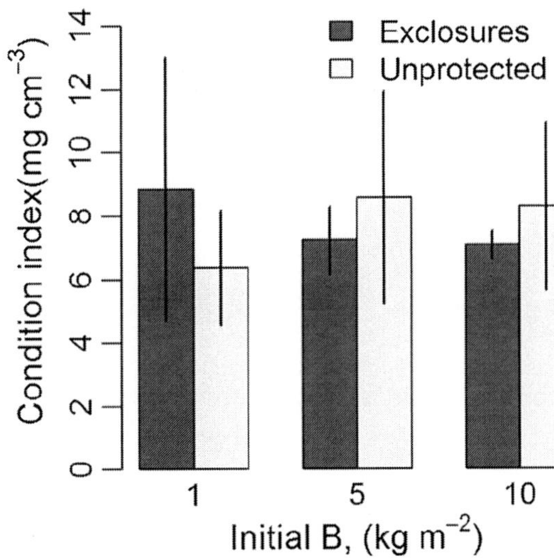


**Figure**. 5 Average condition index (mg $cm^{-2}$) at the end of the experiment, shown per treatment as the initial biomass (B); treatments are not significantly different from each other (two-way ANOVA, $P > 0.05$)

**Table** 2 LME results for the development of attachment strength (*N*) in exclosures and on exposed plots

| Source of variation | *df* | *F* | *P* |
|---|---|---|---|
| Dependent variable: log attachment strength (*N*) | | | |
| Treatment | 1 | 15.44 | 0.001 |
| Time | 4 | 11.87 | <0.001 |

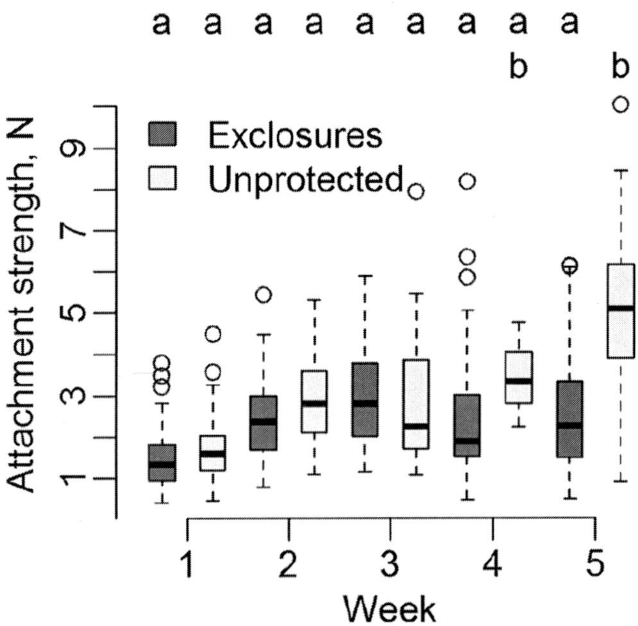


**Figure**. 6 Boxplot of the development of byssal attachment strength (*N* per mussel) of mussels from exclosures and exposed plots over the experimental period; *a* and *b* are statistically different (Tukey HSD, $P < 0.05$)

## DISCUSSION

The present study closely followed mussel performance on an intertidal commercial mussel plot for a 5-week period after seeding. Effects of initial mussel density and crab predation (through use of exclosures) on mussel survival were estimated. Overall losses were extremely high, particularly when compared to earlier tests as detailed below.

## Potential loss factors

The present study confirmed earlier findings that mussel loss is very high shortly after seeding and that mussel loss increases with mussel density (Theisen 1968; Capelle et al. 2014, 2015). Differences between the exclosures and the exposed plots can be used to distinguish between the loss factors. In the exclosures, mussel loss was probably due to mortality associated with handling stress (Calderwood et al. 2014) and intraspecific competition. At the lowest density (1 kg $m^{-2}$), where intraspecific completion was assumed to be low, mortality was still 45 % after 5 weeks. We could not find any studies on mussel bed performance as a function of density by means of exclosures. However, because 1 kg $m^{-2}$ is a low density in common mussel culture practice (Capelle et al. 2014), this 45 % is regarded as background mortality that might show losses following handling stress. At higher densities of 5 and 10 kg $m^{-2}$, mortality was observed to be 71 and 72 %, respectively, during the experiment. This mortality increase can be attributed to intraspecific competition at higher densities. The difference between mussel mortality in the exclosures and the losses on the exposed plots is due to dislodgement, and to predation by birds and shore crabs. This difference was in the range of 27–45 % (average 32.6 %; Table 1), which resulted in a final mussel biomass range of 0.3–0.4 kg $m^{-2}$ on the exposed plots for each treatment.

## Mussel loss on the exposed plots

Exclosures not only protected mussels against crab predation, but also against dislodgement and bird predation. Both bird predation and dislodgement are known factors contributing to loss on mussel beds (Nehls and Thiel 1993; Ens and Alting 1996; Hilgerloh et al. 1997). Herring gulls feed on small mussels (of around 20 mm), and oystercatchers feed on larger mussels (30–45 mm; Meire and Ervynck 1986). These birds are the main predators of mussel seed on intertidal flats (Hilgerloh et al. 1997). The experimental area was embedded in a culture plot on which the same mussel seed was seeded. On that culture plot, mussel farmers were present at low tides every day during the daytime and actively chased the birds away. This activity stopped after the first 3 weeks of the experiment because of the low number of birds observed in the area and subsequently on the culture plot. Exposure of the experimental area during low tides was relatively short and occurred, on average, only 2 h in every 24 h. Such circumstances led to speculation that bird predation might not be a significant loss factor on the exposed plots. The other major mussel predator in this area is starfish (*Asterias rubens*); however, their distribution is limited to the subtidal zone (Saier 2001). Also no starfish were observed on the video. It has been observed that the intertidal mussel beds are destroyed by dislodgement as a result of the effect of severe storms, but they remain relatively stable in sheltered locations (Nehls and Thiel 1993). The experimental site for the present study was located within a sheltered area in the Oosterschelde (Fig. 1), where the wind speed did not peak above 9.2 m $s^{-1}$throughout the duration of the experiment. The dominant direction of the wind during the 5 weeks of the experimental period was south-west. No mussel seed was found in the area around the (seeded) culture plot. These circumstances during the experimental period did reduce dislodgement risk and were important factors in the evaluation of the mussel loss. Thus, only predation by crabs remained as a factor to analyse the primary difference in mussel loss between the exclosures and the exposed plots.

## Mussel loss caused by handling stress

In contrast to a similar experiment that was performed a year earlier on the same culture plot, with 60–85 % loss 4 weeks after relaying on exposed plots, mussel loss was observed to be higher at the exposed plots in the current study with 90–99 % loss (Capelle et al. 2014). Furthermore, in the present study, the mussel loss was also higher than the losses that are generally experienced after seeding (around 50 %) (Theisen 1968; Capelle et al. 2016). According to mussel farmers, such high mussel loss may be due to the possibility that the mussels were kept in the vessel for too long from the time of harvest to seeding, especially because the day was very hot and sunny, and it has been reported that such conditions might cause some kind of stress that decreases survival after relay (Calderwood et al. 2014). It might be a satisfactory explanation for the very high mussel losses (Table 1), because the condition index of the seed, a day after seeding, was high (8.08 mg $cm^{-3}$). In the similar experiment that was performed a year earlier on the same culture plot with mussels from similar origin and size, condition index of mussels at start was only 5.00 mg $cm^{-3}$ (Capelle et al. 2014) and condition index of mussels of a similar size transplanted in Denmark ranged between 3.8 and 5.2 mg $cm^{-3}$ (Dolmer et al. 2012). These losses due to handling stress also explain the total losses (95.6 %), and the resulting loss levels were at the levels that are normally expected after seeding (about 50 %, Table 1).

It has been reported that handling stress also has an impact on byssal thread production shortly after seeding (Calderwood et al. 2014). However, in spite of similar stress experienced by the mussels in each treatment and on the entire culture plot, mussel losses found in the present study were higher than those observed under normal conditions. The results support the suggestion by Calderwood et al. (2014) that mussels should not be kept out of the water for more than 24 h.

## Development of attachment strength

Byssal attachment strength increased to a mean strength of 4.9 ± 0.2 N (SE) $n = 9$ during the fifth week at the exposed plots, while it remained at the same level for the protected mussels (mean 2.5 ± 0.2 N (SE), $n = 8$). Previous studies showed that the presence of shore crabs (effluent) stimulates the creation of byssus (Reimer and Tedengren 1997; Leonard et al. 1999) at a speed that can result in 10 byssus threads after 6 h and, in the field, to attachment strengths (after 30 days) of about 16 N (Côté 1995), which was twice as high as it was when no crabs were present (Leonard et al. 1999). Attachment strength is related to the sum of individual byssal threads (Bell and Gosline 1996), and mussels that are more strongly attached are less likely to be taken by crabs (Lin 1991). However, the byssus production on the exposed plots over time did not increase in the period after seeding, which is when the majority of the losses occur (Capelle et al. 2016). During the first week in the field, the mussels developed an average attachment strength of 1.6 ± 0.1 N (SE). The results of the laboratory experiment indicated that a mussel attachment strength of around 2 N is not sufficient to protect a mussel from crab predation.

Byssal attachment strength, which was measured both in established and in transplanted (after 30 days) mussel aggregations at exposed rocky shores in the study conducted by Leonard et al. (1999), was much higher (about 10–35 N) than it was in the present study. The byssal attachment strength measured after re-attachment under laboratory conditions (Lee et al. 1990; Dolmer and Svane 1994; Babarro and Reiriz 2010) was much lower than that which was reported in the findings by Leonard et al. (1999), and was lower than values measured in the present study. The experimental site at which the present experiment was carried out was a sheltered area of the Oosterschelde. Site conditions are strong determinants of attachment strength (Bell and Gosline 1996), and formation of byssus is costly (8 % of carbon and nitrogen in the total production in summer (Hawkins and Bayne 1985). Sheltered site conditions might, therefore, be more relevant for explaining the development of attachment strength than the presence of shore crabs, and may also play a role in protection against shore crab predation.

## Crab predation

It is estimated that about one-third (average 32.6 %; Table 1) of mussel loss in the 1-month period after seeding can be attributed to crab predation. These results are in line with the Plass-Johnson et al. (2010), who attributed 35 % of the mussel losses to predation and with estimations of Grosholz et al. (2011) who extrapolated results from Beal and Kraus (2002) and from (Beal 2006) where predation losses varied between 13 and 55 %. Murray et al. (2007) estimated that losses as a result of shore crab predation are 9.5 % over 12 months in the Menai Street in Wales, which is substantially lower than in the present study. However, predation rates were based on mean number of crabs per square metre on mussel beds in Wales which was lower than in the present study with on average 0.5 crabs $m^{-2}$ in August (Murray et al. 2007; Kamermans et al. 2009). A peak in foraging crabs can be expected closely after relaying. However, detailed observations from the first weeks after relaying are scarce, but the present study indicates such an effect. Crab abundance was much lower in the second week after seeding than it was in the first week after seeding, probably due to the attraction of the crabs to the smell of dead or damaged mussels during and shortly after seeding. Losses in mussel bottom culture are determined by high losses following seeding (Capelle et al. 2016). Mussels in week 1 and week 2 were between 16 and 18 mm in size. Murray et al. (2007) reported that the number of mussels consumed decreased as the mussel size increased. An individual shore crab can consume 21–36 mussels of 20 mm (minimum size presented) per 24 h depending on the size of the crab (Murray et al. 2007, Kamermans et al. 2009). It is assumed that for mussels of between 16 and 18 mm, the average food intake is around 1.25 mussels per hour, and a similar food intake was observed in the small predation experiment and in Kamermans et al. (2009); thus, the amount of crabs observed would have eaten about 286–1676 (mean ± SD) mussels/$m^2$within the first 2 weeks: 730 (±355) mussels/$m^2$ in the first week and 216 (±229) mussels/$m^2$in the second week. At the exposed plots with the highest density, 7000 mussels/$m^2$ were lost in a period of 5 weeks. With the estimated food intake, it would have taken between four and 24 weeks to consume such a high number of mussels with regard to the number of crabs that was observed. The amount of variation shows that such estimations can only be regarded as a rough estimate; however, it does show that the number of shore crabs observed could explain the observed mussel losses.

## CONCLUSIONS

Based on the present analysis, it can be concluded that mussel losses were high (mean 95.6 %) in a short period (about 1 month) after seeding on an intertidal culture plot, particularly when compared to previous studies. We observed a high and unexpected background mortality, which was the largest relative proportion (45 %) of all losses. Handling stress caused by the culture process could have played a major role in this loss. Moreover, (18 %) higher losses were observed at the higher stocking densities (5–10 kg $m^{-2}$) than at the low mussel density (1 kg $m^{-2}$). Remaining losses (about one-third; 32.6 %) were attributed to crab predation. No correlation was found between the number of shore crabs and the density of mussels. The development in attachment strength of the mussels at the sheltered location was observed to be a slow process and was thus unlikely to serve as protection against predation.

## ACKNOWLEDGMENTS

The authors wish to thank Robbie, Peter, Leon and Sjaak Praet from the company 'N.L. Praet & zn' for their support, the use of their lease site and for the mussel seed. Comments from Neil Bourne and an anonymous reviewer improved the quality of this paper. This work was part of the PRODUS project under contract of the Producers Organisation of the Dutch Mussel Culture and the Ministry of Economic Affairs.

## REFERENCES

1. Babarro JMF, Reiriz MJF (2010) Secretion of byssal threads in *Mytilus galloprovincialis*: quantitative and qualitative values after spawning stress. J Comp Physiol B Biochem Syst Environ Physiol 180:95–104
2. Beal BF (2006) Relative importance of predation and intraspecific competition in regulating growth and survival of juveniles of the soft-shell clam, *Mya arenaria* L., at several spatial scales. J Exp Mar Biol Ecol 336(1):1–17
3. Beal BF, Kraus MG (2002) Interactive effects of initial size, stocking density, and type of predator deterrent netting on survival and growth of cultured juveniles of the soft-shell clam, *Mya arenaria* L., in eastern Maine. Aquaculture 208(1):81–111
4. Bell E, Gosline J (1996) Mechanical design of mussel byssus: material yield enhances attachment strength. J Exp Biol 199(4):1005–1017
5. Beukema JJ (1976) Biomass and species richness on the macro-benthic animals living on the tidal flats of the Dutch Wadden Sea. Neth J Sea Res 10:236–261
6. Brown KM, Aronhime B, Wang X (2011) Predatory blue crabs induce byssal thread production in hooked mussels. Invertebr Biol 130:43–48

7. Burch A, Seed R (2000) Foraging behaviour of *Carcinus maenas* on *Mytilus edulis*: the importance of prey presentation. J Mar Biol Assoc UK 80:799–810
8. Calderwood J, O'Connor NE, Sigwart J et al (2014) Determining optimal duration of seed translocation periods for benthic mussel (*Mytilus edulis*) cultivation using physiological and behavioural measures of stress. Aquaculture 434:288–295
9. Capelle JJ, Wijsman JWM, Schellekens T et al (2014) Spatial organisation and biomass development after relaying of mussel seed. J Sea Res 85:395–403
10. Capelle JJ, Wijsman JWM, Van Stralen MR, et al (2015) Effect of seeding density on biomass production in mussel bottom culture. Manuscript submitted for publication
11. Capelle JJ, Wijsman JW, van Stralen MR et al (2016) Effect of seeding density on biomass production in mussel bottom culture. J Sea Res 110:8–15
12. Caro AU, Escobar J, Bozinovic F et al (2008) Phenotypic variability in byssus thread production of intertidal mussels induced by predators with different feeding strategies. Mar Ecol Prog Ser 372:127–134
13. Côté IM (1995) Effects of predatory crab effluent on byssus production in mussels. J Exp Mar Biol Ecol 188:233–241
14. Crothers JH (1968) The biology of the shore crab *Carcinus maenas* (L.). 2. The life of the adult crab. Field Stud 2:579–614
15. Davies GP, Dare PJ, Edwards DB (1980) Fenced enclosures for the protection of seed mussels (*Mytilus edulis* L.) from predation by shore crabs (*Carcinus maenas* L.). Fisheries Research Technical Report 56
16. Dolmer P, Svane I (1994) Attachment and orientation of *Mytilus edulis* L. in flowing water. Ophelia 40:63–74
17. Dolmer P, Christensen HT, Hansen BW et al (2012) Area-intensive bottom culture of blue mussels *Mytilus edulis* in a micro-tidal estuary. Aquac Environ Interact 3:81–91
18. Ens BJ, Alting D (1996) The effect of an experimentally created mussel bed on bird densities and food intake of the oystercatcher *Haematopus ostralegus*. Ardea 84:493–507
19. Fréchette M, Bourget E (1985) Food-limited growth of *Mytilus edulis* L. in relation to the benthic boundary layer. Can J Fish Aquat Sci 42:1166–1170
20. Freeman AS (2007) Specificity of induced defenses in *Mytilus edulis* and asymmetrical predator deterrence. Mar Ecol Prog Ser 334:145–153
21. Gallagher T, Richardson CA, Seed R et al (2008) The seasonal movement and abundance of the starfish, *Asterias rubens* in relation to mussel farming practice: a case study from the Menai Strait, UK. J Shellfish Res 27:1209–1215
22. Garner YL, Litvaitis MK (2013) Effects of injured conspecifics and predators on byssogenesis, attachment strength and movement in the blue mussel, *Mytilus edulis*. J Exp Mar Biol Ecol 448:136–140
23. Gascoigne JC, Beadman HA, Saurel C et al (2005) Density dependence, spatial scale and patterning in sessile biota. Oecologia 145:371–381
24. Grosholz E, Lovell S, Besedin E et al (2011) Modeling the impacts of the European green crab on commercial shellfisheries. Ecol Appl 21:915–924
25. Hawkins A, Bayne B (1985) Seasonal variation in the relative utilization of carbon and nitrogen by the mussel *Mytilus edulis*: budgets, conversion efficiencies and maintenance requirements. Mar Ecol Progr Ser Oldendorf 25(2):181–188
26. Hilgerloh G, Herlyn M, Michaelis H (1997) The influence of predation by herring gulls *Larus argentatus* and oystercatchers *Haematopus ostralegus* on a newly established mussel *Mytilus edulis* bed in autumn and winter. Helgol Mar Res 51:173–189
27. Kamermans P, Blankendaal M, Perdon J (2009) Predation of shore crabs (*Carcinus maenas* L.) and starfish (*Asterias rubens* L.) on blue mussel (*Mytilus edulis* L.) seed from wild sources and spat collectors. Aquaculture 290:256–262

28. Kangeri AK, Jansen JM, Barkman BR et al (2014) Perturbation induced changes in substrate use by the blue mussel, *Mytilus edulis*, in sedimentary systems. J Sea Res 85:233–240
29. Lee CY, Lim SSL, Owen MD (1990) The rate and strength of byssal reattachment by blue mussels (*Mytilus edulis* L.). Can J Zool 68:2005–2009
30. Leonard GH, Bertness MD, Yund PO (1999) Crab predation, waterborne cues, and inducible defenses in the blue mussel, *Mytilus edulis*. Ecology 80:1–14
31. Lin J (1991) Predator-prey interactions between blue crabs and ribbed mussels living in clumps. Estuar Coast Shelf Sci 32(1):61–69
32. Liu Q-X, Weerman EJ, Herman PMJ, et al (2012) Alternative mechanisms alter the emergent properties of self-organization in mussel beds. In: Proceedings of the royal society B: biological science
33. Mascaró M, Seed R (2001) Foraging behavior of juvenile *Carcinus maenas* (L.) and *Cancer pagurus* (L.). Mar Biol 139(6):1135–1145
34. Meire PM, Ervynck A (1986) Are oystercatchers (*Haematopus ostralegus*) selecting the most profitable mussels (*Mytilus edulis*)? Anim Behav 34(5):1427–1435
35. Murray LG, Seed R, Jones T (2007) Predicting the impacts of *Carcinus maenas* predation on cultivated *Mytilus edulis* beds. J Shellfish Res 26:1089–1098
36. Nehls G, Thiel M (1993) Large-scale distribution patterns of the mussel *Mytilus edulis* in the Wadden Sea of Schleswig-Holstein: Do storms structure the ecosystem? Neth J Sea Res 31:181–187
37. Pinheiro J, Bates D, DebRoy S, et al (2014) nlme: linear and nonlinear Mixed effects models. R package version 3.1-117. http://CRAN.R-project.org/package=nlme
38. Plass-Johnson JG, McQuaid CD, Porri F (2010) Top-down effects on intertidal mussel populations: assessing two predator guilds in a South African marine protected area. Mar Ecol Prog Ser 411:149–159
39. Reimer O, Tedengren M (1997) Predator induced changes in byssal attachment, aggregation and migration in the blue mussel, *Mytulis edulis*. Mar Freshw Behav Physiol 4:251–266
40. R Core Team (2013) R: A language and environment for statistical computing. R Foundation for Statistical Computing, Vienna, Austria. ISBN 3-900051-07-0, URL http://www.R-project.org
41. Saier B (2001) Direct and indirect effects of seastars *Asterias rubens* on mussel beds (*Mytilus edulis*) in the Wadden Sea. J Sea Res 46:29–42
42. Silva ACF, Boaventura DM, Thompson RC et al (2014) Spatial and temporal patterns of subtidal and intertidal crabs excursions. J Sea Res 85:343–348
43. Smallegange IM, Van Der Meer J (2007) Interference from a game theoretical perspective: shore crabs suffer most from equal competitors. Behav Ecol 18:215–221
44. Sneddon LU, Huntingford FA, Taylor AC (1997) The influence of resource value on the agonistic behaviour of the shore crab, *Carcinus maenas* (L.). Mar Fresh Behav Psychol 30:225–237
45. Theisen BF (1968) Growth and mortality of culture mussels in the Danish Wadden Sea. Medd Danm Fiskeri og Havunders NS 6:47–78
46. Whitton TA, Jenkins SR, Richardson CA et al (2012) Aggregated prey and predation rates: juvenile shore crabs (*Carcinus maenas*) foraging on post-larval cockles (*Cerastoderma edule*). J Exp Mar Biol Ecol 432–433:29–36
47. Zuur A, Ieno EN, Walker N et al (2009) Mixed effects models and extensions in ecology with R. Springer, NewYork
48. Zuur AF, Ieno EN, Elphick CS (2010) A protocol for data exploration to avoid common statistical problems. Methods Ecol Evol 1(1):3–14

# Index


## A

Allometric growth, 61, 62, 67, 69, 71, 76

Aquaculture, 1, 7, 27, 28, 29, 306, 307, 308, 310, 312, 374, 375, 377, 393, 394

## B

Bangladesh, 145, 313, 314, 315, 316, 317, 318, 319, 320, 321, 322, 323

Bioeconomy, 95, 96

Biosecurity, 80, 151, 230, 232

Burbot, 61, 62, 72, 76

## C

**Carcinus maenas**, 378, 394, 395

*Chaetoceros*, 4, 7, 15, 16, 17, 18, 19, 30, 83, 168, 269, 278

Community based, 314, 315

## F

Fatty acids, 118, 156, 168

Fisheries, 1, 7, 30, 33, 37, 53, 54, 61, 75, 76, 79, 90, 98, 99, 323, 339, 354, 377, 394

## G

Glucose metabolism, 34

Glycogenolysis, 34, 48

Gut microbiota, 242, 243

## I

Immunostimulation, 80, 93

Integrated multitrophic aquaculture, 96, 97

## M

Morphological characteristics, 62

**Mytilus edulis**, 109, 378, 379, 394, 395

## N

Nutrient composition, 358

Nutrition security, 314

## P

*Penaeus monodon*, 4, 29, 89, 92, 307, 312, 334, 336, 339
Phytoplankton, 30, 156, 180, 202

## R

Rotifers, 357, 358, 366, 368, 370, 373, 374
*Ruditapes decussatus*, 265, 266, 271, 272, 273, 274, 278, 279

## S

Salmonid fish, 34
Sustainability, 1, 80, 96, 114, 147, 155, 220, 234

## T

*Tetraselmis pseudonana*, 4

## V

*Vibrionaceae*, 261, 266, 270, 274, 278

## W

Welfare Transcriptome, 34

## Z

Zero water exchange systems, 80